W9-CUK-342

ACS Symposium Series

M. Joan Comstock, *Series Editor*

Foreword

The ACS SYMPOSIUM SERIES was founded in 1974 to provide a medium for publishing symposia quickly in book form. The format of the Series parallels that of the continuing ADVANCES IN CHEMISTRY SERIES except that, in order to save time, the papers are not typeset but are reproduced as they are submitted by the authors in camera-ready form. Papers are reviewed under the supervision of the Editors with the assistance of the Series Advisory Board and are selected to maintain the integrity of the symposia; however, verbatim reproductions of previously published papers are not accepted. Both reviews and reports of research are acceptable, because symposia may embrace both types of presentation.

Contents

Preface

DEVELOPMENT OF SOPHISTICATED SURFACE ANALYTICAL TECHNIQUES over the past two decades has revived interest in the study of phenomena that occur at the electrode–solution interface. As a consequence of this renewed activity, electrochemical surface science is experiencing a rapid growth in empirical information. The symposium on which this book was based brought together established and up-and-coming researchers from the three interrelated disciplines of electrochemistry, surface science, and metal-cluster chemistry to help provide a better focus on the current status and future directions of research in electrochemistry. The symposium was part of the continuing series on Photochemical and Electrochemical Surface Science sponsored by the Division of Colloid and Surface Chemistry of the American Chemical Society.

This volume contains 36 chapters, the first of which is an overview chapter. The eleven following chapters describe the use of ultrahigh vacuum surface spectroscopic methods in the study of the electrode–solution interface. The next eight chapters report recent advances in the adaptation of new experimental techniques, such as scanning tunneling microscopy, to electrochemical problems. The seven following chapters discuss critical aspects of in situ vibrational spectroscopy. The last nine chapters describe various electrode processes such as the electrochemical reactivity of polymer deposits, CO_2 methanation, and surface organometallic chemistry. The broad but balanced range of important topics discussed by experts here should make this volume invaluable to students and active researchers in electrochemical surface science.

Acknowledgments

The symposium and this volume have benefited from financial support provided by the Division of Colloid and Surface Chemistry of the American Chemical Society, the Petroleum Research Fund, Shell

Development Company, and Texas A&M University. I gratefully acknowledge the cooperation of the authors in meeting the various deadlines.

MANUEL P. SORIAGA
Texas A&M University
College Station, TX 77843

June 30, 1988

Chapter 1

Electrochemistry: The Senior but Underused Area of Surface Science

Manuel P. Soriaga

Department of Chemistry, Texas A&M University, College Station, TX 77843

Electrochemistry ought to be regarded as a vintage area of interfacial science. It has long been known that the application of an external potential at a (conducting) solid-solution interface enhances the reactivities of adjacent chemical species. Catalytic reactions activated by control of potential at the electrode-solution interface thus serves as an alternative to activation by control of temperature at gas-solid interfaces. In fact, on the singular basis of catalytic selectivity, the electrochemical route towards synthesis and catalysis should be favored over the thermal approach because the application of elevated temperatures activates other reaction pathways. For a two-electron reaction, for example, a potential change of only half a volt can lead to a 10^8-fold increase in reaction rate. Assuming an activation energy of 75 kJ mole^{-1}, this enormous rate increase can be effected only at temperatures above 600 K. Such drastic conditions invariably lead to a wide spectrum of thermal excitations which could result in undesirable side-reactions such as indiscriminate bond-breaking.

In addition to the universal concern for catalytic selectivity, the following reasons could be advanced to argue why an electrochemical scheme would be preferred over a thermal approach: (i) There are experimental parameters (pH, solvent, electrolyte, potential) unique only to the electrode-solution interface which can be manipulated to dictate a certain reaction pathway. (ii) The presence of solvent and supporting electrolyte may sufficiently passivate the electrode surface to minimize catalytic fragmentation of starting materials. (iii) Catalyst poisons due to reagent decomposition may form less readily at ambient temperatures. (iv) The chemical behavior of surface intermediates formed in electrolytic solutions can be closely modelled after analogous well-characterized molecular or cluster complexes (1-8). (v) For hydrogenation (or oxidation) reactions, the aqueous solvent functions as a convenient source of hydrogen (or

0097–6156/88/0378–0001$06.00/0

oxygen). (vi) Because electrocatalysis employs a reducing and an oxidizing electrode, there is promise for implementation of the concept of paired electrosynthesis (9).

It is true that electrochemistry has enjoyed tremendous successes in the fields of chemical analysis (sensors) and energy conversion (batteries). It is also a fact that the use of electrochemistry has become widespread in modern-day inorganic chemistry where redox potentials of newly synthesized materials are determined almost as routinely as spectroscopic and crystallographic data. However, despite its historical significance and despite all the apparent advantages enumerated above, electrochemistry remains relegated to the background insofar as large-scale catalytic synthesis is concerned. Technological and engineering complexities, such as slow mass transport in solution and the cost of electricity, are important factors for the persistent neglect of the electrochemical approach. Another equally critical reason is the absence of atomic-level descriptions of processes which occur at the electrode-solution interface. This lack of fundamental information can be attributed to the fact that, until these past two decades, conventional electrochemical methods (10), such as voltammetric, amperometric, coulometric, impedance, and transient measurements, were the only tools readily available for the study of electrode processes; these techniques are severely restricted in the type of information they provide since the results obtained from them are manifestations of only the macroscopic properties of the electrode-solution interface.

The recognition of the lack of molecular-level information on electrocatalytic phenomena and the resolve to have this weakness rectified is now firmly established within the electrochemical and surface science communities. This is evidenced by recent international conferences and workshops devoted solely to electrochemical problems (11-4). It is also well-realized that electrochemistry cannot advance further unless the practitioners in this field begin to view electrode-solution processes from a molecular-level perspective.

The central issue which has to be addressed in any comprehensive study of electrode-surface phenomena is the determination of an unambiguous correlation between interfacial composition, interfacial structure, and interfacial reactivity. This principal concern is of course identical to the goal of fundamental studies in heterogeneous catalysis at gas-solid interfaces. However, electrochemical systems are far more complicated since a full treatment of the electrode-solution interface must incorporate not only the compact (inner) layer but also the boundary (outer) layer of the electrical double-layer. The effect of the outer layer on electrode reactions has been neglected in most surface electrochemical studies but in certain situations, such as in conducting polymers and biological systems, processes within the boundary layer can exert non-trivial influences.

In the surface electrochemist's pursuit of fundamental structure-composition-reactivity correlations, lessons can be and have been learned from the successes achieved in both heterogeneous and homogeneous catalysis. For example, extensive work on fuel cell electrodes (15-24) has accomplished much to show that concepts developed in the study of oxygen reduction by Pt particles in gas-solid reactions are applicable to the electrochemistry of H_2 and O_2. The essential ingredients in experimental strategies which have contributed to the molecular-level understanding of a wide variety of homogeneous catalytic processes include (25-6): (i) the use of pure and well-characterized starting materials, (ii) structural and compositional analysis of homogeneous catalysts, (iii) structural and compositional analysis of important intermediates, (iv) detailed kinetic measurements including the concentration of catalyst precursors, and (v) quantitative analysis of reaction product distributions. These elements, which have now become routine in model studies of heterogeneous catalysis at gas-solid interfaces, are just beginning to be appreciated in surface electrochemical investigations.

Studies of the physics and chemistry at the gas-solid interface have enjoyed tremendous advances from the systematic application of modern surface spectroscopic methods (27), such as low-energy electron diffraction for surface crystallographic determinations, Auger electron spectroscopy for surface elemental analysis, X-ray photoelectron spectroscopy for surface bonding studies, high-resolution electron-energy loss spectroscopy for surface vibrational information, and thermal desorption mass spectrometry for adsorption enthalpy measurements. These powerful techniques have provided revealing glimpses of gas-solid interfacial processes at the atomic level. Unfortunately, they cannot be employed to probe the electrode-electrolyte interface under reaction conditions since surface characterization has to be performed outside the electrochemical cell. Doubts have been raised concerning the validity of correlating structural information acquired in the absence of an electrochemical environment with reactivity data obtained under potential control. To help mollify these concerns, it has been pointed out that fruitful structure-reactivity correlations have been accomplished in homogeneous systems (25-6) despite the fact that structural determinations are done in the solid state while the reactivity measurements are performed in solution. Results from recent studies of *emersed* electrodes (12,28-9) have also dispelled most of the earlier doubts. A recent workshop organized by the National Research Council has recommended that the use of high-vacuum surface physics methods should be continued (12). The utilization of such techniques in surface electrochemistry was initially restricted to only a handful of laboratories (30-2) but the number has now increased. It is particularly encouraging to note that surface science laboratories which previously devoted all efforts to gas-solid interfaces now apportion significant time to the study electrocatalytic phenomena (33-4).

Optical spectroscopic methods not restricted to high-vacuum conditions, such as infrared reflection-absorption spectroscopy (35), ellipsometry (36), and surface-enhanced Raman spectroscopy (37-8), are available for examining intermediates at the electrochemical double-layer under reaction conditions. However, none of these can determine the complete structure and composition of the electrocatalyst-adsorbate interfacial aggregate. The capabilities of laser [as in second-harmonic generation (39-40)] and synchrotron radiation [as in near-edge and extended X-ray absorption fine structure (41)] in elucidating interfacial structures under electrochemical conditions have recently been demonstrated; research in these areas have been intensified. Present vigorous activities in surface electrochemical studies have included the adaptation of new technologies, such as scanning tunneling microscopy (42-3), and the revival of conventional methods, such as Mossbauer spectroscopy (44), nuclear magnetic resonance, and X-ray diffraction (45).

In view of the complexity of heterogeneous systems, none of the above techniques will be able to supply, by itself, a complete atomic-level description of surface phenomena. A multi-technique approach has been perceived by many as most appropriate for fundamental studies in electrochemical surface science (30-2). Since none of the existing electrochemical laboratories are adequately equipped to perform a comprehensive experimental study, collaborative efforts between research groups of different expertise are burgeoning. Easier access to national or central facilities are also being contemplated for experiments which cannot be performed elsewhere. The judicious combination of the available methods in conjunction with the appropriate electrochemical measurements are permitting studies of electrocatalyst surface phenomena unparalleled in molecular detail.

In any scientific endeavor, mere accumulation of empirical information is not enough to guarantee major advances. Efforts have to be exerted to formulate fundamental concepts from the available data. For example, it is no longer sufficient to describe how an adsorbed species is bound to the surface simply in terms of its interfacial orientation. It is essential to proceed beyond such phenomenological illustrations and actually examine the nature of the surface-adsorbate chemical bond. It is in this aspect that concepts from the related areas of coordination and organometallic chemistry have been most meaningful. For example, in the so-called surface-cluster analogy (1-4), bonding in chemisorption systems is modelled after bonding in molecular or cluster complexes. Skepticism to the surface-cluster analogy has been raised because of the widespread knowledge that while metal complexes can be described by discrete orbitals, metal surfaces are usually characterized by band structure. However, what is not known to as many is the fact that the surface band structure can actually be deconvoluted into components of given symmetries with respect to a given site (3-4). In this manner, the surface-adsorbate bond can be described in terms of the overlap of symmetry-adapted surface bands and adsorbate orbitals; this description is exactly the same as that of the metal-ligand bond in monometal or small cluster complexes.

With the experimental and theoretical strategies available today, research in electrochemical surface science has been revived. There is great optimism that much of the mysteries surrounding electrocatalysis will soon be unravelled in molecular detail hitherto unachievable.

Literature Cited

1. Muetteries, E.L. Bull. Soc. Chim. Belg. 1975, 84, 959.
2. Muetteries, E.L. Bull. Soc. Chim. Belg. 1976, 85, 451.
3. Saillard, J.Y.; Hoffman, R. J. Am. Chem. Soc. 1984, 106, 2006.
4. Albert, M.R.; Yates, J.T., Jr. The Surface Scientists's Guide to Organometallic Chemistry; American Chemical Society: Washington, DC, 1987.
5. Rodriguez, J.F.; Bravo, B.G.; Mebrahtu, T.; Soriaga, M.P. Inorg. Chem. 1987, 26, 2760.
6. Rodriguez, J.F.; Harris, J.E.; Bothwell, M.E.; Mebrahtu. T; Soriaga, M.P. Inorg. Chim. Acta. 1988, in press.
7. Soriaga, M.P.; Binamira-Soriaga, E.; Hubbard, A.T.; Benziger, J.B.; Pang, K.-W.P. Inorg. Chem. 1985, 24, 65.
8. Schardt, B.C.; Stickney, J.L.; Stern, D.A.; Frank, D.G.; Katekaru, J.Y.; Rosasco, S.D.; Salaita, G.D.; Soriaga. M.P.; Hubbard, A.T. Inorg. Chem. 1985, 24, 1419.
9. Baizer, M.M.; Lund, H. Organic Electrochemistry; Marcel Dekker: New York. 1983.
10. White, R.E.; Bockris, J.O'M.; Conway. B.E.; Yeager, E. Comprehensive Treatise of Electrochemistry; Plenum Press: NY 1984; Vol 8.
11. Molecular Phenomena at Electrode Surfaces; Symposium sponsored by the Surface and Colloid Chemistry Division of the American Chemical Society. New Orleans, LA, 1987.
12. Characterization of Electrochemical Processes; Special workshop convened by the National Research Council. Denver, CO, 1985.
13. Furtak, T.E.; Kliewer, K.L.; Lynch, D.W., Eds; Non-Traditional Approaches to the Study of the Solid-Electrolyte Interface. North-Holland: Amsterdam, 1980.

14. Hansen, W.N.; Kolb, D.M.; Lynch, D.W., Eds.; Electronic and Molecular Structure of Electrode-Electrolyte Interfaces. Elsevier: Amsterdam, 1983.
15. Ross, P.N. J. Electrochem. Soc. 1979, 126, 78.
16. Stonehart, P.; Zucks, P.A. Electrochim. Acta. 17, 2333.
17. Lundquist, J.T.; Stonehart, P. Electrochim. Acta. 1973, 18, 349.
18. Kunz, H.R. Proc. Electrocatal. Fuel Cell React. 1978, 79, 14.
19. Kunz, H.R.; Gruver, G. J. Electrochem. Soc. 1975, 122, 1279.
20. Stonehart, P.; Kinoshita, K.; Bett, J.S. In Electrocatalysis; Breiter, M.W., Ed.; The Electrochemical Society: Princeton, NJ, 1974.
21. Blurton, K.F.; Greenberg, P.; Oswin, H.G.; Rutt, D.R. J. Electrochem. Soc. 1972, 119, 559.
22. Bett, J.S.; Lundquist, J.; Washington, E.; Stonehart, P. Electrochim. Acta. 1979, 18, 343.
23. Bregoli, L.J. Electrochim. Acta. 1978, 23, 489.
24. Peuckert, M.; Yoneda, T.; Dalla Betta, R.A.; Boudart, M. J. Electrochem. Soc. 1986, 133, 944.
25. Schrauzer, G.N. Transition Metals in Homogeneous Catalysis; Marcel Dekker: New York, 1971.
26. Cotton, F.A.; Wilkinson, G. Advanced Inorganic Chemistry; Wiley: New York, 1980.
27. Somorjai, G.A. Chemistry in Two Dimensions: Surfaces; Cornell University Press: Ithaca, NY, 1981.
28. Hansen, W.N. J. Electroanal. Chem. 1983, 150, 133.
29. Stickney, J.L.; Rosasco, S.D.; Salaita, G.N.; Hubbard, A.T. Langmuir. 1985, 1, 66.
30. Hubbard, A.T. Accts. Chem. Res. 1980, 13, 177.
31. Ross, P.N. In Chemistry and Physics of Solid Surfaces; Vaneslow, R.; Howe, R., Eds.; Springer-Verlag: NY, 1982.
32. Homa, A.S.; Yeager, E.; Cahan. B.D. J. Electroanal. Chem. 1983, 150, 181.
33. Wieckowski, A.; Rosasco, S.D.; Salaita G.N.; Hubbard, A.T.; Bent, B.; Zaera, F.; Somorjai, G.A. J. Am. Chem. Soc. 1985, 107, 21.
34. Stuve, E.M.; Rogers, J.W., Jr.; Ingersoll, D.; Goodman, D.W.; Thomas, M.L.; Paffett, M.T. Chem. Phys. Letters. 1988, in press.
35. Pons, S.; Bewick, A. Langmuir. 1985, 1, 141.
36. Muller. R.H. Adv. Electrochem. Electrochem. Engr. 1973, 9, 227.
37. Van Duyne, R.P. In Chemistry and Physics of the Solid Surface; R. Vaneslow, Ed.; CRC: FL, 1981.
38. Gao, P.; Weaver, M.J. J. Phys. Chem. 1986, 90, 4057.
39. Richmond, G.L. Langmuir. 1986, 2, 132.
40. Campbell, D.J.; Corn, R.M. J. Phys. Chem. 1987, 91, 5668.

41. Gordon, J.G.; Melroy, O.R.; Borges, G.L.; Reisner, D.L.; Abruña, H.D.; Chandrasekhar, P.; Albarelli, M.J.; Blum, L. J. Electroanal. Chem. 1986, 210, 311.
42. Lin, W.; Fan, F.-R.F; Bard, A.J. J. Electrochem. Soc. 1987, 134, 1038.
43. Sonnefield, R.; Hansma, P.K. Science. 1986, 232, 211.
44. Scherson, D.A.; Yao, S.B.; Yeager, E.; Eldridge, J.; Kordesch, M.E.; Hoffman, R.W. J. Electroanal. Chem. 1983, 150, 535.
45. Fleischman, M.; Graves, P.; Hill, I.; Oliver, A.; Robinson. J. J. Electroanal. Chem. 1983, 150, 33.

RECEIVED June 27, 1988

Chapter 2

Molecular Adsorption at Well-Defined Platinum Surfaces

Voltammetry Assisted by Auger Spectroscopy, Electron Energy-Loss Spectroscopy, and Low-Energy Electron Diffraction

Arthur T. Hubbard, Donald A. Stern, Ghaleb N. Salaita, Douglas G. Frank, Frank Lu, Laarni Laguren-Davidson, Nikola Batina, and Donald C. Zapien

Department of Chemistry, University of Cincinnati, Cincinnati, OH 45221–0172

Reviewed here are surface electrochemical studies of organic molecules adsorbed at well-defined Pt(111) electrode surfaces from aqueous solution. Emphasis is placed upon studies of nicotinic acid (NA), pyridine (PYR), and nine related pyridine carboxylic acids. Packing densities (moles per unit area) adsorbed from solution at controlled pH and electrode potential, and measured by means of Auger spectroscopy are discussed. Vibrational spectra of each adsorbed obtained by use of electron energy-loss spectroscopy (EELS) are compared with the IR spectra of the parent compounds. Electrochemical reactivity is studied by use of cyclic voltammetry. Monolayer structures observed by means of low-energy electron diffraction (LEED) are described. Substances studied are as follows: 3-pyridine carboxylic acid (nicotinic acid, NA, "niacin"); pyridine (PYR); 3-pyridylhydroquinone (3PHQ), synthesized here for the first time; 4-pyridine carboxylic acid (isonicotinic acid, INA); 2-pyridine carboxylic acid (picolinic acid, PA); 3,4-pyridine dicarboxylic acid (3,4PDA); and the analogous other pyridine dicarboxylic acids 3,5PDA, 2,3PDA, 2,4PDA and 2,6PDA. Each of the pyridine derivatives is adsorbed at Pt(111) in a tilted vertical orientation; an angle of 70-75° between the pyridyl plane and the Pt(111) surface is typical. Pt-N bonding is evidently the predominant mode of surface attachment of these compounds, although coordination of carboxylate is an important mode of additional surface attachment at positive potentials. EELS spectra display strong O-H stretching vibrations near 3550 cm^{-1} due to carboxylic acids in the meta and para-positions, and weak-to-moderate signals near 3350 cm^{-1} due to ortho-carboxylates.

0097–6156/88/0378–0008$08.25/0

Nicotinic acid and related meta-carboxylic acids display the remarkable characteristic that coordination of the pendant carboxylic acid moieties to the Pt surface is controlled by electrode potential. Oxidative coordination of the carboxylate pendant occurs at positive electrode potentials, resulting in disappearance of the O-H vibration and loss of surface acidity as judged by absence of reactivity towards KOH. Carboxylate in the 4-position of pyridine (as in INA) is virtually independent of electrode potential, whereas strong coordination of ortho-carboxylates to the Pt surface is present at most electrode potentials. Adsorbed pyridine carboxylic acids are stable in vacuum; when returned to solution the adsorbed material displays the same chemical and electrochemical properties as prior to evacuation.

An era of remarkable progress in the study of solid electrode surfaces has begun (1-5). This article offers a brief report of recent progress, arranged as follows: rationales for multi-technique investigation of electrode surfaces; well-defined electrode surfaces as a part of electrochemical research; evidence for the stability in vacuum of layers adsorbed from solution; Auger spectroscopy as a tool for quantitation and characterization of adsorbed layers; electron energy-loss spectroscopy (EELS) for investigation of molecular structure and mode of surface attachment of substances adsorbed at electrode surfaces; low-energy electron diffraction (LEED) as a probe of both substrate surface structure and adsorbed layer long-range order; and voltammetric experiments illustrating the influence of electrode surface crystallographic structure on the electrochemical reactivity of adsorbed molecules.

A substantial added benefit of multi-technique investigations of electrode surfaces is that the same experimental tools required to bring the electrode surface to a well-defined state happen also to be very proficient at revealing the various important changes in the nature of the surface which occur during subsequent chemical or electrochemical treatment or use of the electrode. Also, the diagnostic power of the multi-technique approach is particularly advantageous for investigation of complicated surfaces. Advances in technique have brought many complicated surfaces to a well-defined state (3-6). Although in principle there are many techniques from which to choose, in practice it is advisable to select techniques which work well together, answer the more important questions about the sample, and yield the necessary information without appreciably altering the sample. For example, when the objective is to determine the structure of the all-important first atomic/ionic/molecular layer at a single-crystal surface, LEED is the method of choice (7); X-ray diffraction, EXAFS, neutron diffraction, and atomic/ion scattering, while excellent for certain other purposes, are disadvantageous for this task due to lesser sensitivity, generality and convenience. Auger spectroscopy is preferable for measurement of packing density and elemental composition of surface molecular layers (8), although XPS (ESCA,9) or SIMS (10) might also be employed. Vibrational spectra of molecular or atomic layers are obtainable at high sensitivity and virtually unlimited frequency range by EELS (11). On the other hand, infrared reflection-absorption spectroscopy (IRRAS, 12) offers higher resolution than EELS in a few special cases, such as carbon monoxide, where the signal due to a certain vibration in a monolayer meets practical limits of detection. Surface regular Raman spectroscopy is showing signs of usefulness for molecular layers at smooth surfaces and promises to allow direct comparison between gas-solid and liquid-solid interfacial Raman spectra at least in favorable cases (13). Cyclic voltammetry and chronocoulometry are reliable methods for electrochemical characterization, along with variations based upon the use of thin-layer electrodes (14), although numerous alternative or more complicated methods are available (15). Examination of a surface in the presence of a bulk fluid generally requires sacrifices in sensitivity, resolution, signal-to-noise ratio, interfering signals, and other aspects. Such sacrifices are pointless if not repaid by observation of some otherwise-unobservable phenomenon; repayment is to be expected only in special situations yet to be fully identified. In any event, no one technique measures enough of the properties of a surface to bring it to a well-defined state, and accordingly there is no substitute for a balanced approach to surface characterization based upon Auger, EELS, LEED, IRRAS, and related measurements. Procedures for preparation and characterization of electrode surfaces are discussed in more detail in Reference 5 and references cited therein.

Before proceeding to methods and applications of electrode surface characterization, perhaps we should remind ourselves of the motivations for investigating surface structure and composition. Chemical and electrochemical processes at surfaces depend upon surface structure, composition, mode of bonding and related characteristics. Practical properties such as surface stability, electrical behavior, photochemistry, corrosion, passivation, hydrophilicity, adhesion, mechanical durability and lubricity, to name a few, often vary strongly with surface atomic structure and composition. Surfaces shown to be clean by Auger spectroscopy (or equivalent methods) and shown to have a specific atomic structure by LEED (or equivalent methods) are termed "well-defined surfaces". With apologies for stating the obvious, surfaces not directly characterized as to surface structure and composition are not "well-defined surfaces" even when prepared from an oriented single crystal. This is because surface scientists worldwide have observed that specific crystal planes do not form automatically, even on the surfaces or oriented single-crystals. Instead, contamination, structural reconstruction and disorder are the general rule. Only by repeated and persistent experimentation, guided by direct surface characterization (LEED, Auger, EELS) is the surface brought to a clean, ordered, "well-defined" state.

Reversible electrochemical reactivity is present in the chemisorbed states of some compounds. Characteristics shared by each of these compounds reported to date are as follows:

(a) reversible electroactivity in the unadsorbed state;

(b) attachment to the surface by way of a functional group other than the electroactive center;

(c) and none of the atoms in the electroactive part of the molecular are involved in chemical bonding directly with the surface. Current-potential behavior of two such compounds, 2,5-dihydroxy-4-methylbenzylmercaptan (DMBM) and (3-pyridyl) hydroquinone (3PHQ), is illustrated by Figures 1-A and 1-B, in which only an adsorbed layer is present to participate in the electrode reaction. The reactions are:

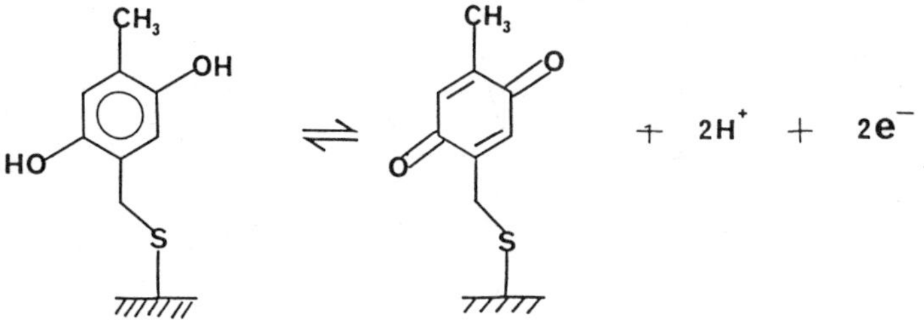

$$\rightleftharpoons \quad + \; 2H^+ \; + \; 2e^- \qquad (1)$$

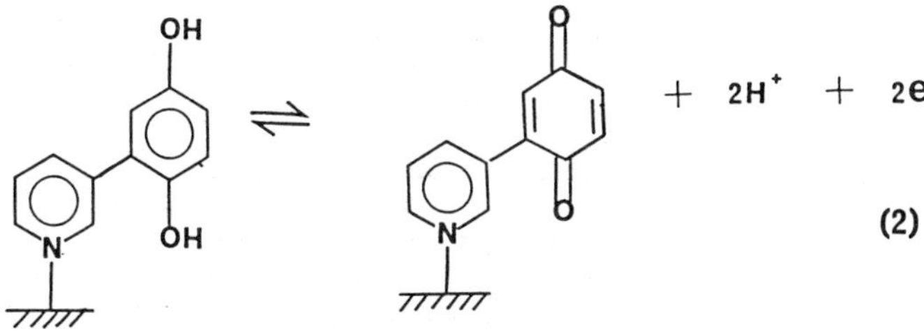

$$\rightleftharpoons \quad + \; 2H^+ \; + \; 2e^- \qquad (2)$$

Prior to recording the current-potential curves in Figure 1-A, the Pt(111) surface was immersed into 0.7mM DMBM, followed by rinsing with pure supporting electrolyte (10mM trifluoroacetic acid). Figure 1-B (solid curve) shows a similar experiment involving immersion into 0.5mM 3PHQ followed by rinsing with 10mM KF (adjusted to pH 4 with HF). The dotted curves in Figures 1-A and 1-B are obtained by a variation of the above procedures in which the adsorbed layer of DMBM or 3PHQ is subjected to one hour in vacuum prior to cyclic voltammetry. The voltammetric results before and after evacuation are virtually identical, demonstrating that DMBM and 3PHQ are retained at the Pt(111) surface in vacuum and are not removed from the surface by rinsing with water. Adsorbed 2,2',5,5'-tetrahydroxybiphenyl (THBP), also, is stable toward evacuation and rinsing, Figure 1-C. Adsorbed hydroquinone (HQ) is typical of numerous horizontally-oriented, adsorbed simple aromatic compounds in being stable toward solution and evacuation (3,4), Figure 1-D.

Auger spectroscopy yields data for molecular layers from which relatively accurate and precise packing densities and elemental analyses are obtained. Two methods are available, allowing verification of precision and self-consistency. The first of these methods is based upon measurement of elemental Auger signals (I_C, I_N, I_O, etc.), while the other method measures the at-

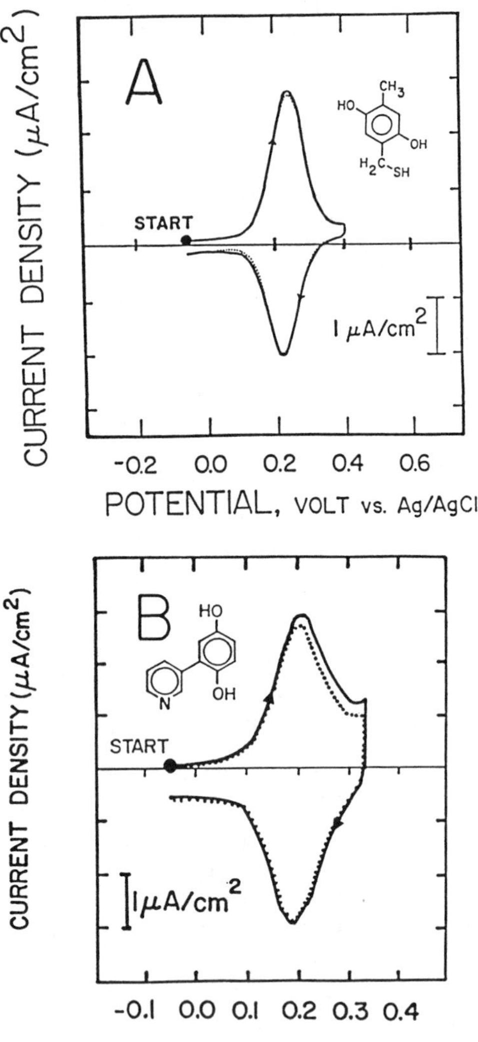

Figure 1. Cyclic voltammetry of adsorbed molecules at Pt(111). Sweep rate, 5 mV/s. A. Solid curve (—): immersion into 0.7 mM DMBM followed by rinsing with 10 mM TFA. Dotted curve (....): as in solid curve, except 1 h in vacuum prior to voltammetry. B. Solid curve (—): immersion into 0.5 mM 3PHQ, followed by rinsing with 1 mM HF. Dotted curve (....): as in solid curve, except 1 h in vacuum prior to voltammetry. *Continued on next page.*

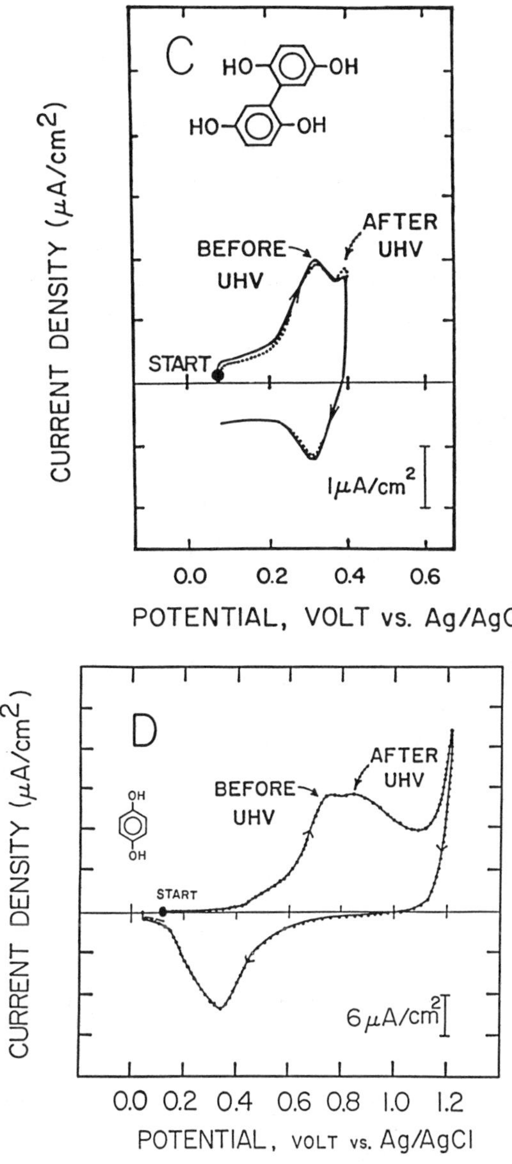

Figure 1. *Continued.* C. Solid curve (—): immersion into 2 mM THBP followed by rinsing with 10 mM TFA. Dotted curve (....): as in solid curve, except 1 h in vacuum prior to voltammetry. (Reproduced with permission from ref. 3. Copyright 1984 Elsevier.) D. Solid curve (—): immersion into 0.03 mM HQ followed by rinsing with 10 mM TFA. Dotted curve (....): as in solid curve, except 1 h in vacuum prior to voltammetry.

tenuation of substrate Auger signal by the adsorbed layer. Packing density measurements by these methods are in excellent agreement. For example, molecular packing density of 3PHQ at Pt(111) (-0.1V, pH4, 0.5mM) is 0.28 nmol/cm^2 based upon I_C/I_{Pt}^o, 0.27 nmol/cm^2 based upon I_{Pt}/I_{Pt}^o, or 0.27 nmol/cm^2 based upon coulometric electrochemical data (16).

Shown in Figure 2-A are the experimental Auger signal ratios I_C/I_{Pt}^o (carbon signal, 272eV/clean surface Pt signal, 161eV), and I_{Pt}/I_{Pt}^o (Pt signal, 161eV, from coated surface/clean surface) graphed versus nicotinic acid (NA) concentration (C_{NA}) at Pt(111). Note that I_C/I_{Pt} increases with C_{NA} due to increases in Γ_{NA}, while I_{Pt}/I_{Pt}^o decreases accordingly. Molecular packing densities obtained from such data using Equations 3 and 4 are shown in Figure 2-B (B_C = 0.377 cm^2/nmol; f=0.70; and K= 0.16 cm^2/nmol) for

$$\Gamma = (I_C/I_{Pt}^o)/[6B_C(1/3+2f/3)] \qquad (3)$$

$$\Gamma = (1-I_{Pt}/I_{Pt}^o)/(9K) \qquad (4)$$

adsorption of NA at -0.2V (vs. Ag/AgCl reference). Packing densities of NA at +0.3V are shown in Figure 2-C. The saturation NA packing density is near 0.38 nmol/cm^2 (43.7 Å/molecule), which falls between the theoretical packing densities of 0.290 nmol/cm^2 (57.2Å/molecule) for a horizontal orientation, and 0.579 nmol/cm^2 (28.7Å/molecule) for the N-attached vertical orientation, based upon covalent and van der Waals radii (17). This is one of several types of evidence pointing to a filled-vertical orientation of adsorbed NA in which the pyridine ring forms an angle of 75° with the Pt(111) surface. A model illustrating the method of calculation of this angle is shown in Figure 3. Pyridine and all of the pyridine carboxylic acids studied to date are oriented near vertically at Pt(111) with most or all of the angles of tilt in the range from 70 to 76°. A brief listing of data will be given in Reference 16. Underlying causes of this recurring orientation will be discussed below after consideration of further types of evidence. The precision, simplicity and self-consistency of NA packing density data based upon Auger spectroscopy is characteristic of data obtained for a wide variety of adsorbed compounds. Auger measurements of this type are a highly recommended starting point for studies of molecular substances at electrode surfaces. (See Table I.)

A striking feature of the EELS spectra of NA adsorbed from acidic solutions (HF/KF) at negative potentials is the pressure of a strong O-H stretching band at 3566 cm^{-1}, Figure 4-A. This band vanishes when the NA layer is rinsed with base, Figure 4-C, and also when adsorption is carried out at positive potentials, Figure 4-B. The EELS spectra are essentially the envelope of the gas-phase or solid IR spectra (lower curves in Figures 4-A through 4-C). Assignments of the EELS spectra of analogy with the IR spectra are given in Table II. The intensity of the O-H peak varies smoothly from maximum to minimum as the electrode potential during adsorption is varied from negative to positive, Figure 5-A. This variation of O-H signal is not due to variation in packing

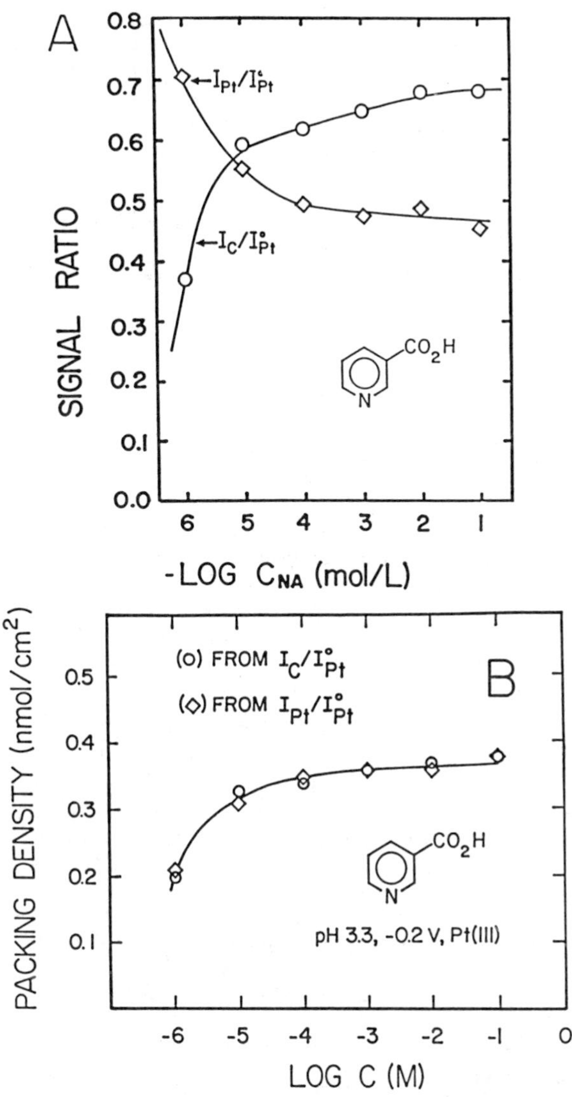

Figure 2. Auger signal ratios and packing densities of NA adsorbed at Pt(111). Experimental conditions: adsorption from 10 mM KF adjusted to pH 7 with HF, followed by rinsing with 1 mM HF (pH 3.3); temperature, 23±1 °C; electron beam at normal incidence, 100 nA, 2000 eV. A. Auger signal ratios: (I_C/I_{Pt}) and (I_{Pt}/I_{Pt}) for NA adsorbed at −0.2 V vs. Ag/AgCl reference. B. Packing density of NA adsorbed at −0.2 V. *Continued on next page.*

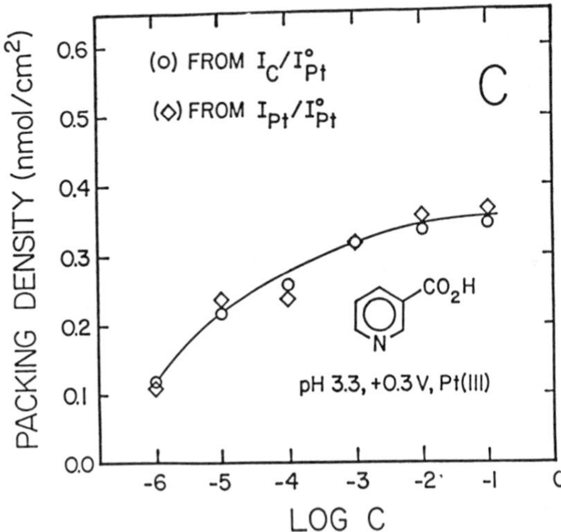

Figure 2. *Continued.* C. Packing density of NA adsorbed at +0.3 V.

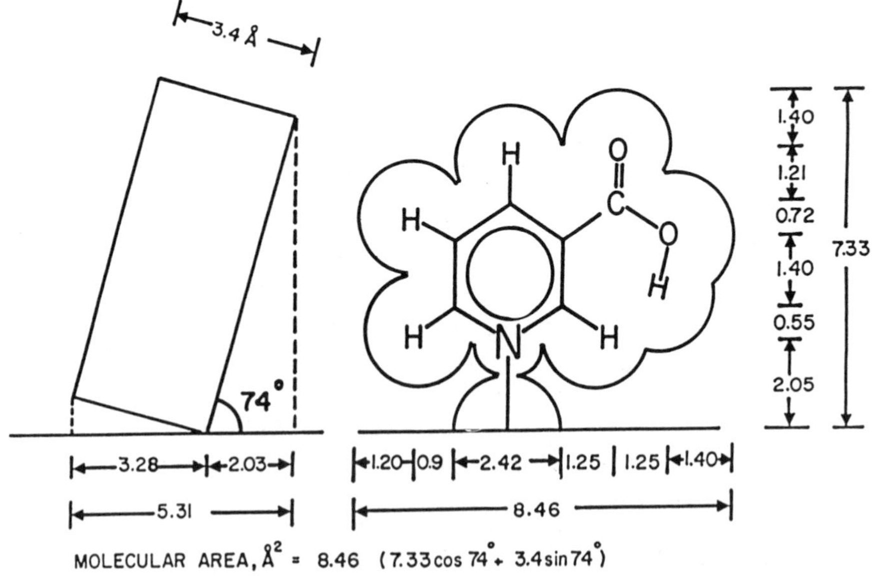

Figure 3. Model illustrating the method of calculation of angle from packing density data.

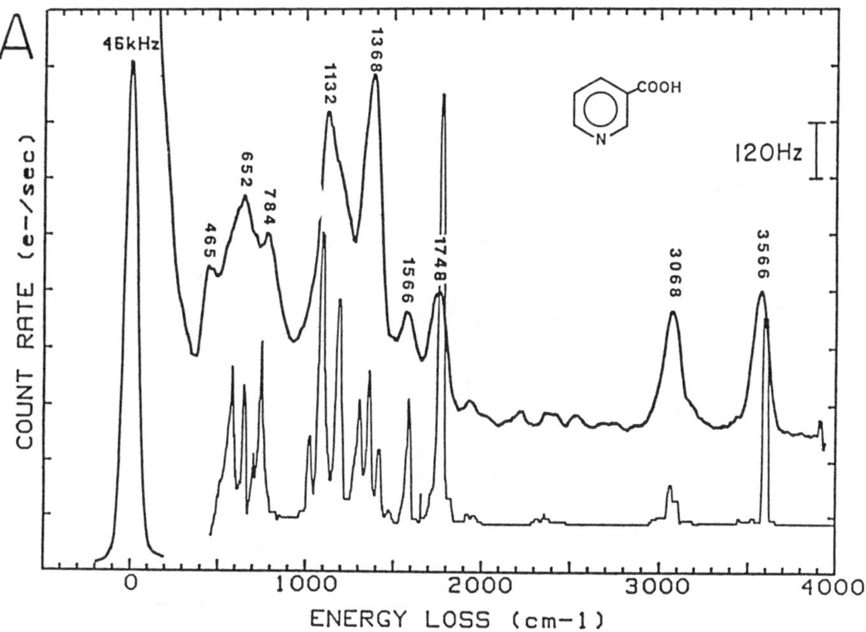

Figure 4. Vibrational spectra of NA. Experimental conditions: adsorption from 1 mM NA in 10 mM KF, pH 3 (A and B) or pH 7 (C), followed by rinsing with 2 mM HF (A and B, pH 3) or 0.1 mM KOH (pH 10 for C); EELS incidence and detection angle 62° from the surface normal; beam energy, 4 eV; beam current about 120 pA; EELS resolution, 10 meV (80 cm^{-1}) F.W.H.M.; IR resolution, 4 cm^{-1}. A. Upper curve: EELS spectrum of NA adsorbed at Pt(111) [pH 3; electrode potential, –0.3 V]. Lower curve in A and B: mid-IR spectrum of NA vapor (18). *Continued on next page.*

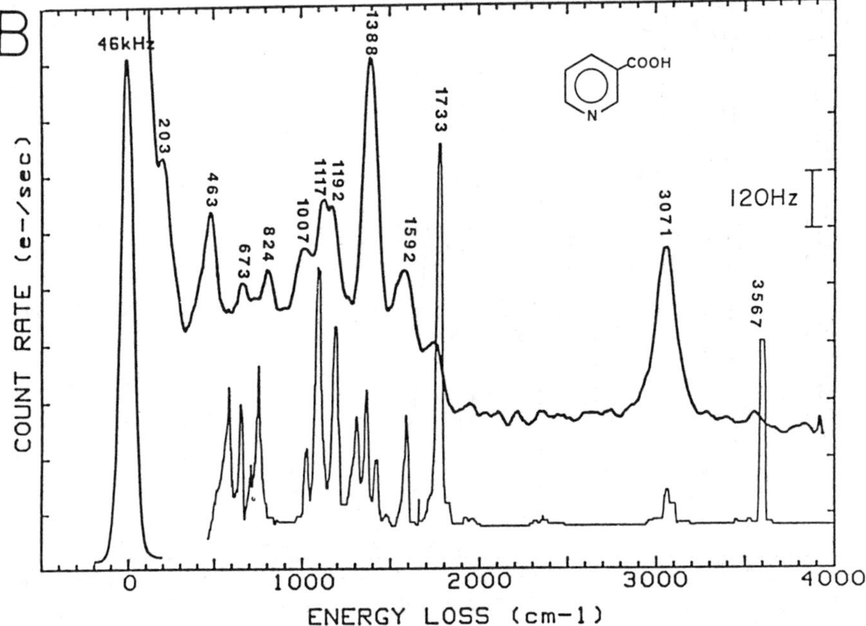

Figure 4. *Continued.* B. Upper curve: EELS spectrum of NA adsorbed at Pt(111). [pH 3; electrode potential, +0.6 V]. *Continued on next page.*

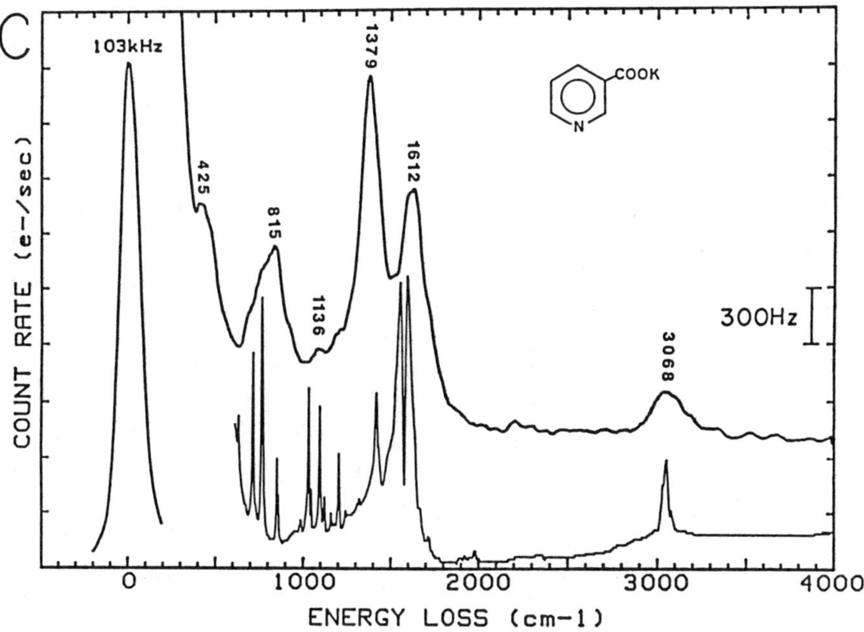

Figure 4. *Continued.* C. Upper curve: EELS spectrum of K^+NA^- adsorbed at Pt(111). [pH 10; electrode potential, –0.3 V].

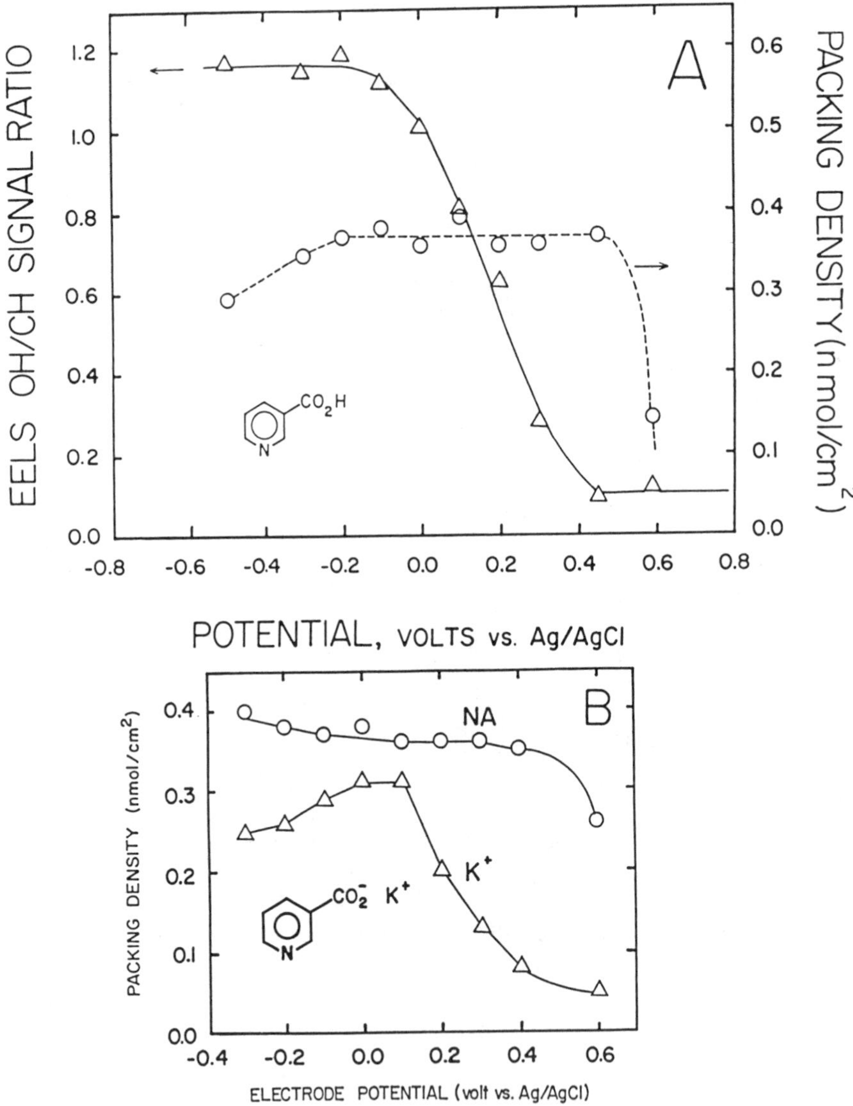

Figure 5. OH–CH signal ratio (EELS) and packing density (Auger) of NA adsorbed at Pt(111) vs. electrode potential. A. Ratio of EELS O–H (3566 cm^{-1}) to C–H (3068 cm^{-1}) peak height. B. Packing density of K^+ ions. Experimental conditions: (A) adsorption from 1 mM NA in 10 mM KF at pH 7, followed by rinsing in 2 mM HF (pH 3); (B) adsorption from 1 mM NA in 10 mM KF at pH 3, followed by rinsing with 0.1 mM KOH (pH 10). EELS conditions as in Figure 4; Auger conditions as in Figure 2.

Table I. Packing Density, Γ(namomol/cm^2)

Ring-to-Surface Angle, Ø (degree)

Compound	Electrode Potential Volt	Γ from I_C/I^o_{Pt}	Γ from I_{Pt}/I^o_{Pt}	Γ_{ec}, from Coulometry	Ø
3PHQ	-0.1	0.28	0.27	0.27	74°
PYR	-0.1	0.46	0.45	----	71°
NA	-0.2	0.36	0.36	----	75°
INA	-0.3	0.42	0.35	----	71°
PA	-0.3	0.28	0.29	----	76°
BA	0.4	0.32	0.35	----	0°
3,5PDA	0.4	0.31	0.29	----	74°
3,4PDA	0.4	0.32	0.26	----	69°
2,3PDA	0.4	0.28	0.26	----	ca 60°
2,4PDA	0.4	0.30	0.27	----	ca 65°
2,5PDA	0.4	0.35	0.29	----	76°
2,6PDA	0.4	0.31	0.27	----	75°

Experimental conditions: 1mM adsorbate in 10mM KF (pH3), followed by rinsing with 2mM HF (pH3).

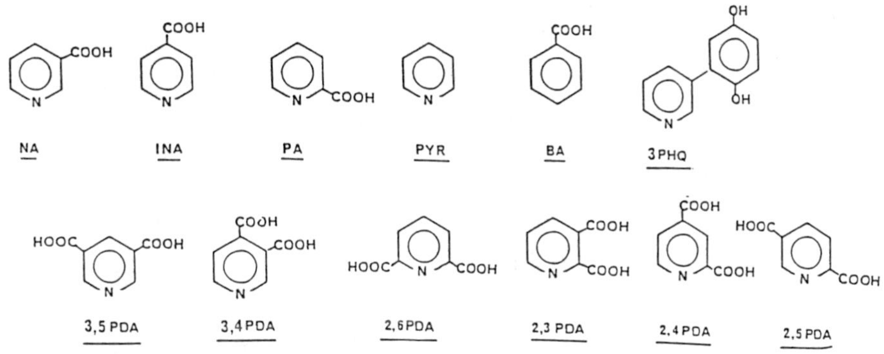

NA INA PA PYR BA 3PHQ

3,5 PDA 3,4 PDA 2,6 PDA 2,3 PDA 2,4 PDA 2,5 PDA

Table II - Assignments of EELS Bands of Adsorbed Nicotinic Acid and Pyridine

Compound	pH/ Electrode Potential	Peak Frequency cm^{-1}	Primary Symmetry Species	Description
NA	3/-0.2	3566	C_S: A'	O-H stretch
		3068	A'	C-H stretch
		1748	A'	C=O stretch
		1566	A'	CC stretch
		1368	A'	CC, CN stretch
		1132	A',A"	C-O stretch; C-H, O-H bend
		784	A',A"	OCO, C-H bend
		652	A',A"	CC, OCO bend
		465	A",A'	ring; Pt-N stretch
NA	3/+0.6	3071	C_S: A'	C-H stretch
		1733	A'	C=O stretch
		1592	A'	CC stretch
		1388	A',A"	CC,CN stretch, C-H bend
		1192	A'	C-H bend
		1117	A'	C-O stretch
		1007	A',A"	ring, C-H bend
		824	A"	OCO, ring bend
		673	A"	OCO, ring bend
		463	A",A'	ring bend; Pt-N stretch
		203	A'	Pt-O stretch
PYR	3/-0.1	3055	C_{2V}: A_1,B_2	C-H stretch
		1537	A_1,B_2	CC stretch
		1471	A_1,B_2	CC, CN stretch
		1250	B_1,B_2	C-H bend
		1134	$A_1;B_2$	C-H bend; X-sens.
		1000	A_1,A_2,B_2	ring; C-H bend; X-sens.
		719	B_1	ring, C-H bend
		416	$B_1;A_1$	ring bend; Pt-N stretch

density, Figures 4 and 5. The disappearance of the O-H vibration with increasing positive potential is due to coordinate covalent bonding of the carboxylic acid moiety with the Pt surface, Equation 5.

At negative potentials in alkaline solutions, adsorbed NA retains K^+ ions, as demonstrated by Auger spectroscopy, Figure 5-B. This retention of K^+ ions is due to interaction of K^+ with the pendant carboxylate moiety and greatly exceeds the amounts expected simply from diffuse double-layer interactions. Potential-dependence of K^+ retention is essentially absent for compounds incapable of potential-dependent carboxylate pendancy (pyridine, picolinic acid, isonicotinic acid and 2,6-pyridine dicarboxylic acid).

Shown in Figure 6-A are EELS spectra of the entire series of pyridine carboxylic acids and diacids adsorbed at Pt(111) from acidic solutions at negative electrode potential. Under these conditions all of the meta and para pyridine carboxylic acids and diacids exhibit prominent O-H vibrations (OH/CH peak ratio near unity). In contrast, at positive potentials only the para-carboxylic acids display pronounced O-H vibrations, Figure 6-B. All of the O-H vibrations are absent under alkaline conditions, Figure 6-C. This situation is illustrated by the reactions of adsorbed 3,4-pyridine dicarboxylic acid:

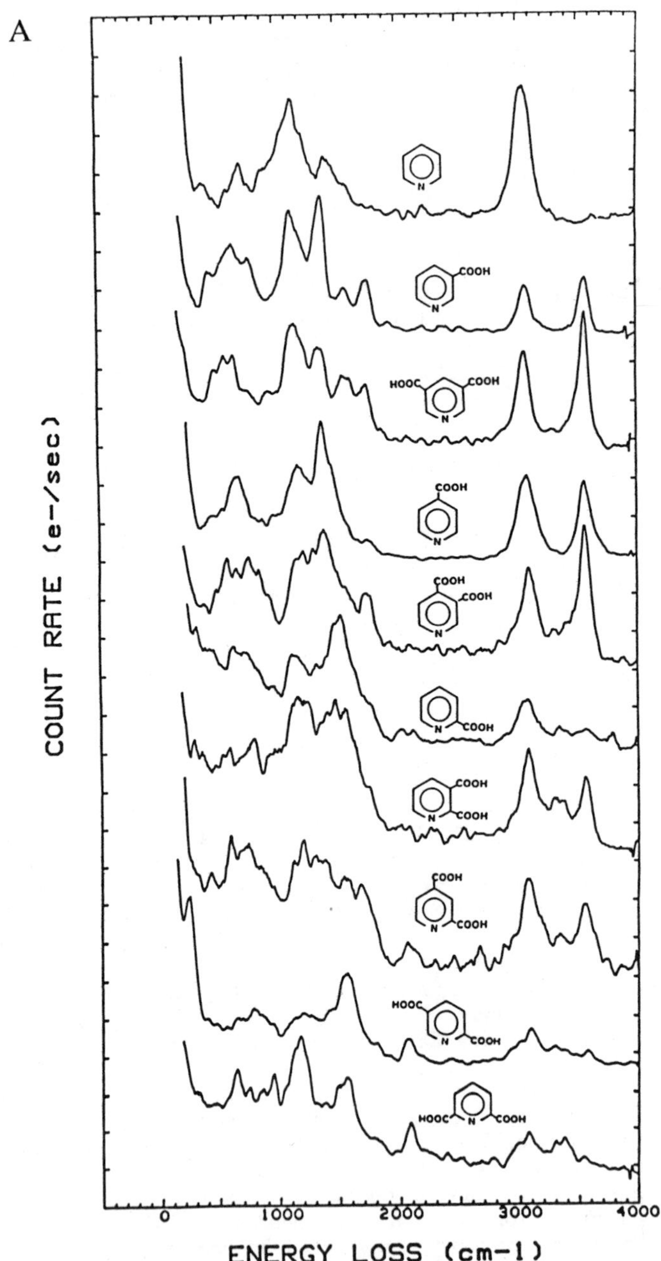

Figure 6. EELS spectra of pyridine carboxylic acids adsorbed at Pt(111). Experimental conditions: (A and B) adsorption from 1 mM NA in 10 mM KF at pH 3, followed by rinsing with 2 mM HF (pH 3); (C) adsorption from 10 mM KF (pH 3), followed by rinsing with 0.1 mM KOH (pH 10); other conditions as in Figure 4. A. Adsorption at –0.2 V vs. Ag/AgCl (pH 3). *Continued on next page.*

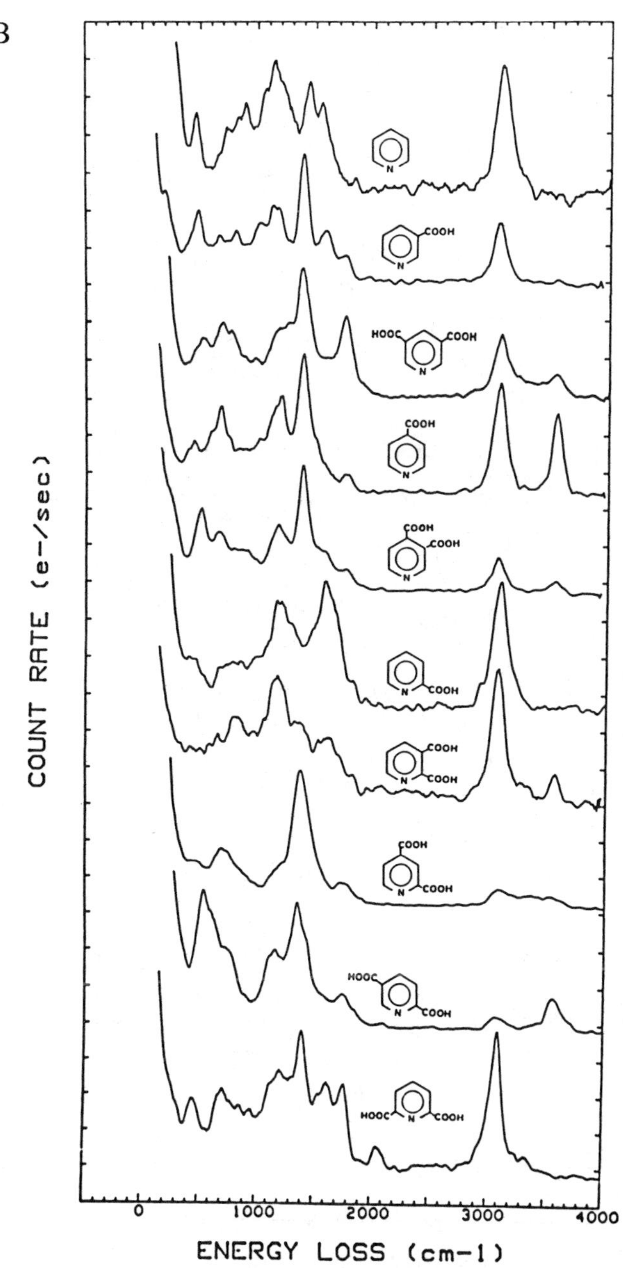

Figure 6. *Continued.* B. Adsorption at +0.6 V (pH 3). *Continued on next page.*

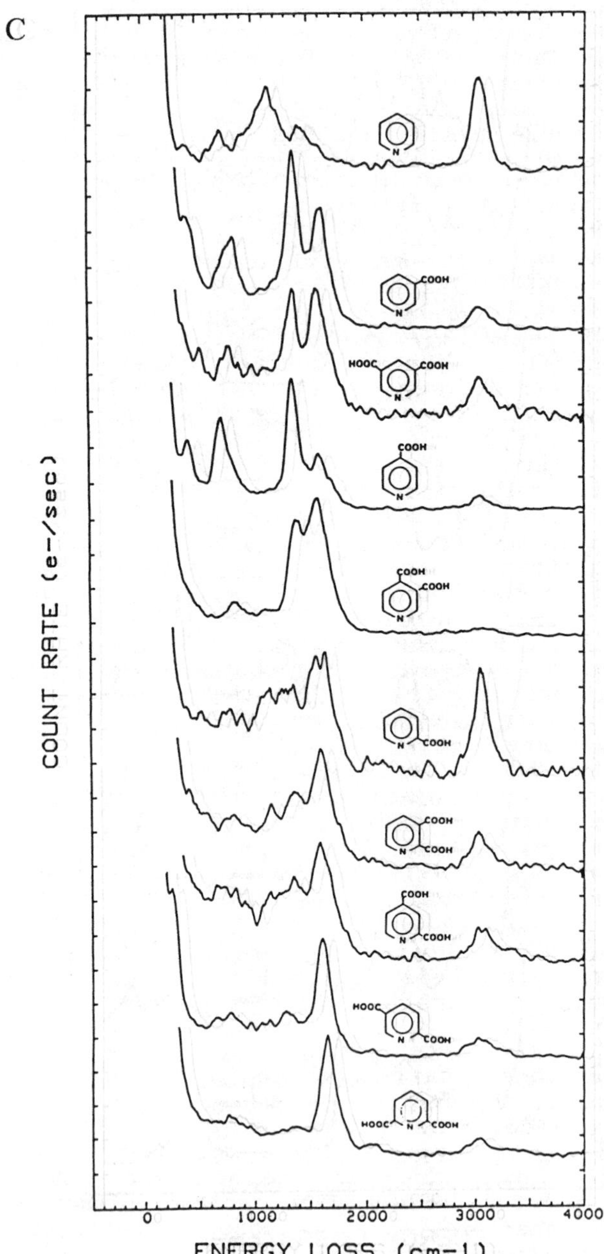

Figure 6. *Continued.* C. Adsorption at –0.2 V (pH 10).

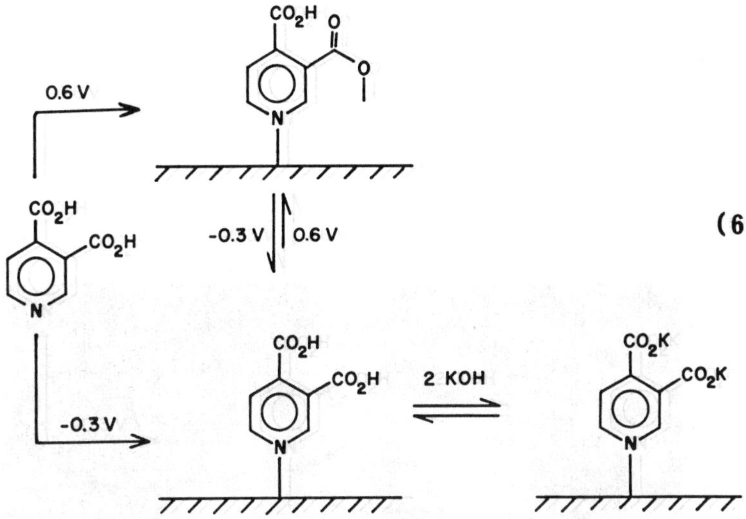

(6)

In summary, para carboxylic acid groups remain active at all potentials in the useful range, while meta carboxylates are complexed with the Pt surface at positive potentials unless sheltered from the surface by ortho-substituents. Ortho-carboxylates are complexed to the Pt surface over most or all of the useful potential ranges.

LEED patterns of pyridine adsorbed at Pt(111) are shown in Figure 7-A. Best clarity of the pattern occurs when the PYR concentration is 1mM. Measurement of the lengths and directions of LEED vectors in Figure 7-A, followed by conversion of the LEED vectors to real space by means of standard formulas, reveals that the structure is approximately (3.3x4.7) with an included angle of 77°. Digital simulation of the LEED pattern demonstrates that the best fit between theory and experiment occurs when the model structure is Pt(111)(3.324x4.738, 77.1^o)R34.0^o-PYR. In matrix notation this is:

$$\begin{vmatrix} a_1 & a_2 \\ b_1 & b_2 \end{vmatrix} = \begin{vmatrix} 1.684 & 2.145 \\ -4.255 & 5.106 \end{vmatrix}$$

The simulated LEED pattern is shown in Figure 7-B. Comparison of the simulated LEED pattern with the observed LEED pattern (lower camera shutter speed) is shown in Figure 7-C. As can be seen, there is good agreement. This PYR layer structure is incommensurate; that is, the mesh vectors of the layer are not exact multiples of the substrate mesh. The nearest commensurate structure would have been ($2\sqrt{3}$x$\sqrt{21}$, 79°)R30^o, Figure 7-D. Although this commensurate alternative is numerically quite similar

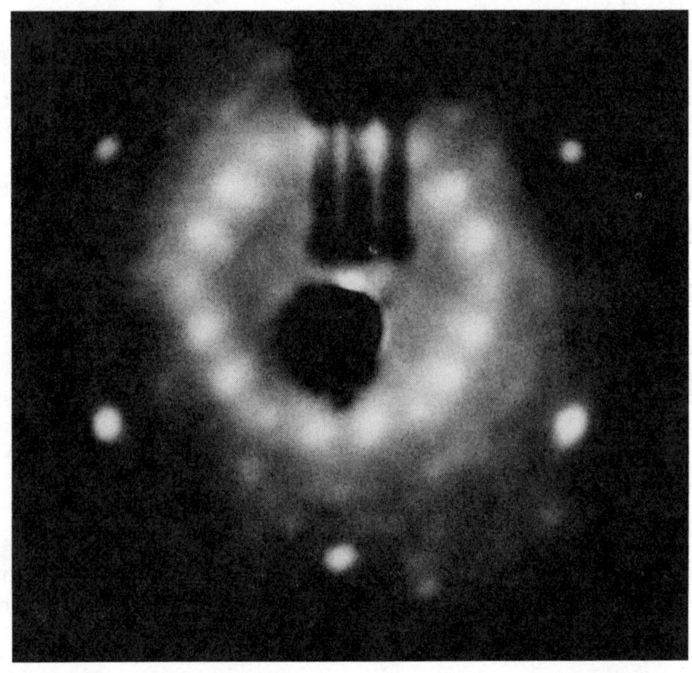

Figure 7. LEED pattern and structure of pyridine at Pt(111). A. LEED pattern of PYR adsorbed at Pt(111), 51 eV. *Continued on next page.*

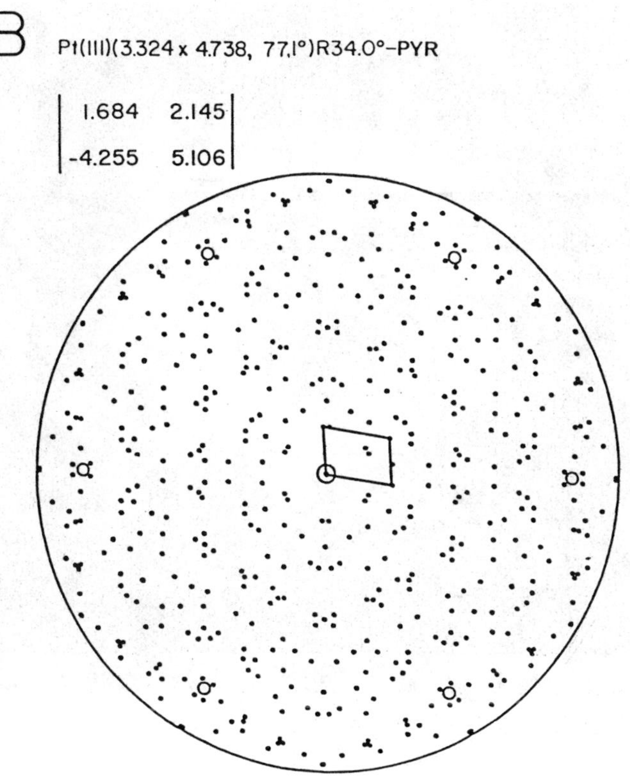

Figure 7. *Continued.* B. Diagram of the LEED pattern in A. *Continued on next page.*

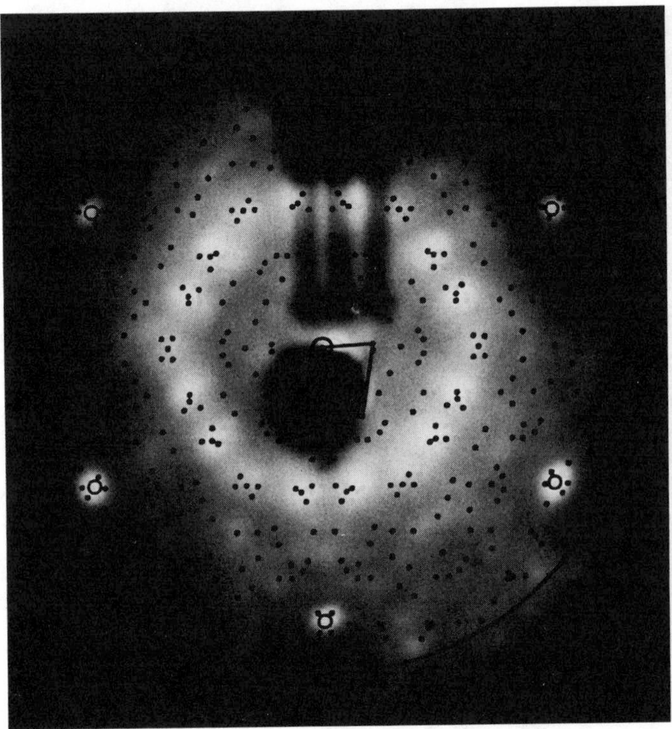

Figure 7. *Continued.* C. Comparison of calculated and observed LEED patterns. *Continued on next page.*

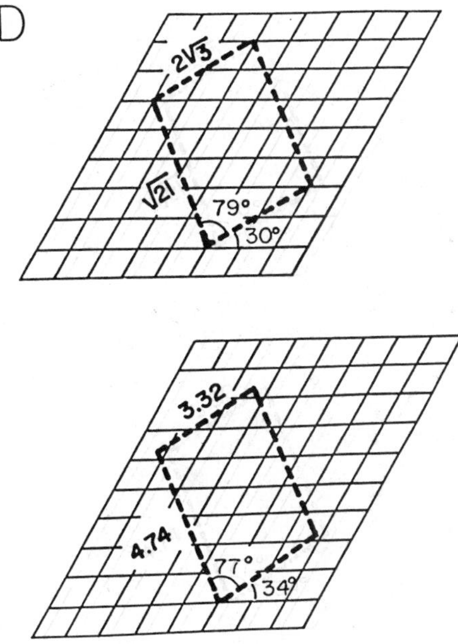

Figure 7. *Continued.* D. Diagrams of observed vs. nearest commensurate meshes. *Continued on next page.*

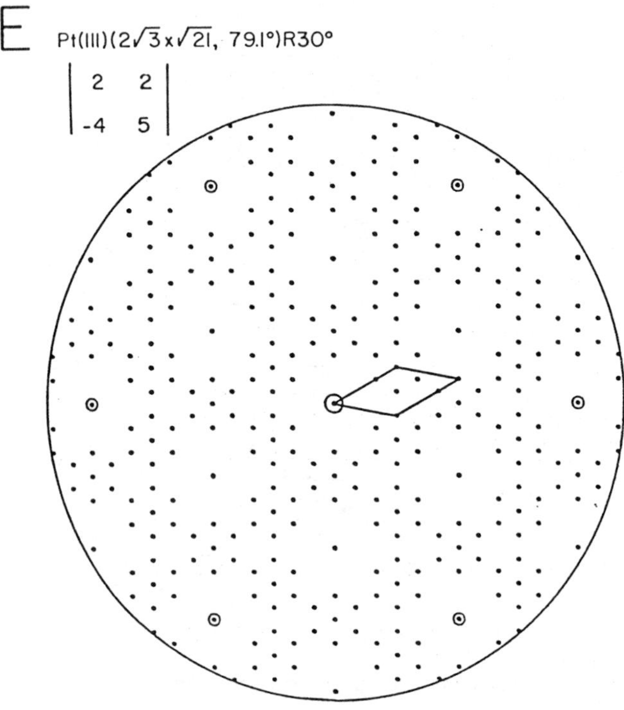

Figure 7. *Continued.* E. Diagram of calculated LEED pattern corresponding to nearest commensurate structure, $(2\sqrt{3}x\sqrt{21}, 79.1°)R30°$. *Continued on next page.*

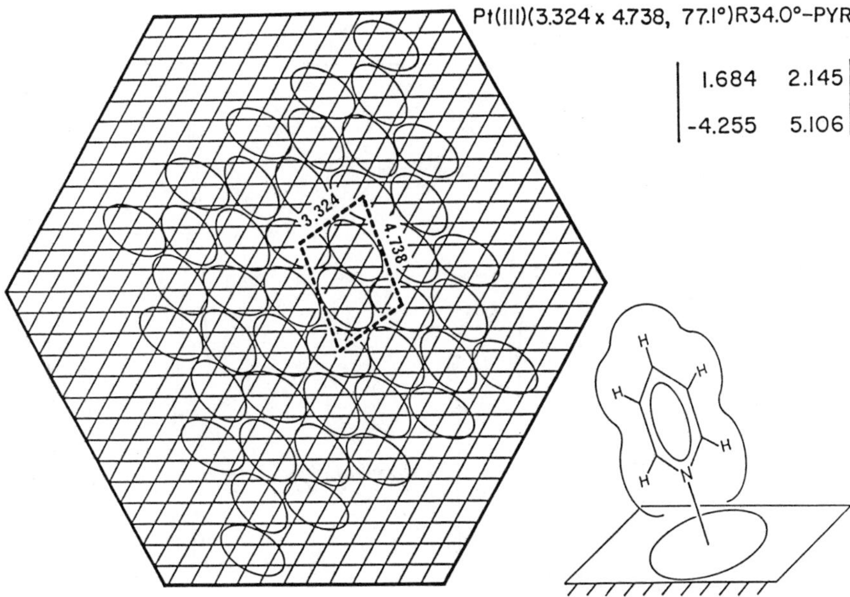

Figure 7. *Continued.* F. Structure of the PYR adsorbed layer: Pt(111)(3.324 x 4.738, 77°)R34°-PYR.

In matrix notation: $\begin{vmatrix} 1.684 & 2.145 \\ -4.255 & 5.106 \end{vmatrix}$

The PYR packing density in this structure is 0.421 nmol/cm^2.

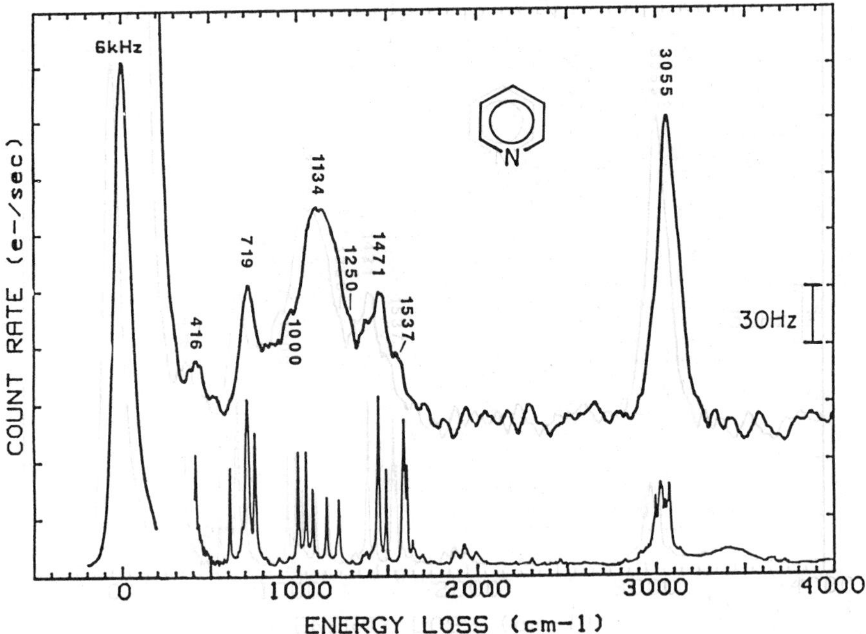

Figure 8. Vibrational spectra of pyridine. Upper curve: EELS spectrum of PYR adsorbed at Pt(111) (pH 3); lower curve: mid-IR spectrum of liquid PYR (18). Experimental conditions: adsorption at −0.1 V from 1 mM PYR in 10 mM KF (pH 3), followed by rinsing with 2 mM HF (pH 3); other conditions as in Figure 4.

to the proposed incommensurate structure, it would have resulted in a very different LEED pattern, Figure 7-E (3.464x4.583, 79°)R30^o or:

$$\begin{vmatrix} 2 & 2 \\ -4 & 5 \end{vmatrix}$$

Auger spectra for PYR indicate that the packing density at the saturation limit is 0.45 nmol/cm^2 based upon (I_C/I^o_{Pt}) or 0.46 nmol/cm^2 from (I_{Pt}/I_{Pt}). A striking characteristic of the PYR isotherms (16) is the virtual constancy of packing density over five order of magnitude in PYR concentration (10^{-4}M to neat PYR, 12M). These data point to a tilted structure with a ring-to-surface angle, $\emptyset=71^o$, Figure 7-F. Based upon three PYR molecules per unit cell, the packing density for this structure is 0.421 nmol/cm^2, in good agreement with experiments. The rigid constancy of PYR packing density and the incommensuracy of the PYR adsorbed layer are indications that the 71° angle of adsorbed PYR results from a balancing of forces between sigma-donor bonding of the nitrogen atom to Pt(111) and pi back-donation from Pt(111) into the aromatic ring. More extensive C-Pt interaction would have resulted in energy setbacks due to decreased aromatic character and decreased nitrogen sigma-donation.

An EELS spectrum of PYR adsorbed at Pt(111) from aqueous solution is shown in Figure 8. Also shown is the mid-IR spectrum of liquid PYR (18). The EELS spectrum of adsorbed PYR is essentially the envelope of the IR spectrum of liquid PYR, with the exception of a peak at 416 cm^{-1} attributable at least in part to the Pt-N bond. Assignments of the EELS peaks based upon accepted IR assignments (19) are given in Table 2. There is also a close correspondence between the EELS spectrum of PYR adsorbed from aqueous solution and the EELS spectra reported for PYR adsorbed at Pt single-crystal surfaces from vacuum (20).

References:

1. Hubbard, A.T. Accounts of Chemical Research 1980, 13, 177.
2. Hubbard, A.T. J. Vac. Sci. Technol. 1980, 17, 49.
3. Hubbard, A.T.; Stickney, J.L.; Soriaga, M.P.; Chia, V.K.F.; Rosasco, S.D.; Schardt, B.C.; Solomun, T.; Song, D.; White, J.H.; Wieckowski, A. J. Electroanal. Chem. 1984, 168, 43.
4. Hubbard, A.T. Chemical Reviews 1988, in press.
5. Hubbard, A.T. in "Comprehensive Chemical Kinetics", Bambord, C.H., Tipper, D.F.H., Compton, R.G., eds., Vol. 28, Chapter 1, (Elsevier, Amsterdam, 1988).
6. Somorjai, G.A. "Chemistry in Two Dimensions: Surfaces", (Cornell University Press, Ithaca, NY, 1981).
7. (a) Duke, C.B. Adv. Chem. Phys 1974, 27, 215;
   (b) Estrup, P.J. in "Characterization of Metal and Polymer Surfaces", L.H. Lee, ed., (Academic Press, NY, 1977), Vol. 1, pp 187ff;
   (c) Somorjai, G.A.; Farrell, H.H. Adv. Chem. Phys. 1971, 20, 215.
8. (a) Chang, C.C. Surface Sci. 1971, 25, 53;
   (b) Hawkins, D. J., "Auger Electron Spectroscopy, A Bibliography, 1927-1975" (Plenum, NY, 1977);
   (c) Somorjai, G.A.; Szalkowski, F.J. Adv. High Temp. Chem. 1971, 4, 137;

(d) Taylor, New Jersey, in "Techniques of Metal Research", R. F. Bunshah, ed. (Wiley-Interscience, NY, 1971), Vol. 7, pp 117ff;
(e) Thompson, M.; Baker, M. D.; Christie, A.; Tyson, J.F. "Auger Electron Spectroscopy" (Wiley, NY, 1985).
9. (a) Betteridge, D. "Photoelectron Spectroscopy: Chemical and Analytical Aspects" (pergamon, NY, 1972);
(b) Carlson, T.A. "Photoelectron and Auger Spectroscopy", (Plenum, NY, 1975);
(c) Delgas, W.N.; Hughes, T.R.; Fadley, C.S. Catalysis Revs. 1970 4, 179;
(d) Herglotz, H.K.; Suchan, H.L. Adv. Coll. Interf. Sci. 1975, 5, 79.
10. Hoenig, R.E. Adv. Mass Spectrom. 1974, 6, 337.
11. (a) Sexton, B.A. Applied Phys. 1981, 126, 1;
(b) Kesmodel, L.L. J. Vacs. Sci. Technol. 1983, A1, 1456.
12. Golden, W.G., in "Fourier Transform Infrared Spectroscopy", Ferraro, J.R. and Bastille, L.J., eds., (Academic Press, Orlando, Florida, 1985), Vol. 4, pp. 315ff.
13. Campion, A. Chem. Phys. Lett. 1987, 135, 501.
14. Hubbard, A.T. Critical Revs. Analytical Chem. 1973, 3, 201.
15. Delahay, P., "Double and Electrode Kinetics" (Wiley, NY, 1965).
16. Stern, D.A.; Laguren-Davidson, L.; Frank, D.G.; Gui, J.Y.; Lin,C.H.; Lu, F.; Salaita, G.N.; Walton, N.; Zapien, D.C.; Hubbard, A.T. J. Amer.Chem.Soc.
17. Pauling, L.C. "The Nature of the Chemical Bond" (Cornell Univ. Press, Ithaca, NY, 1960), 3rd edition.
18. (a) Pouchert, C.J."The Aldrich Library of FTIR Spectra" (Aldrich Chemical Co., Inc., Milwaukee, Wis., 1985).
(b) "The Interpretation of Vapor Phase Spectra" (Sadtler Research Labs., Philadelphia, 1984), Volume 2.
19. Green, J.H.S.; Dynaston, W.; Paisley, H.M. Spectrochim. Acta 1963, 9, 549.
20. (a) Grassian, V.H.; Muetterties, E.L. J. Phys. Chem. 1986, 90, 5600;
(b) Surman, M.; Bare, S.R.; Hoffman, P.; King, D.A. Surface Sci. 1987, 179, 243.

RECEIVED June 29, 1988

Chapter 3

Long-Range Structural Effects in the Anomalous Voltammetry of Platinum(111)

Philip N. Ross, Jr.

Materials and Chemical Sciences Division, Lawrence Berkeley Laboratory, University of California, Berkeley, CA 94720

The history of the observation of anomalous voltammetry is reviewed and an experimental consensus on the relation between the anomalous behavior and the conditions of measurement (e.g., surface preparation, electrolyte composition) is presented. The behavior is anomalous in the sense that features appear in the voltammetry of well-ordered Pt(111) surfaces that had never before been observed on any other type of Pt surface, and these features are not easily understood in terms of current theory of electrode processes. A number of possible interpretations for the anomalous features are discussed. A new model for the processes is presented which is based on the observation of long-period ice-like structures in the low temperature states of water on metals, including Pt(111). It is shown that this model can account for the extreme structure sensitivity of the anomalous behavior, and shows that the most probable explanation of the anomalous behavior is based on capacitive processes involving ordered phases in the double-layer, i.e., no new chemistry is required.

Prior to the publication in 1980 of Clavilier's historic paper (1) reporting anomalous voltammetry of Pt(111), there had been a number of studies of the voltammetry of single crystal Pt electrodes, with some using modern methods of surface analysis (e.g., LEED or RHEED) for characterization of the structure of the crystal prior to immersion in electrolyte (2-6), and all were in qualitative agreement with the seminal work (in 1965) on Pt single crystals by Will (7). By today's standards of surface preparation, Will's procedures for surface preparation were crude, the surface structures were not characterized by use of surface analytical instrumentation (which was neither widely available nor well developed at that time), and he employed extensive potentiodynamic cycling through the "oxide" formation potential region prior to reporting the quasi-steady state voltammetry curve, i.e., the potentiodynamic I-V curve. The studies employing surface analytical methods made a decade or more later were

0097–6156/88/0378–0037$06.00/0

generally regarded as validating the major conclusions of Will's study: 1) that the relative bond strength of hydrogen on the low index surfaces of Pt are (100) > (110) > (111); 2) that the multiple features in the voltammetry curve for polycrystalline Pt in acid electrolyte are associated with adsorption on sites having the different local geometries of the (100), (110) and (111) planes. However, with respect to the quantitative details of the results, there were major discrepancies between different groups, particularly with respect to the charge under the I-V curve in the hydrogen adsorption potential region from Pt(111) surfaces, and to the effect of potential cycling (like that employed by Will) on the amount of this charge and on the detailed shape of the I-V curve (8). It was clear, at least to this author (2,6), that the discrepancies in the results between different groups using UHV methods were caused by the interactive effect between impurities and transformations to the surface structure caused by the potential cycling used to clean the surface of impurities. The use of oxygen atmospheres during transfer of the crystal from UHV to the electrolyte appeared to remove the impurity effect, and produced I-V curves which were essentially indifferent to potential cycling (9). With the publication of these results, the problem of understanding the multiple states of adsorbed on Pt electrodes appeared to be fully resolved, with Will's original site-geometry interpretation upheld and further refined.

There was no precedent in the published or even unpublished history of studies with Pt electrodes (single crystal or otherwise) for the type of I-V curve reported by Clavilier (1) for Pt(111), and his observations might have been dismissed for a time had they not been defended so strongly by Parsons (10). Motivated by subsequent communications with Parsons' on the integrity of Clavilier's results, we accelerated the pace of modifications we were making to the UHV-electrochemistry apparatus in my laboratory, converting the cell to a thin-layer type (11) with the hope of reducing the effect of electrolyte impurities on the voltammetry. With the use of this cell, we were able to produce the first corroboration of the Clavilier type voltammetry with a Pt(111) crystal prepared in UHV and characterized by LEED as a highly ordered (111)-1X1 surface (12). This voltammetry is reproduced in Figure 1a. The sensitivity of this I-V curve to the atomic scale "flatness" of the surface is illustrated by the comparison with the curve 1b, which is the same crystal ion-bombarded but not annealed. If the annealed surface is cycled potentiodynamically through "oxide" formation, the shape of the I-V curve changes rapidly, and is transformed into one resembling Figure 1b, and many others reported previously for "cycled" (111) electrodes (4-7,9). The features which occur in the scans for the annealed (111) surface in the potential range 600-850 mV are the anomalous features first reported by Clavilier. They are anomalous in the sense that they had never been reported before with any Pt electrode, single crystal or polycrystalline, and they occur in a potential region which form thermodynamic considerations (12) indicates they are unlikely to be either adsorbed hydrogen or adsorbed oxygen.

To obtain these startling new results with Pt single crystals, Clavilier employed a novel surface preparation technique that did not require a complex and expensive UHV system, but did require considerable skill and experience. Clavilier's original papers

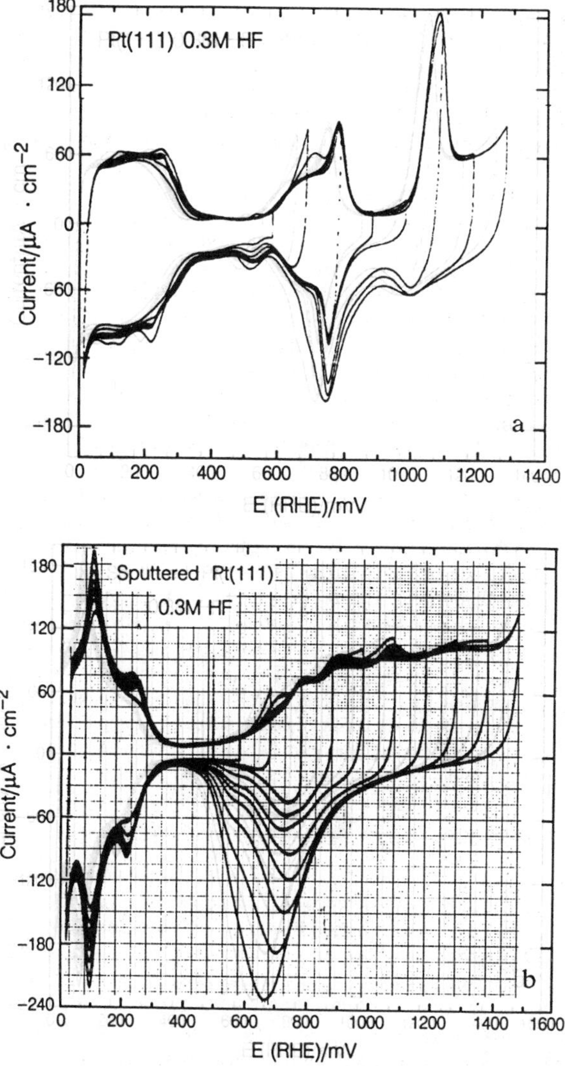

Figure 1. a) Cyclic voltammetry curve recorded from a UHV prepared, well-annealed Pt(111) crystal in dilute HF electrolyte. Crystal was immersed under potential control at 0.6 V (RHE) and swept cathodically to the limiting potential; b) same experiment performed with the same crystal but not annealed after ion bombardment.

should be referred to with regards to the details of the method, as I shall not discuss it further here, and only refer to it subsequently as the "bead method." Because the bead method does not require UHV systems, it can be practiced by a wider number of interested electrochemists, and this has proven to be (and will continue to be) a significant factor in the development of an understanding of the anomalous voltammetry, i.e., what species/processes are involved. A number of laboratories around the world were soon able to use the bead method to reproduce, in virtually every detail, Clavilier's original voltammetry, and have proceeded to use the method in studies of the chemistry of the anomalous features. The particular results with the bead method that I shall refer to in this review are from studies of the effect of anion type and concentration (13-16) and of the effect of atomic scale roughness (observed using step-terrace surface structures) on the potential region and integral charge of the anomalous features (17,18).

In contrast to the successful implementation of the bead method in studying the anomalous features, the contributions from studies with UHV-electrochemical systems has been limited to just a few. Subsequent work from our apparatus following corroboration of Clavilier's results concentrated on the effect of potential cycling through "oxide" formation potential on the surface structure (19), and later on the effect of pH and type of anion (Wagner, F.T.; Ross, P.N., J. Electroanal. Chem., in press) on the anomalous features. Using the system in Yeager's laboratory, Hanson (20) was able to reproduced Clavilier's voltammetry not only for the (111) surface, but also the (100) and (110) surfaces as well. In spite of the relatively small number of contributions to the literature that have come from the UHV-electrochemical systems, they have made and essential validation of the bead method of surface preparation, and have verified the structure sensitivity of the anomalous features inferred from purely electrochemical observations.

Possible Interpretations of the Anomalous Features

Table I gives a summary of the observational data base which I shall use here for a discussion of possible interpretations for the anomalous voltammetry. The discussion here is of necessity brief, and only summarizes the detailed discussion presented in our recent paper (Wagner, F.T.; Ross, P.N., J. Electroanal. Chem., in press). The anomalous features are observed on well-ordered (111) surfaces in a variety of electrolytes over a wide range of pH (0-11), but the potentials at which the features appear and the detailed shapes of the I-V curves vary considerably. Specifically, the potential region (versus RHE) in which the features appear changes with anion concentration in sulphate and chloride electrolytes, but not in fluoride, perchlorate, bicarbonate or hydroxide electrolyte. In sulfate electrolyte, at constant anion concentration the region shifts (versus RHE) with varying pH, while in fluoride, perchlorate, bicarbonate and hydroxide electrolyte it does not. The use of UHV surface analytical techniques has established to a reasonable (but not definitive) extent that adventitious impurities are not involved in the anomalous process, i.e., the only species participating in the chemistry are protons/hydroxyls, water and the anions of the solute. On the basis of the pH and anion concentration dependencies, I agree with the

Table I. Summary of observations of the anomalous features

- o Appearance of anomalous features on Pt(111) requires critical degree of long-range order, e.g., perfectly ordered domains 12-25Å
- o Potential region depends on the type of anion, and anion concentration and pH as follows:

In sulfate and/or chloride	cathodic shift with increasing C_{A-} at constant pH
	no shift vs. SCE at constant C_{A-} with varying pH
In perchlorate fluoride, carbonate and/or hydroxide	no shift with increasing C_{A-} at constant pH
	59 mV/pH vs. SCE at constant C_{A-} with varying pH

- o In sulfate and/or chloride, there is a conservation of charge integrated under the potential region for "normal hydrogen" and the anomalous features with $Q \simeq 280\ \mu C/cm^2$. Roughening causes the normal hydrogen charge to increase with a commensurate decrease in charge for anomalous features.
- o In fluoride, perchlorate and/or hydroxide, there is no simple conservation of charge between the normal hydrogen region and the anomalous features. Roughening causes redistribution of charge both anodic and cathodic to anomalous features.

interpretation put forward by Kolb and co-workers (15) that the anomalous features on Pt(111) in sulfate and chloride acid solutions are associated with processes of specific adsorption of these anions. The identity of the species and processes involved in the anomalous features observed on Pt(111) in sulfate and chloride acid solutions are associated with processes of specific adsorption on these anions. The identity of the species and processes involved in the anomalous features observed in perchlorate, fluoride, bicarbonate and hydroxide electrolyte remains more uncertain than in the cases of the sulfate and chloride electrolytes. Primarily by the process of elimination, i.e., hydrogen adsorption and specific adsorption of anions appear highly unlikely, the process in acidic fluoride and perchlorate electrolytes is most probably related to water dissociation, and the species are most probably surface intermediates of this process, i.e., oxygen-like. In basic electrolytes, it appears likely, by direct analogy with the process in sulfate electrolytes, that specific adsorption of hydroxyl anion is the process there. The specifically adsorbed hydroxyls (pH > 5) may have similar final state molecular configuration on the surface even though they derive from different initial states. It is not clear how these hydroxyl/oxygen-like species differ from the place-exchanged OH formed on the surface at higher potentials (21). It is even more difficult to explain why the molecular configuration of any of these species would be unique to the (111) surface, nor why roughening the (111) surface should cause the process to disappear entirely. In the discussion which follows, a model is presented which attempts to resolve these difficulties within the framework of classical double-layer theory but using new concepts in ordered structures for the double-layer at Pt(111).

Understanding Sensitivity to Long-Range Surface Order

I know of no other examples in the surface science literature in which the adsorption of a small molecule (small in this case referring to the values of molecular/ionic radii of the adsorbate relative to the size of perfectly flat regions on the metal surface) depends critically on the existence of large regions of perfect surface structure. There are some examples of such sensitivity in the adsorption of aromatic molecules, where the "lying-down" configuration, which requires attachment to multiple points on the surface, depends critically on the existence of flat domains of size greater than the molecule itself (see for example the discussion by Hubbard in his chapter in Chemical Reviews, "Electrochemistry at Well-Characterized Surfaces," The American Chemical Society, Washington, D.C., in press). Because these aromatic molecules provided the principal precedent for such structural effects, we spent a great deal of time in my laboratory, perhaps 2-3 man years of effort, pursuing the "impurity hypothesis" based on an adventitious organic impurity. When this hypothesis was found wanting from our studies, we turned to concepts of ordered networks in the double-layer. Fred Wagner, who worked with me on these studies of the anomalies on Pt(111), has pursued the concept of ordered networks in his own studies, as he reports in another paper in this conference. In the discussion which follows, I present a conceptual model for the double-layer at the Pt(111) surface which uses what has been learned about hydrogen bonding in adsorbed water layers from studies of water adsorption on metal surfaces at low temperature (<150 K) and extends the one-dimensional classical models (22) into a three-dimensional model. This model is then used to formulate a mechanism to explain how sulphate anion adsorption would be sensitive to the symmetry of the surface, e.g., (111) vs. (100) or (110), and a critical level of atomic roughness in the surface, e.g., the steps and step-terrace structures.

Fortunately, the large, but diverse literature on the adsorption of water on metals has been collected in a comprehensive manner in the recent review by Thiel and Madey (23). As shown in their review, it is now well established that water forms ordered superlattice structures at low temperatures (e.g., 100°K) on a number of metal surfaces having hexagonal symmetry, i.e., fcc(111) and hcp(001), including and especially Pt(111). The superlattice structures have ($\sqrt{3}$ x $\sqrt{3}$) R-30° symmetry indicated by LEED, and models have been derived showing that such symmetry is expected for long-period ice-like structures. The most detailed and convincing study of these water superlattices was by Doering and Madey (24) for water on Ru(001), the hexagonal surface for the hcp metal. They supplemented their LEED analysis with ESDIAD (electron stimulated desorption-ion angular distribution), which provided direct indication of the orientation of water molecules on the surface, and applied the Bernal-Fowler-Pauling symmetry rules for hydrogen bonding in bulk ice to derive the surface hydrogen bonded network shown in Figure 2. In the first water layer of this bilayer structure, every other water molecule forms a bond to the surface via the oxygen lone pair and two hydrogen-bonds to the oxygen lone pairs of the water molecules in the hydrating part of the bilayer (the puckered positions). Water bilayers not directly bound to the Ru surface were not strongly bound into a regular structure, i.e., the structure is highly ordered only in the first bilayer. There is a close match (within 1.5%) between the Ru and ice lattice parameters, but Thiel and Madey show that this epitaxial relation is not essential in forming the ice-like structure on other metal surfaces, with hexagonal symmetry provided the metal lattice parameter falls within certain limits. The ice-like structures have not been observed on fcc(100), (110) or (111)-stepped surfaces of any metal, even those that form such structures on the (111) surfaces. Thiel and Madey argue that the absence of ice-like structures on these surface is not due to the lack of appropriate epitaxial relations, but to the magnitude of the bond energy of water on the differential surfaces (and at imperfections) in relation to the strength of the hydrogen bond. On numerous fcc(111) surfaces, the metal-water bond energy is equal to twice the OH--O hydrogen bond energy (21 kJ per mole of hydrogen bonds), so that the breaking of the ice symmetry to form an additional metal-water bond (e.g., lowering the water molecules from hydrogen-bonded sites to chemisorption sites) does not result in a net lowering of the surface energy. On the more open fcc surfaces such as (100) or (110), or on "rough" surfaces, the metal-water bond energy is reported to be significantly higher, by 10-20 kJ per mole, so that the ice symmetry is not favored over a structure having a greater number of water molecules bonded to chemisorption sites on the metal surface.

There is no a-priori relationship between these ice-like structures formed on fcc(111) metal surfaces at low temperature and

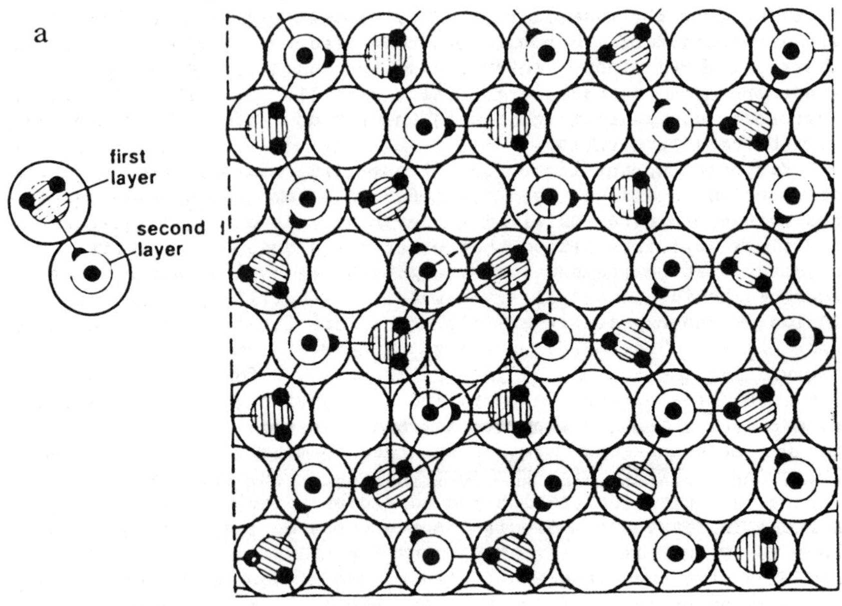

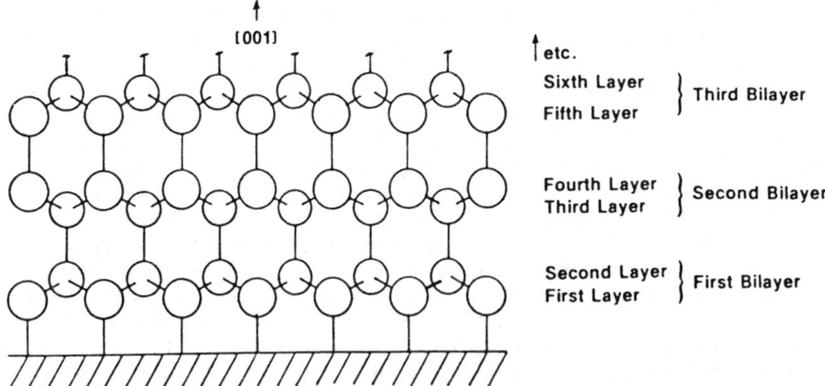

Figure 2. Model for long-period ice-like structures on fcc (111) or hcp (001) surfaces proposed by Doering and Madey (24): a) view perpendicular to fcc(111) surface; b) view parallel to (111) plane.

the structure of water in the double-layer at the same metal surfaces in aqueous solution. One can, however, construct such a relationship starting from the classical model of Bockris and Matthews (25) for the interaction of the hydronium ion with the Pt surface and the classical model (26) of tetrahedral (ice-like) coordination of water about the hydronium ion, and apply (just) the general principles of surface bonding elucidated by Thiel and Madey that give rise to long period ice-like structures, i.e., ice-like structures will occur when the hydrogen bonding forces acting parallel to the surface balance the specific metal-water/hydronium interaction acting perpendicular to the surface. Let us use these principles and classical models to construct a picture of the process of specific adsorption of (bi) sulphate ions on Pt(111) that would explain: 1) why the specific adsorption of (bi) sulphate ion would produce the anomalous features; 2) why the adsorption does not occur to the same extent (or with the same detailed chemistry) on a Pt(100) or (110) surface, nor on a "rough" (111) surface.

To use the general force-balance principle, we need to make some assumptions about the relative strengths of interactions of the various molecules involved: 1) on the atomically flat Pt(111) surface, the $Pt9P_3O^+$ bond is somewhat stronger than the $Pt(H_2O)$ bond (there is some evidence for this in the recent EELS study by Wagner and Moylan (27); 2) the $Pt-HSO_4^-$ bond energy is intermediate between that for $Pt-H_2O$ and $Pt-H_3O^+$; 3) the $Pt-H_2O$ and the $Pt-H_3O^+$ bonds are somewhat structure sensitive, with H_2O (H_3O^+) bonding slightly more strongly at steps and (100) or (110) surfaces; 4) the $Pt-HSO_4^-$ bond is strongly structure sensitive, with the anion bond much more strongly at steps and at (100) and (110) surfaces. Following the classical model of Bockris and Matthews (22), at the potential at the minimum in the voltammetry curve between the charge for "normal" adsorbed hydrogen and the charge from the anomalous features ($\emptyset_{min}$), the inner Helmholtz layer is formed by H_3O^+ hydrated (according to the Boering-Madey argument for the hydrogen bonding in water) into a long-period ice-like bilayer, as depicted in Figure 3. The incorporation of H_3O^+ into a Doering-Madey type ice-like structure is previously suggested by Wagner and Moylan (27) in their study of the coadsorption of water and HF on Pt(111) at low temperature. The orientation of H_3O^+ in such structures is consistent with the enhanced symmetric/asymmetric bend intensity ratio seen in EELS vibration spectra. Chemical intuition and charge compensation suggest the anions form the outer Helmholtz layer by occupying the high symmetry sites in the pure H_2O bilayer, as shown in Figure 4. Note that the anions only partially charge compensate the hydronium ions, so that one could refer to this structure in classical terms as "super-equivalent" hydronium ion adsorption. The structure of successive layers forming the diffuse double-layer would logically become progressively less ordered, since the anion layer (the OHP) becomes the ordering template, but the anions are only weakly hydrated and the ordering force is dissipated. There is ample precedent for the formation of charge compensated bilayers on Pt(111) in this potential region indicated in the studies by Hubbard and coworkers (28-30). As the potential is made anodic to this potential ($\emptyset_{min}$), protons are driven out of the IHP and the anions are drawn towards the metal surface due to the (+) potential field from the Pt surface. The hump in the I-V curve corresponds to the change in

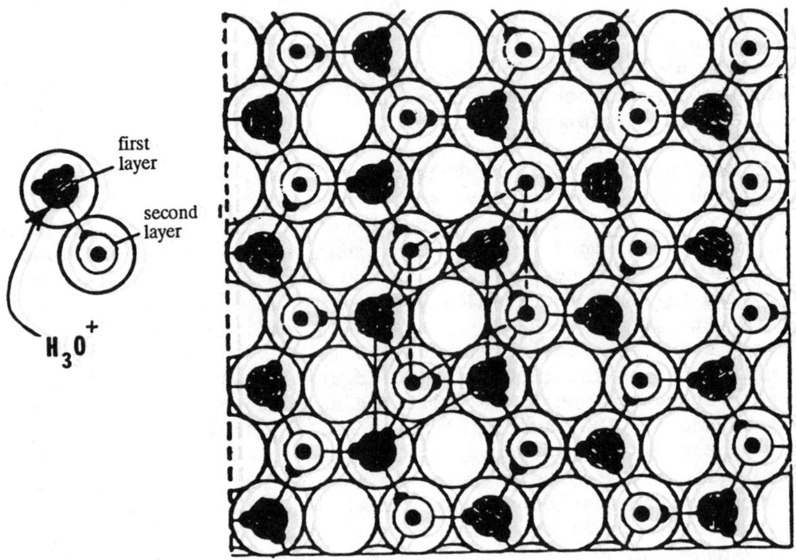

Figure 3. Model for a hydronium ion inner layer on the Pt(111) surface derived from the long-period ice-like structures of water on metals at low temperature.

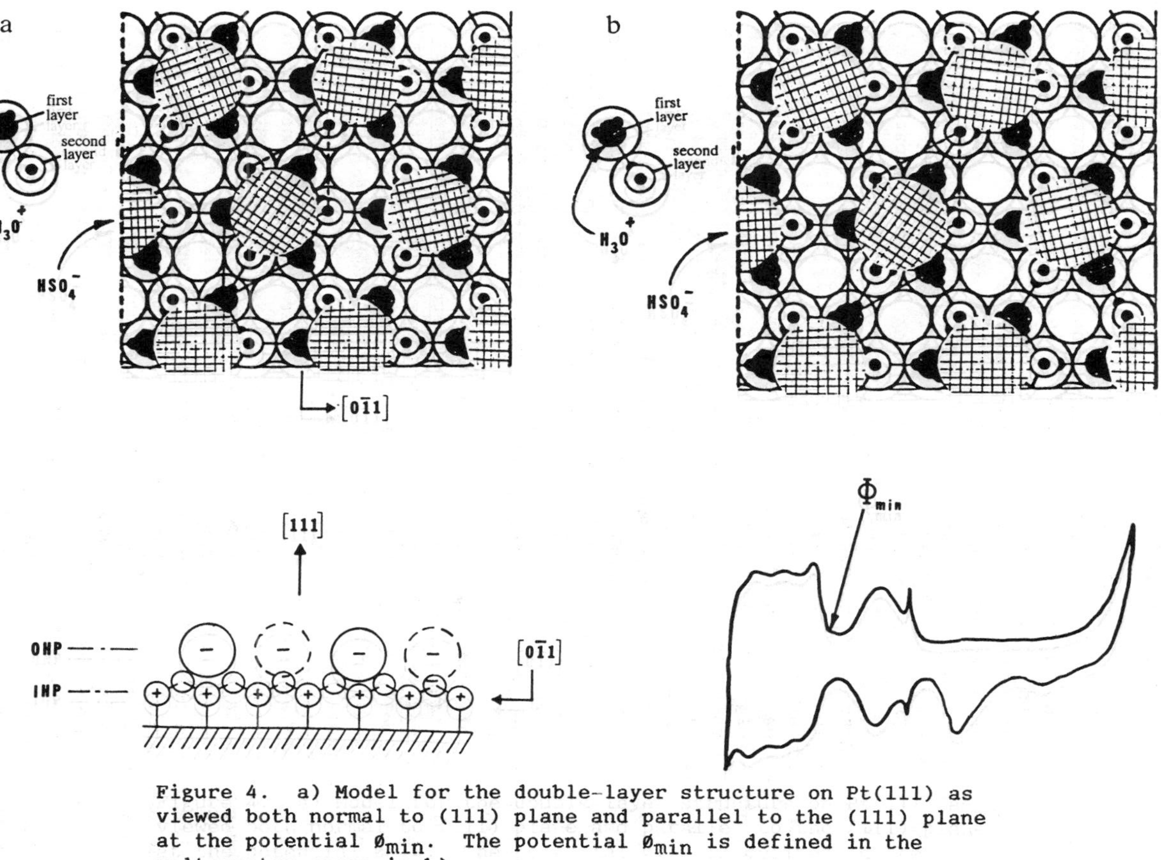

Figure 4. a) Model for the double-layer structure on Pt(111) as viewed both normal to (111) plane and parallel to the (111) plane at the potential ϕ_{min}. The potential ϕ_{min} is defined in the voltammetry curve in b).

capacitance of the inner-layer due to this deprotonation. According to the structural model of Figure 5, the charge under this hump could be as high as 1/3 e^- per Pt atom (or about 85 $\mu C/cm^2$) depending on the interaction of the proton on H_3O^+ with the Fermi sea of the metal at ϕ_{min}. In the limit of strong interaction, this model corresponds to the original Clavilier interpretation of the hump as "hydrogen desorption." As the potential is made cathodic to ϕ_{min}, protons are discharged and adsorb onto the Pt surface in the classical manner (22). However, since classical electrochemical "pictures" of the interface did not depict the structure parallel to the surface, it was not widely appreciated that at any instant in time one-third of the surface atoms are not coordinated to a water molecule, i.e., even if one does not invoke the ice-like structure for water in the double-layer, the water molecules cannot be more densely packed in the plane of the surface than in this structure. Thus, if the bonding of hydrogen to the surface is affected by interaction with water molecules in the double-layer, then it is easily seen from Figure 3 that there are three different energetic states for hydrogen on Pt(111), one having no direct association with a water molecule. It is not merely coincidental, therefore, that a significant fraction of the hydrogen on the Pt(111) surface has exactly the same heat of adsorption in solution as in vacuum (9). Returning again to the capacitive hump anodic to ϕ_{min}, the deprotonation of the inner-layer is followed, at more anodic potentials, by displacement of water from the inner-layer as well and the "contact adsorption" of the (bi)sulphate anion. As shown in Figure 5, the model structure indicates that the sharp spike in the anomalous feature is the contact adsorption of the anion with the Pt surface, 1 anion being adsorbed for every 9 Pt atoms, into a (111)-3 x 3 ordered anionic superlattice.

I cannot emphasize too strongly an important aspect of the model presented here: there are no processes occurring in the anomalous features that do not occur on all Pt surfaces!! What is different about the well-ordered (111) surface from all other Pt surface morphologies is that these processes occur in a much narrower potential region. The narrowing of the potential region where these processes occur is due to the ordered structure of the double-layer at the well-ordered (111) surface, a structure which does not occur on other surface morphologies, as will be discussed in greater detail below. As the structural model presented here indicates, the anomalous processes are all phase transitions, and the capacitive current observed in the I-V curve corresponds to the change in the concentration of ionic charge in the inner layer, from 5.4×10^{14} hydronium ions per cm^2 to 0.2×10^{14} (bi)sulphate anions per cm^2. It is difficult to make the model quantitative at this stage, because the absolute values of the capacitance and the capacitive currents depends on more precise knowledge of the degree of interaction of these ions with the electron sea of the Pt surface. Back-of-the-envelope type calculations suggest the model produces capacitance and capacitive currents consist with experiment. Consider the change in ionic concentrations cited above, and assume the change in "effective charge" in the inner-layer is 50 $\mu C/cm^2$ (the maximum possible value is 115 $\mu C/cm^2$). On a polycrystalline Pt surface, these same capacitive processes occur in the normal "double-layer potential region, from 0.3 V to 0.8 V (RHE). Then, we can approximate the

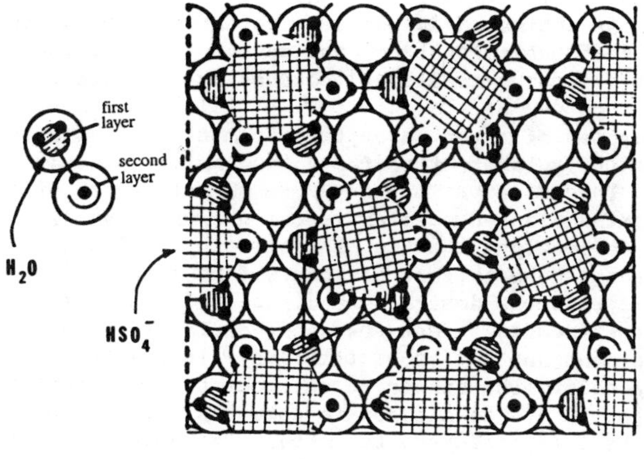

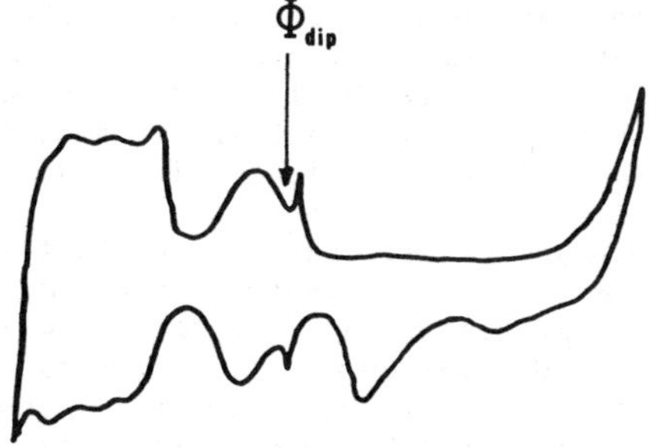

Figure 5. Model showing deprotonation of the inner layer in the potential region between ϕ_{min} and ϕ_{dip}. Protons are transferred to outside the OHP by the conventional Grotthus mechanism.

capacitance as $C = dq/dV = 50/0.5 = 100\ \mu F/cm^2$, and the characteristic current at 0.1 V/s as $10\ \mu A/cm^2$. These are typical of the capacitance and current in the double-layer region observed with polycrystalline Pt. On the well-ordered (111) surface, these same capacitive processes occur over a much narrower potential, about 100 mV, so $C = 50/0.1 = 500\ \mu C/cm^2$ and $i = 50\ \mu A/cm^2$ at 0.1 V/s. Note these estimated capacitive currents are, respectively, characteristic of the current for the sputtered (111) surface (Figure 1b) and for the anomalous features on the annealed (111) surface.

The discussion of anion adsorption from dilute suphuric acid has used the term (bi)sulphate to refer to the anion. The ambiguity is intentional. The local pH in the IHP is much lower than in the bulk and the predominant anion species in the OHP may then be sulphase, rather than the bisulphate anion, which is the predominant anion in the bulk. It is, however, not obvious which of these species is adsorbing, and thus the ambiguous designation.

Why are the anomalous features so specific to the well-ordered (111) surface, and not to (100), (110), or a (111) with a critical level of roughness? The answer to this question, in terms of the model presented here, was alluded to above, but is worth discussing in further detail. The key to the structure sensitivity, indeed to the entire phenomenon itself, is the formation of the ordered hydronium ice-like network in the double-layer. The formation of this network requires that a delicate balance be achieved between relatively weak bond forces, as articulated so elegantly in the discussion of water on metals by Thiel and Madey. The hydrogen bonds are very directional and very sensitive to bond length. The points of attachment of network to the metal atoms must therefore have hexagonal symmetry, a condition which cannot be met on the (100)-1 x 1 (the unreconstructed) surface. As Thiel and Madey have shown, it is possible to construct a ice-like network on an unreconstructed fcc(110) surface like Ag(110), but the Pt(110) surface is reconstructed to a 2 x 1 "missing row" (31) structure which is stable even in acid electrolyte (9). The missing row structure is a sawtooth-type structure, and the vertical articulation of the surface makes it impossible to form the hydrogen-bonded network. In the case of step-terrace structures (17,18), where the anomalous features are seen to depend on a critical length of the (111) terraces, apparently there is a critical ensemble size to form not only the hydronium network inner-layer but the full ordered double-layer. Note in Figures 4-5 that a complete (bi)sulphate 3 x 3 unit cell requires a perfectly flat Pt domain six atoms on a side, which is consistent with the experimental observation of 5 or more as the critical terrace length.

Let us now extend the long-period hydronium ice-like model for the IHP on Pt(111) to explain the observations in electrolytes other than sulphate. In acid chloride, both the observations and the model carry-over directly from the case of sulphate. In fluoride, perchlorate, bicarbonate and hydroxide, in which the anomalous features shift considerably in both potential and appearance (especially in the basic media) from sulphate, another model is needed. Both (bi)sulphate and chloride are large weakly hydrated anions, and in the double-layer model of Figures 4-5, they interact strongly with both the hydronium ions and the Pt surface. The contact adsorption of these anions screens the net (+) image charge on the metal surface from the water molecules in the OHP, so that much higher fields (anodic potential) are required to cause water dissociation and "OH" formation on the surface. Fluoride especially is a small strongly hydrated anion, which does not interact strongly either with hydronium ion or the metal surface, and does not screen the potential field of the metal atoms from the water in double-layer. To account for these fundamental differences in physical properties, it is necessary to construct a double-layer model which retains certain basic elements of Figures 4-5, but is very different in essential features. Because the fluoride ion has weak charge compensating properties, the "super-equivalent" hydronium ion layer cannot form the IHP. Instead, I suggest formation of a Doering-Madey long-period ice-like layer with hydronium ions alternately in the first and second bilayers and insertion of fluoride ions into the more dense 2X2 packing positions in the second bilayer. The latter achieves a nearly charge compensated double-layer at ϕ_{min}, and the view of this structure parallel to the surface would look a great deal like the classical DBM (<u>22</u>) double-layer model. Thus, replacement of (bi) sulphate ion with fluoride ion has two effects on the double-layer structure: a) it reduces the interaction of hydronium ion with the Pt surface because this interaction is intrinsically a cooperative interaction; b) the potential field from the metal is relatively unscreened by the anions. The consequence of these effects are to eliminate the phase transitions in the IHL at ϕ_{min} due to deprotonation and anion adsorption, and instead cause phase transitions at higher potential associated with dissociation of the water molecules in the IHL to form ordered OH structures. As before with sulphate, I argue that <u>no electrode processes take place on Pt(111) in fluoride electrolyte that do not also occur on other Pt surfaces in the same electrolyte</u>. The crucial difference between these processes on Pt(111) and on other Pt surfaces is again the squeezing of the potential region in which these processes take place due to the ordered structures, i.e., transitions from one ordered phase to another in a narrow range of potential (I resist the temptation to give these transitions an order, e.g. first order phase transition, until more is known about them). Potential cycling experiments (Wagner, F.T.; Ross, P.N., <u>J. Electroanal. Chem</u>., in press) that progressively roughen the surface clearly show that of the 80-90 $\mu C/cm^2$ under the anomalous features at 0.6-0.8 V, about 25% of that charge reappears in the "normal" hydrogen region, while the remaining charge is spread anodically into the "oxide" formation potential region. This redistribution of charge with loss of surface order is consistent with the model. Roughening breaks the long-period ice-like structure of the double-layer, which changes the association between water and adsorbed hydrogen producing a redistribution of charge among the three different states of hydrogen on the (111) surface, and some increase in total charge due directly to the increase in the roughness. The "OH" formation process, which could be termed a UPD state of oxygen, moves towards the oxygen Nernst potential due to a lowered free energy of formation on the disordered versus the ordered surface. The process is also no longer a transition between ordered phases, and the charge becomes smeared out in potential.

The Path to Definitive Models of the Anomalous Processes

The path of discovery in the great period of atomic physics from ca. 1920-1940 has become a model for the process of intellectual discovery which social scientists term "the scientific method." Briefly stated, the inquiry begins with the observation of anomalous behavior, e.g., the Stark effect, so termed because the observations are totally new and the results inexplicable in terms of current theory. Others try to duplicate the observations, with usually contradictory results, and a period of controversy ensues until eventually a consensus emerges defining the nature of the phenomena. Hypothetical models are presented which attempt to explain the new behavior, and theorists and experimentalists alternate in disproving one hypothesis and bringing forth a new one, until eventually a model evolves that explains all observations. The purpose of this talk has been to put the study of the anomalous electrochemical behavior of Pt(111) into that same type of intellectual framework. Following a period of controversy concerning the experimental observations, there has emerged a consensus concerning the valid data set that describes the anomalous behavior, which was summarized here in Table I. We are now at the stage of inquiry where hypothetical models are needed to initiate the next stage in the process of discovery. I have attempted to present such a model here. I hope that this model will stimulate theorists to use it as a basis for calculations of free energies of formation of the ordered phases and the corresponding capacitances. These calculations should give rise to new and more refined models, and guide experiments designed to test the new theory. The key to activating this traditional cycle of theory and experiment is entry of theorists, who have so far remained on the sidelines while the experimentalists sort out the valid observations.

Acknowledgments

The author acknowledges the steady support over the last decade for studies on single crystal electrochemistry from the Department of Energy, Assistant Secretary for Conservation and Renewable Energy, Office of Energy Storage and Distribution, under Contract No. DE-AC03-76SF00098.

Literature Cited

1. Clavilier, J. J. Electroanal. Chem. 1980, 107, 205; 1980, 107, 211.
2. Ross, P. N. J. Electroanal. Chem. 1977, 76, 139.
3. Yeager, E.; O'Grady, W.; Woo, M.; Hagans, P. J. Electrochem. Soc. 1978, 125, 348.
4. Hubbard, A.; Ishikawa, R.; Kutakaru, J. J. Electroanal. Chem. 1978, 86, 271.
5. Yamamoto, K.; Kolb, D. M.; Kotz, R.; Lehmpfuhl, G. J. Electroanal. Chem. 1979, 96, 233.
6. Ross, P. N. J. Electrochem. Soc. 1979, 126, 67.
7. Will, F. J. Electrochem. Soc. 1965, 112, 451.
8. Yeager, E. J. Electrochem. Soc. 1980, 128, 160C.
9. Ross, P. N. Surf. Sci. 1981, 102, 463.
10. Parsons, R. J. Electrochem. Soc. 1980, 127, 126C.

11. Ross, P. N., Jr.; Wagner, F. T. In Advances in Electrochemistry and Electrochemical Engineering; Gerischer, H.; Tobias, C. W., Eds.; Wiley: New York, 1985; Vol. 13.
12. Wagner, F. T.; Ross, P. N., Jr. J. Electroanal. Chem. 1983, 150, 141.
13. Clavilier, J.; Faure, R.; Guinet, G.; Durand, R. J. Electroanal. Chem. 1980, 107, 205.
14. Markovic, M.; Hanson, M.; McDougall, G.; Yeager, E. J. Electroanal. Chem. 1986, 214, 555.
15. Al-Jaaf-Golze, K.; Kolb, D.; Scherson, D. J. Electroanal. Chem. 1986, 200, 353.
16. Clavilier, J.; Chauvineau, J. J. Electroanal. Chem. 1984, 178, 343.
17. Love, B.; Seto, K.; Lipkowski, J. J. Electroanal. Chem. 1986, 199, 219.
18. Clavilier, J.; Armand, D.; Sun, S.; Petit, M. J. Electroanal. Chem. 1986, 205, 267.
19. Wagner, F. T.; Ross, P. N. Surf. Sci. 1985, 160, 305.
20. Hanson, M. Ph.D. Thesis, Case Western Reserve University, Cleveland, OH, 1985.
21. Wagner, F. T.; Ross, P. N., Jr. Appl. Surf. Sci. 1985, 24, 87.
22. Bockris, J.; Devanathan, M.; Mueller, K. Proc. Roy. Soc. (London), 1963, A274, p. 55.
23. Thiel, P.; Madey, T. Surf. Sci. Rept. 1987, 7; 211.
24. Doering, D.; Madey, T. Surf. Sci., 1982, 123, 305.
25. Bockris, J.; Matthews, D. Proc. Roy. Soc. (London), 1966, 292, p. 479.
26. Wicke, E.; Eigen, M.; Ackermann, T. Zeit. Phys. Chem. (Neue Folge) 1957, 1, 1195.
27. Wagner, F.; Moylan, T. Surf. Sci. 1987, 182, 125.
28. Salaita, G.; Stern, D.; Lu, F.; Baltruschat, H.; Schardt, B.; Stickney, J.; Soriage, M.; Frank, D.; Hubbard, A. Langmuir 1986, 2, 828.
29. Lu, F.; Salaita, G.; Baltruschat, H.; Hubbard, A. J. Electroanal. Chem. 1987, 222, 305.
30. Stern, D.; Baltruschat, H.; Martinez, M.; Stickney, J.; Song, D.; Lewis, S.; Frank, D.; Hubbard, A. J. Electroanal. Chem. 1987, 217, 101.
31. Ducros, R.; Merrill, R. Surf. Sci. 1977, 55 227.

RECEIVED May 17, 1988

Chapter 4

Gas-Phase Adsorption Model Studies of Electrode Surfaces

J. K. Sass[1] and K. Bange[2]

[1]Fritz-Haber-Institut der Max-Planck-Gesellschaft, Faradayweg 4–6, D–1000 Berlin 33, Federal Republic of Germany
[2]Schott Glaswerke, Postfach 130367, D–6200 Wiesbaden 13, Federal Republic of Germany

Model adsorption experiments in ultrahigh vacuum (UHV.), aimed at simulating the interfacial region between a metal and an electrolyte, are described. It is shown that by measuring the work function change, induced by such a synthetic adsorbate layer, a meaningful comparison to in situ electrochemical data may be achieved. The agreement which has been obtained, by such comparison, for the specific adsorption of the two halides bromide and chloride on Ag{110} is argued to provide definite evidence for the relevance of electrochemical surface science studies. Other recent investigations of interesting coadsorption systems with electrochemical significance, in particular those where non-specific adsorption behaviour would be expected, are briefly reviewed.

Encouraging progress in observing and understanding molecular phenomena at solid-electrolyte interfaces has been made in recent years. Significant contributions to this advancement have come from the application of experimental procedures which were developed for the study of gas-solid interactions in ultrahigh vacuum /1/. These surface science techniques are attractive because they provide a wide range of microscopic information about solid surfaces and adsorbate layers. The drawback is, of course, that they cannot be used directly at electrochemical interfaces and that a certain measure of ambiguity is therefore inherent in such studies.

Considerable effort had to be invested, for example, before contamination levels during transfer of a sample from an electrochemical cell to a vacuum chamber could be adequately assessed and controlled /2/. With this provision, electrode modifications by electrochemical processes can be studied in much greater detail than is possible in situ and many interesting results have been obtained by such experiments. Intrinsic limitations of this transfer method arise, however, with loosely bound

0097–6156/88/0378–0054$06.00/0

species, for example solvent molecules, which may desorb after removal of the sample from solution since the transfer is typically carried out at or near room temperature.

The experimental approach discussed in this article is, in contrast, particularly amenable to investigating solvent contributions to the interfacial properties /3/. Species, which electrolyte solutions are composed of, are dosed in controlled amounts from the gas phase, in ultrahigh vacuum, onto clean metal substrates. Sticking is ensured, where necessary, by cooling the sample to sufficiently low temperature. Again surface-sensitive techniques can be used, to characterize microscopically the interaction of solvent molecules and ionic species with the solid surface. Even without further consideration such information is certainly most valuable. The ultimate goal in these studies, however, is to actually mimic structural elements of the interfacial region and to be able to assess the extent to which this may be achieved.

A key element in considering the properties of the interfacial electric double layer is the distinction between specific and non-specific adsorption. With regard to the feasibility of UHV simulation of interfacial properties this distinction is also very important /4/. Ions which directly contact the electrode surface, such as the halides for example, may be dosed seperately and, because of electronic equilibration, not necessarily in ionic form. Upon subsequent dosing of solvent molecules only lateral motions of the two adsorbates are required for a minimization of the energy. In contrast, the preparation of a fully solvated ion on a metal surface in UHV is clearly a more demanding task. Although solvated ions may be generated in the gas phase there would probably be insufficient intensity in a collimated beam suitable for UHV studies. In sequential dosing of both species, on the other hand, prohibitively large activation barriers for solvent enclosure of the ion may prevent the simulation. Only in one of several coadsorption studies of alkalis and water, for example, was there any evidence for a complete hydration shell around an alkali ion /5/. The major difference in the experimental conditions of this study was that instead of submonolayer alkali deposits multilayers of potassium and water reacted to form bulk potassium hydroxide which to some extent was dissociated and hydrated by additional water. In the only other report, relevant to non-specific adsorption, vibrational spectroscopy indicated the formation of hydrated protons from the interaction of hydrogen fluoride and water /6/. In both of these studies equal amounts of cations and anions were apparently present, however, and it is not yet clear how an excess of hydrated ions might be generated.

For this reason, the emphasis in this article is directed more towards the simulation of specific adsorption and, in particular, the recent encouraging comparison of electrochemical and UHV data for the interaction of bromine and chlorine with Ag{110} /7, 8/. A brief outline of the conclusions emerging from alkali-water coadsorption experiments is given to illustrate basic modes of ion-solvent interaction on metal surfaces and to discuss future directions of this research.

Concepts and Procedures

The consecutive stages of a typical gas-phase-adsorption experiment, in which the simulation of an electrochemical interface is the aim, are schematically illustrated in Figure 1. The depicted sequence illustrates in particular the case when specific adsorption of an ion is expected from electrochemical data. In the first step (Figure 1a) dosing of just the ionic species is performed and its coverage is determined. The next step involves populating nearest neighbour sites of the adsorbed ions with solvent molecules (Figure 1b). From theoretical and experimental work related to heterogeneous catalysis one expects that solvent molecules experience strong perturbations at those sites, and such an effect has indeed been observed for surface interactions between polar molecules and ionic species /3/. With additional solvent adsorption a stage may then be reached where approximately the first adsorbate layer is completed (Figure 1c). This situation is of particular interest for a comparison with electrochemical data because it represents what is commonly considered the inner layer in double layer models of specific adsorption. The deposition of fully solvated ions on top of this first layer, to simulate the outer or diffuse layer of the interface (Figure 1e), is not yet possible, as was pointed out earlier. But at least the presence of an adjacent condensed bulk phase can be simulated by the addition of a few multilayers of solvent in the final step of the experiment (Figure 1d).

Comparing Figure 1d with Figure 1e, it is evident that there are two important features of the electrochemical interface which cannot be reproduced yet by the simulation: the above mentioned ionic excess charge in the diffuse layer and the bulk electrolyte ions with their screening properties. Fortunately, the condition of zero diffuse layer charge can often be extracted from electrochemical data such that the absence of the diffuse layer does not seriously depreciate the purpose of the UHV experiment. Similarly, it may be expected that the structural properties of the inner layer, for a certain composition, do not depend on the electrolyte concentration in the bulk solution phase.

Despite these arguments and the conceptual attractiveness of the procedure which is sketched in Fig. 1 convincing evidence for the relevance of a particular gas phase adsorption experiment can only be obtained by direct comparison to electrochemical data. The electrode potential and the work function change are two measurable quantities which are particularly useful for such a comparison. In both measurements the variation of the electrostatic potential across the interface can be obtained and compared by properly referencing these two values /7/. Together with the ionic excess charge in the double layer, which in the UHV experiment would be expressed in terms of coverage of the ionic species, the macroscopic electrical properties of the interfacial capacitor can thus be characterized in both environments.

Such a comparison has formed the basis, for example, for the assertion that the double layer can be emersed essentially intact from solution /8/. A common ambiguity, although for different reasons, in both emersion and UHV model experiments is the difference in the amount of solvent present either at the emersed or synthesized interface, compared to the in-situ situation. In the UHV the total amount of solvent adsorbed, and its distribution into the first and subsequent layers, can in many instances directly be determined, but this information is difficult to obtain and not yet available for the emersed and the real interface. To gather such missing pieces in the interfacial puzzle is the motivation for the work described in this paper. One important prerequisite for any model of the double layer is, for example, the density of solvent molecules in the inner layer as a function of the charge on the interfacial capacitor.

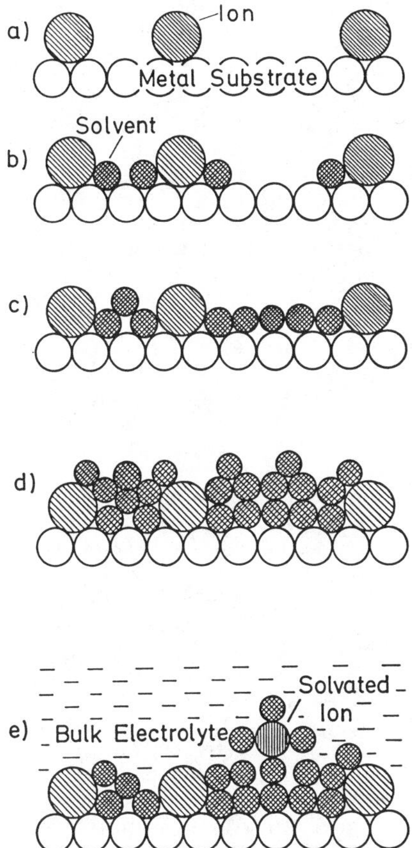

Figure 1. A schematic representation of the synthesis of the electrochemical double layer in UHV: a) adsorption of specifically adsorbed ions without solvent; b) addition of hydration water; c) completion of the inner layer; d) addition of solvent multilayers, e) model for the double layer at an electrode surface in solution.

Coadsorption of Bromine and Water on Ag{110}

Br / Ag {110}

In preparation for the coadsorption experiments it is important to study the nature of the interaction of the ionic species, by itself, with the metal surface. As electronegative species, the halogens are well known to attract charge from the metal substrate and to form a dipole layer which increases the work function. Since the halogens are typically exposed to the metal surface in molecular form in a surface science experiment and the sticking coefficient varies with adsorbate coverage the need for coverage calibration arises. Fig. 2 summarizes work function measurements and LEED results which we have used for this purpose /9/. The p (2x1) and c (4 x 2) LEED patterns, which appear in the coverage ranges indicated in Fig. 2, correlate very well with the concept of a linear increase of the work function ϕ: At about two thirds of the maximum work function change, the p(2x1) LEED pattern, which is first seen at ~ 0.25 monolayers, disappears and, at a slightly higher coverage, is replaced by a c (4x2) pattern. We obtain a saturation coverage of 0.75 for Br on Ag{110} and can assume a linear relationship between work function and Br coverage.

H_2O / Br / Ag{110}

The key issue in simulation experiments, and the most difficult to address, is the transferability of the conclusions drawn from the results obtained in vacuum. In this section we shall therefore examine in some detail a recently proposed procedure /9/ which permits one to directly compare properties of the synthetic interfacial layers prepared in vacuum and those present at in situ electrochemical interfaces. For reasons of space limitations it will not be possible to review how some of the information presented subsequently may be obtained by standard surface science methods /1/.

As pointed out above the comparison between both sets of results relies upon the physical equivalence of the two measureable quantities work function (surface science) and electrode potential (electrochemistry) /7/. This equivalence has been realized in electrochemistry some time ago and has been exploited to analyze measured values of the potential of zero charge /7/ and of work function changes upon emersion of electrodes at fixed potential /8/. In the simulation experiments the approach is quite similar in that one prepares a well-defined composition of the synthetic electrochemical adsorbate layer and then obtains the electrostatic potential drop across it by a work function measurement.

In the previous section the procedure for determining the bromine coverage has been described Obtaining the absolute water coverage after a particular dosing event frequently requires more subtle considerations, in particular when there is no temperature distinction in thermal desorption spectroscopy (TDS) between H_2O molecules directly bound to the Ag {110} surface and those residing in subsequent multilayers /10/. By exploiting the known stoichiometry of the hydroxyl formation and recombination processes on Ag{110} from the reaction of adsorbed oxygen and H_2O /11/ it is, however, possible to cope with this problem, such that the relative water coverages, which may be obtained by TDS, can be converted into absolute coverages of H_2O.

As an additional bonus, the application of TDS also provides information on the amounts of water present in different local environments of the composite double layer, since water bound to bromine desorbs at a higher temperature than that residing at still unmodified areas of the substrate surface /12/. With appropriate analysis of work function data, obtained at different bromine coverages as a function of water coverage, it is therefore possible to identify the particular contribution of surface hydration water and free water in the inner layer to the electrostatic potential drop at the interface /9/.

In Figure 3 the results of such data analysis are shown. At bromine coverages below ~ 0.25, two water molecules are bound to the specifically adsorbed halide ion and the orientation of this surface hydration water (cf. Figure 1) is seen to be such that the normal components of the permanent dipole moment, with smaller contributions from charge transfer and the electronic polarizability /13/, essentially compensate the dipole layer of opposite polarity induced by the ionic halogen adsorbate (see curve b in Figure 3). Of course, it must be realized that this effect of the two adsorbates counterbalancing each other's dipole layer is a pure coincidence and that the important conclusion to be drawn from this result is the invariability of the effective dipole moment of the composite bromide-plus-hydration-water complex at bromide coverages below ~0.25. Provided that this result is of a more general nature, an important building block of the inner layer would have been identified by the UHV simulation experiments.

Following the conceptual outline of the double layer simulation in UHV, sketched in Figure 1, curve c in Figure 3 shows the work function change due to the adsorbed bromide, the surface hydration shell and <u>also</u> the water occupying areas of the Ag{110} surface where the influence of the bromide, which is only short-range, cannot exert itself (cf. Figure 1c). The shape of this curve is not difficult to interpret when we recall that the hydrated surface bromide does not contribute to the work function change and that the "free" water is less tightly bound than the hydration water. Since there is only limited space available in the first layer, i.e. the inner layer in phenomenological models for anion specific adsorption /14/, the free water is gradually displayed from this inner layer and the work function rises because of the concomitant loss of the free H_2O molecules which when present induce a work function decrease /15/.

A somewhat curious effect arises when additional water is dosed on top of this synthetic inner layer, in that the work function is observed to exhibit substantial further decreases. This implies that the water molecules in the multilayers above the inner layer assume some measure of preferential orientation, induced by the presence of the adsorbed bromide in the first layer. This result is probably connected to structure making and structure breaking, or hydrophobic and hydrophilic, properties of soild surfaces, but will not be discussed in detail here.

As pointed out above, the data presented in Fig. 3c correspond to the simulation of the inner layer in halide specific adsorption, for the case of zero diffuse layer charge. From differential capacity measurements equivalent information, i.e. the electrostatic potential drop across the inner layer as a function of halide coverage, can be obtained and used to obtain values for zero diffuse layer charge. In Figure 4 the full line is a representation of such data for bromide specific adsorption on a Ag {110} electrode /16/. The agreement between these measurements and the corresponding work function data taken in UHV is surprisingly good, when recalling that they were obtained at very different temperatures of ~ 100 K and ~ 300 K. Also shown in Figure 4 is a similar comparison, based on recent data /17/, for chloride specific adsorption on Ag{110}. Again agreement is obtained, albeit over a smaller coverage range, between the results for the synthetic inner layer and the one in solution.

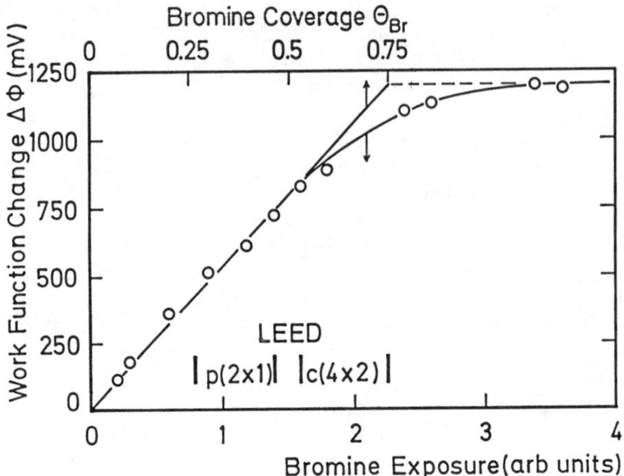

Figure 2. The change in work function (ϕ) of an Ag{110} surface with bromine exposure. Absolute coverages were determined from the observed LEED patterns.

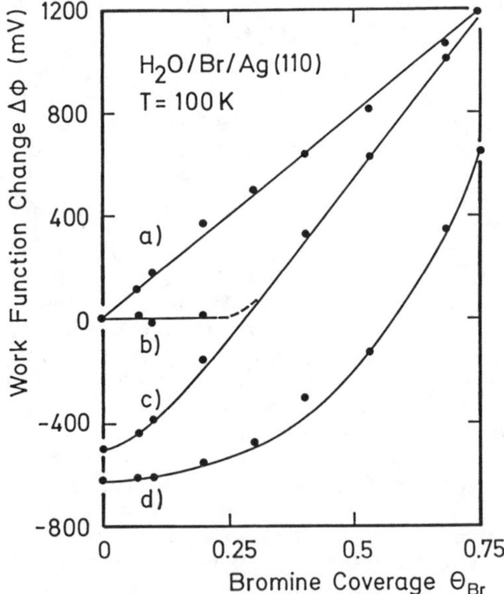

Figure 3. Work function change for a) bromine on Ag{110}; b) bromine and hydration water; c) bromine and water to complete the inner layer; d) bromine and multilayers of water.

It is also of interest that the potential drop across the inner layers are very similar for the two halides, although chloride alone induces a limiting work function increase of ~ 1500 mV which is 300 mV larger than that for bromide alone (cf. Figure 2). This may be due to the fact that in a stronger dipole field (halide plus screening charge) the surface hydration water is more strongly reoriented with its permanent dipole in an antiparallel configuration .

The good agreement between electrochemical and UHV data, documented in Figure 4, is a very important result, because it proves for the first time that the microscopic information which one obtains with surface science techniques in the simulation studies is indeed very relevant to interfacial electrochemistry. As an example of such microscopic information, Figure 5 shows a structural model of the inner layer for bromide specific adsorption at a halide coverage of ~ 0.25 on Ag{110} which has been deduced from thermal desorption and low energy electron diffraction measurements /12/. Qualitatively similar models have been obtained for H_2O / Br / Cu(110) /18/ and also for H_2O / Cl /Ag{110}.

Simulation of Non-Specific Adsorption in Vacuum

Alkali ions in aqueous solution are probably the most typical and most widely studied representatives of non-specific adsorption. The electrochemical term of non-specific adsorption is used to denote the survival of at least the primary hydration shell when an ion is interacting with a solid electrode. As pointed out previously, the generation of such hydrated ions at the gas-solid interface would be of great value because it would provide an opportunity to simulate the charging of the interfacial capacitor at the outer Helmholtz plane or perhaps even in the diffuse layer.

Although quite a few studies of the coadsorption of water and alkalis on metal surfaces in UHV. have been reported /19-21/ the possibility of complete hydration of the alkali adsorbate has not been considered in most cases. The reason is probably that, as yet, all the experimental evidence suggests that the alkali ions are "specifically" adsorbed in such gas-phase simulation experiments, even when an excess of water (several multilayers) is made available. This result is not yet understood, although one should again keep in mind that the simulation experiments are typically performed 150 K below room temperature.

Despite this apparent non-electrochemical behaviour of adsorbed alkalis, in the presence of coadsorbed water, such investigations are definitely of great value to the understanding of the physical phenomena in the double layer, at the microscopic level. For example, by work function measurements in conjunction with coverage determinations of the two coadsorbates one can obtain the initial dipole moment of water molecules interacting with different alkalis. Figure 6 shows a compilation of available data plotted against alkali ion radius It is interesting that the natural orientation of water, with the oxygen towards the metal surface /13/, is gradually reversed when the ionic radius increases. For cesium the perpendicular alignment of the H_2O molecular axis by the alkali-surface-ion plus screening-charge dipolar field appears to be substantially larger than for the opposite orientation on the clean surface. Further experiments are, however, necessary to elucidate more clearly the interaction mechanism for this important coadsorption system. The interesting recent results concerning the generation of hydrated protons on metal surfaces in UHV. /6/, mentioned above, provide encouragement for further attempts to

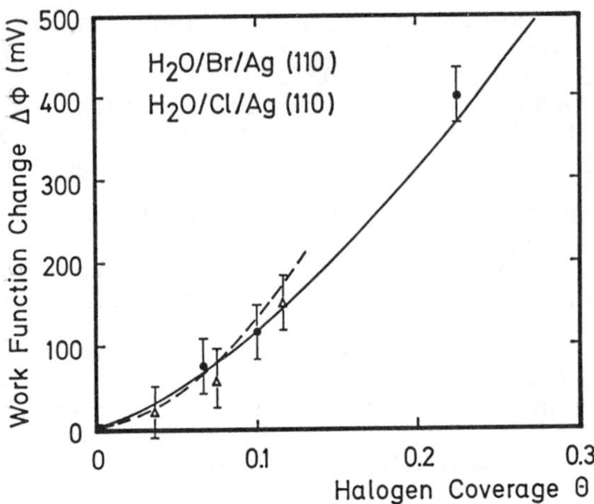

Figure 4. Comparison of UHV and in situ electrochemical data. Solid and dotted lines correspond to the potential drop across the inner layer for the Br/H_2O and Cl / H_2O systems at a Ag{110} electrode. Full circles correspond to the change in work function of a Ag{110} surface with Br and water to complete the inner layer (taken from figure 3 c). Open triangles correspond to similar work function data for the Cl / H_2O system on Ag{110}.

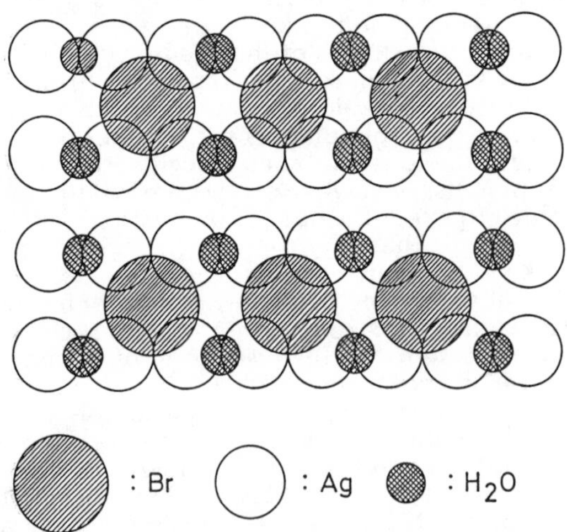

Figure 5. Proposed structure for Br / H_2O adsorption on Ag{110} which is compatible with LEED, $\Delta\phi$ and TDS measurements /12/.

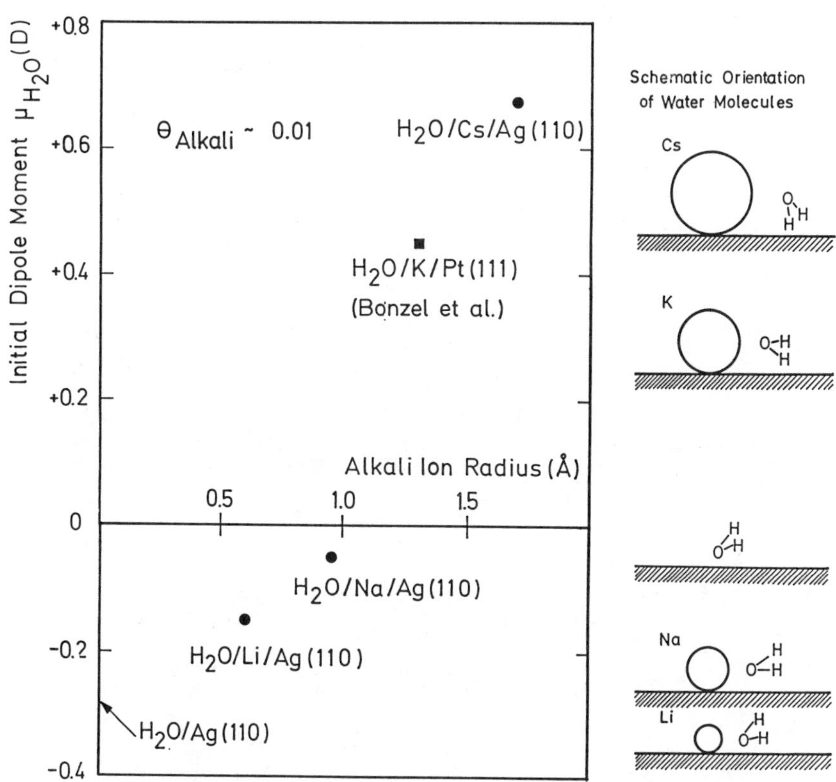

Figure 6. Initial dipole moments and the proposed orientations for surface hydration water on clean and alkali covered Ag{110}. In addition to our results for Ag{110}, data for H_2O/ K /Pt{111}, taken from reference 13, are also included.

prodouce hydrated alkali ions on metal surfaces by adsorption from the gas phase.

Acknowledgments

One of us (J.K.S.) would like to thank M. Docherty, P. Clarke and B. Schlereth for support in the preparation of this manuscript. The generous help of D. Lackey is gratefully acknowledged. This work was supported in part by Deutsche Forschungsgemeinschaft through SFB 6.

References

1. See for example: D.A. King and D.P. Woodruff (Eds.), The Chemical Physics of Solid Surfaces and Heterogeneous Catalysis, Elsevier, Amsterdam, 1981-1988.
2. F.T. Wagner and P.N. Ross, Appl. Surf. Sci. 24, 187 (1985).
3. J.K. Sass, Vacuum 33, 741 (1983).
4. J.K. Sass, K. Kretzschmar and S. Holloway, Vacuum 31, 483 (1981).
5. H.P. Bonzel, G. Pirug and A. Winkler, Chem. Phys. Lett. 133, 116 (1985).
6. F.T. Wagner and T.E. Moylan, Surface Sci. 182, 125 (1987).
7. S. Trasatti, in: H. Gerischer and C.W. Tobias (Eds.), Advances in Electrochemistry and Electrochemical Engeneering, Vol. 10, Wiley, New York, 1977, p. 213.
8. D.M. Kolb and W.N. Hansen, Surface Sci. 79, 205 (1979)
9. K. Bange, B. Straehler, J.K. Sass and R. Parsons, J. Electroanal. Chem. 229, 87 (1987).
10. P.A. Thiel and T.E. Madey, Surf. Sci. Reports 7, 211 (1987)
11. K. Bange, J.K. Sass, T.E. Madey, E.M. Stuve, Surface Sci., 183, 334, (1987).
12. K. Bange, T.E. Madey and J.K: Sass, Surface Sci. 162, 272 (1985)
13. G. Pirug, H.P. Bonzel, J.E. Müller, Phys. Rev. Lett., 58(20), 2138, (1987).
14. R. Parsons, Trans. Faraday Soc. 64, 1638 (1955).
15. R. Döhl-Oelze, Thesis, Technische Universität Berlin, 1986.
16. G. Valetta and R. Parsons, J. Electroanal. Chem. 123, 141 (1981).
17. B. Straehler, J.K. Sass, K. Bange and R. Parsons, Phys. Rev. Lett., submitted for publication.
18. K. Bange, R. Döhl, D.E. Grider and J.K. Sass, Vacuum, 33, 757 (1983).
19. M. Kiskinova, G. Pirug and H.P. Bonzel, Surface Sci., 150, 319 (1985).
20. P.A. Thiel, J. Hrbek, R.A. de Paola and F.M. Hoffmann, Chem. Phys. Lett., 108, 25 (1984).
21. C. Benndorf, C. Nöbl and T.E. Madey, Surface Sci. 138, 292 (1984).

RECEIVED August 24, 1988

Chapter 5

Modeling the Aqueous–Metal Interface in Ultrahigh Vacuum via Cryogenic Coadsorption

Frederick T. Wagner and Thomas E. Moylan

Physical Chemistry Department, General Motors Research Laboratories, Warren, MI 48090–9055

Three stages in the modeling of electrochemical interfaces in ultrahigh vacuum are described, and experimental examples of each stage, verified with a variety of surface spectroscopies, are given. The first stage is the adsorption of water alone, or water and a neutral probe molecule, on a well-defined metal surface. Hydrophilic coadsorption of CO and water on Rh(111) is contrasted with hydrophobic coadsorption on Pt(111), demonstrating that weakly-bound water can modify the adsorption of even strongly-bound, neutral species. The second stage of modeling is the introduction of solvated ionic species into the model double layer. Coadsorption of HF and water yields adsorbed H_3O^+ ions; the solvation stoichiometries of ions in the first monolayer and in subsequent layers are determined. The third stage of modeling is establishment of potential control in UHV. Hydrogen coadsorption is used to deflect the effective potential of the water monolayer below the potential of zero charge. The unique ways in which UHV models can contribute to an improved molecular-scale understanding of electrochemical interfaces are discussed.

Ultrahigh vacuum surface spectroscopies can provide far greater breadth and depth of information about surface properties than can yet be achieved using in situ spectroscopies at the aqueous/metal interface. Application of the vacuum techniques to electrochemical interfaces is thus desirable, but has been plagued by questions of the relevance of the emersed, evacuated surfaces examined to the real electrochemical interfaces. This concern is accentuated by surface scientists' observations that in UHV no molecular water remains on well-defined surfaces at room temperature and above (1). Emersion and evacuation at room temperature may or may not produce significant changes in electrochemical interfaces, depending on whether or not water plays a major role in the surface chemistry.

0097–6156/88/0378–0065$06.00/0

Bulk water can be kept on surfaces in UHV by maintaining a surface temperature below 160 K. A wide range of spectroscopies indicate that water layers formed at these temperatures, even for submonolayer coverages, are extensively hydrogen-bonded and thus represent an associated, condensed phase rather than a collection of isolated adsorbed water molecules (1). We can therefore hope to study fully hydrated species in UHV on liquid nitrogen cooled surfaces. The relevance of such low-temperature interfaces to normal aqueous solutions is attested to by frozen electrolyte experiments (2), in which electrochemistry quite similar to normal, room-temperature aqueous behavior has been observed down to around 150 K. To the authors' knowledge no one has yet succeeded, in a UHV-electrochemical transfer experiment, at cooling a sample during emersion to maintain a layer of bulk water on the surface after evacuation, though several groups are now equipped for the attempt. However, an alternate approach allowing UHV studies of aqueous interfaces is available and forms the subject of this paper. This approach is to grow, from the vapor phase and without removing the sample from the UHV analytical environment, models of electrochemical double layers containing all relevant species. There are three major steps to such modeling: (1) adsorption of water alone, or water plus a neutral probe molecule, (2) addition of solvated ionic species to the UHV interface, and (3) establishment of control over the effective electrochemical potential of the UHV interface. This chapter will describe, and provide an experimental illustration of, each of these steps.

Experimental

All experiments were performed in a stainless steel ultrahigh vacuum chamber equipped for temperature programmed desorption (TPD), high resolution electron energy loss spectroscopy (HREELS), Auger electron spectroscopy (AES), and low energy electron diffraction (LEED) (3-5). Gases were impinged onto the Rh and Pt single crystal samples, cooled to 100 K with liquid nitrogen, through capillary array dosers which allowed background pressures in the 10^{-10} Torr (10^{-8} Pa) range to be maintained during dosing. Water was admitted into the chamber from the vapor above a freeze-pump-thaw cycled sample of the pure liquid held in a glass ampoule. Anhydrous HF, condensed from a commercial anhydrous cylinder (pressurized with He) into a stainless steel tube, was also freeze-pump-thaw cycled to remove volatile impurities.

Step 1 - Water plus a neutral probe molecule, CO

The first step in the UHV modeling of aqueous/metal interfaces is the adsorption of water alone, or water with a neutral probe molecule, on the metal of interest. This step sets up the hydrogen-bonded, high dielectric constant medium which could do much to distinguish aqueous-solid from gas-solid interfaces by providing an environment amenable to ionic species. Extensive UHV studies of water on a variety of surfaces (ably reviewed in Ref. 1) have shown water to be weakly adsorbed, with H_2O-metal and H_2O-H_2O interactions summing to 10-12 kcal/mole. Since water is weakly bound compared to many neutral molecules such as CO, one might expect the presence or absence of water to have little effect on the adsorption chemistry of such strongly bound neutrals. While for CO on Pt(111) the expectation of small water effects is borne out, on Rh(111) attractive CO-water interactions significantly modify the adsorption chemistry, suggesting possible ramifications for electrocatalysis (5).

TPD of CO plus water on Rh(111) and Pt(111). Figure 1 shows how increasing predoses of CO on Rh(111) continuously shift the thermal desorption of 1/2 monolayer of D_2O to higher temperatures, indicating net attractive interactions between water and CO on this surface. Figure 2 gives the results of the analogous experiments on Pt(111); on this surface even the lowest CO preexposures shift water desorption to lower temperatures, indicating net repulsive interactions between water and CO. On both surfaces the only desorbing species were water and CO; thus no evidence for an irreversible water gas shift reaction was seen. The lack of H_2 and/or CO_2 desorption, taken in the context of the known thermal behavior of functional groups, also argues against surface reaction to form adsorbed species such as formyl, formate, or carbonate in the temperature range below which water desorbs ($T \lesssim 210$ K).

HREELS of CO plus water on Rh(111). More specific information about the actual species present after coadsorption is provided by high resolution electron energy loss spectroscopy (HREELS), which yields vibrational data roughly analogous to those obtained by surface infrared spectroscopy. Figure 3 shows Rh(111) vibrational spectra for one-half monolayer of D_2O with increasing CO predoses (5). The D_2O-alone spectrum (Figure 3a) is characterized by a librational (hindered rotational) mode at 580 cm^{-1}, the D-O-D scissoring mode at 1190 cm^{-1}, and O-D stretches around 2480 cm^{-1} (4). On Rh(111) low coverages of CO alone give a single C-O stretch at 2020-2070 cm^{-1} ascribed to CO bound atop a single Rh atom. For higher coverages of CO alone a second peak, ascribed to CO in a two-fold bridge site, grows in at 1830 cm^{-1}, but its height never exceeds 70% of that of the atop CO peak. In the presence of water small CO coverages yield not the 2020 cm^{-1} atop peak seen in the absence of water, but rather a new peak around 1620 cm^{-1}. Even at high CO coverage, the $CO+D_2O$ spectrum is distinct from the CO-alone spectrum, as with water present the intensity of the 1830 cm^{-1} (bridged) peak exceeds that of the 2040 cm^{-1} (atop) peak, and a shoulder on the low-frequency side of the 1830 cm^{-1} peak indicates that some of the CO is still in the form which produced the 1620 cm^{-1} peak at low CO coverage.

We believe that the ~1620 cm^{-1} peak arises from CO bound in a three-fold hollow site, by analogy to rhodium cluster compounds yielding a CO stretch at 1600-1685 cm^{-1} (6-7) and to the 1655-1700 cm^{-1} CO stretch observed for CO+benzene/Rh(111) (8), where LEED intensity analysis has placed the CO in the three-fold hollow. A HREELS peak around 1620 cm^{-1} could also arise from reaction of CO and water to produce formate, formyl, or carbonate, all of which should have vibrational modes in this frequency range. However, none of the other modes expected for such species were observed, and the lack of H_2 or CO_2 desorption products also argue against the presence of such species. XPS data are also consistent with a water-induced shift in CO binding site. Addition of water to low coverages of CO on Rh(111) shifts the XPS O 1s signal due to CO from 531.7 to 530.0 eV binding energy, as expected for a shift from an atop to a multiply-bound site (9). Thus, water, though weakly bound to Rh(111), shifts the much

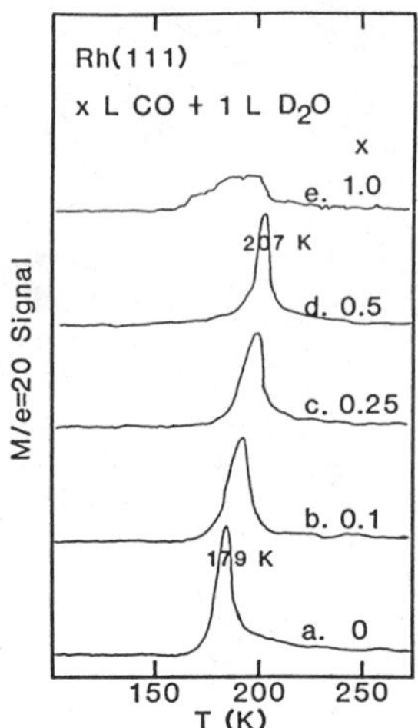

Figure 1. D_2O TPD from Rh(111) of varying exposures of CO followed by 1 L D_2O. (Reproduced with permission from Ref. 5. Copyright 1988 Elsevier.)

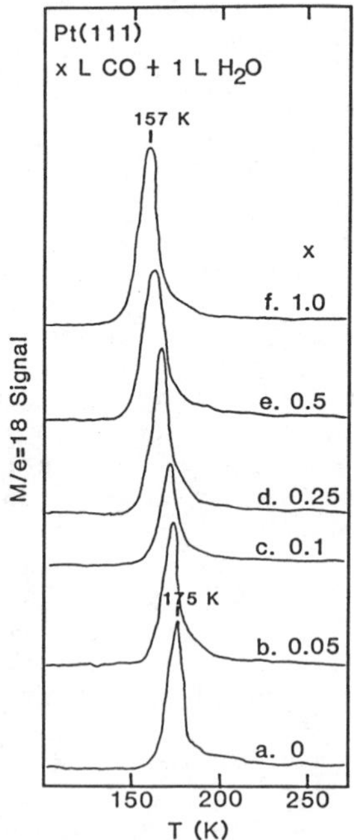

Figure 2. H_2O TPD from Pt(111) of varying exposures of CO followed by 1 L H_2O. (Reproduced with permission from Ref. 5. Copyright 1988 Elsevier.)

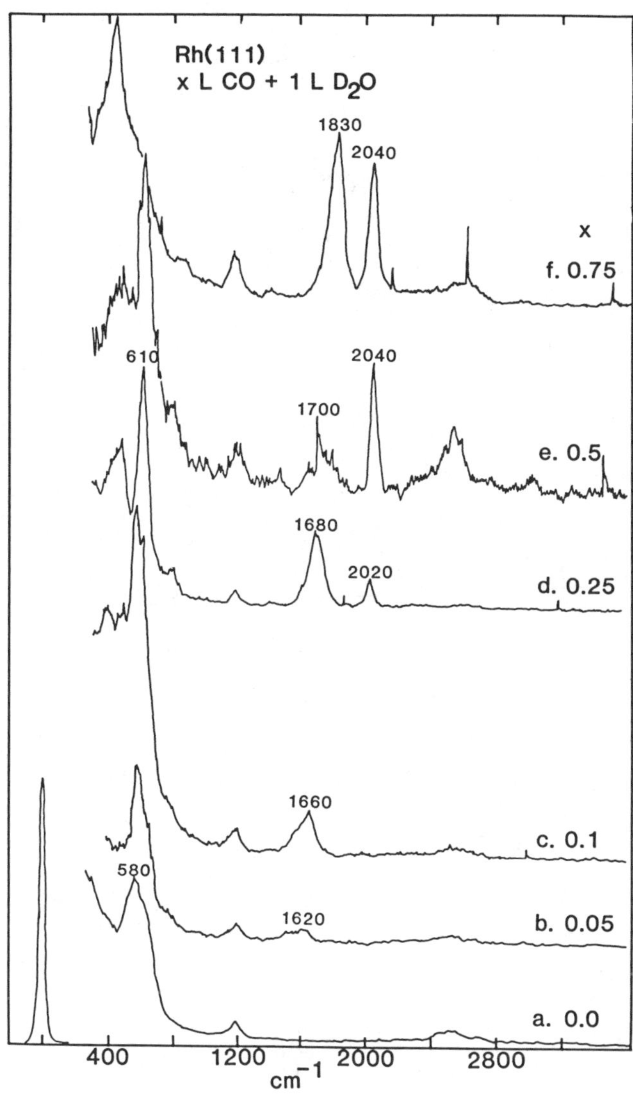

Figure 3. Rh(111) HREELS spectra of a.) 1 L D_2O, b. -f.) varying exposures of CO followed by 1 L D_2O. (Reproduced with permission from Ref. 5. Copyright 1988 Elsevier.)

more strongly-bound CO molecule to a site not populated in the absence of water.

Coadsorption on Rh(111) also modifies the HREELS features due to water. As shown in Figure 3, a substantial sharpening and shift to 610 cm^{-1} of the pure D_2O band at 580 cm^{-1} (due to the superimposed D_2O libration and Rh-O stretch) correlates with the growth of the ~1650 cm^{-1} CO mode. The fact that the vibrational spectra of both species are modified by coadsorption suggests adsorption of water and CO onto adjacent sites, a form of hydrophilic coadsorption.

H_2O+CO/Rh(111) produced features analogous to those shown for D_2O+CO/Rh(111). A partial superposition of the downshifted CO stretch and the H-O-H scissor mode at 1620 cm^{-1} makes the effect less well-resolved for H_2O than for D_2O, but we in fact observed the CO downshift first with H_2O.

HREELS of CO plus water on Pt(111). Figure 4 shows HREELS spectra for CO and a half monolayer of water on Pt(111). The water-alone spectrum (Figure 4a), exhibits a libration at 690, scissor at 1620, and OH stretch at 3440 cm^{-1}, as well as additional features at 970, 1020, and 1950 cm^{-1} previously ascribed to a minority water species (4). On Pt(111) metal coadsorption produces not the single low frequency three-fold peak seen for these CO and water coverages on Rh(111), but rather two peaks at 2080 and 1830 cm^{-1}, corresponding to the atop and two-fold CO species seen on both surfaces for higher coverages of CO alone. Increasing water additions to a constant low coverage of CO on Pt(111) decrease the atop/bridge peak height ratio in the same way that increasing the coverage of CO alone would do. On Pt(111) the HREELS features due to water are unchanged by the presence of CO. These observations indicate that water and CO adsorb onto separate patches on the surface, in a form of hydrophobic coadsorption. Water condenses into hydrogen-bonded islands, as indicated by the low O-H stretching frequency. CO spreads to cover the rest of surface, giving a phase similar to that for CO alone, but with a coverage normalized to the water-free, not total, surface area. CO-CO repulsions, which have been well documented on Pt(111) (10), produce a surface pressure within the CO patches which bears upon the edges of the water islands. It is this lateral pressure which causes water to desorb from Pt(111) at lower temperatures in the presence of coadsorbed CO.

Hydrophilic versus hydrophobic coadsorption. The contrast between the hydrophilic and hydrophobic coadsorption seen on Rh(111) and Pt(111), if confirmed under normal electrochemical conditions, might be of electrocatalytic importance. On Rh(111), where net attractive CO-H_2O interactions produce a mixed phase in which CO is displaced to a three-fold binding site which is not occupied in the absence of water, CO and water appear to occupy adjacent binding sites. Such thorough mixing of the oxygen source (water) and the intermediate [or poison] (CO) should improve electrooxidation rates for $C_xO_yH_z$ fuels (11). On Pt(111), where net repulsions cause condensation of CO and water into separate patches, reaction between the adsorbed species could occur only at the boundaries between patches, and one would expect slower kinetics.

The different coadsorption chemistries on these two structurally and chemically similar surfaces illustrate the importance of the

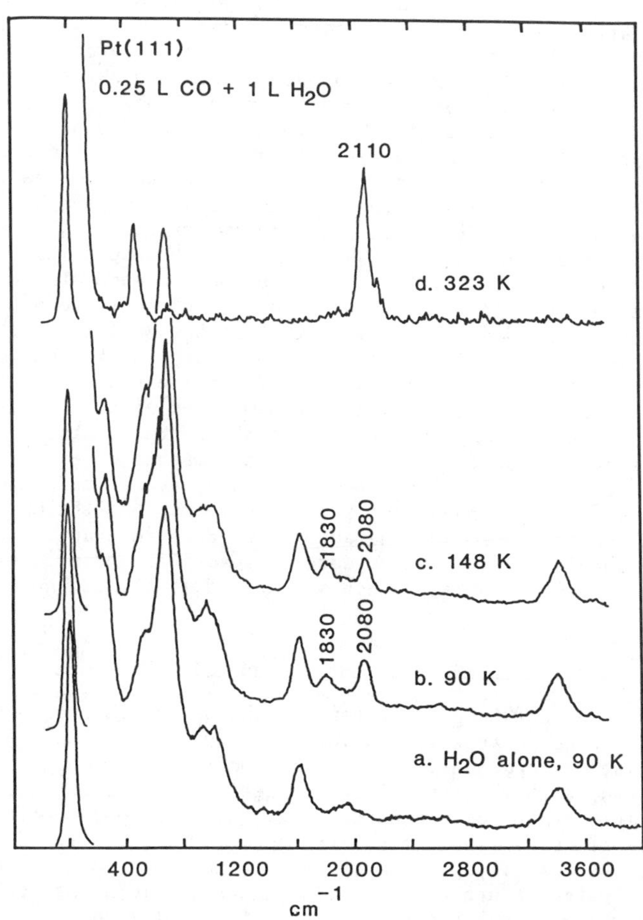

Figure 4. Pt(111) HREELS spectra for a.) 1 L H_2O at 90 K (does not change upon annealing); 0.25 L CO followed by 1 L H_2O: b.) at 90 K, c.) after flash to 148 K (nothing desorbed), d.) after flash to 323 K (water desorbed). (Reproduced with permission from Ref. 5. Copyright 1988 Elsevier.)

subtle balance of the many different lateral interactions present in multicomponent adsorption systems. Both true lateral interactions and the site-to-site variations in metal-adsorbate interaction strengths will influence the coadsorption geometry. The valence electrons in Pt have more d-character, and are therefore more localized, than those in Rh. One would then expect a greater site-to-site variation in metal-adsorbate interactions on Pt than on Rh, as has been noted for NO adsorption (12). Moving CO and H_2O off of their optimum one-component binding sites to optimize CO-H_2O interactions would thus exact a greater metal-adsorbate free energy penalty on Pt than on Rh. There is also some evidence that the strength of water-water interactions is different on these two surfaces (4).

Thus, although water is weakly bound on surfaces, its presence can, but need but not in all cases, alter the balance of forces which determines the adsorption state of much more strongly bound species. Further work is needed to determine whether the Rh(111) or the Pt(111) case is more typical, and also to determine the extent to which the cryogenic coadsorption experiments are representative of real aqueous environments. In the meantime, the electrochemist seeking information from UHV gas-phase adsorption studies to help interpret his results should be aware that the presence of water can alter adsorption chemistry even for strongly-bound neutral molecules which are not good hydrogen-bond donors or acceptors. And the UHV surface scientist hoping to provide gas-adsorption data relevant to electrochemistry should consider doing water coadsorption experiments to check for surprising effects such as those demonstrated here for Rh(111).

Step 2 - Addition of ionic species to the UHV model: HF + H_2O

All electrochemical techniques measure charge transferred across an interface. Since charge is the measurable quantity, it is not surprising that electrochemical theory has been founded on an electrostatic basis, with chemical effects added as a perturbation. In the electrostatic limit ions are treated as fully charged species with some level of solvation. If we are to use UHV models to test theories of the double layer, we must be able to study in UHV the weakly-adsorbing systems where these ideal "electrostatic" ions could be present and where we would expect the effects of water to be most dominant. To this end, and to allow application of UHV spectroscopic methods to the pH effects which control so much of aqueous interfacial chemistry, we have studied the coadsorption of water and anhydrous HF on Pt(111) in UHV (3). Surface spectroscopies have allowed us to follow the ionization of the acid and to determine the extent of solvation both in the layer adjacent to the metal and in subsequent layers.

TPD of HF alone, and of HF plus water, on Pt(111). Anhydrous HF, in which hydrogen bonding is extensive, shares many properties with water, including a broad temperature range of stability of the liquid and the high dielectric constant of the liquid. This parallelism of behavior extends to the interactions with the Pt(111) surface. Like water, HF is molecularly adsorbed on Pt(111). Each substance, adsorbed alone, gives a single desorption peak for coverages up to the monolayer and a second desorption peak at lower temperature for subsequent layers. However, HF desorbs at 50 K lower temperatures (130 and 126 K for the monolayer and multilayer, respectively, versus 179 and 168 K for water), probably due to the fewer hydrogen bonds per molecule formed by HF.

Coadsorption of HF and water has no effect on the water desorption peaks, but stabilizes part or all of the HF to higher temperatures, as shown by Figure 5. As long as at least 5 molecules of water per HF molecule are added to the surface (up to monolayer coverage, or 8 H_2O/HF for subsequent layers) no HF desorbs until water starts to leave the surface around 170 K, peaking at 180 K. As long as at least 1 molecule of water is initially present per HF, no HF desorption will occur until 150 K, peaking at 162 K. If more HF than H_2O molecules are present initially, some HF will desorb in a peak at 136 K, near the temperature at which HF alone desorbs. Coadsorption thus can yield HF desorption at three peaks, one not stabilized vs. HF alone, one stabilized by 30 K, and one stabilized by 50 K, i.e., to the water desorption temperature.

HREELS of the H_2O + HF system. The nature of the interaction stabilizing HF on the surface is made clear by the HREELS spectra of Figure 6. As the concentration of HF in the water layer is increased a new peak around 1150 cm^{-1} (and several smaller peaks) first increases and then, as the HF/H_2O ratio exceeds 1, decreases in intensity. By analogy to vibrational spectra of acid hydrates of known structure (13-16), this peak is identified as the symmetric bending mode of the pyramidal H_3O^+ ion. We have observed the same peak upon coadsorption of water and other, stronger, mineral acids. The reaction

$$H_2O + HF \longrightarrow H_3O^+ + F^-$$

can proceed further to the right in these cryogenic experiments than in room temperature aqueous solution because the low temperature decreases the importance of the unfavorable entropy of solvation of the fluoride ion (17).

Solvation stoichiometries. HREELS is sensitive to internal modes of ion cores, but details of solvation would produce at most secondary effects on the vibrational spectra. However, the identification of the ionization reaction, coupled with the TPD data outlined above, allow us to quantify the solvation. The HF desorption peak coincident with the water peak at 180 K, seen for molecular ratios of $H_2O/HF \geq 5$ in the monolayer and ≥ 8 in subsequent layers, corresponds to the recombination of H_3O^+ and F^- ions which had been fully solvated. Screened by solvating water molecules, these ions cannot combine to form HF until the solvation water desorbs. The figure of 8 for layers subsequent to the monolayer agrees with the sum of the primary solvation numbers measured in room temperature aqueous solution for the H^+ [4] and F^- [4±1] ions (18). This agreement, in combination with the frozen electrolyte work of Stimming's group (2), gives hope that the cryogenic aqueous environment accessible in UHV may be relevant to standard electrochemistry. The lesser number of solvation waters seen in the first monolayer suggests that even these classically "non-specifically adsorbed" ions lose some of their solvation waters upon adsorption, at least on Pt(111), a surface which, when well ordered, gives notoriously strange voltammetry (19-20). It also

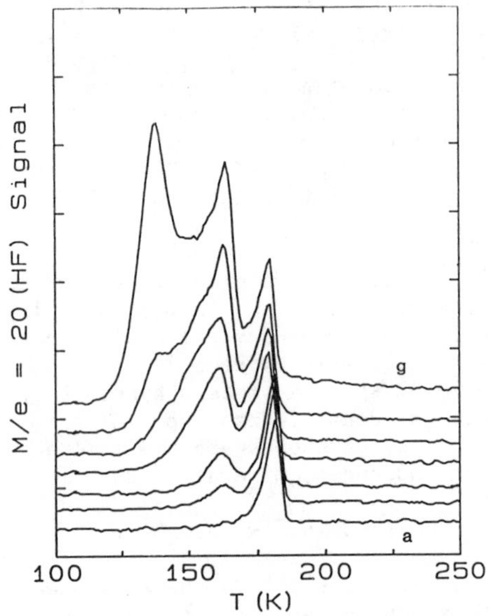

Figure 5. HF thermal desorption signals, normalized to H_2O coverage, for HF/H_2O monolayer ratios of: a.) 0.12, b.) 0.15, c.) 0.20, d.) 0.42, e.) 0.56, f.) 0.75, and g.) 1.28. (Reproduced with permission from Ref. 3. Copyright 1987 Elsevier.)

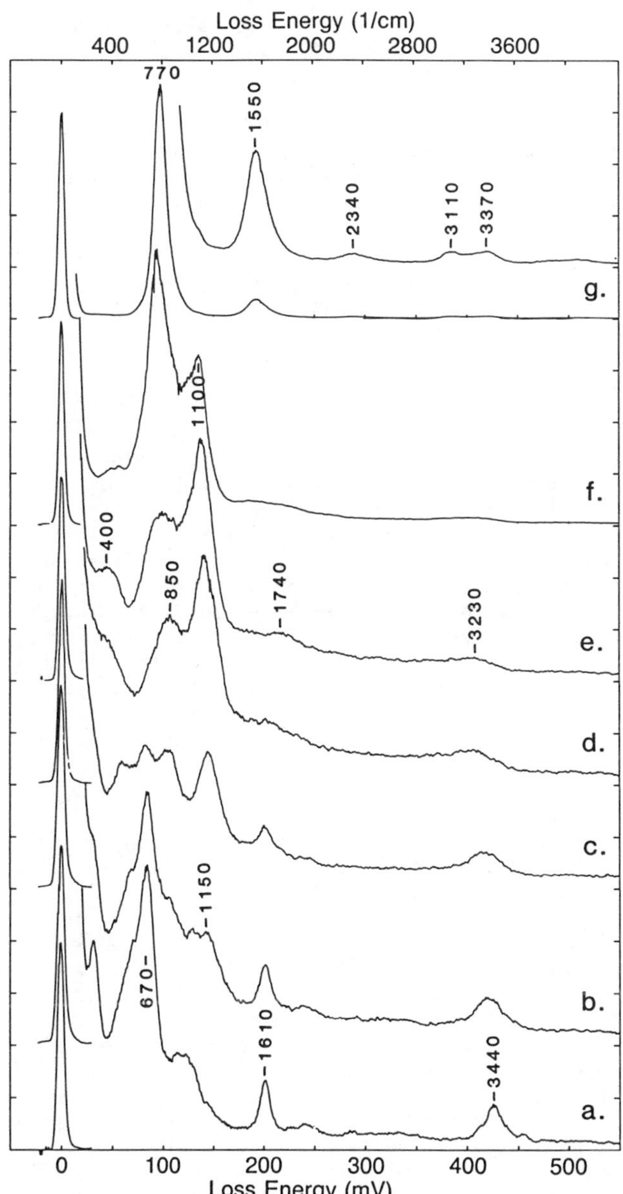

Figure 6. HREELS spectra for H_2O and HF adsorption and coadsorption on Pt(111) at 100 K, HF/H_2O ratio increasing upwards. a.) 1 monolayer (ML) H_2O, x100; b.) 1ML H_2O + 0.05ML HF, x100; c.) 1ML H_2O + 0.12 ML HF, x100; d.) 1 ML H_2O + 0.5 ML HF, x100; e.) 1 ML H_2O + 1 ML HF; f.) 1 ML HF + 0.25 ML H_2O, x100; g.) 1 ML HF, x16.5 and x 110. (Reproduced with permission from Ref. 3. Copyright 1987 Elsevier.)

draws into question the concept of a hydration sheath separating the surface from the solvated ions (21), which would yield a higher number of waters per HF in the first ionic layer than in subsequent layers.

The HF desorption peak around 160 K corresponds to the fractional distillation of HF from the monohydrate phase $[H_3O^+][F^-]$, which has been well established in bulk studies (22). The bulk work has not identified any water-rich distinct phases. The 136 K HF desorption peak seen for $HF/H_2O>1$ is due to the loss of excess molecular HF from the layer.

Coadsorption of HF and H_2O has shown that we can grow in UHV an, if not yet demonstrably the, ionic aqueous environment with controlled pH and composition.

Step 3 - Control of effective electrode potential: $H_2 + H_2O$

The unique aspect of electrochemistry lies in the ability to change the electrode potential and thus concentrate an applied perturbation right at the interface. Electric fields of 10^7 V/cm can be generated electrochemically with a half-lemon, scraped zinc (since 1983) penny, and copper wire as opposed to the massive Van de Graaff generator and electric power plant required for non-electrochemical approaches to the same field strength. If UHV models are to provide useful molecular-scale insight into electrochemistry, some means of controlling the effective electrode potential of the models must be developed.

One approach to potential control in UHV lies in chemical poising, roughly analogous to pinning an inert but catalytic electrode at a given potential by immersing it in a solution containing controlled concentrations of both members of a redox couple. To apply this approach in UHV we supply both members of the redox couple as species adsorbed, in controlled quantities, from the gas phase. We then allow equilibration to occur.

The close relation between the work function, measurable in UHV, and the electrochemical potential has been experimentally demonstrated by Hansen and Kolb (23-24). Bange et al. (25) have presented a direct comparison between UHV and electrochemical data (the latter corrected for diffuse layer effects) on the $Br_2+H_2O/Ag(110)$ system. They found that coadsorption of water with Br brought the UHV data into good agreement with electrochemical measurements. We consider here the $H_2+H_2O/Pt(111)$ system (in which molecular H_2 is never present as an adsorbed species). Because of the extreme reversibility of hydrogen electrochemistry on Pt, the effective electrode potential should be a single-valued function of the amount of hydrogen added to the water layer. Our attempts to measure work functions were frustrated by a persistant 10^8 Ω path to ground on our sample manipulator at the time. Instead, we will infer the electrode potential, relative to the potential of zero charge, from spectroscopic examination of the types of species present on the surface after equilibration.

Figure 7 compares electrochemical and UHV data for hydrogen adsorption on Pt(111). The solid curve, a, is the cathodic sweep from a cyclic voltammogram in 0.3 M HF (20) at 25 mV/s, with a constant double-layer charging current subtracted. The voltammogram plots current I (charge/time) versus potential V for a constant negative sweep rate ($dV/dt=-|C|$). The current in this potential region is generally ascribed to the discharge of hydronium ions to form

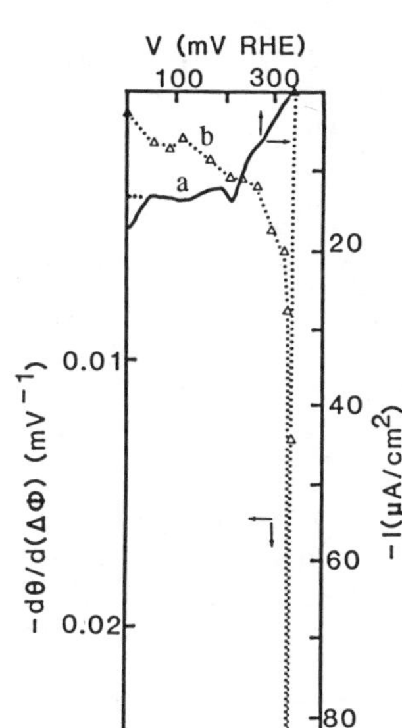

Figure 7. Comparison of (a, solid) electrochemical and (b, dashed) UHV measurements of the H_{ads} coverage/potential differential versus potential on Pt(111). a.) cathodic sweep (25 mV/s) voltammogram in 0.3 M HF from Ref. 20, constant double layer capacity subtracted. b.) $d\theta/d(\Delta\Phi)$ versus $\Delta\Phi$ plot derived from $\Delta\Phi$ versus θ plot of Ref. 26. Potential scales aligned at zero coverage. Areas under curves correspond to a.) 0.67 and b.) 0.73 H per surface Pt atom.

adsorbed hydrogen atoms, with one electron passed per hydrogen atom adsorbed. The current therefore equals the time derivative of the hydrogen coverage θ ($-I=d\theta/dt$). Given the constant potential sweep rate, the current is directly proportional to the potential derivative of coverage ($-I=-|C|d\theta/dV$). Since, once diffuse layer effects are corrected for, a change in electrochemical potential V is equivalent to a change in work function $\Delta\Phi$ (referenced to the clean surface), the current can also be thought of as the work function derivative of hydrogen coverage ($-I=-|C|d\theta/d(\Delta\Phi)$), as noted on the left-hand vertical axis of Figure 7. This interpretation of the voltammogram allows direct comparison with UHV data for gas-phase hydrogen adsorption.

The work function change $\Delta\Phi$ of Pt(111) in UHV as a function of hydrogen coverage θ has been reported by Christmann et al. (26). Exchanging their x and y axes and taking the first derivative yields a $-d\theta/d\Delta\Phi=f(\Delta\Phi)$ curve, shown as a dashed line, b, in Figure 7, analogous to the voltammetric data shown as the solid line, a. Though the same quantities are plotted in the two curves, the functional forms are quite different, indicating that water and/or other double layer components do play major roles in determining the coverage/potential relationship for hydrogen electrosorption on Pt(111). UHV models of this system should contain, at the least, water coadsorbed with the hydrogen.

TPD and HREELS of H_2 + H_2O on Pt(111). Despite the intimations of water effects given by Figure 7, coadsorption of hydrogen has little if any effect on the thermal desorption of water. However, as noted in the case of HF+H_2O, the lack of a change in water TPD does not necessarily indicate a lack of significant interactions. Figure 8 shows the effects of an increasing H_2 predose on the HREELS spectrum of a monolayer of water adsorbed at 90 K and then flashed to 150 K (Wagner, F.T. and Moylan, T.E. Surface Sci., submitted). For H_2 doses below 0.65 L no new peaks (in addition to the water-alone modes) are seen. This is not surprising in light of the very weak HREELS features produced by H alone on Pt(111) (27). Above 0.65 L a new peak grows in at 1150 cm^{-1}, similar to the peak seen for water coadsorption with HF and other mineral acids; we again assign this peak to the symmetric bending mode of H_3O^+. Unlike the case for HF + H_2O, no reaction is seen at 100 K. Figure 8 thus gives evidence for a kinetically hindered reaction

$$H_{ads} + H_2O_{ads} \longrightarrow H_3O^+{}_{ads} + e^-{}_{metal}$$

which proceeds at 150 K when the initial hydrogen coverage exceeds ~20% of saturation, but not at lower hydrogen coverages. Note that both reactants and products are formally reduced one electron with respect to water on an uncharged metal surface.

The above reaction yields an adsorbed cation which has weak specific interactions with the surface. The presence of this cation in the absence of charge-balancing anions (such as F^- in the HF+H_2O case) thus indicates that the surface is at an effective potential below the potential of zero charge (pzc), i.e., the countercharge resides in the metal. The lack of reaction for initial hydrogen coverages below ~20% of saturation suggests that the effective potential of the water-only monolayer lies above the pzc, i.e., where the

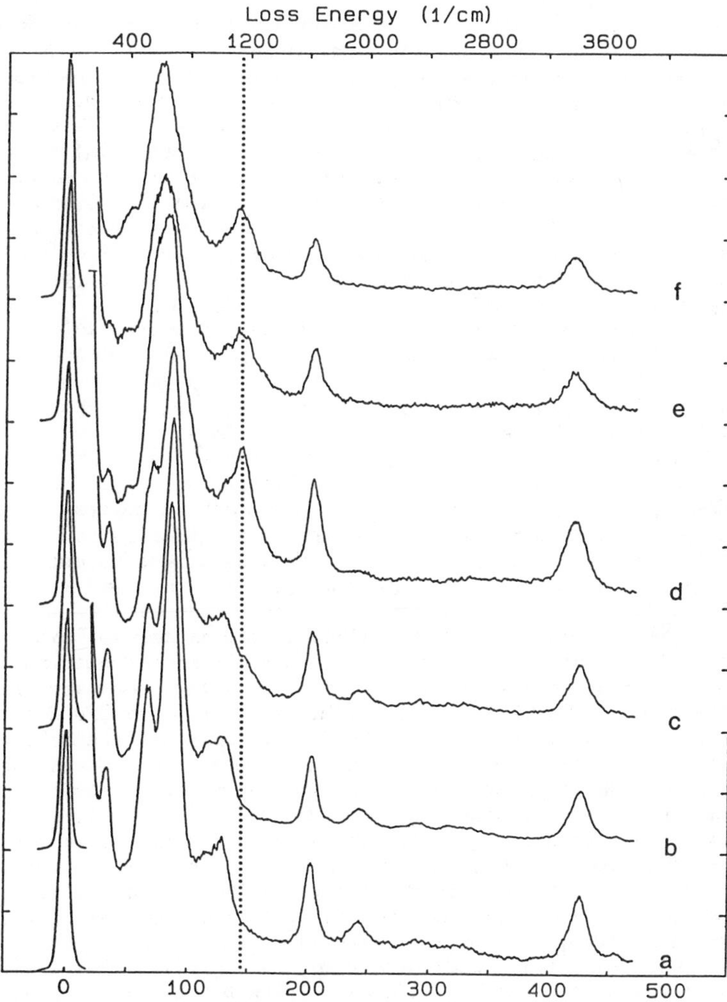

Figure 8. HREELS spectra at 90 K of coadsorbed layers flashed to 150 K showing effect of increasing H_2 predose, all x100: a.) 1.4 L H_2O; b.) 0.3 L H_2 + 1.4 L H_2O; c.) 0.65 L H_2 + 1.4 L H_2O; d.) 1 L H_2 + 1.4 L H_2O; e.) 2.5 L H_2 + 1.4 L H_2O; f.) 10 L H_2 + 1.4 L H_2O.

net charge on the metal is positive. To form an equilibrium layer including adsorbed cations one must add sufficient reducing agent to shift the potential first down to, and then past, the pzc.

Comparison with room-temperature electrochemistry

The UHV data suggest that 20% of saturation hydrogen atom coverage must be added to the water monolayer to shift the potential below the pzc. Electrons can be added to the interface either by adsorption of the reducing agent hydrogen or by passage of charge through an external circuit; after equilibration both methods of reduction should produce the same interface provided the same number of electrons are added. To form one hydrogen atom by reduction of bulk water (or $H_3O^+_{aq}$), one electron must be added to the electrode from the external circuit. The gas-phase addition of 20% of the 0.73 H/Pt UHV saturation hydrogen converage is thus equivalent to the addition of 35 $\mu C/cm^2$ of cathodic charge to the Pt(111)/0.3 M HF interface.

If one integrates the area between the voltammogram in Figure 7a and the double-layer corrected zero current line, starting from the onset of hydrogen adsorption at 0.34 V RHE, one reaches 35 $\mu C/cm^2$ at a potential of 0.22 V RHE. To the as yet undetermined extent that the low temperature UHV data is representative of real electrochemical conditions, this result suggests that the pzc of Pt(111) lies at 0.22 V RHE. It should be noted that this quantitative comparison between UHV hydrogen dosage and electrochemical charge is not affected by the percentage of H_{ads} and ($H_3O^+_{ads}$ + e^-_{metal}) pairs on the equilibrated surface; what matters is the total number of reducing equivalents added to the water + metal system.

The proposed position of the pzc lies just at the anodic end of the unusual broad, flat section (peculiar to well-ordered Pt(111)) of the hydrogen electrosorption region. This flat section would then correspond to the coadsorption of H_3O^+ and H; repulsions between charged H_3O^+ groups could help explain why a structurally uniform surface can produce such a broad electrosorption peak. The sloping section of the voltammogram above 0.22 V would correspond to adsorption of H alone. A somewhat similar interpretation was advanced by Bewick and Russell (28), based on EMIRS data, for the low and high potential main peaks in the voltammogram of polycrystalline Pt in acid. They ascribed the low potential peak to hydrogen strongly associated with water, while the high potential hydrogen gave no evidence for strong association.

A water-alone monolayer potential above the pzc is in accordance with an absolute work function measurement for the water monolayer on Pt(111) of 4.8 eV (29). Comparing this to the hydrogen electrode (4.7 eV below vacuum (30) for the normal hydrogen electrode NHE) corrected by 7x0.059 V for a nominal pH 7 yields a water-alone monolayer potential of +0.5 V vs. RHE at pH 7. This lies 0.3 V above our proposed pzc of 0.2 V RHE. This relatively high apparent potential of the water monolayer has been noted previously (Sass, J.K., private communication), and has raised concern about the relevance of the UHV monolayer to real electrochemical conditions, since most electrochemical measurements of the pzc of polycrystalline Pt have been closer to 0.2 V than to 0.5 V (31). By showing that the water monolayer lies above, not at, the pzc, the present H_2+H_2O data remove part of the apparent discrepancy between the electrochemical and UHV results. If future UHV work function data show a large (~0.3 V) decrease in the water monolayer work function upon addition of small (<20% saturation) amounts of hydrogen, all of the apparent discrepancy could be quantitatively accounted for.

Conclusions: Roles for cryogenic UHV model double layers

We have given examples of how the effects of an aqueous environment can be studied in UHV, how solvated ionic species can be added to that environment, and how the effective electrode potential can be controlled. Thus, the essential features of the electrochemical double layer can be reproduced in UHV at low surface temperatures. However, a basic question remains: why should the electrochemist be interested in <u>model</u> double layers when the real thing can be so simply generated by plunking an electrode into a liquid electrolyte? UHV models offer two advantages: (1) control of interfacial parameters far more precise than that which can be achieved in liquid, and (2) detailed analysis of fully solvated species by a host of complementary surface spectroscopies, all applicable to the exact same surface. In the standard electrochemical experiment we control the electrode potential and the bulk concentration of electrolyte species. The concentrations of all adsorbed species are not directly controlled and are often difficult or impossible to measure, and the possibility of uncontrolled surface impurities is always lurking in the background. In the UHV model we directly control the amount of adsorbed species by controlled dosing from the gas phase; the dependent variable (the potential) is readily measurable as the work function. After a given composition of an electrochemical interface is postulated, we can prepare that composition in UHV, transfer it to an electrochemical cell, and see if it behaves as expected. Working with cryogenic model double layers grown in UHV, we have available "in situ" (i.e., with double layer and solvent present) measurements of surface structure, elemental composition, electronic structure, oxidation state, and molecular connectivity (through vibrational and ion spectroscopies). All of these techniques can be applied to the same surface, while in situ techniques for normal spectroelectrochemistry require a myriad of specialized, often mutually exclusive, electrode forms and cell designs.

Nevertheless, the cryogenic UHV models remain <u>models</u>, the relevance of which to real electrochemistry remain incompletely established. We can generate <u>an</u> extensively hydrated ionic environment in UHV, but is it the correct one? Grounds for optimism on this point have been demonstrated in Ref. 25 and discussed less concretely in Ref. 3. Frozen electrolyte work (2) has demonstrated considerable mechanistic continuity between normal electrochemistry and electrochemistry in a perchloric acid hydrate at temperatures down to 150 K, but how general is this result?

Attainment of complete answers will require coordinated application of (1) solvated, cryogenic UHV models, (2) true in situ spectroscopies under realistic conditions, and (3) UHV-electrochemical transfer experiments. As non-UHV techniques for generating clean, well-ordered electrode surfaces are developed (19) and validated with UHV work (20), high-quality in situ spectroelectrochemical data on single crystals will provide direct points of comparison for UHV model studies. UHV-electrochemical transfer experiments provide a means of applying vacuum surface spectroscopies to real electrochemical, albeit emersed and evacuated, interfaces. A growing body of data (23-24) shows that, at least in some cases, emersion and evacuation do not destroy the dominant chemistry of the interface. Cryogenic UHV model work, by identifying in which systems water effects are important and in which they are not, can help establish the cases where ex situ UHV spectroscopic data are truly representative of the original electrochemical interface. Synergistic use of these three approaches should provide the hard, molecular-scale, data needed to determine the true balance between chemical and electrostatic effects at electrochemical interfaces, and to refine our previous, largely electrostatics-based, views of the double layer.

Literature Cited

1. Thiel, P. A.; Madey, T. E. Surface Sci. Rept. 1987, 7, 211.
2. Frese, U.; Stimming, U. J. Electroanal. Chem. 1986, 198, 409.
3. Wagner, F. T.; Moylan, T. E. Surface Sci. 1987, 182, 125.
4. Wagner, F. T.; Moylan, T. E. Surface Sci. 1987, 191, 121.
5. Wagner, F. T.; Moylan, T. E.; Schmieg, S. J. Surface Sci. 1988, 195, 403.
6. King, R. B.; Ackermann, M. N. J. Organomet. Chem. 1973, 67, 431.
7. Eisenberg, R.; Hendriksen, D. E. In Advances in Catalysis; Eley, D. D.; Pines, H.; Weisz, P. B., Eds.; Academic: New York, 1979; Vol. 28, p 87.
8. Mate, C. M.; Somorjai, G. A. Surface Sci. 1985, 160, 542.
9. DeLouise, L. A.; Winograd, N. Surface Sci. 1984, 138, 417.
10. Poelsema, B.; Verheij, L. K.; Comsa, G. Phys. Rev. Letters 1982, 49, 1731.
11. Gilman, S. J. Phys. Chem. 1964, 68, 70.
12. Banholzer, W. F.; Park, Y. O.; Mak, K. M.; Masel, R. I. Surface Sci. 1983, 128, 176; R.I. Masel, private communication.
13. Taylor, R. C.; Vidale, G. L. J. Am. Chem. Soc. 1956, 78, 5999.
14. Savoie, R.; Giguere, P. A. J. Chem. Phys. 1964, 41 2698.
15. Ferriso, C. C.; Hornig, D. F. J. Chem. Phys. 1964, 41, 2698.
16. Falk, M.; Giguere, P. A. Can. J. Chem. 1957, 35, 1195.
17. Myers, R. T. J. Chem. Educ. 1976, 53, 17.
18. Bockris, J. O'M.; Reddy, A. K. N. Modern Electrochemistry; Plenum: New York, 1977; p 131.
19. Clavilier, J.; Faure, R.; Guinet, G.; Durand, R. J. Electroanal. Chem. 1983, 150, 141.
20. Wagner, F. T.; Ross, P. N. Jr. J. Electroanal. Chem. 1983, 150, 141; J. Electroanal. Chem., submitted.
21. Ref. 18, chap. 7.
22. Cady, G. H.; Hildebrand, J. H. J. Am. Chem. Soc. 1930, 52, 3843.
23. Hansen, W. N.; Kolb, D. M. J. Electroanal. Chem. 1979, 100, 493.
24. Kolb, D. M. Z. Phys. Chem. (N.F.) 1987, 154, 179.
25. Bange, K.; Straehler, B.; Sass, J. K.; Parsons, R. J. Electroanal. Chem. 1987, 229, 87.
26. Christmann, K.; Ertl, G.; Pignet, T. Surface Sci. 1976, 54, 365.
27. Baro, A. M.; Ibach, H.; Bruchmann, H. D. Surface Sci. 1979, 88, 384.
28. Bewick, A.; Russell, J. W. J. Electroanal. Chem. 1982, 132, 329.
29. Fisher, G. B. GMR-4007; GM Research Labs: Warren, MI, 1982.
30. Gomer, R.; Tryson, G. J. Chem. Phys. 1977, 66, 4413.
31. Frumkin, A. N. Petry, O. A. Electrochim. Acta 1970, 15, 391.

RECEIVED August 24, 1988

Chapter 6

Aromatic Polymerization on Metal Surfaces

Jay B. Benziger, N. Franchina, and G. R. Schoofs

Department of Chemical Engineering, Princeton University, Princeton, NJ 08544

Thin polymer films have been prepared by surface catalysis in ultrahigh vacuum and electrochemical deposition from solution. These two routes of synthesis result in poly(thiophene), poly(aniline) and poly(3,5-lutidine) films that have similar infrared spectra. These polymer films are highly orientationally ordered; the rings are perpendicular to the surface in poly(thiophene) and poly(3,5-lutidine) films, and the phenyl rings are parallel to the surface in poly(aniline). Both the catalytic and electrochemical polymerization reactions appear to result from electron donation from the aromatic species to the surface producing a radical cation that leads to electrophilic addition.

A variety of aromatic compounds can be polymerized electrochemically to form a thin polymer film[1-4]. Many films involving heteroatom aromatics can be made electrically conducting or insulating through oxidation and reduction[5-8]. There has been a great deal of interest in films of poly(aniline), poly(thiophene) and poly(pyrrole) because they can be made electrically conducting, are easily synthesized, and are fairly robust . Polymer films of aniline, thiophene and pyrrole are readily formed on platinum electrodes by anodic oxidation of the monomer from a solution of a polar organic solvent, such as acetonitrile[1,2]. As formed these polymer films have associated anions and are conducting. They may be removed from solution stored in air and used in a wide variety of electrolyte solutions[7,9].

Polymer films may also be synthesized by direct chemical routes[10]. These chemical synthesis routes generally involve some form of electrophilic substitution using Grignard reagents or other modified precursors. The purity and electrical properties of the films produced from the chemical route are generally poorer than the films produced electrochemically, but the chemical synthesis

0097–6156/88/0378–0083$06.00/0

routes provide insight into how the films are formed electrochemically. Our recent investigations of the interactions of heteroatom species with well defined nickel surfaces indicated that monolayers of heteroatom aromatic compounds adsorbed on atomically clean nickel single crystals polymerize[11,12]. Furthermore, it was found that several of these polymers were orientationally ordered relative to the surface. This suggests a third route for synthesizing aromatic polymer films, and suggests that it may be possible to produce highly ordered films.

In this paper we describe the preparation of thin polymer films by surface catalysis and anodic deposition. The results indicate that both synthesis routes produce orientationally ordered films that have similar infrared spectra. It is also shown that thin ordered films of poly(thiophene) have different electrochemical behavior than the fibrous films that are electrically conducting.

Experimental

Polymer films were produced by surface catalysis on clean Ni(100) and Ni(111) single crystals in a standard UHV vacuum system [12,13]. The surfaces were atomically clean as determined from low energy electron diffraction (LEED) and Auger electron spectroscopy (AES). Monomer was adsorbed on the nickel surfaces circa 150 K and reaction was induced by raising the temperature. Surface species were characterized by temperature programmed reaction (TPR), reflection infrared spectroscopy, and AES. Molecular orientations were inferred from the surface dipole selection rule of reflection infrared spectroscopy. The selection rule indicates that only molecular vibrations with a dynamic dipole normal to the surface will be infrared active [14], thus for aromatic molecules the absence of a C=C stretch or a ring vibration mode indicates the ring must be parallel the surface.

Polymer films were produced by anodic deposition by potentiostatic deposition onto a platinum electrode. Deposition was done from 1 M solutions of the monomer in 1M $LiClO_4$ in acetonitrile. The films were characterized by cyclic voltammetry and reflection infrared spectroscopy in an apparatus described elsewhere [15].

Results

Thiophene. The first example of surface catalyzed polymerization we observed was the formation of thiophene oligimers on a clean Ni(111) surface [11]. The TPR results summarized in Figure 1 indicated molecular fragments with five and more carbons (up to C_8 fragments were detected) from thiophene reacting on Ni(111). These intermediates decomposed over a wide temperature range, approximately 400 K. The existence of products with more than four carbons indicated that dimers and other oligimers were being formed. Furthermore, the broad temperature range for dihydrogen evolution resulting from decomposition is suggestive of a wide range of surface species. The broad range of decomposition temperatures was also observed by Zaera et al for thiophene reacting on Ni(100)[16] , and by Lang and Masel on Pt (111), (100) and (210) [17].

A series of infrared spectra taken after adsorption and subsequent heating of thiophene on Ni(111) are shown in Figure 2. At the initial adsorption temperature of 170 K the infrared spectrum

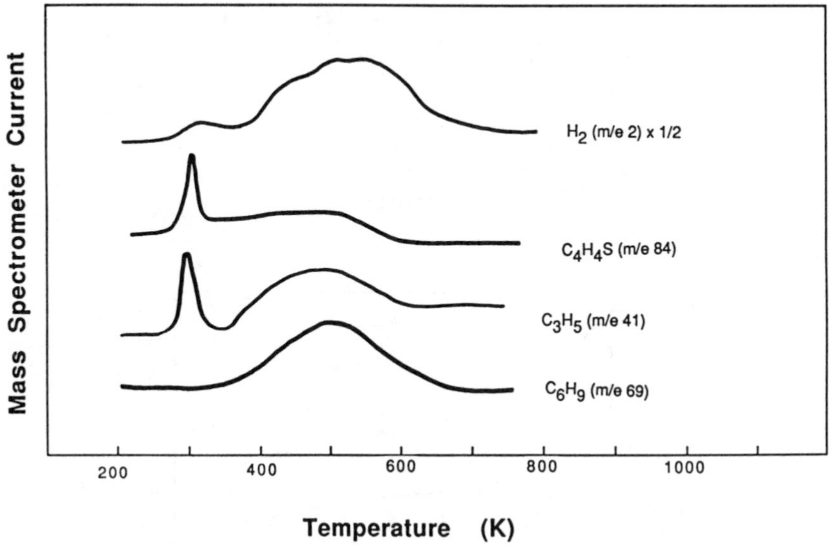

Figure 1. Temperature Programmed Reaction Spectra of Thiophene on Ni(111)

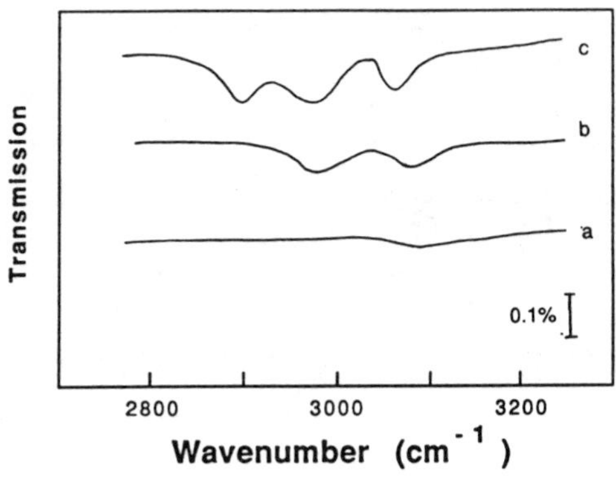

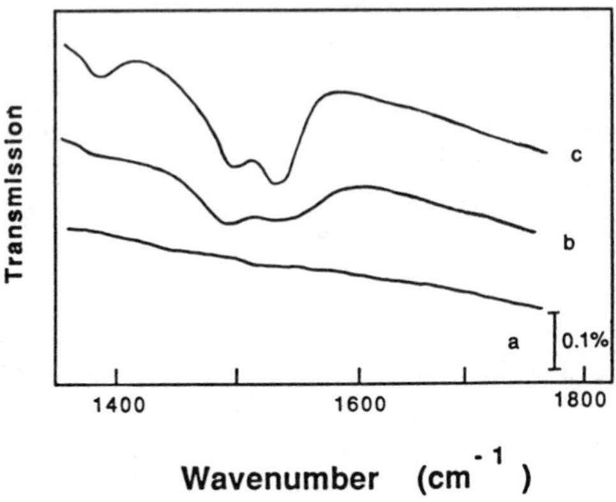

Figure 2. Reflection infrared spectra of thiophene films (a) thiophene adsorbed on Ni(111) at 170 K (b) thiophene adsorbed on Ni(111) at 170 K heated to 320 K (c) electrochemically grown poly(thiophene) on Pt

was featureless. After heating to 320 K the infrared spectrum showed the presence of CH stretching modes between 2800 and 3100 cm^{-1}, and C=C stretching modes at 1520 and 1495 cm^{-1}. These results indicate that the thiophene was with the ring parallel to the surface below 200 K. At 320 K the thiophene had reacted to form a molecular entity with vibrational modes normal to the surface. We suggest these changes result from the surface polymerization of thiophene. At the low temperature the thiophene cannot react, and it pi-bonds to the surface. We have carried out extended Huckel calculations for thiophene adsorbed on a 19 atom Ni cluster and found that a flat configuration with the thiophene ring 2.6 Å above the surface was the most stable configuration [see ref. 18 for details of the computational technique]. The flat orientation was also observed by Sexton for thiophene on Cu(100) [19], which is consistent with our observations that aromatic molecules tend to bond flat on metal surfaces.

As the temperature was raised the thiophene reacted, with both structural and orientational changes observed. The TPR results show the formation of oligimeric species and the infrared spectra indicate the polymer must have CH and C=C bonds perpendicular to the surface. Zaera et al [16] have reported vibrational spectra taken with electron energy loss spectroscopy for thiophene reacting on a Ni(100) surface. Their spectra agree with the infrared spectra reported here, extending the frequency range to below 1000 cm^{-1}. Between 90 and 185 K they observe a change in the EEL spectrum indicating a surface reaction of thiophene. The resulting absorption bands are compared to the absorption bands of the two isomers of poly(thiophene) in Table I.

Table I

Vibrational Bands of Thiophene Derivatives

C_4H_4S/Ni(100)*	poly(2,5-thienylene)**	poly(2,4-thienylene)**	
3025	3090		C-H stretch
1450	1490 1408	1475	C=C stretch
1340	1230	1320	ν cycle
1150		1170	ν cycle
1070	1052		C-H in plane bend
850	788	820 730	C-H out of plane bend

* from ref. 16
** from ref. 20

There is remarkable agreement between the bands from thiophene reacting on Ni(100) with a mixed isomeric form of poly(thiophene). Furthermore, Zaera et al observed that the absorption bands all gradually diminished in intensity between 220 and 600 K with the evolution of hydrogen, which is characteristic of polymer degradation. Zaera et al attributed their results to a metallocycle on the surface. We believe that surface polymerization of thiophene is a more reasonable explanation of the experimental results for the following reasons: (i) C_{5+} desorption products were observed; (ii) the broad temperature range for decomposition; (iii) the agreement of the ir bands between thiophene on Ni(100) and the isomers of poly(thiophene); and (iv) Ni catalysts are known to catalyze polymerization of thiophene in the absence of hydrogen [21]. The absence of sulfur containing oligimers among the desorption products is difficult to account for, and may be the result of the poly(thiophene) bonding to the surface through the sulfur atoms.

Poly(thiophene) films have also been formed on a Pt foil potentiostatically from 1 M thiophene in 1M $LiClO_4/CH_3CN$ at 1.74 V vs. Ag/AgCl(1M). The films were then removed from the thiophene solution and placed in 1M $LiClO_4/CH_3CN$ and cycled between 0 V to 2.0 V vs. Ag/AgCl to test for the presence of polymer films. At deposition potentials below 1.74 V no film deposition was detected, at higher potentials the films deposition rate was more rapid and not readily controlled. For short deposition times (< 5 s) an orange-gold film was deposited on the Pt electrode. Assuming a two electron transfer per thiophene molecule grafted onto the surface the density of the gold film is ~ 2×10^{15} molecules/cm^2. Cycling this film in electrolyte caused no change in the color of the film. There is also very little doping (incorporation of perchlorate anion) into the film; the thin orange-gold film can be doped to <5% with the perchlorate ion. Increasing deposition times resulted in thicker black films. Cycling these films in electrolyte showed these films could be oxidized (doped) and reduced (undoped) as indicated by the voltammogram. In the reduced form the films appeared a dark red. These thicker films could be doped with perchlorate ion, with the doping level increasing with increasing film thickness up to a maximum doping level of approximately 25%.

A reflection infrared spectrum of the anodically deposited orange-gold film is shown in Figure 2. The electrochemically deposited poly(thiophene) film has an infrared spectrum that is very similar to that obtained from the film formed catalytically on Ni(111), suggesting a common structure. Some differences in the ratio of peak intensities are seen which may be due to differences in the isomeric forms of the polymers. It was also found that the catalytically prepared film can only be grown to monolayer thickness. Adsorption of thiophene on a Ni(111) surface with a poly(thiophene) film results in thiophene desorption during temperature programmed heating, no additional polymerization was detected. Reflection spectra of the thick black films formed electrochemically could not be obtained as they lacked sufficient reflectivity.

Surface polymerization of thiophenes was found to be affected by both surface impurities and substituent groups on the thiophene ring. The reactions of thiophene on a Ni(111) surface with sulfur impurities was examined [11]. The sulfur inhibited the polymerization so that the only reactions observed were desorption of thiophene with a small amount of decomposition. Thiophene polymerization was also found to be inhibited on dirty platinum electrodes. It was observed that without rigorous cleaning of the platinum electrode before immersion into the thiophene solution higher potentials were required to obtain thiophene polymerization, and the rate of polymer formation was greatly reduced. No polymerization of 2,5-dimethylthiophene was observed either catalytically on the Ni(111) surface [11], or electrochemically on a Pt electrode.

Aniline. Aniline black is a well known polymer of aniline formed by electrophilic addition[3,4]. Numerous investigators have formed poly(aniline) films by anodic deposition of Pt and other electrode materials. We have examined the interaction of aniline with clean Ni(111) and Ni(100) surfaces in ultrahigh vacuum and found aniline to form an orientationally ordered, thermally stable polymer film. Electrochemically prepared poly(aniline) films also show the high degree of orientational ordering.

The evidence for catalytic polymerization of aniline is a combination of TPR, AES and infrared results. The TPR results for aniline on Ni(111) are summarized in Figure 3. After adsorption of aniline at 170 K aniline and dihydrogen desorb at 240 K. Dihydrogen, ammonia, HCN and various hydrocarbon products are found to evolve at 1000 K. These results indicate a thermally very stable surface species. The Auger spectra did not change with temperature between 300 K and 800 K, with the carbon lineshape being indicative of a molecular entity [22], and a molecular coverage of aniline based on the C(KLL)/Ni(LMM) and N(KLL)/Ni(LMM) ratios of 1.5 x 10^{15} molecules/cm^2. The infrared spectrum of aniline adsorbed on Ni(111) at 165 K shown in Figure 4 is almost identical to liquid aniline, indicating a frozen multilayer. After heating to 400 K the infrared spectrum shows no ring vibration modes in the 1350-1850 cm^{-1} range, and no CH stretches around 3000 cm^{-1}; an NH stretch at 3200 cm^{-1} is evident in the spectrum.

The AES results indicate that the aniline coverage is more than two times greater than the maximum coverage based on van der Waals radii. The TPR results show this species is too stable to be a condensed multilayer. Hence, we conclude that aniline polymerized forming a very stable polymer layer. In addition, the absence of infrared bands corresponding to C=C stretches or ring vibrations indicated that the poly(aniline) film was formed with the phenyl rings parallel to surface. The infrared results also indicated that the poly(aniline) film had N-H bonds which were oriented perpendicular relative to the surface. Almost identical results were obtained for the interaction of aniline with a Ni(111) surface.

Poly(aniline) films were prepared electrochemically on a Pt electrode potentiostatically at 0.85 V vs Ag/AgCl for 5 s from a 1 M aniline in 1 M $LiClO_4/CH_3CN$ solution. The films were a faint blue-green color. Thicker films were formed at greater potentials. Even the very thin films could be oxidized and reduced. After removing the electrode from the aniline solution and placing it in electrolyte solution the film could be oxidized and reduced. Oxidation peaks occurred at 0.2 V and 0.48 V, scanning was kept below 1 V to avoid polymer degradation [7,8]. Infrared spectra of the electrochemically prepared poly(aniline) were taken after emersion from the electrolyte solution at 0.8 V, 0.35 V and 0 V, the spectra are shown in Figure 4. At all three potentials the ir spectra show only weak C=C stretches or ring vibrations in the 1350-1850 cm^{-1} range, indicating that the poly(aniline) is formed with the phenyl ring nearly parallel to the surface. In the NH stretching region (circa 3200 cm^{-1}) one can see that the reduced film has an NH stretching band. After the first oxidation the NH stretch decreases in intensity and shifts to lower frequency, and after the second oxidation the NH stretch vanishes. From the infrared results the reduced film from electrochemical deposition appeared to be almost identical to the catalytically formed film. As was observed with thiophene polymerization, the catalytic polymerization of aniline would produce only a monolayer while the electrochemical polymerization could produce thick films.

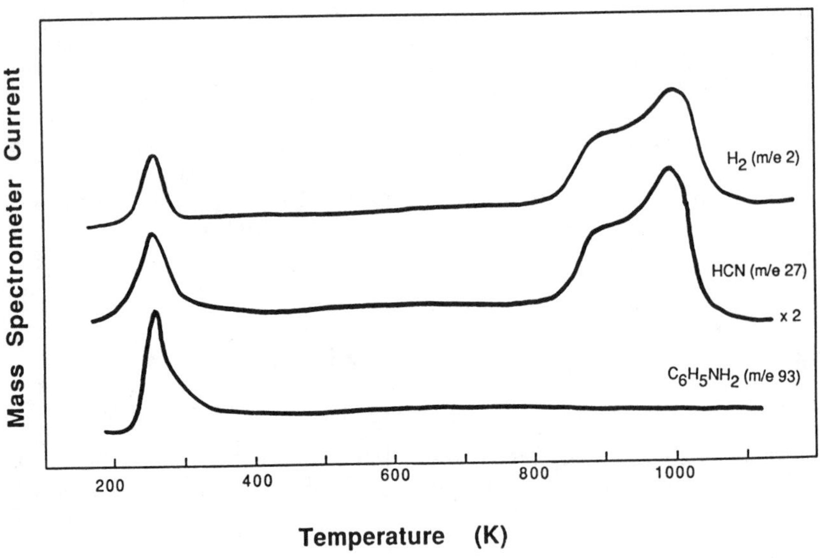

Figure 3. Temperature Programmed Reaction Spectra of Aniline on Ni(111)

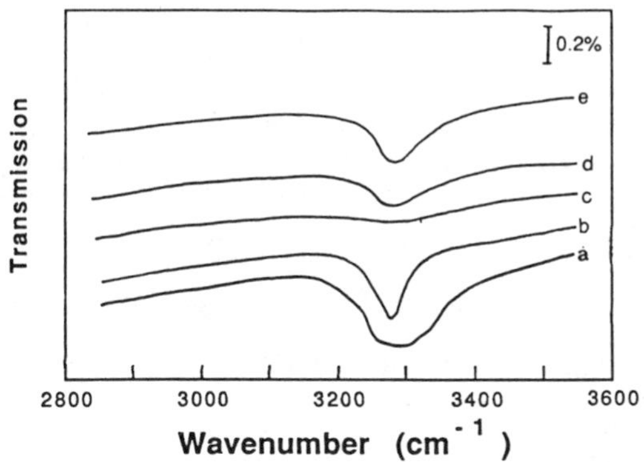

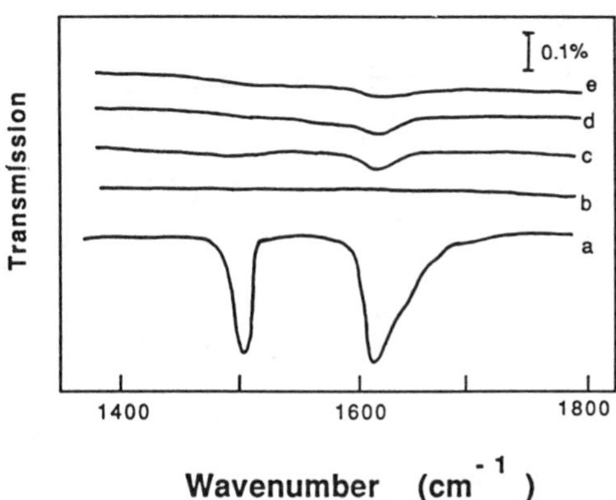

Figure 4 . Reflection infrared spectra of aniline films (a) aniline multilayer adsorbed on Ni(111) at 165 K (b) poly(aniline) monolayer on Ni(111) at 400 K (c) electrochemically grown poly(aniline) on Pt oxidized at 0.8 V vs. Ag/AgCl(1M) (d) electrochemically grown poly(aniline) on Pt reduced at 0.35 V vs. Ag/AgCl(1M) (e) electrochemically grown poly(aniline) on Pt reduced at 0.0 V vs. Ag/AgCl(1M)

Pyridine. Pyridine and its methyl substituted derivatives (picolines and lutidines) were found to polymerize electrochemically and, under certain circumstances, catalytically. This behavior was not expected because usually pyridine undergoes electrophilic substitution and addition slowly, behaving like a deactivated benzene ring. The interaction of pyridine with a Ni(100) surface did not indicate any catalytic polymerization. Adsorption of pyridine below 200 K resulted in pyridine adsorbing with the ring parallel to the surface. The infrared spectrum of pyridine adsorbed at 200 K showed no evidence of either ring vibrations or CH stretches (Figure 5). Desorption of molecular pyridine occurred at 250 K, and above 300 K pyridine underwent a substitution reaction forming an α-pyridyl species on the surface [12,23]. Prominent ring modes in the infrared spectrum indicated that the ring was oriented perpendicular to the surface. The α-pyridyl subsequently reacted forming a variety of species including pyridine, HCN, and dihydrogen that desorbed at 475 K. There was evidence of a small amount of a more stable surface species that decomposed above 600 K suggesting the possibility of oligimerization on the surface, but this was a minor reaction path.

Methyl substituents on the pyridine ring had a profound impact on the reactivity of the pyridine ring. 2,6-Lutidine did not react to any appreciable extent on Ni(100) [12]. The infrared spectrum of 2,6-lutidine showed no C=C stretches and ring vibrations, but did show CH stretches. This indicated adsorption occurred with the ring parallel to the surface, and the CH stretches resulted from the methyl groups. The presence of the methyl groups inhibited attack at the α-carbon and resulted in the lutidine desorbing without reaction. In contrast, 3,5-lutidine adsorbed with its ring perpendicular to the surface. The infrared spectrum for 3,5-lutidine adsorbed on Ni(100) at 180 K shown if Figure 6 is very similar to liquid lutidine. The C=C stretches and ring modes are all very prominent. During temperature programmed heating no lutidine was found to desorb from the surface. Dihydrogen was evolved in a series of desorption peaks between 300 and 600 K, along with small amounts of methane and ammonia. The infrared spectra showed a gradual loss of most of the absorption bands, but the C=C stretching band at 1590 cm^{-1} persisted to above 750 K. The TPR results showed a mixture of hydrocarbons, ammonia and hydrogen cyanide all were evolved at 900 K from the decomposition of a stable surface species. Auger electron spectra indicated the C(KLL) lineshape was that of a molecular entity to above 800 K. The molecular coverage after adsorption at 180 K, based on the C(KLL)/Ni(LMM) and N(KLL)/Ni(LMM) ratios, was approximately 0.7×10^{15} cm^{-2}; and this coverage remained nearly constant up to 800 K. This is approximately the maximum coverage based on close-packed van der Waals radii. These results suggest the formation of a surface polymer of lutidine, which displays comparable thermal stability to poly(aniline). The poly(lutidine) is different from the poly(aniline) in the orientation of the rings, as infrared spectroscopy shows poly(lutidine) has the rings perpendicular to the surface.

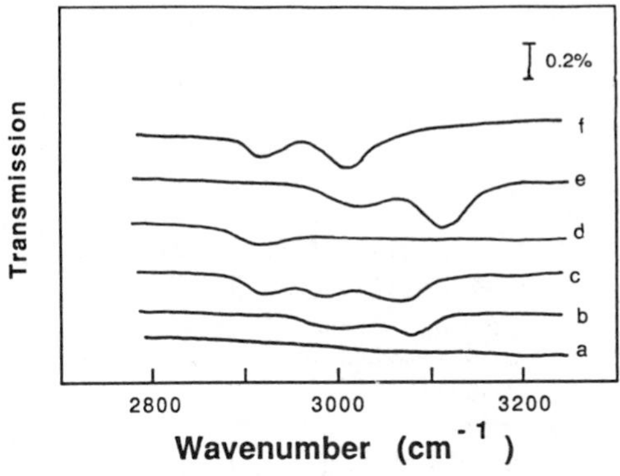

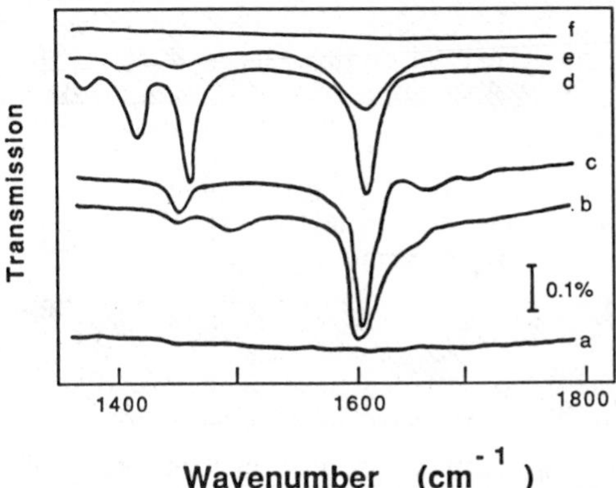

Figure 5. Reflection infrared spectra of pyridine films (a) pyridine adsorbed on Ni(111) at 180 K (b) pyridine adsorbed on Ni(111) at 320 K (c) electrochemically grown poly(pyridine) on Pt (d) 3,5-lutidine adsorbed on Ni(111) at 180 K (e) 3,5-lutidine polymer on Ni(111) at 750 K (f) 2,6-lutidine adsorbed on Ni(111) at 180 K

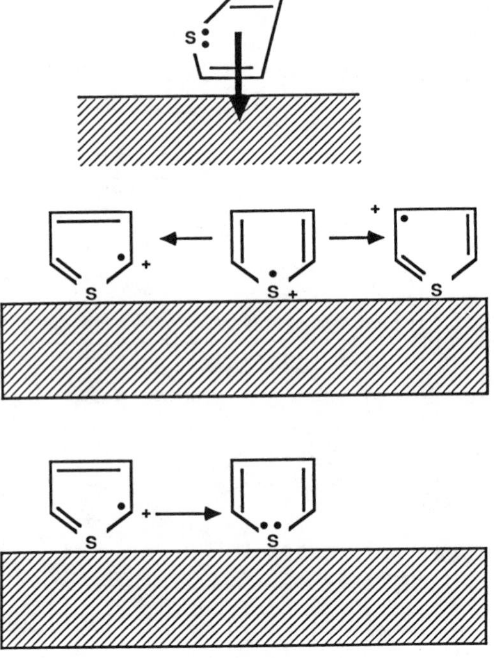

Figure 6. Electrophilic polymerization of thiophene. Initial π-bonding, formation of radical cation, electrophilic addition.

Pyridine was found to polymerize on a Pt electrode from a solution of 1 M pyridine in 1 M $LiClO_4/CH_3CN$ at potentials above 0.8 V vs Ag/AgCl. A colorless film was formed, but it could be oxidized and reduced when placed in plain electrolyte solution. The infrared spectrum of the electrochemically formed poly(pyridine) film is shown in Figure 5. It displays a very intense, narrow band at 1500 cm^{-1} indicative of C=C stretches that are perpendicular to the surface. 3,5 Lutidine also was polymerized on a platinum electrode under the same conditions, and its infrared spectrum is similar to that for the surface catalyzed poly(lutidine). The C=C stretching band for the poly(lutidine) is at slightly higher frequency than that observed for poly(pyridine), and also is broader. We also attempted to polymerize 2,6-lutidine electrochemically but found that at potentials of less than 2 V no polymerization of the 2,6-lutidine was observed.

Pyrrole. Pyrrole is the nitrogen analog of thiophene and is expected to behave in much the same fashion. Electrochemically the polymerization of pyrrole is well established [2]. Catalytic routes for polymerization of pyrrole do not appear to be as well established as those for thiophene. Our work and the work of others suggests that the surface catalytic polymerization of pyrrole is also less likely than for thiophene. Temperature programmed reactions of pyrrole on Ni(100) showed different types of reactions occurred at different heating rates [24]. At heating rates < 10 K/s pyrrole decomposed on Ni(100) to adsorbed carbon and nitrogen, and hydrogen which recombined and desorbed as dihydrogen. At higher heating rates, > 50 K/s, pyrrole appeared to polymerize, forming a surface species that decomposed above 570 K yielding H2, HCN, and various hydrocarbon products. Infrared spectroscopy also reflected these differences in reaction pathway caused by different heating rates. The infrared spectra for pyrrole adsorbed on Ni(100) indicate that after adsorption at 170 K the molecule adsorbed with its ring parallel to the surface as no C=C, or CH modes were apparent in the infrared spectrum. After heating to 400 K at a rate of 8 K/s the infrared spectrum remained featureless, whereas after heating to 400 K at 48 K/s small features were evident at 1550 cm^{-1} corresponding to the C=C stretching, and at 3075 and 3360 cm^{-1} resulting from CH and NH stretches respectively. These features are weak, consistent with the TPR results showing that the surface concentration of oligimer is low. Pyrrole was also found to reorient on Pt(111) circa 300 K [25].

A poly(pyrrole) film was deposited on a Pt electrode from potentiostatic conditions at 0.8 V vs Ag/AgCl. The film was colorless, its presence was verified by oxidation and reduction of the film in plain electrolyte solution. The infrared spectrum of the electrochemically prepared poly(pyrrole) is similar to the catalytically prepared films indicating the two films are structurally similar.

Discussion

The results suggest that there is a remarkable similarity in the structure of thin films formed by anodic deposition on a Pt electrode to those formed catalytically on Ni surfaces in ultrahigh vacuum. The reactions that form these polymers are both suggested to be electrophilic addition reactions. In the case of electrochemical anodic deposition an electron is pulled from the aromatic ring producing a radical cation, shown schematically for thiophene in Figure 6. The radical cation would attack an aromatic ring as an electrophile, resulting in chain propagation. In the case of a clean metal surface empty states act as electron acceptors, again producing a radical cation that will act as an electrophile. Work function measurements of aromatic molecules on Pt surfaces have shown that the surface acts as an acceptor for the π electrons in the aromatic ring[26,27]. The electron donation of the aromatic molecules can be understood in terms of energy level matching arguments [18].

The catalytic polymerization reaction is distinguished from its electrochemical counterpart by the extent of polymerization. Catalytic surface polymerization appears to be limited, whereas one can grow arbitrarily thick films electrochemically. This difference is due to the available sink for electrons to propagate the reaction. On an open circuit metal surface, so the reaction is self-limited by the number of electron acceptors. This may be due to either the availability of surface states, or the polymer film hindering electron transfer across the layer as it grows. Electrochemically there is a continual drain of electrons, limited only by the conductivity through the growing film, so the film may be grown arbitrarily thick. There are also activation barriers associated with both electrochemical and surface catalyzed reactions. In the electrochemical route a sufficient potential must be applied to allow electrons to flow from the adsorbate into the solid. The activation barrier associated with the catalytic path is that the electrons are thermally activated to go from the π-orbitals in the aromatic to the empty surface states.

The activation associated with the catalytic path also appears to cause orientational changes in a variety of the molecules. This can also be accounted for by the electrophilic nature of the reactions. In the case of thiophene, pyrrole and pyridine the highest occupied molecular orbitals are the π-orbitals. The lone electron pair lies at lower energies. At low temperatures adsorption occurs through a multi-center bond between the π-orbitals of the aromatic species and the metal d-orbitals, and results in the ring being parallel to the surface. Promotion of an electron from the lone electron pair to bond with the empty metal d-states above the Fermi level results in a perpendicular configuration as the lone electron pair is in the plane of the ring. The electrochemical deposition process effectively reduces the activation barrier for bonding between the surface states and the lone electron pair. By decreasing the potential one effectively lowers the lumos of the surface reducing the activation barrier and making the perpendicular configuration more favorable. We have carried out extended Huckel calculations for pyridine adsorbed on a 17 Ni atom cluster modelling adsorption on a Ni(100) surface. For the normal d-level of the Ni atoms the Fermi level of the cluster is at -8.22 eV and the parallel mode of pyridine binding is 24 kJ/mol more stable than the perpendicular mode; however when the Ni d-level decreased so the Fermi level of the cluster is -9.11 eV the perpendicular mode of bonding to the surface becomes more stable. This appears to explain why thiophene, pyrrole and pyridine all readily polymerize anodically at room temperature with the ring perpendicular to the surface.

The methyl groups on the pyridine ring result in a major difference in the reactivity of lutidines. In 3,5-lutidine the methyl groups act as electron donors tending to increase the stability of the π-bonds, and activating the ring for electrophilic attack at the α-positions. The MOs in 3,5-lutidine show the π-levels pushed to lower energy relative to pyridine, while the lone pair is unaffected so there is a greater preference for bonding through the lone electron pair. The methyl groups also push the ring away from the surface when bonding in the parallel orientation, also contributing to the preference for the perpendicular orientation. In contrast the methyl groups in 2,6-lutidine sterically hinder bonding with the lone electron pair on the nitrogen and also do not activate the ring for electrophilic reactions. Thus one sees little reactivity and only a parallel bonding configuration. These effects were manifested both catalytically and electrochemically.

Aniline polymerized differently from the other heteroaromatics as it retained a parallel bonding configuration for the rings. The NH2 group activated the ring for electrophilic attack leading to the polymerization reactions. Unlike the other molecules the lone electron pair on the nitrogen in aniline lies almost co-planar with the p-orbitals of the phenyl ring. This results in the ring parallel, or nearly parallel to the surface independent of the reaction conditions.

Conclusions

Aromatic molecules can be polymerized catalytically on clean metal surfaces, or electrochemically to produce oriented polymer films. Initial adsorption of aromatic molecules occurs by electron donation from the aromatic molecule to the surface. This electron donation creates radical cations that can polymerize. Molecular orientation in the films depends on the stable bonding configuration of the radical cation. Thiophene, pyridines, and pyrrole all polymerize with the ring substantially perpendicular to the surface, whereas aniline polymerizes with the phenyl rings parallel to the surface. The catalytically prepared films terminate at monolayer coverage due to availability of surface states to act as electron acceptors, while thick electrochemically prepared films can be prepared by providing a drain for the electrons.

Acknowledgments

The authors wishes to thank the NSF and the AFOSR for providing financial support for this work.

Literature Cited

1. Tourillon, G; in *Handbook of Conducting Polymers*, Skotheim, T.A., ed.; Dekker, New York, 1986.
2. Diaz, A.F.; Logan, J.A.; *J. Electroanal. Chem.* **1980**, *111*, 111.
3. Mohilner, D.M.; Adams, R.N.; Argersinger, W.J., Jr.; *J. Am. Chem. Soc.* **1982**, *84*, 3618.

4. Macdiarmid, A.G.; Chiang, J.C.; Richter, A.F.; Epstein, A.J.; *Synthetic Metals* **1987**, *18*, 285.
5. Tourillon, G.; Garnier, G.; *J. Phys. Chem.* **1983**, *87*, 2289.
6. Genies, E.M.; Vieil, E.; *Synthethic Metals* **1987**, *20*, 97.
7. Paul, E.W.; Ricco, A.J.; Wrighton, M.S.; *J. Phys. Chem.* **1985**, *89*, 1441.
8. Mermilliod,N.; Tanguy, J.; Hoclet, M.; Syed, A.A.; *Synthetic Metals* **1987**, *18*, 359.
9. Tourillon, G.; Garnier, F.; *J. Electrochem. Soc.* **1983**, *130*, 2042.
10. Kovacic, P.; McFarland, K.N.; *J. Polym. Sci. Polym. Chem.* **1979**, *17*, 1963.
11. Schoofs, G.R.; Preston, R.E.; Benziger, J.B.; *Langmuir* **1985**, *1*, 313.
12. Schoofs, G.R.; Benziger, J.B.; *J. Phys. Chem.* **1988**, *92*, 741.
13. Benziger, J.B.; Preston, R.E.; Schoofs, G.R.; *Appl. Optics*, **1987**, *26*, 343.
14. Greenler, R.G.; *J. Chem. Phys.* **1966**, *44*, 310.
15. Pang, K.P.; Benziger, J.B.; Soriaga, M.P.; Hubbard, A.T.; *J. Phys. Chem.* **1984**, *88*, 4583.
16. Zaera, F.; Kollin, E.B.; Gland, J.L.; *Langmuir* **1987**, *3*, 555.
17. Lang,J.F.; Masel, R.I.; *Surface Science* **1987** , *183* , 44.
18. Myers, A.K.; Benziger, J.B.; *Langmuir* **1987**, *3*, 414.
19. Sexton, B.A.; *Surface Science* **1985**,*163*, 99.
20. Yamamoto, T.; Sanechika, K.; Yamamoto, A.; *Bull. Chem. Soc. Japan* **1983**, *56*, 1497.
21. Bonner, W.A.; Grimm, R.A.; In "The Chemistry of Organic Sulfur Compounds"; Kharasch, N; Meyers, C.Y. Eds.; Pergamon Press: Oxford, 1966; Chapter 2.
22. Haas, W.T.; Grant, J.T.; Dooley, D.J.; *Proc. 2nd Intern. Symp. on Adsorption/Desorption Phenomena*; Ricca, F. Ed.; Academic Press: London, 1973.
23. Wexler, R.M.; Tsai M.-C.; Friend, C.M.; Muetterties, E.L.; *J. Am. Chem. Soc.* **1982**, *104*, 2034.
24. Schoofs, G.R.; Benziger, J.B.; *Surface Science* **1987** , *192* , 373.
25. Tourillon, G.; Raaen, S.; Skotheim, T.A.; Sagurton, M.; Garrett, R.; Williams, G.P.; *Surface Science* **1987**, *184*, L345.
26. Gland, J.L.; Somorjai, G.A.; *Surface Science* **1974**, *41*, 387.
27. Gland, J.L.; Somorjai, G.A.; *Surface Science* **1973**, *38*, 157.

RECEIVED May 17, 1988

Chapter 7

Surface Chemistry of Copper Electrodes

Preliminary Studies

John L. Stickney, Charles B. Ehlers, and Brian W. Gregory

School of Chemical Science, University of Georgia, Athens, GA 30602

Initial studies of the reactivity of the low index planes of copper exposed to oxygen, water vapor, neat water, HCl gas, and aqueous 1mM HCl, have been performed. An ultrahigh vacuum surface analysis instrument, directly coupled to an antechamber, used for high pressure and solution experiments, was used to investigate the surface reactivity of a copper single crystal. This crystal was oriented on three faces, each to a different low index plane. Low energy electron diffraction and Auger electron spectroscopy were used to study the surfaces before and after each reaction. The resulting structures and compositions were consistent with previous gas phase studies in the literature, and evidenced significant stability for the copper surface in aqueous solutions at open circuit.

This paper describes equipment, procedures and results for investigation of transition metal surface reactivity. Specifically, the surface reactivity of copper single crystals was examined under conditions relevant to the electrochemistry and corrosion of copper.

Copper is the primary metal used in construction of electronic equipment. Preparation of copper components often involves electroplating, electropolishing, or exposure to humid air. The extensive use of copper justifies a systematic examination of its surface reactivity.

Initial studies, described here, involved the use of an ultrahigh vacuum (UHV) surface-analytical instrument coupled to an antechamber. The antechamber allows experiments in solution and electrochemical treatments without transfer of samples outside of the system's controlled atmosphere. Focusing on the chemistry of copper surfaces in aqueous environments suggests the importance of studying the initial stages of surface reactivity with oxygen and water. Electrochemical experiments involve electrolytes; thus their surface reactivity should be studied as well.

Oxygen adsorption on the low-index planes of copper is one of

0097–6156/88/0378–0099$06.00/0

the most extensively studied surface systems, and has been partially reviewed by Spitzer and Luth (1,2). From an electrochemical standpoint, water adsorption is of great interest, although copper is much less reactive to water than to oxygen at room temperature (3-12). Water is adsorbed at or above room temperature only in the presence of catalytic amounts of adsorbed oxygen (4-6,8,10,11), and its adsorption involves decomposition to oxygen adatoms (11). Previous surface studies of the adsorption of electrolytes, or electrolyte analogues, have mostly involved Cl_2 (Stickney, J.L.; Ehlers, C.B.; Gregory, B.W.; Langmuir, submitted.) " S.E.G." The adsorption of HCl from both the gas phase and from solution onto the low-index planes of copper has been recently studied in this author's lab (S.E.G.). In general, halogens and hydrogen halides adsorb strongly on copper surfaces, forming adlayers of halogen atoms at low exposures, < 1000L. High exposures of halogens result in formation of thin films of copper halides (13).

Experimental. The apparatus used in these experiments is shown in Figure 1. The instrument consists of an analysis chamber containing optics for examination of the sample surface by low energy electron diffraction (LEED), a hemispherical electron analyzer for Auger electron spectroscopy (AES), and a quadrupole mass spectrometer for thermal desorption spectroscopy (TDS). Attached to the analysis chamber is an antechamber (Figure 2) into which the sample was transferred for experiments involving high pressures, up to ambient, or solutions. The antechamber allows introduction of high pressures of reactant gases while leaving the analysis chamber at UHV pressures, 10^{-9} to 10^{-11} Torr. A gate-valve interlock connects a stainless steel bellows-sealed compartment to the antechamber. This compartment contains an electrochemical H-cell made of Pyrex glass, complete with a glass frit separating the auxiliary electrode compartment.

Extraction and translation of the sample to the antechamber were performed with a linear and rotational feedthrough assembly (Figure 2). The rotational feedthrough was used both to rotate the sample between crystal faces for analysis and to lock the sample holder onto the extraction rod for transfer into the antechamber. The sample holder and X-Y-Z-R manipulator, located in the analysis chamber, were connected with four contacts: one chromel and one alumel (for the thermocouple), and two high current connectors for sample heating. These four contacts were set inside a 1-1/4" I.D. high-precision bearing. This bearing provided rotation (which was necessary to examine all three faces of the crystal), room for the electrical contacts, and clearance for extraction of the crystal.

Sample. The sample used in the experiments described here was a single crystal of copper oriented and polished on three faces. Each of the three faces was oriented to a different low-index crystallographic plane; the (100), the (111) or the (110). These three faces were parallel to a vertical axis, allowing the study of each, in turn, by rotation about this axis (Figure 2). The low-index planes could thus be studied without removing the sample from the UHV system. More importantly, however, experiments involved equivalent treatment of the three faces at each stage. Initial ion bombardment took place in a cage, similar to an ion gauge, in which the sample

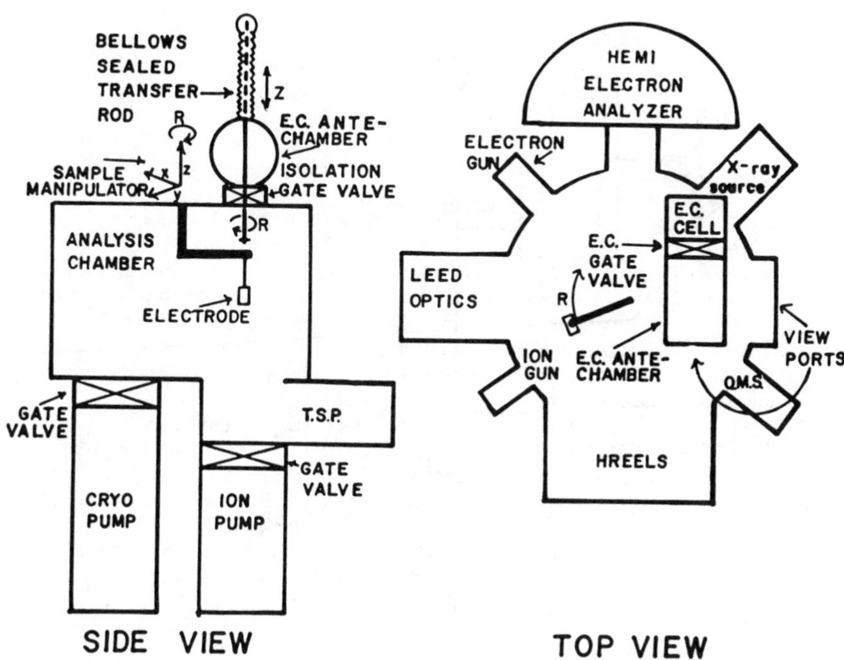

Figure 1. Diagram of the surface analysis instrument.

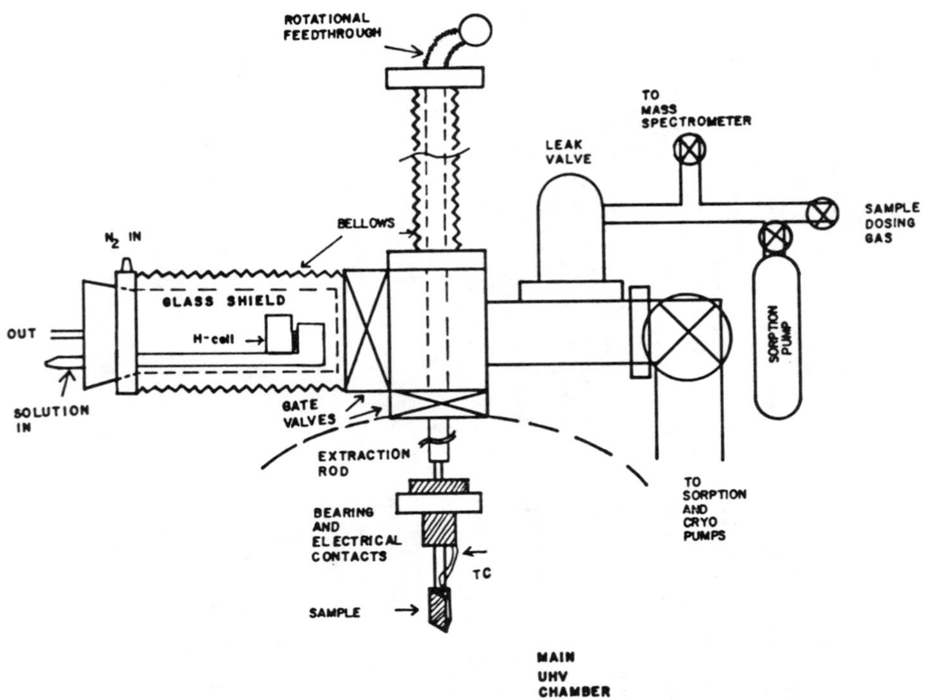

Figure 2. Diagram of the electrode, antechamber, and dosing manifold.

acted as the collector in 10^{-5} Torr of argon. All three faces were bombarded at the same time, which prevented one from becoming contaminated while the others were cleaned. Gas dosing was performed dynamically but without a gas dosing tube so that equivalent exposures resulted on all surfaces. Gas composition, partial pressure, and total gas exposure were monitored with a quadrupole mass spectrometer (UTI). A Pyrex sorption pump was included in the dosing manifold (Figure 2). The sorption pump and high purity gas source were used to rinse the manifold multiple times prior to gas introduction, which resulted in higher dosent purity. Immersion of the sample in solution also resulted in equivalent exposures on the three faces. Solutions were prepared with pyrolytically distilled water (14) and reagent grade chemicals.

Sample pretreatment consisted first of ion bombardment which removed the top monolayers of the sample surface, followed by annealing at 550°C for 15 minutes by resistance heating. Subsequent AES spectra were taken and monitored for signs of contamination. LEED was used to characterize the regularity of the single crystal surfaces, insuring the reproduciblity of the sample surface prior to each experiment.

Before solution experiments were run, the sample was transferred to the antechamber. Ultrahigh purity (UHP) gases, argon and nitrogen (Matheson), were used to bring the antechamber to ambient pressure (Figure 3b). Subsequently, the gate valve was opened (Figure 2), and the H-cell inserted. With the H-cell in position, the sample was lowered in and solutions were delivered from pressurized bottles exterior to the instrument.

Pump-down of the antechamber, following solution experiments, involved sorption pumping and cryo-pumping resulting in the pressure decreasing from ambient to 10^{-9} Torr in 5 minutes. The resulting sample surfaces were subsequently examined by both LEED and AES after transfer back to the analysis-chamber.

Results and Discussion

Oxygen Adsorption. Initial exposure of the copper crystal to oxygen resulted in formation of structures previously reported in the literature (1-3,15). On Cu(100), structures formed were the same as those described by Lee and Farnsworth (15) as early as 1965. Exposures less than 500L at crystal temperatures above 100°C resulted in formation of an oblique structure, characterized by groups of four spots near the (1/2, 1/2) positions. Exposures above 500L at room temperature resulted in the initial formation of a Cu(100)(/2X/2)R45-O structure. There is some disagreement in the literature as to whether this pattern represents a separate structure or whether it is an enhancement of a subset of spots characteristic of a Cu(100)(/2X2/2)R45° structure, observed at higher exposures (Figure 3c). The extra beams resulting from the 2/2 pattern vary considerably in intensity as the oxygen coverage is changed. High exposures, above $2x10^{4}$ L resulted in multiple layers of copper oxidation (Figure 3d) and a disordered LEED pattern.

Studies of oxygen adsorption on Cu(111) have been reported in the literature to a lesser degree than studies of Cu(100) or Cu(110). This is primarily due to the lack of formation of an ordered structure on Cu(111) (2,17,13) at low exposures. Oxygen adsorbs on

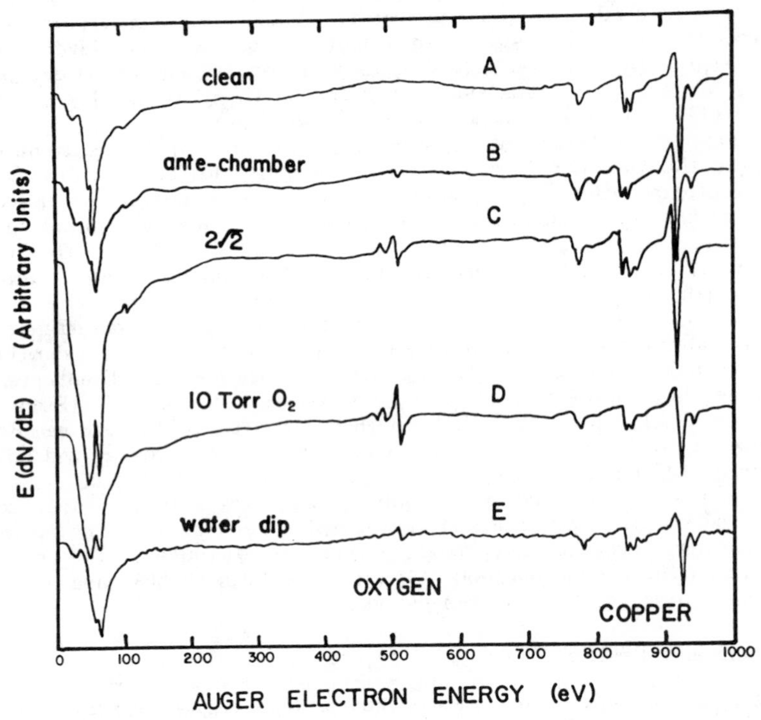

Figure 3. Auger spectra of copper surfaces after various stages of oxidation.

Cu(111) to a lesser extent than on the other two low-index planes. This adsorption results in formation of an ordered structure above 10^3 L at room temperature, which is displayed as an oblique unit cell [ ]. This pattern has been previously reported (16). It is interesting that it has not been observed by other workers, possibly because of the extreme O_2 exposures required and the faintness of the pattern.

Ion scattering experiments by Niehus (17) on Cu(111) with adsorbed oxygen indicate that the copper surface undergoes a reconstruction. Room temperature work function measurements evidenced no change upon adsorption of oxygen, which was interpreted as incorporation of oxygen. Alternatively, those authors (2) suggested that the absence of a work function change could be interpreted as evidence for reconstruction, in line with the ion scattering work (17). These results are consistent with Pritchard and co-workers (13), who account for the [ ] pattern by proposing a reconstruction of the top layer of copper to form an adlayer having (/2X/2)R45° symmetry.

Oxygen adsorption on Cu(110) proceeded initially with formation of a (2X1) structure (1,18-22). Quantitation of the oxygen Auger current indicated significantly less oxygen present when the (2x1) was initially formed, compared with a coverage of 0.5, which is expected for an oxygen structure with this symmetry. The low oxygen coverage, responsible for initial formation of the (2X1), was especially evident in direct comparisons with oxygen coverages and structures on the other two faces. Ion scattering experiments (18,21,22) and X-ray adsorption (XAS) studies (19,20) suggest that the surface undergoes a reconstruction upon oxygen adsorption, resulting in a missing row structure where every other row of <100>-packed surface-copper atoms are removed. Oxygen atoms reside in the leftover short bridge sites (18,20). It thus appears that reconstruction was catalyzed by a low coverage of oxygen, much lower than the 5.4×10^{14} atoms/cm^2 expected for 1/2 coverage. Increasing the oxygen coverage above 0.5 formed a c(6X2) pattern, reported elsewhere (1,22), which was optimum near unit coverage. Uncertainty as to whether the surface remains reconstructed (18-22) prevents simple interpretation of the c(6X2) pattern.

Water Adsorption. As mentioned in the introduction, water is not strongly adsorbed on copper at room temperature (3-12). In the present studies, no adsorption was observed until extremely high exposures of water (>10^7 L). At these levels, the partial pressure of oxygen in the water vapor was significant and probably responsible for the presence of adsorbed oxygen. The same structures observed with pure oxygen, mentioned previously, were observed in the present case.

The water used in these experiments was made by pyrolytic distillation, a process involving saturation of the distilling water vapor with oxygen and passing it over heated platinum (14). Extensive sparging with UHP N_2 was performed prior to experiments, removing most of the oxygen. Subsequent experiments will involve a final boil in the absence of oxygen in order to decrease the O_2 content further.

Suspension of the crystal above liquid water for 2 minutes resulted in oxidation to the extent observed in the presence of 10

Torr of water vapor or oxygen exposures over 1,000L; the 2/2 LEED pattern was observed on Cu(100) and the c(6X2) was observed on Cu(110). There was increased diffuse intensity in these patterns over those resulting from exposure to pure oxygen, but the symmetry was distinct. No elements, other then oxygen and copper, were evidenced by Auger spectroscopy.

Immersion of the copper crystal into water resulted in a 25% increase in the oxygen Auger current on Cu(111) and Cu(110), while it was relatively unchanged on Cu(100) (Figure 3e). The 2/2 pattern was still present on Cu(100) after emersion while the other low-index planes displayed diffuse LEED patterns with spots visible only at energies above 100 eV. The facts that spots were visible above 100 eV and the minor increase in oxygen Auger current suggest that the oxidation was no more than two monolayers thick. This also indicates significant stability for the copper surfaces at open circuit in water. Subsequent studies will involve potential control and examination of copper surface stability vs. potential. Supporting electrolytes will be required in these studies.

Electrolyte Adsorption. Initial studies of electrolyte adsorption have been performed using HCl. HCl adsorption in the gas phase and from solution was investigated and is described in a separate paper (S.E.G.). HCl gas adsorbs on copper with an initial sticking coefficient close to 1. This adsorption rate decreases by 2 to 3 orders of magnitude after 1/2 monolayer of adsorption. Ordered structures were observed on all three of the low index planes at HCl exposures of 20L or less: Cu(100)(/2X/2)R45°-Cl, Cu(111)(/3X/3)R30°-Cl and Cu(110)c(2X2)-Cl. These patterns resemble those formed by Cl_2 adsorbed on other transition metals and with other hydrogen halides and halogens. That is, in general, these patterns result from structures which are approximately hexagonally close-packed layers of adsorbed halogen atoms (1,2).

A series of transition structures were observed with LEED on the Cu(110) surface at exposures exceeding 10L. These structures indicated that the Cl atoms on the surface were undergoing compression in the <110> direction, close packing at their van der Waals diameter. The final ordered structure was a Cu(110)(3X2)-Cl at a chlorine coverage of 2/3. HCl exposures exceeding 1000L result in formation of bulk copper chlorides, evidenced by increased diffuse intensity in the LEED pattern, decreased copper Auger currents, and increased chlorine Auger currents. This stage of halogen adsorption has been extensively investigated by Sesselmann and Chuang on polycrystalline Cu (13).

A crystal supporting preadsorbed Cl adlattices formed with HCl gas, as described previously, was immersed in water for one minute to investigate the stability of the Cl adlayers. Subsequent examination of the emersed (removed) crystal evidenced a partial loss of chlorine, about 1/3, and uptake of oxygen on both Cu(111) and Cu(110). There was no detectable oxygen uptake on the Cu(100) surface.

Ordered LEED patterns were evident on all three surfaces. A Cu(100)(/2X/2)R45°-Cl was observed with LEED, the same pattern present on the surface before immersion, but with a slight increase in spot diffuseness. A 28% decrease in Cl Auger current was observed and was responsible for the degradation in LEED pattern clarity. It

Figure 4. LEED pattern resulting from immersion of a Cu(111)(/3x/3)R30°-Cl structure into water. Cu(111)(/3x/3)R30° split pattern, 60 eV.

Figure 5. LEED pattern resulting from immersion of a Cu(110)c(2x2)-Cl structure into water. Cu(110)c(4x2)-Cl,O pattern, 61 eV.

appears that the remaining Cl was stable under the present conditions and dense enough to prevent oxidation of the Cu(100) surface.

Chlorine adsorbed on Cu(111) was less stable, decreasing by 37% during immersion. Initially sharp /3 spots, before immersion, were enlarged to form triangular spots after immersion (Figure 4). The adsorption of significant amounts of oxygen, as well as the decrease in Cl, may account for the significant diffuse intensity present in the patterns, but the triangular shape of the fractional index spots is indicative of a Cl structure reported previously by Goddard and Lambert. Lambert et.al. reported formation of a /3 "split spot" pattern by exposure of Cu(111) to Cl_2 gas at 100°C. They proposed a Cu(111)(6/3X6/3)R30°-Cl structure at a coverage of 0.45. The present structure (Figure 4) has less Cl than was present in the initial /3 structure, optimally 0.33. The presence of oxygen on the surface, as well, may substitute for lost Cl, but the pattern is better explained by a different structure.

A /3 split spot pattern was observed in this author's lab after immersion of a clean Cu(111) surface into 1mM HCl (S.E.G.). This pattern consisted of well-resolved triplets of spots at the /3 positions, similar to the patterns resulting from iodine adsorption on copper UPD (24) and on Ag(111) (25). The structures proposed in these cases (24,25) more fully explain the observed "split" pattern and Cl coverage. It was proposed that a reconstruction of the surface to one having /3 local structure, occurred and that this was responsible for the absence of intensity anywhere but at the /3 positions. This /3 local symmetry existed inside a larger unit cell, defined by a series of phase boundaries, which accounts for the "split spot" spacings and orientations. It appears that the Cu(111)(/3X/3)R30°-Cl surface began reconstructing to form a /3 split structure upon immersion in water but due to loss of Cl and oxygen incorporation, did not form the sharp split pattern (S.E.G.).

The initially present Cu(110)c(2X2)-Cl structure, converted to a c(4X2) with considerable diffuse intensity and streaks in the <110> direction (Figure 5). This LEED pattern was not observed in experiments involving HCl adsorption on Cu(110), described elsewhere (S.E.G.). A Cu(110)c(4X2) with sharp spots, and without the diffuse background and streaking, was observed after heating an extensively oxidized, 10 Torr of O_2, Figure 3d, Cu(110) surface to 550°C and cooling. Since this Cu(110) surface contained significant amounts of both Cl and oxygen, proposing a structure would be premature. This result will be further examined in a series of experiments, presently underway, involving copper electrode potential dependence of surface structure and composition.

Literature Cited

1. Spitzer, A.; Luth, H. Surface Sci. 1982, 118, 121-35.
2. Spitzer, A.; Luth, H. Surface Sci. 1982, 118, 136-44.
3. Sexton, B. A. J. Vac. Sci. Technol. 1979, 16, 1033-36.
4. Au, C. T.; Breza, J.; Roberts, M. W. Chem. Phys. Lett. 1979, 66, 340-3.
5. Au, C. T.; Roberts, M. W. Chem. Phys. Lett. 1980, 74, 472-4.
6. Spitzer, A.; Luth, H. Surface Sci. 1982, 120, 376-88.
7. Mariani, C.; Horn, K. Surface Sci. 1983, 126, 279-85.

8. Bange, K.; Grider, D.; Sass, J. K. Surface Sci. 1983, 126, 437-43.
9. Anderson, S.; Nyberg, C.; Tengstal, C. G. Chem. Phys. Lett. 1984, 104, 305-11.
10. Spitzer, A.; Ritz, A.; Luth, H. Surface Sci. 1985, 152/153, 543-49.
11. Spitzer, A.; Luth, H. Surface Sci. 1985, 160, 353-61.
12. Nyberg, C.; Tengstal, C. G.; Uvdal, P.; Anderson, S. J. Elec. Spec. and Rel. Phenom. 1986, 38, 299-307.
13. Sesselmann, W.; Chuang, T. J. Surface Sci. 1986, 176, 32 & 67.
14. Conway, B. E.; Angerstein-Kozlowska, H.; Sharp, W. B. A.; Criddle, E. E. Anal. Chem. 1973, 45, 1331.
15. Lee, R. N.; Farnsworth, H. R. Surface Sci. 1965, 3, 461.
16. Judd, R. W.; Hollins, P.; Pritchard, J. Surface Sci. 1986, 171, 643-53.
17. Niehus, H. Surface Sci. 1983, 130, 41-9.
18. Bronckers, R. P. N.; De Wit, A. G. J. Surface Sci. 1981, 112, 133-52.
19. Dobler, U.; Baberschke, K.; Haase, J.; Puschmann, A. Phys. Rev. Lett. 1984, 52, 1437-40.
20. Dobler, U.; Baberschke, K.; Vvedensky, D. D.; Pendry, J. B. Surface Sci., 1986, 178, 679-85.
21. Yarmoff, J. A.; Cyr, D. M.; Huang, J. B.; Kim, S.; Williams, R. S. Amer. Phys. Soc. 1986, 33, 3856-68.
22. Gruzalski, G. R.; Zehner, D. M.; Wendelken, J. F. Surface Sci. 1984, 147, L623-9.
23. Goddard, P. J.; Lambert, R. M. Surface Sci. 1977, 69, 180.
24. Stickney. J. L.; Rosasco, S. D.; Hubbard, A. T. J. Electrochem. Soc. 1984, 131, 260.
25. Salaita, G. B.; Lu, F.; Laguren-Davidson, L.; Hubbard, A. T. J. Electroanal. Chem. 1987, 229, 1.

RECEIVED May 17, 1988

Chapter 8

Molecules at Surfaces in Ultrahigh Vacuum

Structure and Bonding

J. Somers, T. Lindner, and A. M. Bradshaw

Fritz-Haber-Institut der Max-Planck-Gesellschaft, Faradayweg 4–6, D–1000 Berlin 33, Federal Republic of Germany

Using spectroscopic techniques based on photoabsorption and photoelectron emission it is possible to obtain information on the geometry and electronic structure of adsorbed molecules as well as of adsorbed molecular fragments resulting from simple heterogeneous reactions. In fact, the correct assignment of the electronic energy levels of such species using photoemission is often only possible once the molecular orientation is known.

Our understanding of the energetics and dynamics of molecular adsorption on single crystal metal surfaces under ultra-high vacuum (uhv) conditions is presently limited by the meagre amount of information available on the geometric structure and electronic energy levels of surface molecules. For example, only in the case of a few simple systems such as diatomics - and then usually CO - are the adsorption site and molecular orientation known with any accuracy. Structural analysis via low energy electron diffraction (LEED) is restricted to ordered overlayers. Whilst many adsorbed molecules form regular two-dimensional arrays, it is by no means the rule. For molecular fragments formed as intermediates in heterogeneous reactions (e.g. formate, methoxy, cyano) it is certainly the exception. Moreover, LEED theory at its present state of development has difficulties in coping with polyatomic molecules. There are, however, a number of techniques which do not require the existence of an ordered array and which still give structural information in a direct or indirect way. One possibility is to measure the angular distribution of ions desorbed as a result of electron- or photon-induced electronic transitions. The technique is normally referred to as ESDIAD and is described by Madey in another chapter of the volume. Another possibility is to apply selection rules in photoemission to establish molecular orientation [1] but, as we shall see below, it is not possible to simultaneously determine structure and assign levels. A new development in recent years has been the application of structural tools based on photoabsorption and photoionisation: x-ray absorp-

0097–6156/88/0378–0111$06.00/0

tion fine structure [2] and photoelectron diffraction [3]. Since they both require a continuously tunable source of soft x-rays, usually in the region 200 - 1000 eV photon energy, their application is dependent on the availability of synchrotron radiation and suitable grazing-incidence grating monochromators [4]. (This spectral range covers the core levels of C, N and O; 1s (or higher) levels of heavier elements are accessible at higher photon energy with crystal monochromators.) The measurement of the polarisation dependence of the near edge x-ray absorption fine structure (NEXAFS) is a particularly straightforward technique, if applied correctly, and figures prominently in this article.

The most direct technique for probing surface electronic structure is undoubtedly photoemission, or photoelectron spectroscopy. In a reasonable approximation, referred to as Koopmans' theorem, the measured ionisation potentials can be equated with the ground state orbital energies. When the molecule is only weakly adsorbed a comparison with the photoelectron spectrum of the corresponding gas phase species normally suffices to assign the various adsorbate-induced features. For this purpose, laboratory line sources (HeI, etc.) are usually adequate. Where a strong perturbation of the energy levels of the molecule takes place, assignment can only be carried out using selection rules preferably in conjunction with polarised light. This is particularly true of molecular fragments for which no comparison with gas phase species is possible. For the application of photoemission selection rules prior knowledge of the orientation of the molecule is necessary. Synchrotron radiation in the photon energy range 10 - 50 eV (the so-called normal-incidence region) is the obvious choice for such investigations because of its tunability and intrinsically high polarisation.

In the investigations of molecular adsorption reported here our philosophy has been to first determine the orientation of the adsorbed molecule or molecular fragment using NEXAFS and/or photoelectron diffraction. Using photoemission selection rules we then assign the observed spectral features in the photoelectron spectrum. On the basis of Koopmans' theorem a comparison with a quantum chemical cluster calculation is then possible, should this be available. All three types of measurement can be performed with the same angle-resolving photoelectron spectrometer, but on different monochromators. In the next Section we briefly discuss the techniques. The third Section is devoted to three examples of the combined application of NEXAFS and photoemission, whereby the first - CO/Ni(100) - is chosen mainly for didactic reasons. The results for the systems CN/Pd(111) and HCOO/Cu(110) show, however, the power of this approach in situations where no *a priori* predictions of structure are possible.

Photoionisation and photoelectron emission

Below the photoionisation threshold a core electron in a free molecule can be excited into empty anti-bonding molecular orbitals (m.o.'s) as well as into Rydberg states. These transitions are observable as sharp features directly below the corresponding absorption edge (carbon K, oxygen K etc.). Above the photoionisation threshold further transitions into higher-lying anti-bonding virtual m.o.'s will occur, their broad spectral features being superimposed on the background continuum absorption. All these features associated with the existence of the molecular potential constitute the near edge x-ray absorption fine structure (NEXAFS) or x-ray absorption near edge structure (XANES). At higher energies above the edge further very weak structure may be visible due to a scattering phenomenon. This is the *extended* x-ray absorption fine structure (EXAFS) resulting from interference between the emitted photoelectron wave and waves backscattered from other atoms in the molecule. (The excitation into anti-bonding m.o.'s above the threshold may also be viewed as EXAFS-type scattering resonances. Hence the term "shape" resonance: the energy and width of the features depends on the shape of the molecular potential.) When the molecule is placed on the surface, several factors come into play. Firstly, the experiment has to be performed quite differently. Since the substrate is usually a compact solid surface, a measurement of the absorption spectrum in transmission is no longer feasible. As shown schematically in Figure 1a, the yield of Auger electrons, or the high energy portion of the secondary electrons (partial electron yield) are measured, both signals being proportional to the number of excited core electrons in the immediate surface region. Upon adsorption the near edge resonances may be shifted and their degeneracy lifted or they may even disappear as a result of the formation of the bond to the substrate. New resonances (so-called substrate resonances) may be observed. Since the chemisorption process invariably results in a molecule with a single, fixed orientation and since the excitations are subject to dipole selection rules, the transitions in NEXAFS are polarised. This forms the basis for the determination of molecular orientation.

Since the surface atoms are likely to be relatively strong backscatterers the *extended* fine structure will be dominated by the scattering from the substrate. This relatively weak modulation at higher energy is referred to as the surface EXAFS, or SEXAFS. An analysis of the SEXAFS can give further structural information, most valuably, on the adsorption site (for further details, see Ref. [2]).

In photoelectron spectroscopy, shown schematically in Figure 1b, the electrons emitted as a result of their excitation into states above the photoionisation threshold are analysed according to their kinetic energy. The energy balance is given simply by $h\nu = E_B + E_K$, where E_B is the energy of a bound level and E_K the kinetic energy of the photoemitted electron. The technique is often referred to as XPS (x-ray photoelectron spectroscopy) or ESCA (electron spectroscopy for chemical analysis), the latter being the original acronym proposed by K. Siegbahn. If the photon energy is low ($h\nu < 50$ eV), such that essentially only valence electrons are excited, then we normally refer to UPS (ultra-violet photoelectron spectroscopy) or simply photoemission. At these low photon energies the photoionisation cross-section for valence (or outer shell) electrons is also highest. For a molecule on a surface the primary excitation is a dipole transition from a molecular orbital, modified by the interaction with the substrate, into a continuum

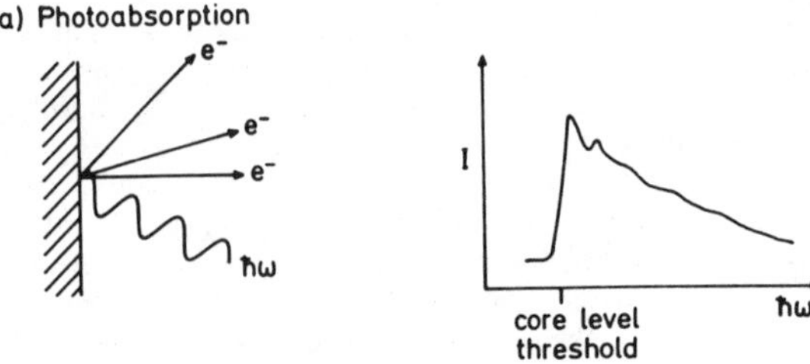

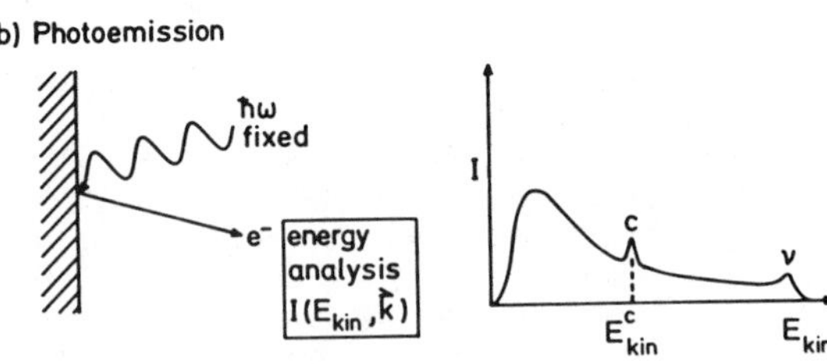

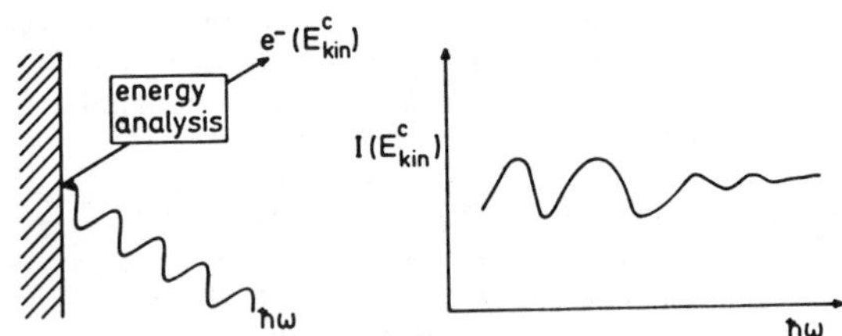

Figure 1: Schematic representation of the three techniques (a) x-ray photoabsorption (NEXAFS/SEXAFS), (b) photoelectron spectroscopy (photoemission) and (c) photoelectron diffraction.

level. The same photoabsorption selection rules thus apply as in NEXAFS. Whereas in NEXAFS, however, the final state is a bound or continuum level associated with the molecule itself and thus of (potentially) known symmetry, the final state continuum wavefunction in photoemission must be selected in an angle-resolved experiment. When the molecular orientation is also known, the symmetry of the initial state can be determined. How this procedure can be used in assignment will become clear in the examples below.

In photoemission there is an interesting experimental variation known as constant initial state (CIS) spectroscopy, in which the photon energy and the analysed kinetic energy are scanned simultaneously. For the excitation out of a given occupied level, a whole range of final states is then sampled. This approach is particularly useful in band structure studies of solids when the momentum, or wavevector, of the final state is also know, i.e. when the emission direction is also defined. By the same token, a differential partial cross-section is measured in such an experiment on a free molecule. If the photoelectron current is angle-integrated, meaning that all the photoemitted electrons are collected, then the photoabsorption spectrum, or partial photo-ionisation cross-section, should be obtained. Putting the molecule (or an atom) on a surface and taking a CIS spectrum of a core level introduces, in the same way as in SEXAFS, scattering from the substrate (Figure 1c). The resulting interferences, which depend on the adsorption site and orientation modulate the differential cross-section; the phonomenon is normally referred to as photoelectron diffraction (PED). It has recently been shown that the modulations in intensity are of the order of 50% of the background signal in experiments on the C and O 1s levels [5]. (Photoelectron diffraction effects are also observed when the photoelectron emission angle is varied at fixed photon energy [6]. It is thus sensible to distinguish between *energy-scanned* and *angle-scanned* PED.) By performing model calculations for a given structure and emission direction and comparing with experiment, information on adsorption site and orientation is obtained. How does the technique relate to SEXAFS? By varying the photon energy in PED the scattered electron intensity is distributed between the various final state manifolds corresponding to different emission directions. If it were possible to angle-integrate over the whole photoelectron current, then we would recover the SEXAFS signal. Again, it is essentially the difference between a partial cross-section and a differential partial cross-section. Before concluding this section it should perhaps be noted that SEXAFS has so far been more successful in surface structural studies than photoelectron diffraction, largely because multiple scattering effects can be neglected. On the other hand, the potential of PED has not yet been fully realised; we show some PED data below which illustrate this point.

<u>Some examples</u>

<u>Ni{100}-CO</u>. Figure 2 shows the NEXAFS at the C and O 1s edges for CO adsorbed on a Ni{100} surface at half-monolayer coverage [7]. This surface concentration corresponds to the formation of an

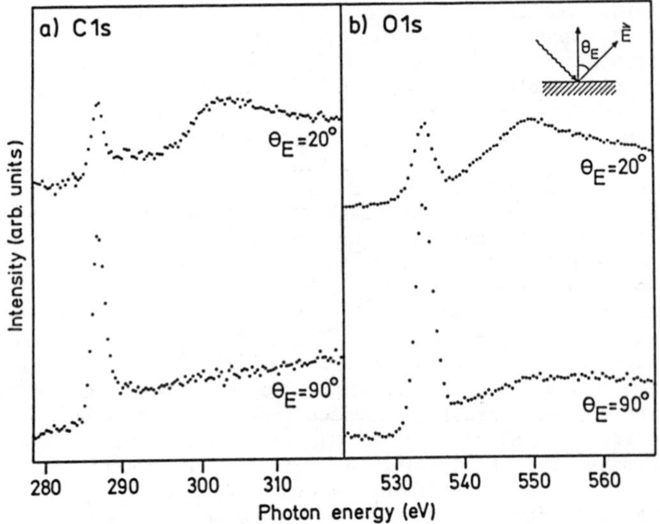

Figure 2: NEXAFS at the (a) carbon and (b) oxygen 1s edges for Ni{100}(√2x√2)R45°-CO.

ordered overlayer designated as $(\sqrt{2}x\sqrt{2})R45°$ in the standard notation. In each case spectra for two angles of the electric vector relative to the surface normal, θ_E = 20° and 90°, are shown. Two distinct resonances can be seen, corresponding to transitions into the bound 2π level and into the 6σ-derived level in the continuum. We note the difference in linewidth between these two features. Equally pronounced is the different polarisation dependence in each case. At θ_E = 90° the σ resonance is virtually absent and the π resonance strong. At θ_E = 20° the integrated intensity of the σ resonance is larger than that of the π. Starting from the dipole matrix element $\langle f|\tau|i\rangle$ for an electronic transition, it is easy to show [8] that for an isolated, oriented diatomic the σ resonance is polarised along the molecular axis and the π resonance in a plane perpendicular to that axis. The corresponding angular dependences are given by $I(\sigma) \sim P\cos^2\alpha$ and $I(\pi) \sim 1 - P\cos^2\alpha$, where α is the angle between the C-O axis and the **E** vector and P the degree of polarisation of the incident light. If the molecule is adsorbed with its axis perpendicular to the surface, which we might expect from all the previous studies on this system (and there have been many!), the effective symmetry either remains $C_{\infty v}$ or is reduced to C_{4v} or C_{2v} depending on the adsorption site [9]. For present purposes it is important to remember that the symmetry axis remains and that the simple polarisation dependences above still hold. Should the molecule be tilted, the symmetry will drop to C_s and the degeneracy of the π orbital will be lifted. Strictly speaking, it is then no longer possible to determine the angle of the C-O axis relative to the surface. This is because the σ resonance and the a' component of the π resonance are no longer polarised parallel and perpendicular to the molecular axis. They are now polarised in the single symmetry plane; the $\pi(a'')$ component is polarised perpendicular to that plane. In practice, however, it is generally found that the σ resonance is unaffected by the lowering of symmetry. Even the π resonance may not be strongly affected; the test for this of course is the observation of a measurable splitting.

A simple procedure presents itself as a result of the above discussion. We fit calculated polarisation dependences to the experimental data, also allowing for the possibility of a random azimuthal orientation. If the molecule appears to be adsorbed perpendicular to the surface, the use of the $\cos^2\alpha$ formulae is retrospectively justified. If a tilt angle emerges, then its value might possibly be incorrect, unless only the σ resonance has been used in the analysis. The results from the oxygen edge for Ni{100}$(\sqrt{2}x\sqrt{2})R45°$-CO are shown in Figure 3. The π intensities normalised to the edge jump have been plotted as a function of θ_E, the angle between the **E** vector and the surface normal. The data show that the molecular axis is perpendicular to the surface to within 10°. The integration procedure requires that the form of the background under the resonance is known. We refrain, however, from discussing background and normalisation effects here, although they do give rise to some serious problems. We note for example that the σ resonance does not go to zero for θ_E = 90° in the oxygen data as it does in the C data, which is probably due to an underlying substrate resonance. For this reason the π resonance was used for the analysis; since the molecule is clearly oriented normal to the surface, this presents no problem. The σ data at the C edge could have been used but are less reliable because of the poorer S/N ratio (and are, incidentally, incomplete!).

Figure 4 shows two angle-resolved photoelectron spectra for CO adsorbed on the same surface in the ($\sqrt{2}$x$\sqrt{2}$)R45° structure [10]. The corresponding spectrum of the free molecule in the valence region [11] contains three features due, in order of increasing ionisation energy, to the 5σ, 1π and 4σ orbitals (see inset). On the surface only two distinct adsorbate-derived features are observed, as in spectrum (a). That this results from overlapping 1π and 5σ features can be shown in the following way [12]. Since the molecular axis is oriented perpendicular to the surface, the point group remains $C_{\infty v}$ or, if the influence of the surface is felt [1], is reduced to C_{4v} or C_{2v}. (In fact the on-top site is occupied, giving C_{4v}. The arguments that follow are also valid for the bridge site, however.) If the photoelectron detector is placed in the rotational axis or in a mirror plane of the system, the final state wave function must be symmetric with respect to the corresponding symmetry operation. This constraint which is placed on the final state $|f\rangle$ enables us to orient the **E** vector such that the dipole matrix element is zero for a particular initial state $|i\rangle$. This is the essence of the photoemission selection rules referred to above. Figure 4 is a case in point. It can be easily shown for the geometry of spectrum (a) that both a_1 and e states (σ- and π-derived, respectively) are allowed. In the geometry $\mathbf{E} \perp \mathbf{k}_\parallel$ of spectrum (b), where $\mathbf{k}_\parallel$ is the component of photoelectron momentum parallel to the surface, a_1 states are forbidden. This is often referred to as the "forbidden geometry" experiment. If the interaction with the surface is weak (effective point group $C_{\infty v}$) and multiple scattering unimportant in the final state, this will also hold when the azimuthal direction defined by $\mathbf{k}_\parallel$ no longer correspond to a symmetry plane common to the adsorbate and the surface [1,13]. In spectrum (b) the 5σ- and 4σ-derived orbitals disappear and we are able to locate the 1π orbital. Since the separation of the 1π and 4σ features is roughly that of the free molecule, it is clear that the formation of the chemisorption bond has resulted in a relative shift of the 5σ feature by about 3 eV: its ionisation energy becomes greater than that of 1π. Using Koopmans' theorem we can then compare this result with orbital energy diagrams resulting from calculations performed at various levels of sophistication [14]. These generally show that as the Ni-CO distance decreases in the end-on configuration all the levels shift to lower energies, but only the 5σ and 1π orbitals are significantly perturbed by the bonding with the substrate. The energy of the 5σ changes most strongly with considerable donation of charge from this orbital into the metal. The metal gives back charge into the C-O antibonding 2π* level just as in the donor-acceptor model of ligand bonding in coordination chemistry. This concept of charge donation can be misleading and it is perhaps more useful to think in terms of a new set of molecular orbitals for the adsorbate/substrate complex. The 2π* thus overlaps with metal states of appropriate symmetry and forms an occupied bonding level largely of metal character situated energetically somewhere in the Ni d band. The corresponding anti-bonding level remains largely CO

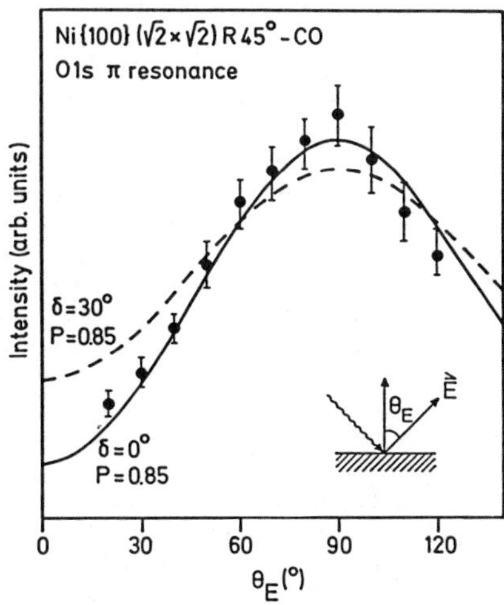

Figure 3: Polarisation dependence of the π resonance intensity at the O 1s edge for Ni{100}(√2x√2)R45°-CO. Degree of photoionisation, P, of the incident radiation is taken to be 0.85 for the calculated curve. δ is the angle of tilt of the C-O axis relative to the surface normal. The analysis assumes no reduction in symmetry from $C_{\infty v}$.

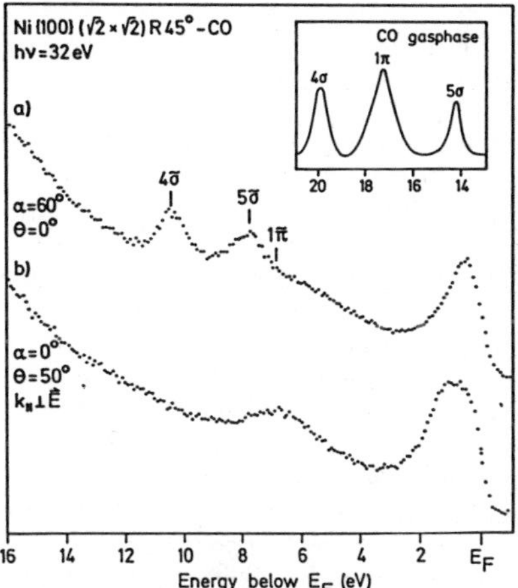

Figure 4: Angle-resolved photoelectron spectra for Ni{100} (√2x√2)R45°-CO. hν = 32 eV (a) angle of incidence, α = 60°, polar angle of emission, θ = 0°. (b) α = 0°, θ = 50°, $\mathbf{E} \perp \mathbf{k}_{\parallel}$. Inset: Gas phase photoelectron spectrum of CO at the same photon energy. After [10] and [11].

$2\pi^*$ in character and is the unoccupied orbital giving rise to the π resonance in NEXAFS. It can also be probed with inverse photoemission.

Two further remarks should be made at this juncture. The first concerns Koopmans' theorem, which turns out in some cases not to be such a trivial approximation. In most photoemission studies to date it has simply been assumed that, as far as the *relative* energies are concerned, the ionisation potentials can be compared directly with calculation, i.e. that relaxation indeed plays a role but the shift is the same for all orbitals. When intramolecular relaxation and image charge screening are dominant this may be a reasonable approximation. In the presence of so-called charge transfer screening (the "unfilled orbital" mechanism), which appears for example to affect the levels of σ symmetry in adsorbed CO, this may not be the case. The second point concerns dispersion. Due to lateral interactions in ordered overlayers, overlap of orbitals on adjacent molecules leads to band formation. Energy dispersion $\varepsilon(\mathbf{k}_\parallel)$ is then observable in the experimental spectrum: as the polar emission angle is varied, and thus $\mathbf{k}_\parallel$, the energy ε of a feature will change. This effect can be very pronounced in strongly-bound atomic overlayers with bandwidths of 2 eV or more, but is also observed for molecular adsorbates. In fact, it was first observed for the latter in the system Ni{100}-CO [15].

Pd{111}-CN. The usual bonding geometry for an adsorbed diatomic molecule is the end-on configuration where the molecular axis is perpendicular to the surface, as in the case of Ni{100}-CO described above. This observation is consistent with the behaviour of CO, NO or N_2 as ligands in co-ordination chemistry. By the same token we would perhaps expect a surface CN species also to be "terminally" bonded via the C atom as is normally found in cyano complexes. Surface vibrational spectroscopy has, however, indicated that surface CN formed by the decomposition of C_2N_2 on Pd and Cu surfaces is adsorbed in a lying-down configuration [16]. This result has since been confirmed by NEXAFS [17] and has led to a new consideration of the photoemission data from adsorbed CN [18].

The N 1s NEXAFS from the system Pd{111}-CN is shown in Figure 5a at two angles $\theta_E = 20°$ and 90° in an analogous way to Figure 2. It is immediately apparent, however, that CN differs considerably in its orientation. For CO on Ni{100} the π resonance is at a maximum for the **E** vector parallel to the surface; the σ resonance peaks at $\theta_E = 20°$ as the **E** vector approaches the surface normal. For CN on Pd{111} not only do the features hardly change in relative intensity, but a splitting of the π resonance is also observed (apparent only as a weak shoulder on the low energy side of the resonance at the relatively low resolution of Fig. 5). This indicates a C_s point group. By consideration of the polarisation dependence of the two components, it can be shown that the a" component dominates at practically all angles. This is particularly useful because the $\sigma/\pi(a")$ intensity ratio can then be used for the analysis, thus avoiding the need for normalisation (but not solving the background problem!). Allowing for a random azimuthal orientation of molecules the expected ratio is given by $I(\sigma)/I(a") \sim [1 + P\cos^2\theta_E]/[1 - P\cos^2\theta_E]$. As expected, this does not give

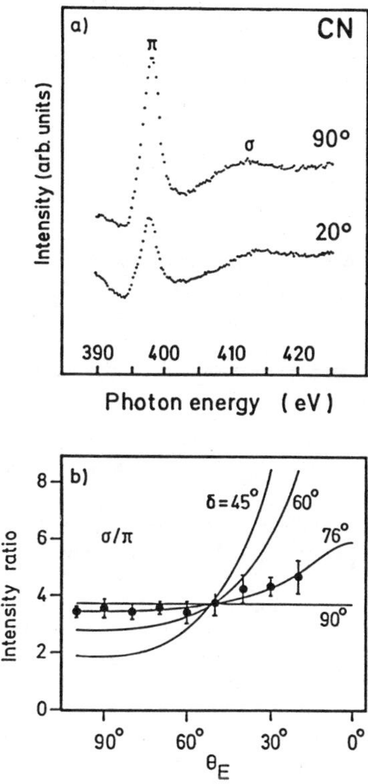

Figure 5: (a) N 1s NEXAFS from the system Pd{100}-CN at θ_E = 20° and 90° (b) Polarisation dependence of the σ/π intensity ratio. P = 0.85. After [17].

any information on the tilt of the CN axis. If we assume, however, that the σ resonance is unaffected by the reduction in symmetry, i.e. the effective point group remains $C_{\infty v}$, we can replace the numerator by $P(\sin^2\delta\sin^2\theta_E + 2\cos^2\delta\cos^2\theta_E) + (1-P)\sin^2\delta$, where δ is the angle of tilt relative to the surface normal. For a parallel-bonded species this reduces to $1-P\cos^2\theta_E$ and the intensity ratio becomes constant. As the experimental data of Figure 5b show, the ratio is not quite constant and a best fit to the calculated curve is obtained for $\delta = 76°$ (± 5°). Since an angle of 14° relative to the plane is rather unlikely on a close-packed surface, we suspect that the molecule is indeed parallel-bonded and that some other source of error is responsible for the deviation. The most likely candidate is again a substrate scattering feature under the σ resonance: Note the small shift in the energy of the σ resonance between θ_E = 90° and θ_E = 20° in the spectra of Figure 5a.

Although the C-N axis lies parallel, or nearly parallel, to the metal surface, it is only the knowledge of the appropriate point group which helps us further in the application of the photoemission selection rules. Since the C_s group has a single symmetry plane and the molecules show a random azimuthal orientation, it is not possible, however, to perform the analysis as above for CO. In fact, the intrinsically low symmetry prevents conclusive experiments involving particular emission directions for defined orientations of the **E** vector. The spectra in Figure 6 taken for different polar and azimuthal emission angles at hν = 35 eV show four adsorbate-derived features at 5.2, 6.1, 7.2 and 9.1 eV. Some limited success was obtained with the same forbidden geometry experiment as above for Ni{100}-CO, although this is, strictly speaking, not applicable in the present case. At θ = 70° a considerable reduction in the intensity of peaks 1 and 3 is obtained on going from $\mathbf{E} \parallel \mathbf{k}_\parallel$ (c) to $\mathbf{E} \perp \mathbf{k}_\parallel$ (d). Emission from a' orbitals is only forbidden for $\mathbf{E} \perp \mathbf{k}_\parallel$ when the **E** vector is perpendicular to the single symmetry plane. For randomly oriented symmetry planes the reduction in intensity in this "forbidden" geometry should, at the most, only be factor of 0.5. How many levels are expected and what is their symmetry? From the NEXAFS we know that the lowest unoccupied level(s) are 2π-derived. The 5σ- and 1π-derived orbitals are thus filled and the molecule is adsorbed essentially as CN^-. The chemisorption bond appears to derive its strength mainly from the interaction of this anionic species with its image charge. In this situation we might expect a shifting and splitting of the orbitals to be the main effect rather than a rehybridisation. The four levels are then expected to correspond to 4σ, 5σ, 1π(a') and 1π(a"). On the basis of the "forbidden" geometry experiment and other, qualitative arguments a preliminary assignment has been made in which peaks 1 and 3 are attributed to the 5σ- and 1π(a')-derived levels, respectively. Peak 2, which is the strong feature in normal emission, appears to be 1π(a")-derived. The most certain assignment is peak 4 at 9.1 eV which is 4σ-derived. Because of the low symmetry, mixing between 4σ, 5σ and 1π(a') is also possible. Unfortunately, the cluster calculations that exist assume the end-on adsorption geometry and a comparison with calculation is at present not possible.

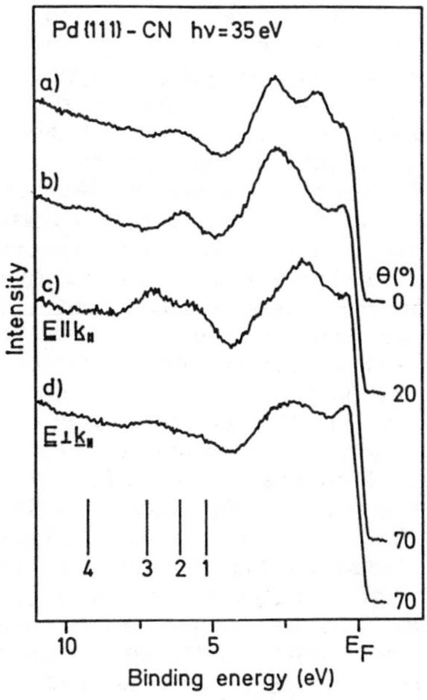

Figure 6: Angle-resolved photoemission spectra for Pd{111}-CN. hν = 35 eV. (a) α = 70°, θ = 0° (b) α = 70°, θ = 20° (c) α = 0°, θ = 70°, **E** ∥ $\mathbf{k}_\parallel$ (d) α = 0°, θ = 70°, **E** ⊥ $\mathbf{k}_\parallel$. After [18].

Cu{110}-HCOO. The decomposition of formic acid on metal and oxide surfaces is a model heterogeneous reaction. Many studies have since shown that it proceeds via a surface formate species. Thus on Cu{110} adsorbed formic acid is found at low temperature. On heating to ~ 270 K deprotonation occurs, giving rise to the surface formate, which in turn decomposes at ~ 450 K with evolution of H_2 and CO_2. In previous studies, particularly with vibrational spectroscopy, it had been demonstrated that the two C-O bonds are equivalent and that the symmetry is probably C_{2v} [19]. A NEXAFS study by Puschmann *et al.* [20] has subsequently shown that the molecular plane is oriented perpendicular to the surface and aligned in the <110> azimuth.

Figure 7 shows NEXAFS spectra at the oxygen edge from the system Cu{110}-HCOO with the **E** vector aligned in (a) the <110> azimuth and (b) the <100> azimuth. The π-type resonance corresponds to excitation into the $2b_2$ orbital and is expected to be polarised perpendicular to the molecular plane. There are actually two σ-type resonances, corresponding to $7a_1$ and $5b_1$; at the C edge they can actually be resolved [21]. In the <100> azimuth (Figure 7b) we note that the intensity of the π resonance varies drastically as the angle between the **E** vector and the surface normal is changed. In <110> this effect is hardly observed, indicating that the molecular plane is perpendicular to the surface and aligned in the <110> azimuth, i.e. parallel to the atom rows of the {110} surface. (The measurable intensity in the π resonance when the **E** vector is oriented in the <110> azimuth is due to incomplete polarisation of the incident radiation.) This qualitative conclusion is supported by comparing the θ_E = 90° spectra in Figures 7a and 7b (full lines). Here we observe a drastic change in the π/σ intensity ratio as the **E** vector (parallel to the surface) is moved around from the <110> to the <100> azimuth. A quantitative analysis of the polar angular dependence in the <100> azimuth indicates that the molecular plane is indeed perpendicular to the surface, the accuracy of the determination being ± 10° [20].

Puschmann *et al.* also carried out a SEXAFS analysis giving an adsorption site shown as A in the inset to Figure 8. The adsorption site of the formate species on copper surfaces in general has, however, proved to be controversial [22,23]. Most recent photoelectron diffraction data has indicated that it is the same site on both Cu{100} and Cu{110} and that the "aligned bridge site" (B) is most likely. We show the corresponding photoelectron diffraction data in Figure 8 without any calculated curves or any discussion of the level of theory-experiment agreement since this will be the subject of a forthcoming paper [24]. It has already been mentioned above that PED differs from SEXAFS in that a differential partial cross section is measured. Whilst this results in a smaller absolute signal in PED, the actual intensity modulations are about an order of magnitude greater. In addition, the nature of the structural information sampled differs and the technique is sensitive to both distance *and* real space direction of the scatterers. The adsorption site information is thus more directly involved in establishing the observed interferences than in SEXAFS, although the precision obtained for bond lengths may be lower. Nevertheless, even at the level of qualitative comparison,

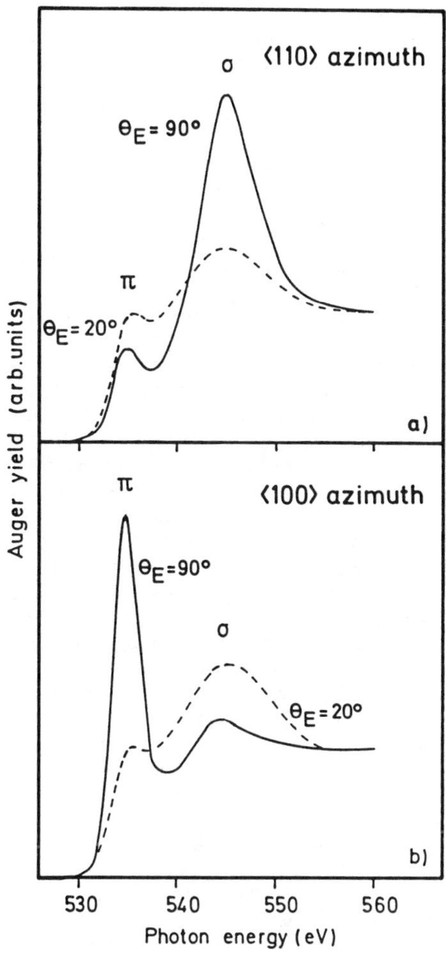

Figure 7: O 1s NEXAFS from the surface formate species on Cu{110}. The **E** vector is aligned in (a) the <110> azimuth and in (b) the <100> azimuth. After [20].

the data shown in Figure 8 from the two Cu surfaces are so similar that they cannot be reconciled with two completely different adsorption sites as suggested by the original SEXAFS studies [20,22].

Having established the orientation of the formate species on the surface we can proceed to examine the photoemission data. Figure 9 shows the effect of deprotonation of adsorbed formic acid which occurs on warming the surface to above ~ 270 K. Spectrum (a) can be assigned by comparison with the photoelectron spectrum of the free molecule. The formate species also gives rise to four spectral bands and, since the number of expected orbitals is the same, it is tempting to assume a one-to-one correspondence, allowing of course for the change in symmetry from C_s to C_{2v}. The application of selection rules proves, however, that such an assignment is incorrect [25].

Figure 10 shows three spectra at hν = 25 eV with the **E** vector parallel to the surface and aligned along the ⟨110⟩ azimuth, i.e. oriented in the molecular plane of the formate species. Spectrum (b) was obtained at normal photoelectron emission, for which the selection rules tell us that *only* levels belonging to b_1 in C_{2v} will be observed. This immediately assigns two features in the spectrum at 4.8 eV and 9.6 eV below E_F. By moving the detector off-normal into the ⟨100⟩ azimuth (spectrum (a), $\mathbf{E} \perp \mathbf{k}_\parallel$) emission from a_2 states should be observed as well. Whereas peak 3 remains in the same place peak 1 shifts slightly to lower binding energy indicating that it also contains a level of a_2 symmetry. Similarly, by moving the detector off-normal into the ⟨110⟩ azimuth (spectrum (c), $\mathbf{E} \parallel \mathbf{k}_\parallel$) a_1 and b_1 states are expected. Under these conditions peak 1 moves up in binding energy, as does peak 3. In addition, peak 4 is observed. Thus three a_1 states are also present. Peak 2 is only visible with $\mathbf{E} \perp \mathbf{k}_\parallel$ for non-normal incidence (not shown in Figure 10), indicating that it belongs to b_2. By performing further confirmatory experiments at other orientations of the **E** vector, in particular when it is aligned in the ⟨110⟩ azimuth, a complete assignment is possible. Peak 1 contains three bands due to $1a_2(\pi)$, $4b_1(\sigma)$ and $6a_1(\sigma)$ at 4.7, 4.8 and 5.1 eV below E_F, whereas peak 2 consists only of $1b_2(\pi)$ at 7.8 eV. Peak 3 contains $3b_1(\sigma)$ and $5a_1(\sigma)$ at 9.6 and 9.7 eV; peak 4 is due to $4a_1(\sigma)$ at 13.0 eV. These measured ionisation energies have been compared with HF-SCF calculations for the formate ion [26] as well as with an INDO Cu{110}-HCOO cluster calculation [27]. The relative orbital energies from the latter, semi-empirical treatment are in reasonable agreement with the measured binding energies although the absolute values, as expected, are way out. The important result from this calculation is the correct assignment of the photoelectron spectrum (via Koopmans' theorem), in particular that three levels are expected in the first band and only one in the second. An analysis of the percentage formate character in the adsorbate-derived orbitals reveals that the $1a_2$, $4b_1$ and $6a_1$ orbitals are most strongly involved in the chemisorption bond. Relative to the formate ion, surface formate has both lower σ and π populations but the σ population difference is the greater. The π donation occurs mainly via the $1a_2$ orbital; the strongest σ donor

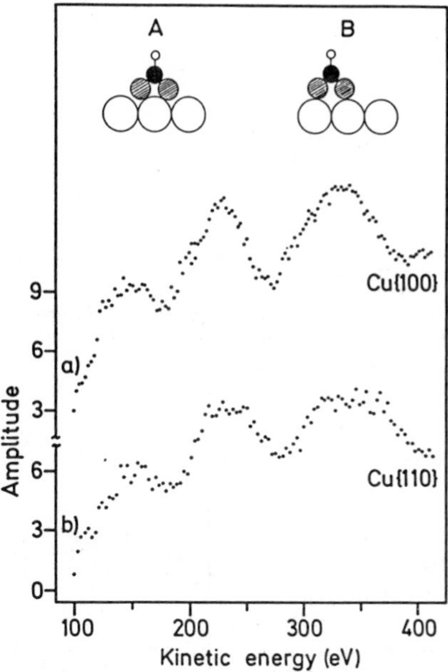

Figure 8: Photoelectron diffraction data (normal emission) for the surface formate species on (a) Cu{100} and (b) Cu{110}. Insets: A) The aligned atop site and B) the aligned bridge site. After [5].

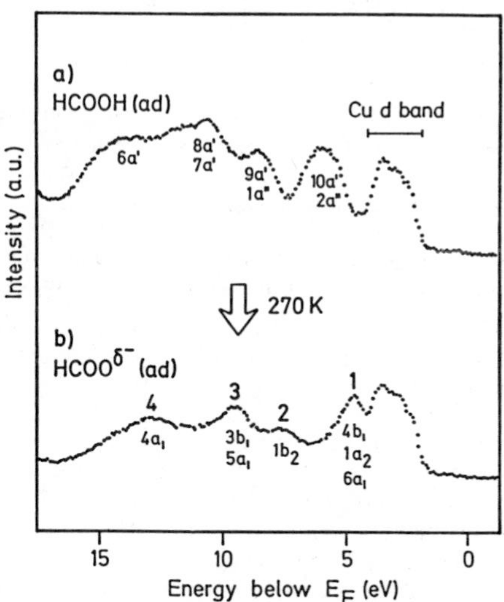

Figure 9: (a) Photoelectron spectrum of adsorbed formic acid on Cu{110} and (b) corresponding spectrum after formation of the formate species above ~ 270 K. hν = 25 eV, α = 60°, **E** ‖<110>, θ = 60° in the <100> azimuth. After [25].

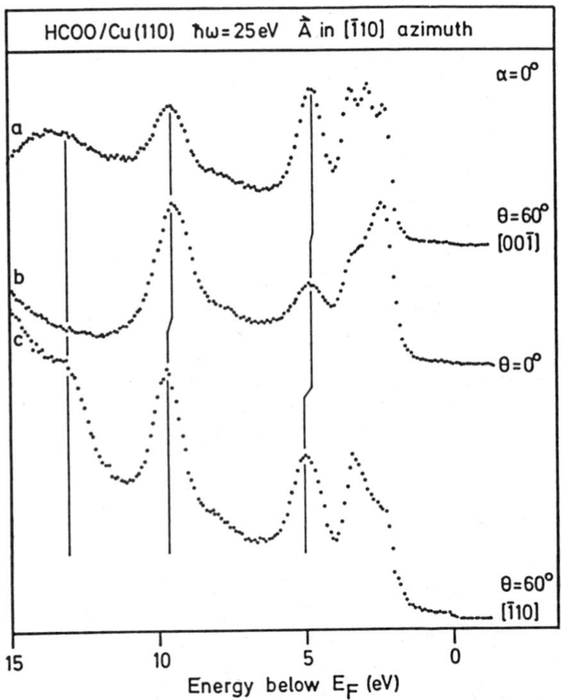

Figure 10: Angle-resolved photoelectron spectra from the system Cu{110}-HCOO for three different emission angles. hν = 25°, **E** ‖<110>. After [25].

is the $4b_1$ orbital. Classical back-bonding is neglible because the empty $2b_2$ orbital is too high in energy.

Concluding remarks

The three examples show how the approach outlined in the first two sections can be successfully implemented to obtain information on the structure and bonding of molecules and molecular fragments adsorbed on surfaces. The system Ni{100}-CO is perhaps already too well investigated and correspondingly documented to induce any surprise. Moreover, its properties are exactly those that might have been predicted on the basis of coordination chemistry. Not so the system Pd{111}-CN, where the parallel, or nearly parallel, adsorption geometry is confirmed and the consequences for the bonding scheme are investigated. Here, the lack of an appropriate quantum chemical calculation for adsorbed CN is a serious drawback to further understanding of this interesting system. Perhaps the textbook example of the interplay between structural tool, photoemission and theory is provided by the surface formate species. We are concerned here with a polyatomic surface intermediate in a heterogeneous reaction, the properties of which cannot be predicted by reference to those of a free molecule. The almost complete characterisation of the formate species on Cu{110} using a variety of surface techniques, not least those based on photoabsorption and photoionisation phenomena, might be regarded as an important advance in surface chemistry.

Acknowledgments

This work has been supported financially by the Deutsche Forschungsgemeinschaft through the Sonderforschungsbereich 6-81 and by the Fonds der Chemischen Industrie. We also acknowledge the contribution of our colleagues H. Conrad, R. Hemmen, P. Hollins, D. Kilcoyne, R. Klauser, M. Kordesch, M. Surman, G. P. Williams and D. P. Woodruff who have been associated with some of the work reviewed in this article.

Literature Cited

1. Richardson, N. V.; Bradshaw, A. M. In Electron Spectroscopy: Theory, Techniques and Applications; Baker, A.; Brundle, C. R., Eds.; Academic Press: London, 1981; Vol. 4, p. 153.
2. Stöhr, J. In Chemistry and Physics of Solid Surfaces;Vanselow, V. R.; Howe, R., Eds.; Springer Ser. Chem. Phys; Springer-Verlag: Berlin, Heidelberg 1984; Vol. 35, p. 231; Haase, J. Appl. Phys. 1985, A38, 181.
3. Barton, J. J.; Robey, G. W.; Shirley, D. A. Phys. Rev. 1986, B34, 778 ; Woodruff, D. P. Surface Sci. 1986, 166, 377.
4. Johnson, R. L. In Handbook on Synchrotron Radiation; Koch, E.-E., Ed.; North Holland: Amsterdam; 1983; Vol. 1, p.173.
5. McConville, C. F.; Woodruff, D. P.; Prince, K. C.; Paolucci, G.; Cháb, V.; Surman, M.; Bradshaw, A. M. Surface Sci. 1986, 166, 377; Woodruff, D. P.; Kilcoyne, A. L. D.; McConville, C.

F.; Lindner, Th.; Somers, J.; Surman, M.; Bradshaw, A. M. Vacuum, in press.
6. Fadley, C. S. Phys. Scripta 1987, T17, 39.
7. Somers, J.; Hollins, P.; Lindner, Th.; Bradshaw, A. M. unpublished results.
8. Somers, J. S.; Lindner, Th.; Surman, M.; Bradshaw, A. M.; Williams, G. P.; McConville, C. F.; Woodruff, D. P. Surface Sci. 1987, 183, 576.
9. Klauser, R.; Spieß, W.; Bradshaw, A. M.; Hayden, B. J. Elect. Spec. Rel. Phen. 1986, 38, 187 and references therein.
10. Klauser, R.; Surman, M.; Lindner, Th.;Bradshaw, A. M. Phys. Scripta 1987, 35, 561.
11. Plummer, E. W.; Gustafsson, T.; Gudat, W.; Eastman, D. E. Phys. Rev. 1977, A15, 2339.
12. Allyn, C. L.; Gustafsson; T.; Plummer, E. W. Sol. State Commun. 1977, 24, 531.
13. Scheffler, M.; Bradshaw, A. M. In The Chemical Physics of Solid Surfaces and Heterogeneous Catalysis;King, D. A.; Woodruff, D. P., Eds.; Elsevier: Amsterdam 1983; Vol. 2, p. 165.
14. There are many of these for Ni{100}-CO, we cite here an early one and a recent one. Cederbaum, L. S.; Domcke, W.; von Niessen, W.; Brenig, W. Z. Phys. 1975, B21, 381; Sung, S.-S.; Hoffmann, R. J. Am. Chem. Soc. 1985, 107, 578.
15. Horn, K.; Bradshaw, A. M.; Jacobi, K. Surface Sci. 1978, 72, 719.
16. Kordesch, M. E.; Stenzel, W.; Conrad, H. Surface Sci. 1987, 186, 601.
17. Somers, J.; Kordesch, M. E.; Lindner, Th.; Conrad, H.; Bradshaw, A. M.; Williams, G. P. Surface Sci. 1987, 188, L693; Spectrochimica Acta 1987, 43A, 1561.
18. Somers, J.; Kordesch, M. E.; Hemmen, R.; Lindner, Th.; Conrad, H.; Bradshaw, A. M. Surface Sci., in press.
19. Hayden, B. E.; Prince, K. C.; Woodruff, D. P.; Bradshaw, A. M. Phys. Rev. Lett. 1983, 51, 475.
20. Puschmann, A.; Haase, J.; Crapper, M. D.; Riley, C. E.; Woodruff, D. P. Phys. Rev. Lett. 1985, 57, 2598.
21. Somers, J.; Lindner, Th.; Bradshaw, A. M., to be published.
22. Stöhr, J.; Outka, D.; Madix, R. J.; Döbler, U. Phys. Rev. Lett. 1985, 54, 1256.
23. Crapper, M. D.; Riley, C. E.; Woodruff, D. P. Surface Sci. 1987, 184, 121.
24. Woodruff, D. P.; McConville, C. F.; Kilcoyne, A. L. D.; Lindner, Th.; Somers, J.; Surman, M.; Paolucci, G.; Bradshaw, A. M. Surface Sci., in press.
25. Lindner, Th.; Somers, J.; Bradshaw, A. M.; Williams, G. P. Surface Sci. 1987, 185, 75; Hofmann, P.; Menzel, D. Surface Sci., in press.
26. Peyerimhoff, S. D. J. Chem. Phys. 1967, 47, 349.
27. Rodriguez J. A.; Campbell, C. T. Surface Sci. 1987, 183, 449.

RECEIVED July 12, 1988

Chapter 9

Kinetics of Ethylidyne Formation on Platinum(111) Using Near-Edge X-ray Absorption Fine Structure

F. Zaera[1], D. A. Fischer[2], R. G. Carr[3], E. B. Kollin[4], and J. L. Gland[4]

[1]Department of Chemistry, University of California, Riverside, CA 92521

[2]National Synchrotron Light Source, Brookhaven National Laboratory, Upton, NY 11973

[3]Stanford Synchrotron Radiation Laboratory, P.O. Box 4349–Bin 69, Stanford, CA 94305

[4]Corporate Research Science Laboratory, Exxon Research & Engineering Company, Annandale, NJ 08801

The kinetics of ethylidyne formation from chemisorbed ethylene on Pt(111) surfaces was studied by using near-edge x-ray absorption fine structure (NEXAFS). The feasibility of using this technique for studying rates of reactions was successfully tested for saturation and sub-monolayer initial coverages of the reactants. Ethylidyne formation rates are first order in ethylene coverage over the entire range studied. A substantial isotope effect was observed; normal ethylene reacts about twice as fast as fully deuterated ethylene. We obtained activation energies of 15.0 and 16.7 Kcal/mole for C_2H_4 and C_2D_4, respectively. Hydrogen thermal desorption spectra (TDS) were obtained for both isotopic isomers. They also yielded first order kinetics and a noticeable isotope effect. Analysis of the hydrogen desorption peak shapes assuming a single rate limiting reaction results in values for the activation energy about 3 Kcal/mole higher than those obtained from the isothermal NEXAFS kinetic experiments. However, these differences can be explained by including a hydrogen recombination step in the interpretation of the TDS data.

The chemisorption and thermal decomposition of ethylene over platinum (111) surfaces have been extensively studied by several groups using a range of modern surface science techniques (1-3). Chemisorption at low temperatures is molecular, with the

0097–6156/88/0378–0131$06.00/0

carbon-carbon bond axis parallel to the surface and a carbon-carbon bond length of 1.49A (4,5). Thermal desorption experiments indicate that dehydrogenation occurs in a stepwise fashion starting around room temperature (2). After the loss of one hydrogen, ethylene rearranges to form a new moiety, ethylidyne ($\equiv CCH_3$), in which the carbon-carbon bond stands perpendicular to the surface in a three-fold hollow site with the α carbon bonded to three platinum atoms and three hydrogens bonded to the β carbon (6,7). Additional dehydrogenation above 450K results in the formation of hydrocarbon fragments with C_xH stoichiometry, where x has a value between one and two.

Although the structures of chemisorbed ethylene and ethylidyne have been carefully established, the mechanism by which ethylene transforms into ethylidyne remains a mystery. Understanding the reactivity of these adsorbed hydrocarbons is particularly important in view of their role during the catalytic hydrogenation of ethylene. We have recently shown (8,9) that the steady state catalytic hydrogenation of ethylene over Pt(111) at room temperature and atmospheric pressures occurs not on the bare metal surface but in the presence of either an ethylidyne layer or a related hydrocarbon fragment. In order to explain these results, we have proposed a mechanism in which hydrogen atoms may be transferred from the platinum surface to ethylene molecules weakly chemisorbed on a second layer on top of the strongly bonded carbonaceous moieties. We suspect that the intermediate(s) involved in the conversion of ethylene into ethylidyne are closely related to those intervening in the mechanism for ethylene hydrogenation described above.

In this report we present NEXAFS results for the kinetics of ethylidyne formation. Previous data is scarce and comes mostly from thermal desorption (TDS) experiments (2) . The only reported study of isothermal rates of reactions for this system was done by Ogle et. al. using secondary ion mass spectrometry (SIMS) (10). However, due to the difficulties in calculating ion yields in SIMS, quantitation of the data is not very reliable, and their work was not conclusive. We have determined here that the reaction of chemisorbed ethylene to form ethylidyne is first order in ethylene coverage. A noticeable isotope effect was observed, with activation energies of 15.0 and 16.7 Kcal/mole for C_2H_4 and C_2D_4 respectively. These values are smaller than those calculated from TDS, but the differences can be reconciled by including the recombination of hydrogen atoms on the surface in the interpretation of the thermal desorption experiments.

Experimental

The NEXAFS experiments were performed at the Stanford Synchrotron Radiation Laboratory, beamline I-1. This line is equipped with a grasshopper monochromator, 1200 lines/mm, as described elsewhere (11). The entrance and exit slits were set at 15μm, yielding a resolution of $\Delta E/E=8 \times 10^{-6}E$ (E in eV); for light of 300 eV photon energy it resulted in a linewidth of about 0.7 eV. We estimate the total photon flux under those conditions to be on the order of 1×10^9 photons/sec. at 300 eV and for a ring current of 50 mA.

The end station consisted of a ultra-high vacuum chamber pumped with an ion and a Ti sublimation pump to a base pressure in the 10^{-10} torr range. It was equipped with a cylindrical mirror analyzer for Auger electron spectroscopy (AES), LEED optics, and sputtering and gas dosing facilities. The sample (a Pt(111) crystal prepared using standard procedures) was mounted on a manipulator for alignment with respect to the incident beam, and could be cooled to 130K and resistively heated above 1200K. Temperatures were measured with a type K thermocouple spotwelded to the edge of the crystal. The sample was cleaned by Ar ion bombardment followed by annealing to 1100K or by oxygen treatment at 900K until no impurities were detected by AES. All spectra reported here were taken at saturation coverages, that is, after exposure to 10L of ethylene at temperatures below 150K ($1L=1x10^{-6}$ torr.s, pressure corrected for ion gauge sensitivity).

NEXAFS spectra were collected by using a partial electron yield (PEY) detector. This detector consisted of a channel-plate, about 5 cm in diameter, located about 5 cm from the sample and at an angle of about 45° below the synchrotron plane. A set of grids were placed in front of the channel-plate in order to apply a retarding field of ~-200 Volts, so most of the secondary electrons emitted by photon excitation could be rejected, reducing the corresponding background signal and the noise level. Photon energies were calibrated using the strong absorption feature at 291.0 eV in the incoming beam due to carbon deposited on the beam line optics (12).

Results and Discussion

NEXAFS spectra were taken at glancing and normal incidence angles for both ethylene and ethylidyne. Ethylene spectra were taken after dosing at low temperatures, while ethylidyne data were acquired after flashing to 300K for 60 seconds. We obtained spectra identical to those previously reported (13). Figure 1 shows the spectra obtained for ethylene and for ethylidyne at normal incidence. The spectrum for ethylidyne displays a peak at 285.8 eV while no features are seen in that energy range for chemisorbed ethylene. We carried out our kinetic experiments by following the absorption signal at 285.8 eV photon energy and normal incidence.

The rate of ethylidyne formation was measured by recording the partial electron yield signal as a function of time after setting the crystal temperature to a preestablished value. An example of the results is shown in Figure 2 for the conversion of C_2D_4 at 264K. A jump in the signal due to ethylidyne formation is clearly observed immediately after heating the sample (t = 0). The figure illustrates some typical characteristics of these experiments. First, the background signal is on the order of $8x10^3$ counts per second (c/s), while the signal due to saturation of ethylidyne is about $2.5x10^2$ c/s, yielding a signal-to-background ratio of about 3% (note that each point represents counting for 10s). The absolute noise, on the other hand, is close to statistical (below 25 c/s for 10s counting time per point), so the signal-to-noise is better than 10%. These values assure the feasibility of the kinetic experiments even for starting coverages below 10% of saturation.

Similar experiments were done at several temperatures starting with either normal or fully deuterated ethylene chemisorbed on

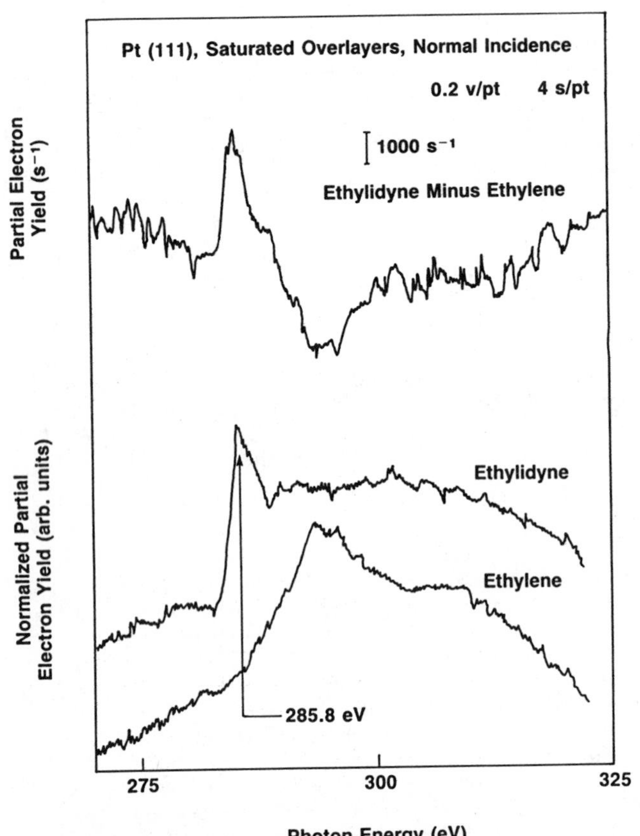

Fig. 1. NEXAFS spectra of ethylene (T=90K) and ethylidyne (T=300K) chemisorbed on Pt(111) for normal incidence. The difference between the spectra is also shown to indicate the maximum for ethylidyne at 285.8 eV photon energy.

Pt(111) surfaces. The data was then normalized and plotted in a semilogarithmic fashion (Figure 3). These plots yielded straight lines, indicating that the rate of reaction is first order with respect to ethylene coverage. The linearity of the graphs break down in the first few seconds of each run because of the time needed for the crystal temperature to equilibrate. It is also worth noticing that in some cases good quality data was obtained for coverages down to 3-5% of saturation.

Rate constants were calculated from the slopes of plots such as those shown in Figure 3. The results for both C_2H_4 and C_2D_4 are presented in Ahrrenius plots in Figure 4. Each rate was accurate to better than 20%. It is clear from the data that the ethylidyne formation displays a noticeable isotope effect,since the rates for normal ethylene are about twice as fast as those for the deuterated compound. We calculated values for the activation energy of 15.0 ± 1.0 and 16.7 ± 1.0 Kcal/mole for C_2H_4 and C_2D_4, respectively. A compensation effect is also seen in the preexponential factors: $3.6x10^{10}$ versus $3.5x10^{11}$ s^{-1}.

Previous H_2 TDS results have shown that the first peak at 297K, due to ethylidyne formation, does not shift in temperature at different initial ethylene coverages, indicating that the reaction rate is first order (2). Based on these results it was proposed that a C-H bond breaking step is limiting for that process. The authors reported values of 18.4 Kcal/mole and $4x10^{13}$ s^{-1} for the activation energy and preexponential factor respectively for normal ethylene. Since these values are considerably higher than those obtained here by NEXAFS, we decided to repeat and extend the thermal desorption experiments.

H_2 (2 amu) and D_2 (4 amu) TDS from saturation C_2H_4 and C_2D_4 are shown in Figure 5. The spectrum for normal ethylene is qualitatively equal to that reported by Salmeron et. al. (2) , although the peaks occur at slightly higher temperatures in our case. The hydrogen partial pressure resulting from H_2 desorption after ethylidyne formation peaks at 305K. For deuterated ethylene the deuterium desorption peaks at 316K, indicating a slower reaction rate in this case. Assuming that the decomposition of ethylene is a one step process, analysis of the TDS peak shapes (14) yields first order kinetics, with activation energies of 18.0 and 19.8 Kcal/mole and preexponential factors of $7.7x10^{12}$ and $4.7x10^{13}$ s^{-1}. The values for normal ethylene are, within the experimental errors, the same as those reported by Salmerón et. al.

In order to understand the TDS spectra better, we did a computer simulation of the thermal desorption peaks by integrating the rate equations for all the processes involved. Although we believed that a C-H bond breaking is the limiting step in most of the TDS temperature range, hydrogen atom recombination is a second order reaction and therefore may affect the overall reaction rate at the tails of the TDS peaks. This is indeed what we found in our calculations. We calculated TDS profiles using our NEXAFS results for the kinetic parameters of the ethylidyne formation step, and hydrogen recombination was simulated using data from the thermal

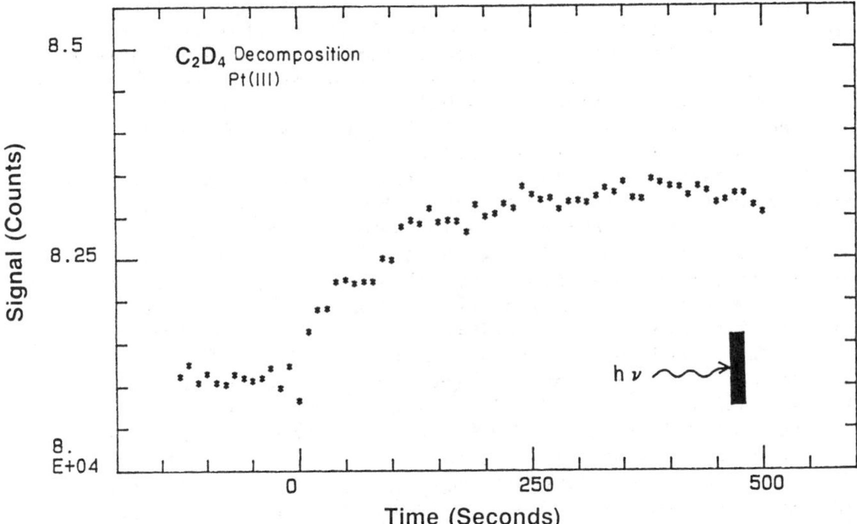

Fig. 2. Time dependence of the absorption signal following a jump in surface temperature from 130 to 264K for C_2D_4 chemisorbed on Pt(111). The photon energy was 285.8 eV, incident normal to the surface.

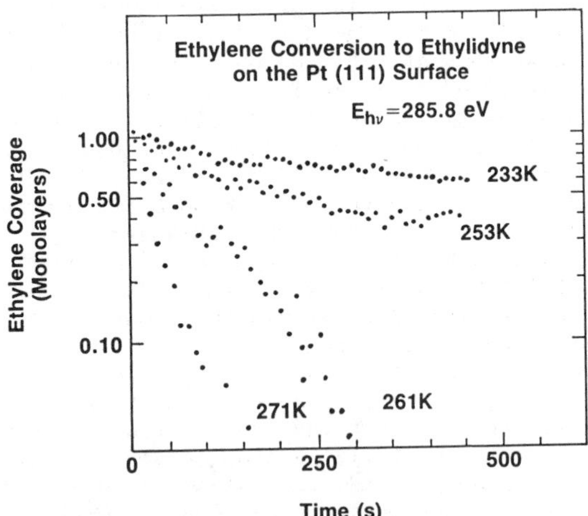

Fig. 3. Time dependence of the conversion of normal ethylene adsorbed on Pt(111) to ethylidyne at four different temperatures.

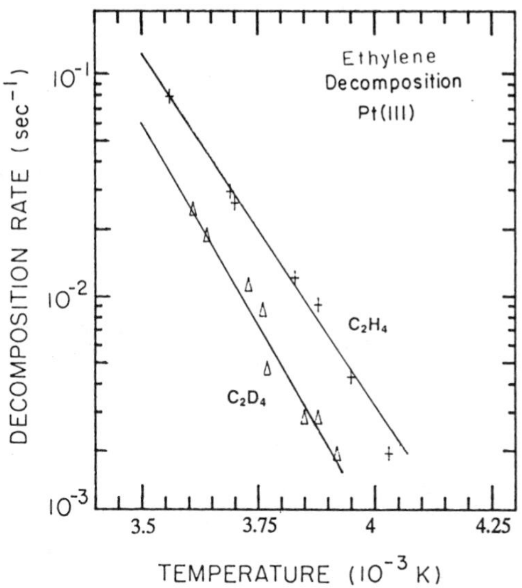

Fig. 4. Temperature dependence of the rate constants for ethylidyne formation from C_2H_4 (+) and from C_2D_4 (Δ).

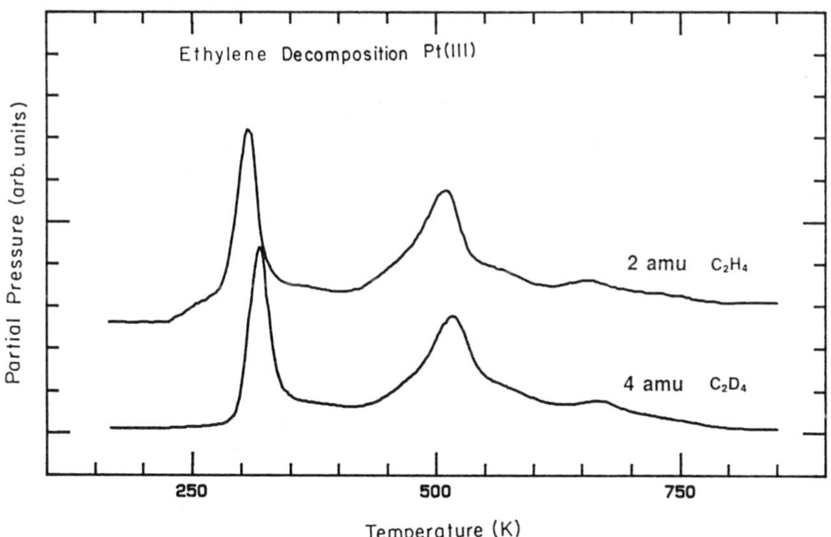

Fig. 5. H_2 (2 amu) and D_2 (4 amu) thermal desorption spectra from chemisorbed C_2H_4 and C_2D_4 on Pt(111), respectively. Heating rate = 10K/s.

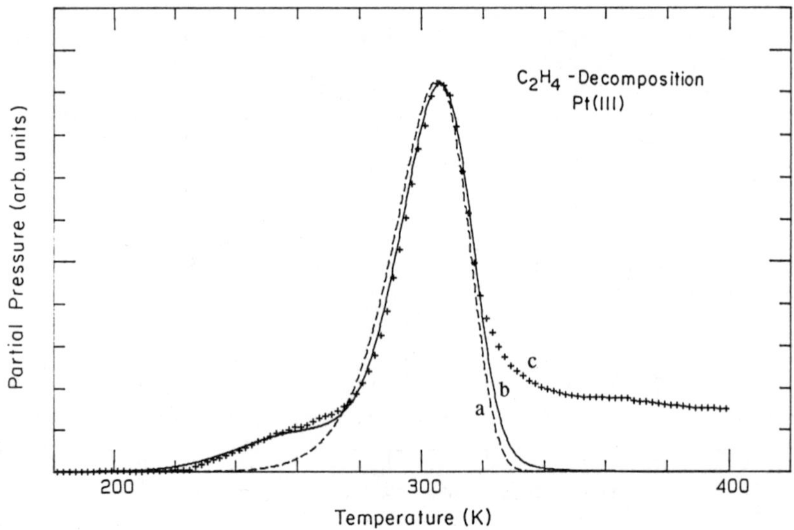

Fig. 6. Computer simulation for hydrogen TDS from chemisorbed ethylene on Pt(111): (a) First order process only, with activation energy = 15.0 Kcal/mole (dashed line); (b) same as (a) but including a hydrogen recombination step (solid line); (c) experimental data (crosses). See details in text.

desorption spectra reported by Christmann et. al. (15). The results are shown in Figure 6. We can see that if only the C-H bond breaking is considered, the simulated TDS peak is much broader than the experimental result (dashed line versus crosses). The shapes are also clearly different, since the experimental data has a sharper leading edge that the simulated peak. However, if hydrogen recombination is included in the calculations, the fit is greatly improved (solid line). It must be pointed out that these simulations use no adjustable parameters. There is some uncertainty in the values for hydrogen recombination, since activation energy values up to 19 Kcal/mole have been reported for hydrogen on Pt(111) (16). Also, hydrogen desorption may be modified by ethylene coadsorption. We decided to fit our parameters to the hydrogen TDS data from clean Pt(111) from ref. 15 (Ea = 9 Kcal/mole) and use those values for the simulations of ethylene decomposition, but we found that the results did not depend very strongly on the actual numbers used as long as hydrogen desorption was faster than C-H bond breaking at the peak of the TDS. We tried values for the activation energy for hydrogen desorption between 7 and 14 Kcal/mole and obtained no significant changes in the overall simulated TDS. Finally, the low temperature peak around 250K could also be simulated by assuming an initial coverage of about 0.15 of a monolayer of coadsorbed hydrogen (from background gases).

Conclusions

We have measured the kinetics of ethylidyne formation from chemisorbed ethylene over Pt(111) surfaces. The rates of reaction display a first order dependence on the ethylene coverage. There is an isotope effect, since the reaction for C_2H_4 is about twice as fast as for C_2D_4. We obtain values for the activation energy of 15.0 and 16.7 Kcal/mole for the normal and deuterated ethylene, respectively. These values are lower than those obtained from TDS experiments, but the differences can be reconciled by taking into account the hydrogen recombination when analyzing the thermal desorption data.

These experiments also show the value of NEXAFS as a technique for following the kinetics of surface processes. We have shown that experiments can be tailored so a specific reaction can be studied, even if gas evolution is not involved. This represents an advantage over thermal desorption experiments, where several steps may be required in order to desorb the products to be detected. Another advantage of NEXAFS is that rates are measured isothermally, so the kinetic parameters can be determined with accuracy. Finally, NEXAFS is not a destructive technique, so we need not to worry about modifying the surface compounds while probing the system, as would be the case with other techniques such as Auger electron spectroscopy.

Literature Cited

1. Baro, A.M.; Ibach, H. J. Chem. Phys. 1981, 74$16C, 4194.
2. Salmerón, M; Somorjai, G.A. J. Phys. Chem. 1982, 86, 341.

3. Zaera, F; Somorjai, G.A. In Hydrogen in Catalysis: Theoretical and Practical Aspects; Paal, Z.; Menon, P.G., Eds.; Marcel Dekker, New York, 1987.
4. Demuth, J.E. IBM J. Res. Dev. 1978, 22, 265.
5. Hiett, P.J.; Flores, F.; Grant, P.J; March, N.M.; Martin-Rodero, A; Senatore, G. Surf. Sci. 1984, 140, 400.
6. Kesmodel, L.; Dubois,L; Somorjai, G.A. J. Chem. Phys. 1979, 70, 2180.
7. Skinner, P; Howard, M.W.; Oxton, I.A.; Kettle, S.F.A.; Powell, D.B.; Sheppard, N. J. Chem. Soc. Faraday Trans. 1981, 2, 77, 1203.
8. Zaera, F.; Somorjai, G.A. J. Am. Chem. Soc. 1984, 106, 2288.
9. Godbey, D.; Zaera, F.; Yates, R.; Somorjai, G.A. Surf. Sci. 1986, 167, 150.
10. Ogle, K.M.; Creighton, J.R.; Akhter,S; White, J.M. Surf. Sci. 1986, 169, 246.
11. Brown, C.; Bachrach, R.Z.; Lien, N. Nucl. Instrum. Methods 1978, 152, 73.
12. Stöhr, J; Jaeger, R. Phys. Rev. B 1982, 26, 4111.
13. Koestner, R.J.; Stöhr, J.; Gland, J.L.; Horsley, J.A. Chem. Phys. Letts. 1984, 105, 332.
14. Chan, C.-M.; Aris, R.; Weinberg, W.H. Appl. of Surf. Sci. 1978, 1, 360.
15. Christmann, K.; Ertl, G.; Pignet, T. Surf. Sci. 1976, 365.
16. Poelsema, B.; Mechtersheimer, G.; Comsa, G. Surf. Sci. 1981, 111, 519.

RECEIVED May 17, 1988

Chapter 10

Low-Energy Electron Diffraction, X-ray Photoelectron Spectroscopy, and Auger Electron Spectroscopy

Lead Underpotential Deposition on Silver Single Crystals

Michael E. Hanson[1] and Ernest Yeager

Case Center for Electrochemical Sciences and the Chemistry Department, Case Western Reserve University, Cleveland, OH 44106

The underpotential deposition of lead has been examined on LEED-characterized single crystal silver surfaces with 0.1 M HF as the electrolyte using a special ultra-high vacuum-electrolyte transfer system. Each of the low index surfaces has a characteristic voltammetry curve with multiple adsorption and desorption UPD peaks. LEED studies of the UPD layers indicate unique superlattices which are highly dependent on the coverage as well as the particular single crystal surface. The UPD layers have also been examined with AES and XPS. These indicate that under some conditions lead in oxidized form is also present on the surface after the electrochemical measurements, thus complicating the interpretation of the LEED patterns.

The underpotential deposition (UPD) of metals on foreign metal substrates is of importance in understanding the first phase of metal electrodeposition and also as a means for preparing electrode surfaces with interesting electronic and morphological properties for electrocatalytic studies. The UPD of metals on polycrystalline substrates exhibit quite complex behavior with multiple peaks in the linear sweep voltammetry curves. This behavior is at least partially due to the presence of various low and high index planes on the polycrystalline surface. The formation of various ordered overlayers on particular single crystal surface planes may also contribute to the complex peak structure in the voltammetry curves.

In order to gain more insight into the dependence of the UPD process and structure of the layer on the crystal structure of the substrate, the UPD of lead has been studied on silver crystal surfaces using linear sweep voltammetry. Low energy electron diffraction (LEED) has been used to examine the initial substrate surface as well as the UPD layers as a function of the potential

[1]Current address: General Electric Lighting, Nela Park, Cleveland, OH 44112

0097-6156/88/0378-0141$06.00/0

at which the electrode was emersed from the electrolyte. The transfer of the single crystal surfaces from the ultra-high vacuum (UHV) environment where they were prepared into the electrochemical environment, as well as their return to the UHV for post electrochemical examination with LEED, X-ray photoelectron spectroscopy (XPS) and Auger electron spectroscopy (AES), was carried out with the special system developed in the authors' laboratory (Figure 1).

Experimental

A detailed description of the experimental procedure is reported elsewhere (1). Silver single crystals were obtained from Metal Crystals Ltd., Cambridge, England, with surfaces corresponding to the (100), (110) and (111) planes, oriented to better that 1°. These surfaces were mechanically polished with diamond paste, gradually decreasing the grit to 1 μm and then chemically polished using previously reported procedures (1,2). The Ag crystals were then transferred into vacuum chamber A (Figure 1) where the surface was argon ion sputtered and then thermally annealed at 600^oC at 10^{-10} Torr.

The surface purity was checked with AES and XPS, and the surface morphology was confirmed with LEED. For the samples used in the work herein reported, no impurities were observed in the AES or XPS prior to transfer into the electrochemical environment. The crystal was then transferred into vacuum chamber B (Figure 1) by means of a transfer manipulator and the chamber was back-filled with ultra pure argon. A quasi-thin-layer cell with a 2 mm. gap was then formed with an α-H(Pd) counter electrode and Pb reference electrode. Various linear sweep voltammetry measurements were then carried out using 0.1 M HF + 3.0 mM PbF_2 as the electrolyte. Following these measurements, the single crystal electrode was separated from the electrolytic solution at a particular controlled potential and any remaining electrolyte on the silver electrode was removed if necessary by touching the edge of the crystal with a piece of deashed filter paper mounted on the counter-reference manipulator assembly. The filter was precleaned with ~20% by wt. HF followed by extended washing in triply distilled water. Chamber B was then re-evacuated and the single crystal electrode returned to chamber A at a pressure in the 10^{-10} Torr range in about 5 minutes. LEED, XPS and AES measurements were then carried out.

Results and Discussion

Linear Sweep Voltammetry

Each of the low index Ag single crystals displayed mutually unique voltammetry curves with multiple adsorption/desorption peaks (Figure 2). The nominal features of these curves are similar to those obtained by other authors for Ag single crystal surfaces in HF or $HClO_4$ using both UHV and non-UHV methods (4-7).

Voltammetry curves obtained by Vitanov et al. (8) on electrochemically grown Ag(111) surfaces with an ultra low step density

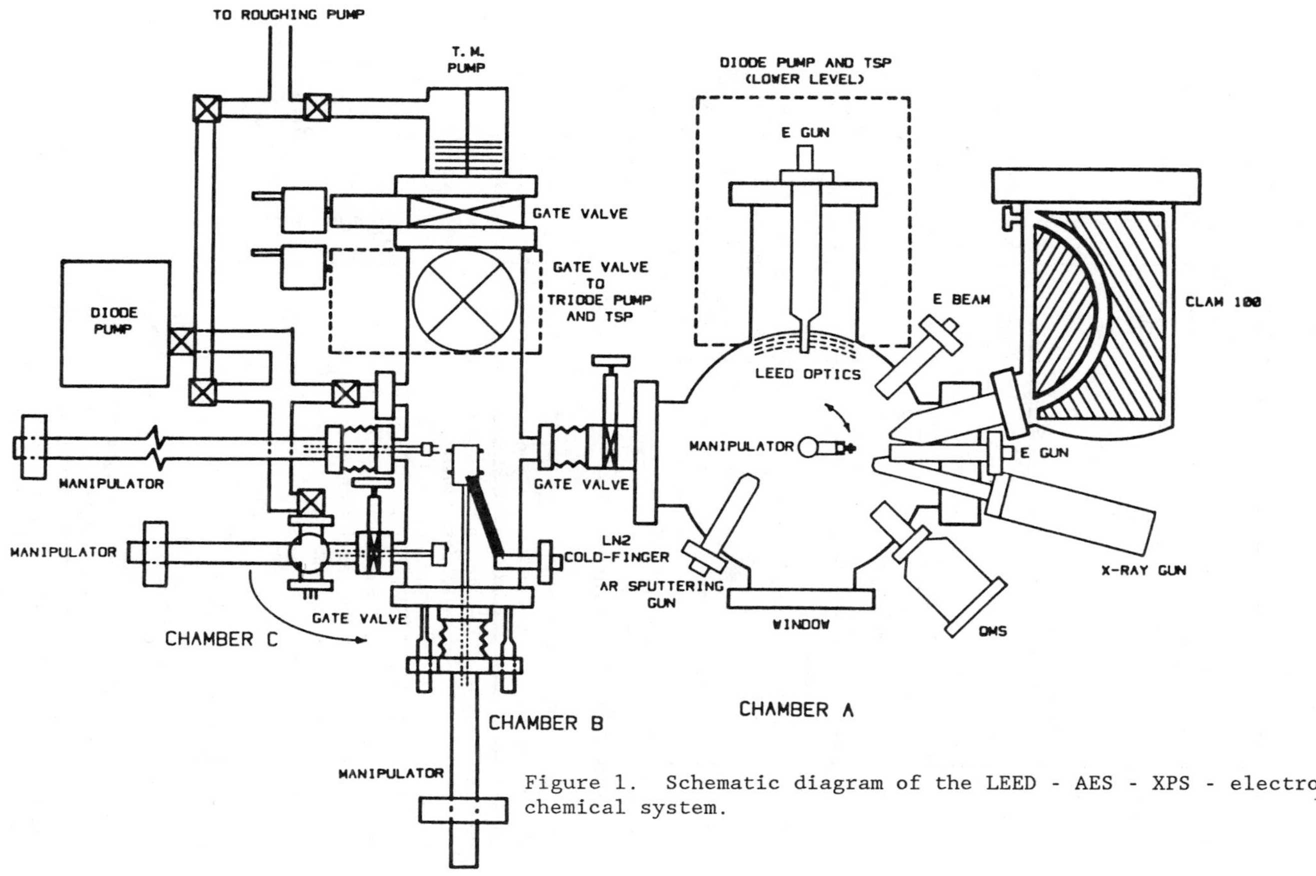

Figure 1. Schematic diagram of the LEED - AES - XPS - electrochemical system.

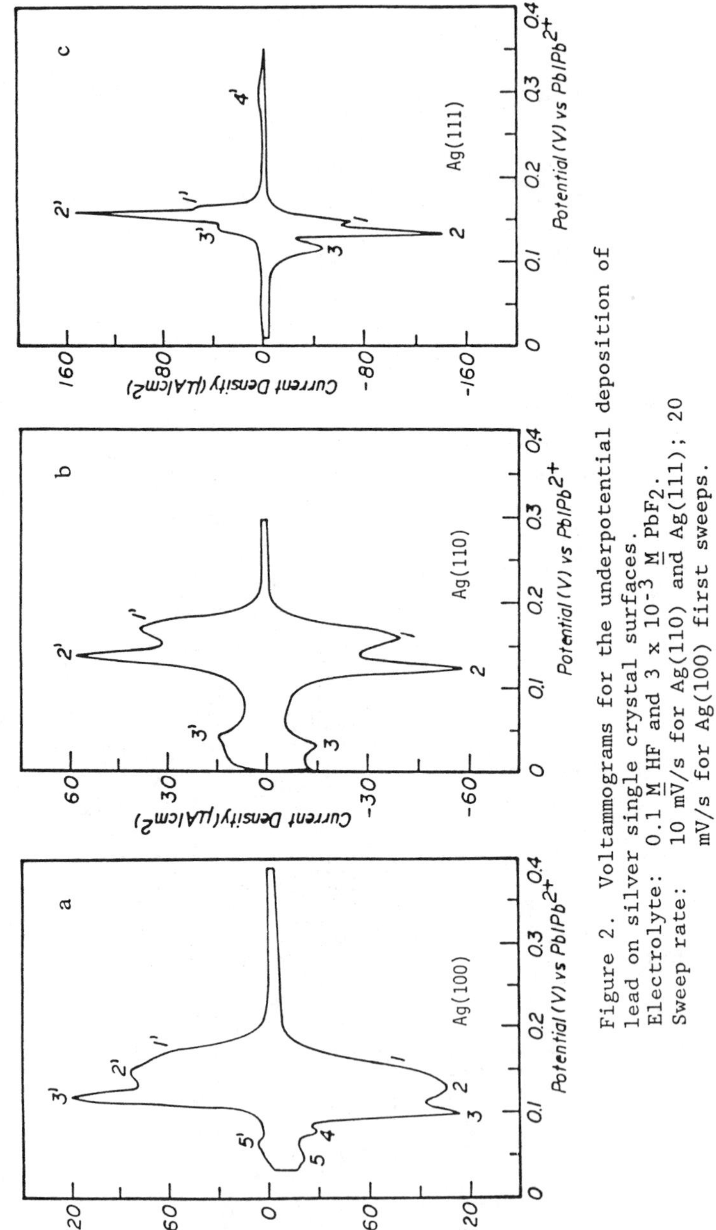

Figure 2. Voltammograms for the underpotential deposition of lead on silver single crystal surfaces.
Electrolyte: 0.1 M HF and 3×10^{-3} M PbF_2.
Sweep rate: 10 mV/s for Ag(110) and Ag(111); 20 mV/s for Ag(100) first sweeps.

display only one prominent adsorption/desorption peak. These authors report that the shape of those voltammetry curves did not change with extended sweeping within the UPD potential range or with extended polarization at a potential more cathodic than the adsorption peak potential. On Ag(111) surfaces with a higher step density, they propose that structural transformations occur in a partial adsorption layer causing the voltammetry curve to display two additional adsorption/desorption peaks, similar to peaks 1-1′ and 3-3′ in the present work (Figure 2c). The magnitude and facility of this effect appears to be proportional to the initial step density of the Ag(111) substrate, and the adsorbate structural transformation results in the growth of additional steps.

A similar effect was observed in our work and in the work of others (5), where voltammetry curves changed after extended cycling, particularly if the cathodic sweep was reversed before the full Pb deposition coverage. The observed "cathodic memory effect" may be due to the proposed structural transformation phenomenon and subsequent step density growth, initially facilitated by a high step density on a UHV-prepared or chemically polished (6) Ag(111) substrate. Post electrochemical LEED analysis on Ag(111)-Pb(UPD) surfaces provided additional evidence of a step density increase during Pb underpotential deposition, which will be discussed later in this text. (See Figure 3.)

Post Electrochemical Electron Spectroscopy

Pb UPD films were emersed under potential control at different coverages and analyzed by XPS, LEED and AES. XPS spectra in the Pb(4f) region of many of these films (Figures 4 and 5) show 4f electron binding energies higher than expected for Pb in the metallic state (Figure 6). Chemical shifts are more consistent with a PbF_2 standard (Figure 5c). XPS and AES frequently detected O,F and possibly C, although not consistently (Figure 7). Clear detection of the C(KVV) peak in Figure 7 is obstructed by one of the Ag peaks.

After the loss of potential control, the Pb UPD layer may react with residual electrolytic solution to generate H_2 and the Pb may be oxidized to the divalent state yielding a layer of PbO, $Pb(OH)_2$ or PbF_2. This is more likely to occur with UPD of Pb on Ag than for example on Au since the UPD potentials are more cathodic on Ag. Thus, for $\Delta E_p = 0.15$ V (see Figure 2a) and $E^o(Pb/PbF_2) = -0.344$ V in the reaction:

$$Pb(UPD) + 2HF = PbF_2 + H_2 \quad (1)$$

$$E = 0.344 - 0.15 - 0.296 \cdot \log \frac{[a_{PbF_2}][P_{H_2}]}{[a_{HF}][a_{Pb(UPD)}]} \quad (2)$$

$$= 0.344 - 0.15 - 0.296 \cdot \log(Q)$$

At equilibrium (E=0), this expression predicts log(Q) = 6.5 (where Q is the equilibrium constant for equation 1. This calculation

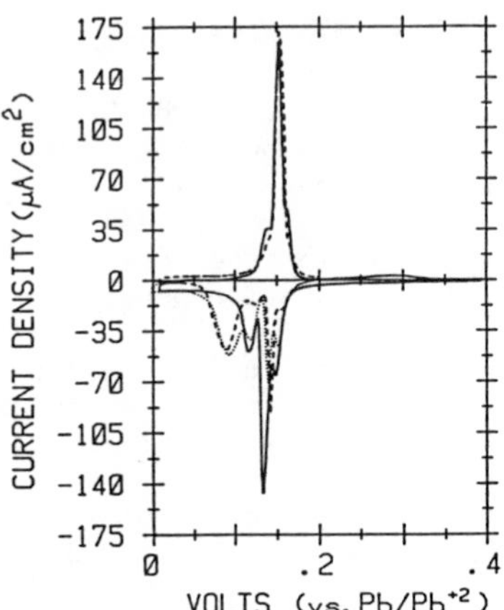

Figure 3. Summary of voltammetry curves of Ag(111) in 0.1 M HF and 3 x 10^{-3} M PbF_2; Sweep rate = 10 mV/s.
——Before cathodic memory effect.
·····After cathodic memory effect, obtained by closing the cathodic window and re-opening in 10 mV increments.
-----Partially recovered curve, obtained by closing the anodic window and re-opening in 10 mV increments.

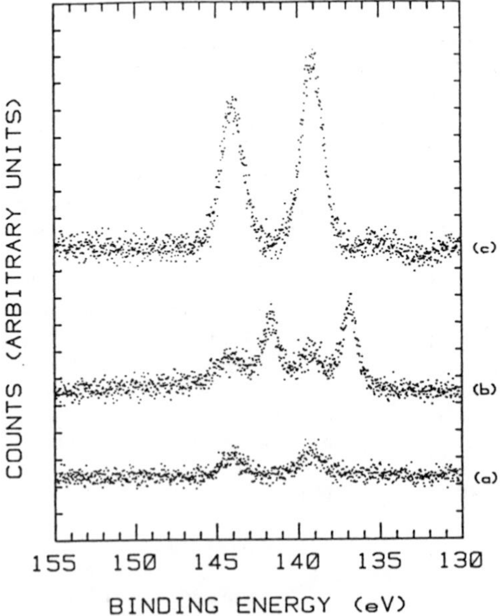

Figure 4. Pb(4f) XPS spectra of Pb UPD films on Ag(110). Magnesium X-ray anode ($h\nu$ = 1253.6 eV).
(a) Deposition charge = 166 $\mu C/cm^2$, θ_{Pb} = 0.31
(b) Deposition charge = 264 $\mu C/cm^2$, θ_{Pb} = 0.49
(c) Deposition charge = 384 $\mu C/cm^2$, θ_{Pb} = 0.71

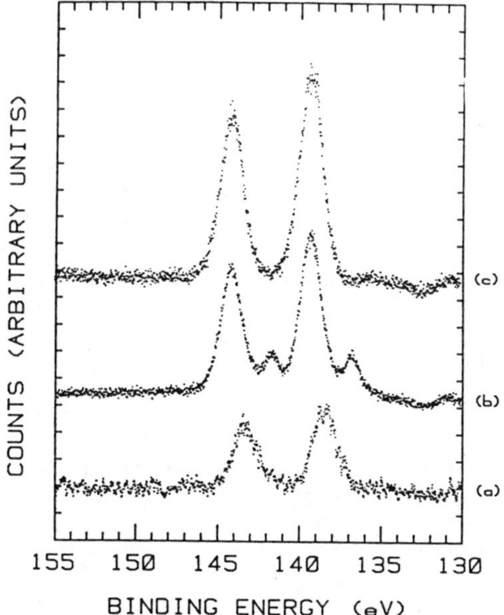

Figure 5. Pb(4f) XPS spectra of Pb UPD films on Ag(111). Magnesium X-ray anode ($h\nu$ = 1253.6 eV).
(a) Deposition charge = 244 $\mu C/cm^2$, θ_{Pb} = 0.55
(b) Deposition charge = 345 $\mu C/cm^2$, θ_{Pb} = 0.78
(c) PbF_2 standard

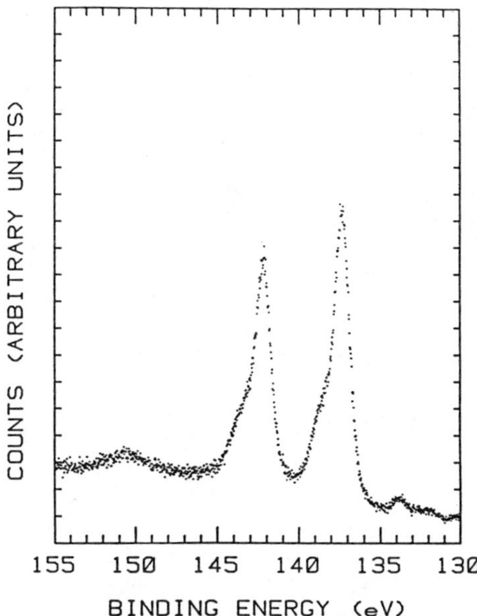

Figure 6. Pb(4f) XPS spectrum of sputtered Pb foil. Magnesium X-ray anode ($h\nu$ = 1253.6 eV).

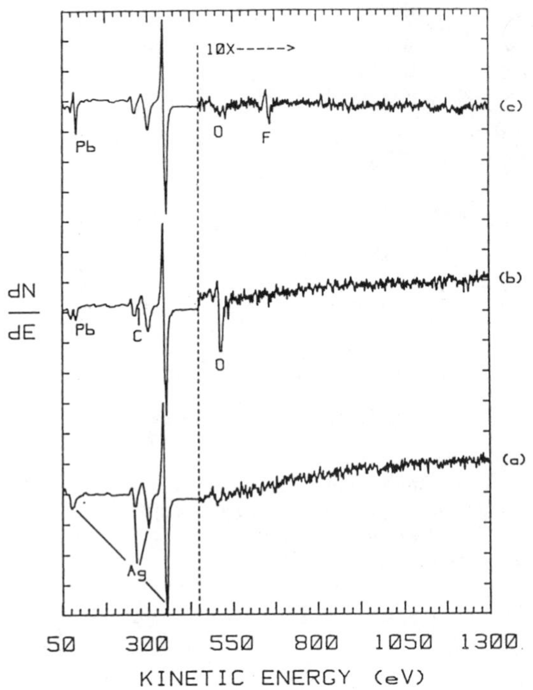

Figure 7. AES spectra of Pb UPD films on Ag(111). $E_p = 2500$ eV, $I_p = 10\mu A$, $V_m = 5$ V.
(a) Clean Ag(111), before electrochemistry.
(b) Deposition charge = 244 $\mu C/cm^2$, $\theta_{Pb} = 0.55$
(c) Deposition charge = 345 $\mu C/cm^2$, $\theta_{Pb} = 0.78$

indicates that the oxidation of the Pb(UPD) is thermodynamically favorable, although low hydrogen overvoltage is required to make this reaction kinetically feasible.

These XPS results cause great uncertainty as to the chemical state of emersed Pb(UPD) films on Ag and validate questions raised about partial desorption of UPD films after potential control is lost (3,9). Nevertheless, XPS chemical shifts of the Pb(4f) peaks show a dependence on the emersion potential; i.e. the Pb coverage before emersion. Some emersed films also displayed at least two different chemical states of the UPD Pb (Figures 4b an 5b).

The significance of the post-electrochemical LEED studies is very uncertain because of the partial oxidation of the UPD-Pb layers. Nonetheless the LEED measurements were carried out. Superlattice geometries, as judged by LEED, were dependent on the Pb(UPD) coverage at emersion. Figures 8 and 9 show LEED photographs and possible real-space illustrations for UPD Pb on Ag(100) at low and high emersion coverages. At a low coverage (Figure 8), the Pb appears to adsorb in site-specific registry of the Ag(100) substrate. At high coverage (Figure 9), the emersed film shows hexagonal symmetry with two equiprobable orthogonal domains in which substrate registry is not obeyed.

Pb superlattice geometry was also found to be dependent on substrate geometry. Figure 10 shows a LEED photograph and possible real-space mdoel of an emersed partial coverage Pb(UPD) film on Ag(111). Three equiprobable hexagonal superlattices are present, rotated -11°, 0° and +11° relative to the hexagonal substrate lattice. The poor quality of the LEED pattern does not permit a clear observation of the scattering centers in the holes of these honeycomb structures. The very high background intensity and broad spot sizes confirm that the step and defect density of the substrate increased during electrochemistry.

Summary

UHV-prepared Ag(100), Ag(110) and Ag(111) each displayed a unique Pb UPD voltammetry curve. Observations of the voltammetry curves on Ag(111) suggest low-coverage Pb may place-exchange with the substrate at step sites, causing an increase in the density of step sites upon further cycling. LEED studies of Pb UPD films detected unique superlattice geometries which were dependent upon the coverage of Pb at emersion and the crystallographic orientation of the Ag substrate. XPS of emersed UPD Pb films infer that much of the Pb is in an oxidized state, e.g. PbO or PbF_2. This may be the result of a reaction of the UPD Pb with residual electrolyte on the emersed surfaces to form H_2 once potential control is lost, since the Pb on Ag is a relatively cathodic UPD system. It is uncertain but likely that this oxidation alters the geometry of the UPD Pb films as they exist on the electrochemical environment under potential control.

This study illustrates the complications that may occur in post-electrochemical studies of UPD layers using UHV techniques, particularly when the layers are thermodynamically unstable on open circuit. Fortunately considerable progress has been reported including papers at this symposium in the use of *in-situ* techni-

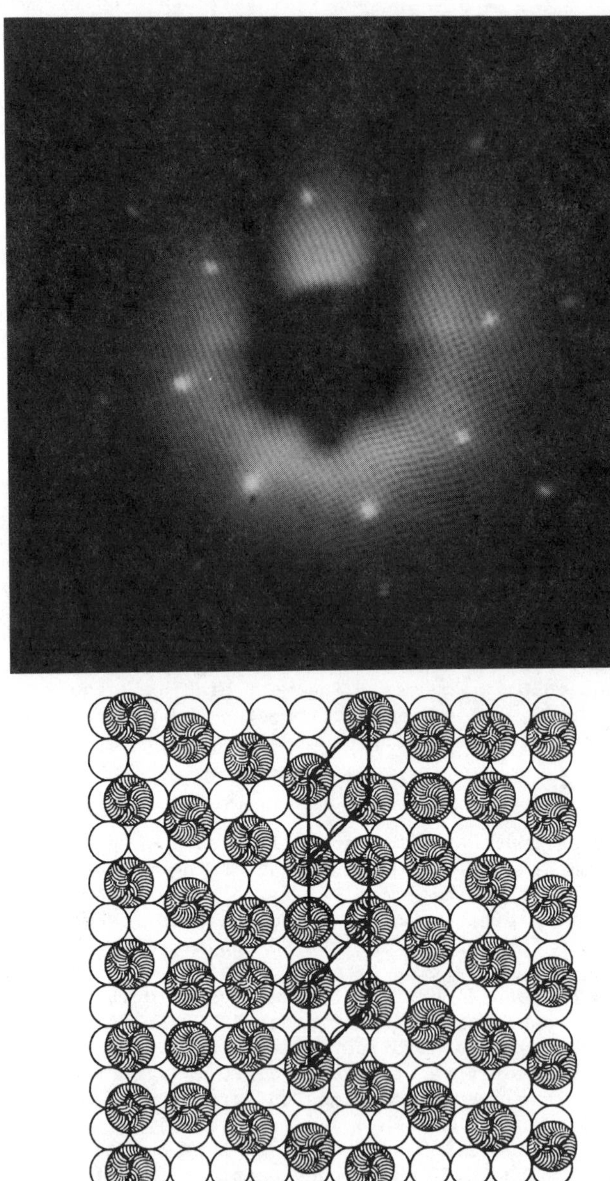

Figure 8. LEED pattern and real-space model of partial coverage UPD Pb on Ag(100). Deposition charge density before emersion = 53 $\mu C/cm^2$, θ_{Pb} = 0.14, Ep = 27 eV.

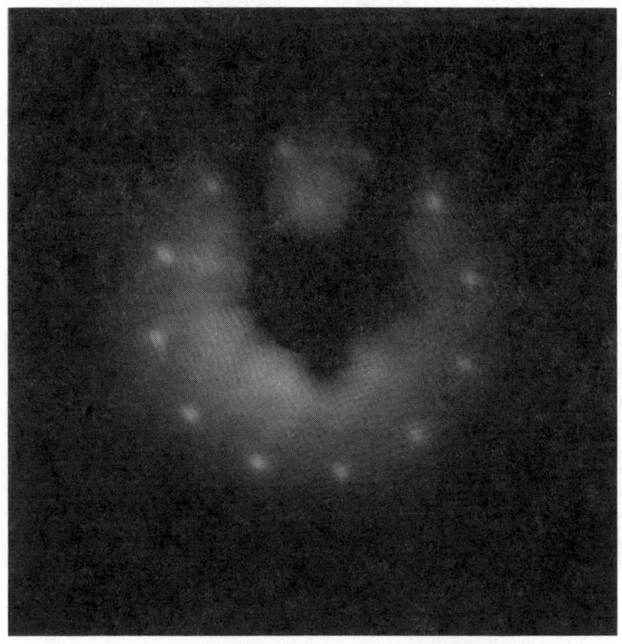

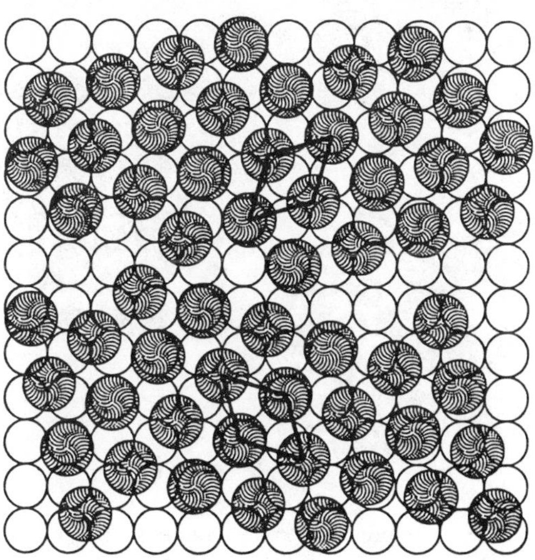

Figure 9. LEED pattern and real-space model of high coverage UPD Pb on Ag(100). Deposition charge denisty before emersion = 328 $\mu C/cm^2$, θ_{Pb} = 0.86, Ep = 27 eV.

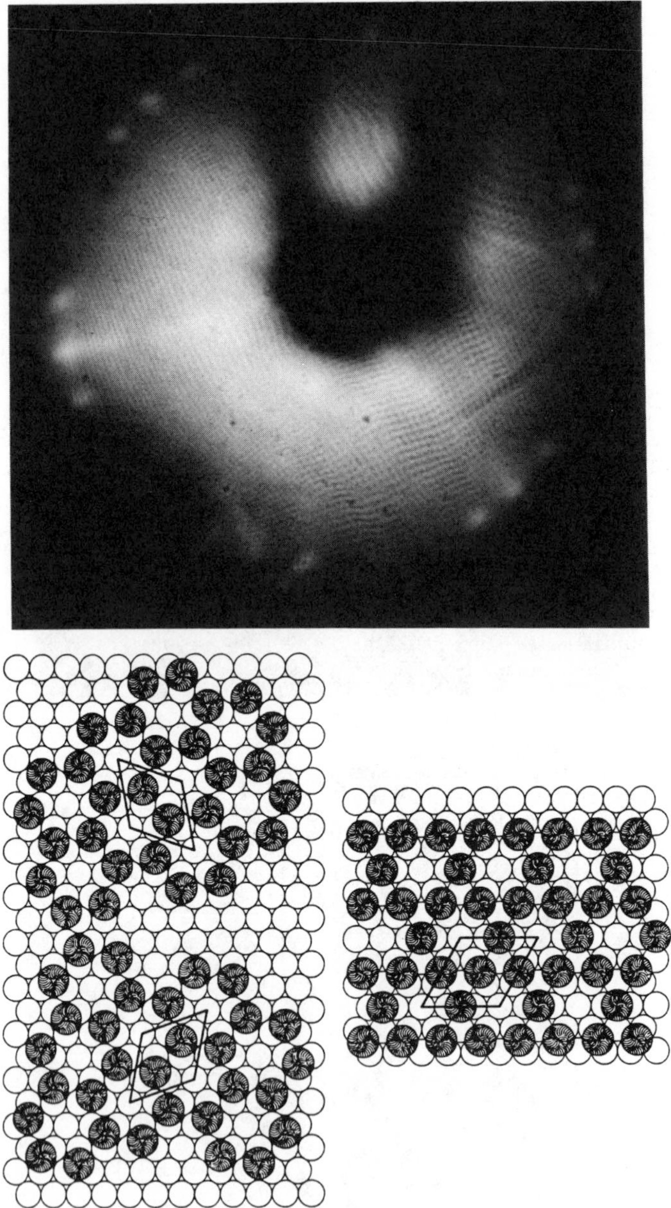

Figure 10. LEED pattern and real-space model of partial-coverage UPD Pb on Ag(111). Deposition charge density before emersion = 244 $\mu C/cm^2$, θ_{Pb} = 0.55, Ep = 25 eV.

ques for the study of the structure of layers on electrode surfaces using synchrotron radiation in low angle X-ray scattering, EXAFS (10) and X-ray standing wave techniques (11).

Acknowledgment

The authors acknowledge very helpful discussions with Dr. R. Adzic of the Institute of Electrochemistry, Belgrade, concerning the underpotential deposition of lead on single crystal silver substrates after chemical polishing. The authors also acknowledge support of the research by the U.S. Office of Naval Research.

References

1. M. E. Hanson, Ph.D. Thesis, Case Western Reserve University, Cleveland, Ohio (1985).
2. R. R. Adzic, M. E. Hanson and E. Yeager, J. Electrochem. Soc., 131, 1730 (1984).
3. L. Laguren-Davidson, F. Lu, G. N. Salaita and A. T. Hubbard, Langmuir, 4, 224 (1988).
4. W. J. Lorenz, E. Schmidt, G. Staikopv and H. Bort, 12th Faraday Symposium, Electrocrystallization, Nucleation and Phase Formation, Southampton, England (1977).
5. H. Siegentaller and K. Juttner, Electrochim. Acta, 24, 109 (1979).
6. K. Juttner, Electrochim. Acta, 31, 917 (1986).
7. A. Bewick and B. Thomas, J. Electroanal. Chem., 84, 127 (1977).
8. T. Vitanov, A. Popov, G. Staikov, E. Bodevski, W. J. Lorenz and E. Schmidt, Electrochim. Acta, 31, 981 (1986).
9. B. C. Schardt, J. L. Stickney, D. . Stern, A. Wieckowski, D. C. Zapien and A. T. Hubbard, Langmuir, 3, 239 (1987).
10. J. G. Gordon, O. R. Melroy, M.G. Samani, M. F. Toney and G. L. Borges, 194th American Chemical Society National Meeting, Symposium on Photochemical and Electrochemical Surface Science, New Orleans, paper 186 (1987).
11. H. D. Abruna, loc. cit., paper 185.

RECEIVED July 14, 1988

Chapter 11

Structural, Catalytic, Electronic, and Electrochemical Properties of Strained-Copper Overlayers on Ruthenium(0001)

D. W. Goodman and E. M. Stuve[1]

Sandia National Laboratories, P.O. Box 5800, Albuquerque, NM 87185

The chemical behavior of monolayer coverages of one metal on the surface of another, i.e. Cu/Ru, Ni/Ru, Ni/W, Pd/W, has recently been shown to be dramatically different from that seen for either of the metallic components separately. These chemical alterations, which modify the chemisorption and catalytic properties of the overlayers, have been correlated with changes in the structural and electronic properties of the bimetallic system. The films are found to grow in a manner which causes them to be strained with respect to their bulk lattice configuration. Earlier studies have addressed the adsorption of CO and H_2 on copper overlayers on Ru(0001) as well as the measurement of the elevated pressure kinetics of the methanation, ethane hydrogenolysis, and cyclohexane dehydrogenation reactions. In addition, unique electronic interface states have been identified for this bimetallic system. A comparison of electrochemical (EC) and ultrahigh vacuum (UHV) deposition methods has been made via a series of co- characterization experiments in which Cu was deposited in one environment (UHV or EC) and characterized in the other (EC or UHV). Both EC stripping and temperature programmed desorption (TPD) in vacuum of the Cu show that the difference in the heat of desorption between multilayer and monolayer Cu is ~6 kcal/mole and identical for both methods.

[1]Visiting Associated Western Universities summer faculty; permanent address: Department of Chemical Engineering, BF–10, University of Washington, Seattle, WA 98195

0097–6156/88/0378–0154$06.00/0

It has long been recognized that the addition of impurities to metal catalysts can produce large effects on the activity, selectivity, and resistance to poisoning of the pure metal (1). For example, the catalytic properties of metals can be altered greatly by the addition of a second metal (2). A long-standing question regarding such bimetallic systems is the nature of the properties of the mixed metal system which give rise to its enhanced catalytic performance relative to either of its individual metal components. These enhanced properties (improved stability, selectivity and/or activity) can be accounted for by one or more of several possibilities. First, the addition of one metal to a second may lead to an electronic modification of either or both of the metal constituents. This electronic perturbation can result from direct bonding (charge transfer) or from a structural modification induced by one metal upon the other. Secondly, a metal additive can promote a particular step in the reaction sequence and, thus, act synergistically with the host metal. Thirdly, the additive metal can serve to block the availability of certain active sites, or ensembles, prerequisite for a particular reaction step. If this "poisoned" reaction step involves an undesirable reaction product, then the net effect is an enhanced overall selectivity. Further, the attenuation by this mechanism of a reaction step leading to undesirable surface contamination will promote catalyst activity and durability.

The present studies are part of a continuing effort (3-12) to identify those properties of bimetallic systems which can be related to their superior catalytic properties. A pivotal question to be addressed of bimetallic systems (and of surface impurities in general) is the relative importance of ensemble (steric or local) versus electronic (nonlocal or extended) effects in the modification of catalytic properties. A complete understanding of surface impurity effects (including alloying) in catalysis will likely include components of both electronic and ensemble effects, the relative importance of each to be assessed for a given reaction and reaction conditions. An emphasis of our research has been in the area of addressing and partitioning the importance of these two effects in the influence of alloying and surface additives on surface reactions.

Experimental

The studies described other than the combined ultrahigh vacuum (UHV)-electrochemical (EC) studies involve the use of an experimental apparatus of the type described in references 14 and 15. This device consists of a surface analysis chamber vacuum interlocked to a microcatalytic reactor. Both regions are of ultrahigh vacuum construction and capable of ultimate pressures of less than 2 times 10^{-10} torr. In the surface analysis chamber, techniques such as Auger spectroscopy (AES), X-ray photoelectron spectroscopy (XPS), ultraviolet photoelectron spectroscopy (UPS), low energy electron diffraction (LEED), and temperature programmed desorption (TPD) are available for sample preparation and for sample characterization before and after reaction. The sample support assembly allows the metal single-crystal catalyst to be transferred in vacuo from the surface analysis region to the microcatalytic reactor. In the microcatalytic reactor, reaction kinetics can be measured for single crystal catalysts having surface areas of less than 1 cm^2 at reactant pressures up to several atmospheres.

The single crystal catalysts, ~1 cm in diameter and 1 mm thick, are typically aligned within 0.5° of the desired orientation. Thermocouples are generally spot-welded to the edge of the crystal for temperature measurement. Details of sample mounting, cleaning procedures, reactant purification, and product detection techniques are given in the related references. The catalytic rate normalized to the number of exposed metal sites is the specific activity, which can be expressed as a turnover frequency (TOF), or number of molecules of product produced per metal atom site per second.

The electrochemical experiments were conducted in an apparatus consisting of an electrochemical cell attached directly to a UHV system and has been described in detail elsewhere (16). The transfer between UHV and the EC was accomplished via a stainless steel air lock vented with ultra-pure Ar. Differentially pumped sliding teflon seals provided the isolation between UHV and atmospheric pressure. The sample was mounted on a polished stainless steel rod around which the teflon seals were compressed. All valves in the air lock were stainless steel gate valves with viton seals. Details of the electrochemical cell and conditions are contained in reference 16. Electrochemical potentials are referred to a saturated calomel electrode (SCE).

Results and Discussion

Structural, Catalytic, and Electronic Properties. Interest in bimetallic catalysts has increased steadily because of the commercial success of these systems, which allow an enhanced ability to control the catalytic activity and selectivity by tailoring the catalyst composition (17-29). A key point in these investigations, as with the studies involving other impurities, has been determining the relative importance of ensemble and electronic effects in defining catalytic behavior (29-32). It is advantageous to simplify the problem by utilizing models of a bimetallic catalyst, which have been made by deposition of metals on single-crystal substrates in a UHV environment. A combination that has been studied extensively in supported catalyst research is copper on ruthenium (Cu/Ru). The immiscibility of copper in ruthenium circumvents the complication of determining the three-dimensional composition. Furthermore, distinct TPD features exist for the first monolayer of copper relative to multilayer coverages (3). Thus the copper coverage can be calibrated accurately by TPD measurements as discussed in Section II.

A comparison of CO desorption from ruthenium (6), and from multilayer (10 ML) and monolayer copper covered ruthenium is shown in Figure 1. The CO coverage is at saturation. The TPD features of the 1 ML copper (peaks at 160 and 210 K) on ruthenium are at temperatures intermediate between those found for adsorption on surfaces of bulk ruthenium and copper, respectively. This suggests that the copper monolayer is perturbed electronically and that this perturbation is manifested in the bonding of CO. An increase in the desorption temperature relative to bulk copper indicates a stabilization of the CO on the copper monolayer which suggests an electronic coupling of the CO through the copper to the ruthenium.

Model kinetic studies of these Cu/Ru(0001) catalysts have been performed for methanation (7), cyclohexane hydrogenolysis (9), and cyclohexane dehydrogenation (10) reactions. For the first two reactions, copper serves as an inactive diluent, blocking sites on a one-to-one basis. The latter reaction is quite different. Figure 2 shows the effect, caused by addition of copper onto ruthenium on the rate of dehydrogenation of cyclohexane to benzene. The overall rate of this reaction increases by approximately an order of magnitude at a copper coverage of 0.75 ML. This translates into a specific rate enhancement for ruthenium of ~40. At higher coverages, the rate decreases to an activity approximately equal to that of copper-free ruthenium.

The rate enhancement for cyclohexane dehydrogenation observed for submonolayer copper deposits may result from changes in the geometric (6) and the electronic (8) properties of the copper overlayer relative to bulk copper. Alternatively, the two metals may catalyze different steps of the reaction cooperatively. For example, dissociative H_2 adsorption on bulk copper is unfavorable because of an activation barrier of approximately 5 kcal/mol (33). In the Cu/Ru system, ruthenium may function as a reservoir for atomic hydrogen, which is accessible via spillover to neighboring copper. Kinetically controlled spillover of hydrogen from ruthenium to copper (5) is consistent with the observed optimum reaction rate at an intermediate copper coverage.

The unique chemical behavior seen in CO adsorption and for certain catalytic reactions is mirrored in unique physical and electronic properties. For example, the adsorption and growth of copper films on the Ru(0001) surface have been studied (3,6,34-41) by work function measurements, LEED, AES, and TPD. The results from recent studies (3,6,8) indicate that for submonolayer depositions at 100 K the copper grows in a highly dispersed mode, subsequently forming two-dimensional islands pseudomorphic to the Ru(0001) substrate upon annealing to 300 K. Pseudomorphic growth of the copper during the first monolayer indicates that the copper-copper bond distances are strained by almost 6% beyond the equilibrium bond distances found for bulk copper. Thermal annealing to 600 K of copper films at coverages in excess of 2 ML results in the agglomeration of copper into three-dimensional islands. The particles formed expose primarily Cu(111) surfaces and partially uncover the underlying ruthenium surface. This is the origin of the residual activity of the copper films at coverages greater than 1 ML.

In terms of the electronic properties, recent angle-resolved photoemission (ARUPS) studies (8) also reveal unique structure, an interface state that is related to the altered bonding of copper films intimate to ruthenium. The ARUPS data, shown in Figure 3, are using HeI photon radiation (21.2 eV) at normal incidence and at an electron emission angle corresponding to the excitation of electrons from states of a particular symmetry character with respect to the crystal structure. The spectra correspond to the energy distribution of photoelectrons, and are shown as a function of the coverage of copper overlayer. The zero of electron binding energy is the

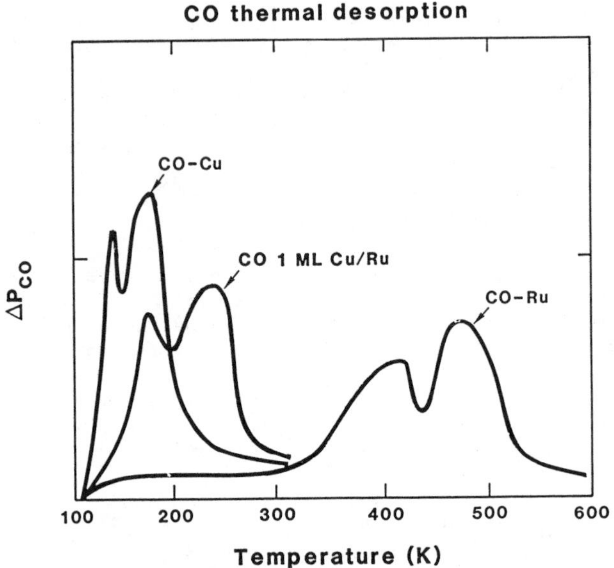

Figure 1. TPD results for CO adsorbed to saturation levels on clean Ru(0001), on multilayer Cu, and on a 1 ML Cu covered Ru(0001). (Data from ref. 6.) (Reprinted with permission from ref. 42. Copyright 1986 Annual Reviews, Inc.)

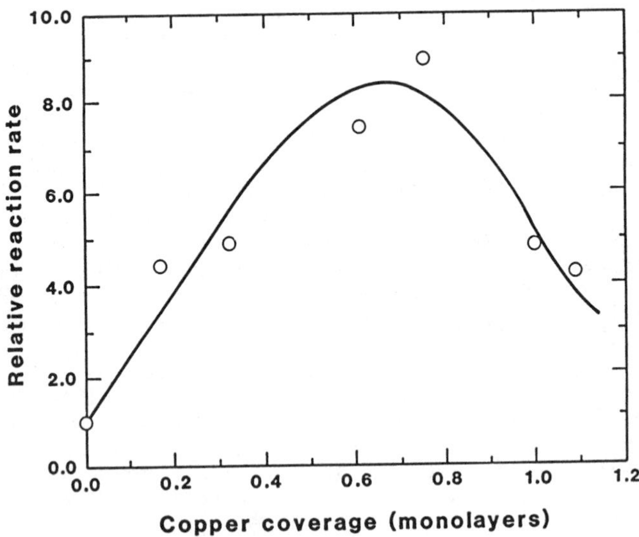

Figure 2. Relative rate of reaction vs. surface Cu coverage on Ru(0001) for cyclohexane dehydrogenation to benzene. P_T = 101 Torr. H_2/cyclohexane = 100. T = 650 K. (Data from ref. 10.) (Reprinted with permission from ref. 42. Copyright 1986 Annual Reviews, Inc.)

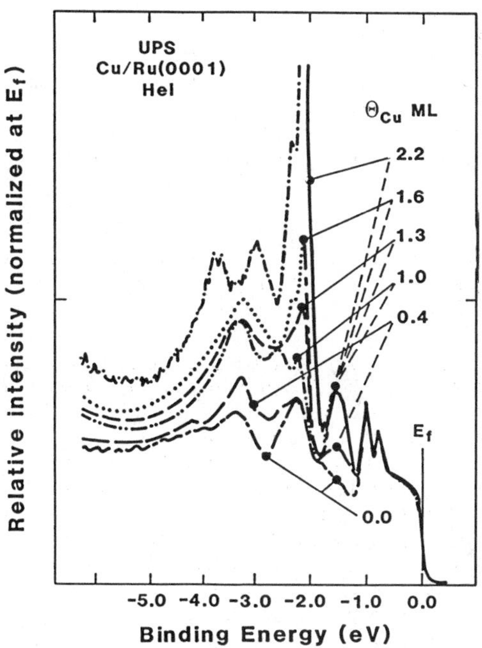

Figure 3. ARUPS energy distribution curves taken with HeI radiation at normal incidence and an electron emission angle of 52° shown as a function of copper coverage. The intensity of the various curves has been normalized at the Fermi level E_f. The individual curves are matched to their corresponding copper coverages in monolayers by the solid lines and the saturation behavior of the interface state at approximately −1.5 eV is identified by the dashed lines. (Data from ref. 8.) (Reprinted with permission from ref. 43. Copyright 1987 American Association for the Advancement of Science.)

abrupt rise in photoemission intensity, at the Fermi level. Overlayer copper does not alter appreciably the ruthenium electronic structure at binding energies less than ~1 eV except for an intensity attenuation. Hence, the spectra can be normalized so that they have equal intensity at the Fermi level. The fingerprint of bulk copper is the narrow d-band peak just below 2 eV for copper coverages greater than 1 ML. Below this coverage the overlayer copper causes an increase in intensity in narrow regions centered at ~1.5 and ~3.6 eV. These states remain unchanged in energy and relative intensity for copper overlayer of a dozen monolayers. These two features exemplify the behavior of interface states, which are states that are not seen for either component of a bimetallic surface alloy but which exist because of the abrupt change in electronic properties at the interface.

The interface nature of the states shown in Figure 3 have been corroborated in recent theoretical calculations (8). The calculated energy position of these states is in close agreement with the experimental results. Their orbital character has been determined as well. These results indicate that the interface states at about 1.5 eV are of appropriate character and energy to be active in environmental chemical reactions involving sub- and monolayer copper films on ruthenium.

Vacuum Desorption and Electrochemical Properties Figures 4a-4d shows a series of TPD spectra of Cu from Ru(0001) (16) corresponding to three regions of Cu coverage----low, medium, and high. The TPD traces were independent of the deposition parameters such as evaporation rate or temperature of the anneal subsequent to evaporation. In the low coverage region (Figure 4a) the initial buildup of Cu can be seen. The initial stage of Cu growth is indicated by the appearance of an approximately zero order desorption peak and, following the notation of Christmann and co-workers (34), is noted as β_2. This state reflects Cu coverages up to approximately one monolayer (one Cu atom per surface Ru atom). The saturated β_2 state has a desorption maximum at ~1210 K.

Higher Cu exposures (Figures 4b, 4c, and 4d) cause the appearance of a second binding state, β_1, with a desorption maximum at a temperature below that of the β_2 state. The kinetics of the desorption process of the β_1 state are approximately zero order, indicating that the rate of the desorption is independent of the Cu concentration on the surface. The general adsorption behavior and peak temperatures in Figure 4 are completely in agreement with the work of Christmann, et al. (34).

The evolution of the Cu desorption first entails filling of the β_2 state followed by filling of the β_1 state. As the β_1 state grows, the β_2 state remains essentially unchanged. This is consistent with the assignment of the β_2 state to Cu-Ru interactions and the β_1 state to three-dimensional Cu-Cu interactions (34). The ordering of these TPD features imply Cu-Ru interactions which are ~6 kcal/mole more stable than Cu-Cu interactions (16). The growth mechanism of the Cu overlayer in either case of preparation is suggested by the TPD results which are consistent with a Frank-van der Merwe or layer-by-layer mechanism. Furthermore, the filling of the β_2 TPD peak provides an accurate calibration point for a single

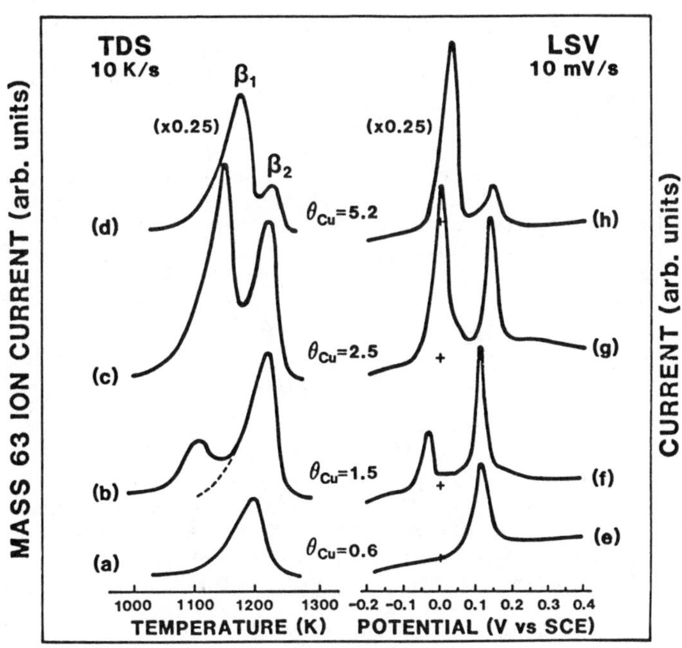

Figure 4. Left, copper TPD from a Ru(0001) surface at copper coverages corresponding to a, 0.6 ML, b, 1.5 ML, c, 2.5 ML, and d, 5.2 ML. (Data from ref. 3.) Right, electrochemical stripping curves as a function of Cu coverage (determined by AES) for vacuum deposited Cu on Ru(0001). (Data from ref. 16.)

monolayer of Cu which, assuming purely epitaxial growth, is 1.59 X 10^{15} Cu atom/cm^2 (35).

A Ru(0001) sample, with vacuum deposited Cu, has been characterized by cyclic voltammetry by transferring to an electrochemical cell (16). Figures 4e-4h shows the anodic stripping curves for four different coverages of Cu. A single stripping peak was observed at +110 mV for 0.6 ML Cu and shifted to +145 mV for 5.2 ML Cu. This peak represents the removal of the first monolayer of Cu or Cu in direct contact with the Ru surface. The curve for 5.2 ML Cu shows an additional peak at -20 mV for the stripping of multilayer Cu. The splitting between the multilayer and monolayer stripping peaks of 140 mV is equivalent to an absolute energy difference of 6.4 kcal/mole, essentially the same value found for the difference in apparent activation energy for monolayer versus multilayer Cu removal by TPD techniques described above. Auger spectra recorded after electrochemical stripping and removal of the electrode at +400 mV showed no signs of Cu remaining on the surface.

There is a good correlation between determination of the Cu coverage by UHV (AES or TPD) and electrochemical methods (16). Figure 5 compares the Cu coverage measured by AES and CV. The CV data were calculated from the integrated charge under the monolayer and multilayer stripping peaks assuming that the stripping reaction was

$$Cu_{ads} \text{ ----> } Cu^{2+}_{ads} + 2\ e^-$$

The error in the CV determinations was estimated to be ±25%. That the line does not go through the origin and its slope deviates from unity is attributed to ambiguities in determining the proper baseline for integrating the charge under the CV peaks.

A typical Auger spectrum of the sample following electrodepostion of Cu is shown in Figure 6 (16). For this experiment, the sample was immersed in a solution of 0.2 M $HClO_4$ and 0.96 mM Cu^{2+} at +400 mV, cycled to -100 mV, and then back to +40 mV at a sweep rate of 20 mV s^{-1}. The sample was held at +40 mV for 2 min and emersed under potential control. It was not rinsed before transfer to vacuum; the small droplet which remained on the electrode was allowed to evaporate in vacuum. Under these conditions, one would expect that the Cu coverage was approximately one monolayer. This is approximately the case as the Cu/Ru ratio in Figure 6 is 0.048 corresponding to about one monolayer of Cu (see Figure 5).

Conclusions

These studies have shown that:

1. Cu deposited onto Ru(0001) at 100K grows in 2-d islands via a Frank-van der Merwe (layer by layer growth) mechanism up to 2 ML. The island sizes but not the basic growth mode are altered by a post-deposition anneal at 900 K.
2. The structure of the first monolayer of Cu is pseudomorphic with respect to the Ru(0001) substrate whereas successive Cu layers grow epitaxially with a Cu(111) structure.

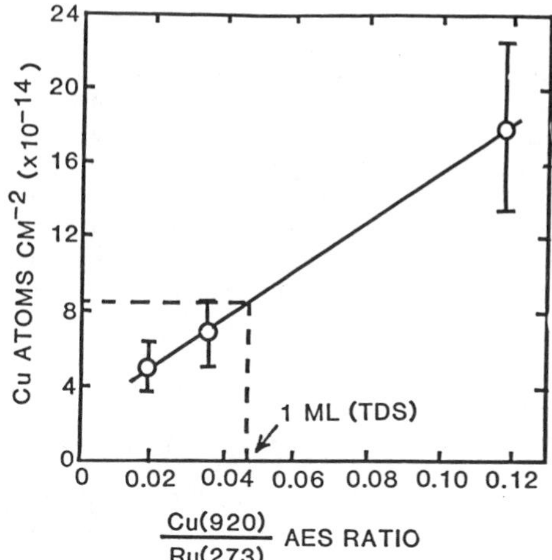

Figure 5. Correlation of electrochemical vs. Auger determination of Cu coverage. The electrochemical measurements were taken from the area under the Cu stripping peaks. (Data from ref. 16.)

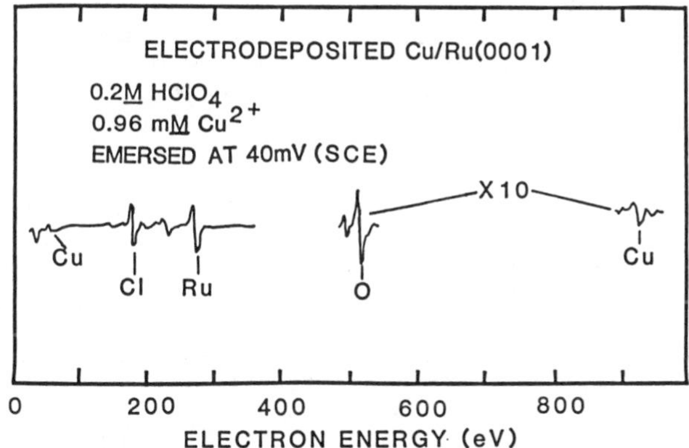

Figure 6. Auger spectrum after electrodeposition of Cu on Ru(0001) from an electrolyte of 0.2 M $HClO_4$ and 0.96 mM Cu^{2+}. The sample was emersed without rinsing at 40 mV (SCE). (Data from ref. 16.)

3. The adsorptive properties of 1 ML of Cu on Ru(0001) toward CO is markedly different than either corresponding to CO on Ru(0001) or CO on bulk Cu(111).
4. The addition of Cu to Ru(0001) results in a dramatic enhancement of the rate of cyclohexane dehydrogenation, despite the fact that Cu is much less active for this reaction than is Ru.
5. ARUPS results have identified unique electronic interface states for the Cu/Ru(0001) system. These states are not present in either metal separately but exist because of the abrupt change in properties at the interface.
6. The difference in desorption energy between multilayer and monolayer Cu is essentially the same for Cu desorption from Ru in vacuum or by electrochemical stripping.

Acknowledgment

We acknowledge with pleasure the partial support of this work by the Department of Energy, Office of Basic Energy Sciences, Division of Chemical Sciences. This work, performed at Sandia National Laboratories, was supported by U.S. Department of Energy Contract No. DE-AC04-76DP00789.

References

1. Imelik, B.; Naccache, C.; Coudurier, G.; Praliaud H.; Meriaudeau, P.; Gallezot, P.; Martin, G. A.; Vedrine, J. C., Eds. Metal-Support and Metal-Additive Effects in Catalysts; Elsevier, 1982.
2. Sinfelt, J. H. Bimetallic Catalysts: Discoveries, Concepts, and Applications; John Wiley & Sons: New York, 1983.
3. Yates, J. T., Jr.; Peden, C. H. F.; Goodman, D. W. J. Catal. 1985, 94, 576.
4. Goodman, D. W.; Yates, J. T., Jr.; Peden, C. H. F. Surf. Sci. 1985 164, 417.
5. Goodman, D. W.; Peden, C. H. F. J. Catal. 1985, 95, 321.
6. Houston, J. E.; Peden, C. H. F.; Blair, D. S.; Goodman, D. W. Surf. Sci. 1986, 167, 427.
7. Peden, C. H. F.; Goodman, D. W. I&EC Fundamentals, 1986, 25, 58.
8. Houston, J. E.; Peden, C. H. F.; Feibelman, P. J.; Hamann, D. R. Phys. Rev. Lett. 1986, 56, 375.
9. Peden, C. H. F.; Goodman, D. W. J. Catal. 1987, 104, 347.
10. Peden, C. H. F.; Goodman, D. W. J. Catal. 1986, 100, 520.
11. Berlowitz, P. J.; Goodman, D. W. Surf. Sci. 1987, 187, 463.
12. Goodman, D. W.; Peden, C. H. F. J. Chem. Soc. Faraday Transactions I, 1987 83, 1967.
13. Goodman, D. W.; Kelley, R. D.; Madey, T. E.; Yates, J. T., Jr. J. Catal. 1980, 63, 226.
14. Goodman, D. W. J. Vac. Sci. Tech. 1982, 20, 522.
15. Goodman, D. W. Accts. Chem. Res. 1984, 17, 194.
16. Stuve, E. M.; Ingersoll, D.; Rogers, J. W., Jr.; Goodman, D. W. Chem. Phys. Letts. 1988.
17. Schwab, G. M. Disc. Faraday Soc. 1950, 8, 166.
18. Dowden, D. A. J. Chem. Soc. 1950, 242.
19. Hall, W. K.; Emmett, P. H. J. Phys. Chem. 1959, 63, 1102.

20. Sachtler, W. M. H.; Dorgelo, G. J. H. J. Catal. 1965, 4, 654.
21. Ertl, G; Kuppers, J. J. Vac. Sci. Technol. 1972, 9, 829.
22. Sinfelt, J.; Carter, J. L.; Yates, D. J. C. J. Catal. 1972, 24 283.
23. Williams, F. L.; Boudart, M. J. Catal. 1973, 30, 438.
24. Stephan, J. J.; Ponec, V.; Sachtler, W. M. H. Surf. Sci. 1975, 47, 403.
25. Helms, C. R.; Yu, K. Y.; Spicer, W. E. Surf. Sci. 1975, 52, 217.
26. Burton, J. J.; Helms, C. R.; Polizzotti, R. S. J. Chem. Phys. 1976, 65, 1089.
27. Silverman, E. M.; Madix, R. J. J. Catal. 1979, 56, 349.
28. Ponec, V. Surf. Sci. 1979, 80, 352.
29. Bond, G. C.; Turnham, B. D. J. Catal. 1976, 45, 128.
30. Betizeau, C.; Leclercq, G.; Maural, R.; Bolivar, C.; Charcosset H.; Frety, R.; Tournayan, L. J. Catal. 1976, 45, 179.
31. Galvagno, S.; Parravano, G. J. Catal. 1979, 57, 272.
32. De Jongste, H. C.; Ponec, V.; Gault, F. G. J. Catal. 1980, 63, 395.
33. Balooch, M.; Cardillo, M. J.; Miller, D. R.; Stickney, R. E. Surf. Sci. 1975, 50, 263.
34. Christmann, K.; Ertl, G.; Shimizu, H. J. Catal. 1980, 61, 397.
35. Shimizu, H.; Christmann, K.; Ertl, G. J. Catal. 1980, 61, 412.
36. Vickerman, J. C.; Christmann, K.; Ertl, G. J. Catal. 1981, 71, 175.
37. Shi, S. K.; Lee, H. I.; White, J. M. Surf. Sci. 1981, 102, 56.
38. Richter, L.; Bader, S. D.; Brodsky, M. B. J. Vac. Sci. Techn. 1981, 18, 578.
39. Vickerman, J. C.; Christmann, K. Surf. Sci. 1982, 120, 1.
40. Vickerman, J. C.; Christmann, K.; Ertl, G.; Heiman, P.; Himpsel, F. J.; Eastman, D. E. Surf. Sci. 1983, 134, 367.
41. Bader, S. D.; Richter, L. J. Vac. Sci. Technol. 1983, A1, 1185.
42. Goodman, D. W. Annu. Rev. Phys. Chem. 1986, 37, 425.
43. Goodman, D. W.; Houston, J. E. Science 1987, 236, 403.

RECEIVED August 26, 1988

Chapter 12

Implications of Double-Layer Emersion

Wilford N. Hansen and Galen J. Hansen

Physics Department, UMC–4415, Utah State University, Logan, UT 84322

The intimate relationship between double layer emersion and parameters fundamental to electrochemical interfaces is shown. The surface dipole layer (χ_S) of 80% sat. KCl electrolyte is measured as the difference in outer potentials of an emersed oxide-coated Au electrode and the electrolyte. The value of +0.050 V compares favorably with previous determinations of χ_S. Emersion of Au is discussed in terms of UHV work function measurements and the relationship between emersed electrodes and absolute half-cell potentials. Results show that either the accepted work function value of Hg in N_2 is off by 0.4 eV, or the dipole contribution to the double layer (perhaps the "jellium" surface dipole layer of noble metal electrodes) changes by 0.4 V between solution and UHV.

Emersion of an electrode from electrolyte with its double layer intact is now a widely accepted phenomenon and technique. Not only is it a phenomenon which deserves careful consideration and study, but also a process which opens up a new set of experimental methods to the study of the electrochemical double layer. Electrode emersion involves the careful removal of an electrode from electrolyte under potentiostatic control, usually hydrophobically (1-5). When fairly concentrated electrolyte parts ("unzips") from the electrode surface during hydrophobic emersion, the double layer remains essentially intact on the electrode surface and no electrolyte outside the double layer remains. This phenomenon is not due to the presence of organics or other impurities as some have suggested. The emersion process works well with rigorously clean electrode surfaces (5).

Emersion involves fundamental aspects of condensed matter surface science and electrochemistry, and its consideration offers new insight into these fundamentals. For example, when a new solid or liquid surface is made the atoms or molecules may rearrange at the surface to form a surface dipole layer. This certainly happens

0097–6156/88/0378–0166$06.00/0

for aqueous electrolyte. Yet as an electrode plus double layer "unzip" from the solution during emersion, the double layer species (including surface water) are held tightly on the electrode surface and probably cannot rearrange to form a new outer dipole layer. Therefore the outer potential of the emersed electrode could well be the same as the inner potential of the electrolyte. This would represent a major breakthrough in measurement capability and in our understanding of the double layer. Even if this situation is only approximately true (to say ±20 mV) for special known cases, it is still very significant. Consider the following points. (a) Classical double layer structure theories could be tested in new ways. Present implications are that theories concerning the role of water may need to be altered. (b) Surface dipole layers of various electrolytes (χ_S potentials) could be directly measured. To date these have been considered unmeasurable (6-7). (c) Absolute half-cell potentials could be determined more directly, and either the electrolyte inner potential ϕ_S or outer potential ψ_S could be used as the reference state, with the difference clearly known. (d) If, in addition, there is no dipole change as the electrode plus double layer is placed in UHV, or if that change is a measurable constant, the absolute half-cell potential can be directly measured from photoelectron spectroscopy data without reference to other data such as the work function of mercury. (e) Needed reference states for liquid ESCA (8) could be easily realized.

Emersion into Ambient Gas

As with other subtle phenomena, such as the quantum mechanical behavior of electrons in metals, we seek a self-consistent picture which will permit a descriptive explanation of the observations. In Figure 1 are shown the outer potential of an electrode (ψ_M) emersed from electrolyte and the outer potential of the electrolyte itself (ψ_S) as a function of electrode potential, both measured with respect to the reference electrode Fermi level and with the same Kelvin probe. The electrode in this case is gold coated with a layer of TiO_2 that is much thicker than the gold film itself and determines the electrochemical behavior. This is one of the most ideally polarizable electrode we have found even though its capacity is large, about ten percent that of gold. It is also very hydrophobic. The electrode is rotating so that it continuously emerges hydrophobically from an 80% saturated KCl solution. The probe was held 1-2 mm from the surface of the rotating electrode or the electrolyte, and the probe reading outside the electrode was taken about two seconds after emersing from the electrolyte. This time lag causes the open presentation (hysteresis) of the upper curve, and the true reading is through the center of the open curve. Measuring the outer potential of the rotating emersing electrode and the electrolyte is discussed in more detail elsewhere (9-13). The 50 mV difference between ψ_M and ψ_S is probably due mainly to the surface dipole of the solution, i.e. $\chi_S = \psi_M - \psi_S$ if ψ_M is equal to the solution inner potential ϕ_S. In any case there is remarkable tracking over the potential range for which the double layer is stable to discharge.

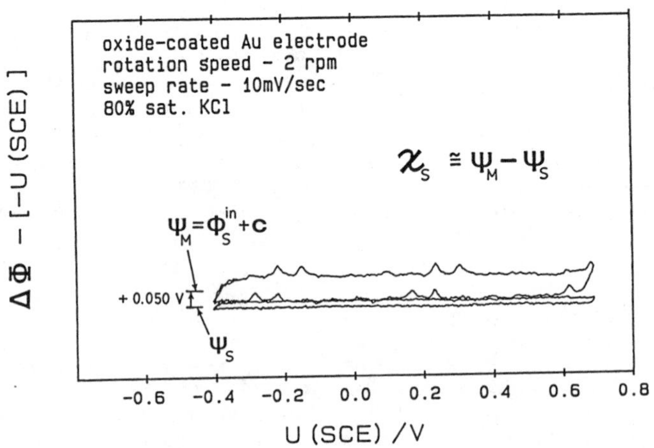

Figure 1. The upper curve is a cyclic emersogram of a rotating oxide-coated Au electrode (outer potential ψ_M vs electrode potential U), measured with a Kelvin probe. (Here ψ_M is plotted as ψ_M minus the ref. electrode Fermi level.) The lower curve is the electrolyte cyclic potentiogram (ψ_S - (-U) vs U), by the same Kelvin probe. Data indicate that $\psi_M - \psi_S \cong 50$ mV at all times.

While the above measurement was done with care, and repeated several times, it depends on the electrolyte surface being clean. We have found that with care the solution can be kept clean so that its χ_S remains constant to within the precision of our instrumentation (±5 mV). The generality of the phenomenon should be tested by repeating the experiment with more electrodes which emerse hydrophobically and give stable emersed double layers. The experiment can also be repeated with various electrolytes. But this involves the question of salt bridge potentials if the SCE with its salt bridge is to be used with various solutions. We have made measurements with a Kelvin probe over solutions at various concentrations of the common salts NaCl, NaF, LiF, KCl, and Na_2SO_4. In all cases reference was to the lead wire of an SCE which contacted the solution via a capillary. All solutions gave a common ψ_S within 10 mV. In the case of KCl there was no change in ψ_S (within 10 mV) for solutions from 10^{-3} M up to 80% sat. Perchloric acid up to one molar was also tested and gave larger deviations, due to either a change in χ_S or more likely a salt bridge potential drop. The deviation was still less than 50 mV. The salt data indicate that χ_S values for all aqueous salt solutions differ by less than 10 mV. The data also indicate that salt bridge potentials using a saturated KCl bridge are less than ten millivolts.

As seen in Figure 1, ψ_M tracks ϕ_S and might be approximately equal to it. If equal, then χ_S = +50 mV. On the other hand, if we knew χ_S from independent data it would tell how close ψ_M is to ϕ_S. We now discuss the data from this point of view. A good argument can be made for taking a value of χ_S = 50 mV for concentrated aqueous KCl. This is obtained by a rationalization of results of Randles (11), Gomer and Tryson (12), and Farrell and McTigue (13). We take the carefully determined value of $\chi(H_2O)$ = 25 ± 10 mV and add another 25 mV because of the ionic strength of concentrated KCl. To the extent these data are correct, the double layer of Figure 1 does indeed emerse such that $\psi_M \cong \phi_S$ over the whole emersion potential range!

Emersion into UHV

Now consider the subject from the point of view of emersion into ultra-high vacuum (UHV). This process makes possible a whole new set of measurements, including photoelectron work function determination at room temperature, identification of much-less-than-monolayer amounts of atomic and molecular species (including water), and the ability to perform experiments with super cleanliness throughout (1,14-19). All this is beautifully illustrated in the work of Koetz and coworkers (19). We have taken some of their data and replotted it for our own purposes as shown in Figure 2. The work function of a gold electrode emersed at various potentials is shown. The straight line through the points is placed by us after a least squares fit. The slope is 1.00 ± 0.02. The root mean square deviation of the points from the line is 0.07 eV. Within this apparent accuracy the work function tracks the potential of emersion exactly, even into the oxide formation region! A systematic constant error is highly unlikely as measurements of photoelectron

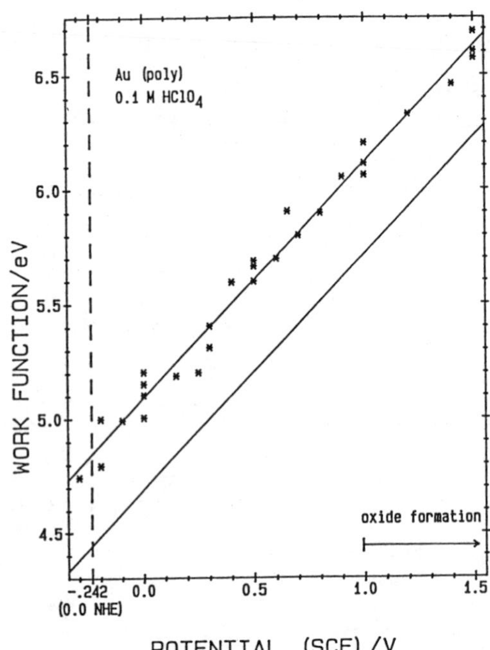

Figure 2. Work function of polycrystalline Au electrode emersed from 0.1 M $HClO_4$ as a function of emersion potential. The work function of the clean metal surface was 5.2 eV (19). If the NHE absolute half-cell potential (with respect to ϕ_S) is 4.45 V, the bottom line is equal to the solution inner potential, ϕ_S. If it is 4.85 V, the upper curve through the points is equal to ϕ_S. As always with WF, the Fermi level is taken as zero.

onsets and the Fermi level (used for WF determination) in photoelectron spectroscopy are easily determined to an accuracy within the observed point scatter. Incidently, this shift in the electrode work function with emersion potential is tied directly to the "electrochemical shift" of the XPS core level spectra of emersed double layer species shown in the earlier pioneering work by D'Agostino and others in this lab (15).

These results are remarkable! Coupled with other results for silver and platinum (19) they show that the emersed electrode work function can be independent of electrode material (even oxide coated) and electrolyte. The ψ_M tracks ϕ_S one-to-one over a large potential region, even after placement in UHV. The apparatus used allowed for emersion and placement in UHV without exposure to air at any time.

The data of Figure 2 imply that either (a) the surface dipole layer of the gold electrode shifts by a systematic 0.4 volts during emersion and placement in UHV, or (b) the accepted work function of mercury in dry nitrogen (Φ(Hg) = 4.50 eV) is off by 0.4 eV. Let us explain. First, this scenario is based on an NHE absolute half-cell potential of 4.45 V. This is equivalent to the value given by Trasatti and officially recommended by I.U.P.A.C. (20). It is also the value implied by the data of McTigue et.al. (13), and is the value obtained by this laboratory (21). However, the value is based on the work function of pure Hg in N_2 of 4.50 eV. Some theoretical treatments also get this value for the NHE without reference to the WF of Hg (22). The lower line in Figure 2 represents the work function of the emersed electrode, Φ_M, which would be measured if the outer potential of the electrode remained equal to the outer potential of the solution (assuming χ_S is negligible), and if the value of 4.45 V for the NHE absolute half-cell potential is correct so that Φ_M = 4.45 when the Fermi level of the electrode equals that of the NHE (-0.242 V vs SCE). It is quite clear that in vacuum, the measured work function has a systematic shift from this line of +0.4 eV. Either the dipole surface potential is shifted 0.4 V or the absolute NHE half-cell potential used here is wrong. If the latter is true, it is because Φ(Hg) is off by 0.4 eV.

Either explanation of the 0.4 eV systematic shift has profound implications. If the accepted Φ(Hg) is off by the 0.4 eV, then many results based on Φ(Hg) are also in error. This would also imply that drastic rethinking on the problem of work functions in gas ambient is in order. Also, the absolute NHE half-cell potential would actually be 4.85 V. Perhaps even more significantly, it would mean that the in situ double layer remains completely intact on the electrode surface without modification during or after emersion and placement in UHV. On the other hand, if the surface dipole layer of the emersed electrode shifts by the 0.4 V during or after emersion, it must require some special mechanism. The mechanism must be somewhat insensitive to metal electrode type (since the data in Ref. 19 for Au, Ag, and Pt all fall on nearly the same line) and to some electro-oxidation. Notice that the 0.4 eV shift occurs at all potentials on the curve, including the pzc (pzc $\leq$ 0.1 V vs SCE). Therefore the change is not due to changes in ionic charge or to double layer collapse, since at the pzc there is no double layer in the usual sense. (The shift due to any change in the ionic double layer will increase away from the pzc, rather than remaining constant like that in Figure 2). There is also a problem if we consider the 0.4 eV shift to be due to removal of oriented water dipoles. One would think that the orientation and hence the potential drop due to the water layer would depend on electrode potential. On the other hand much of the double layer water may lie between discrete ions where there is very little field. But it is difficult to imagine a constant 0.4 V dipole layer of water which is independent of potential, noble metal, and surface oxidation. Perhaps a better guess is that the "jellium" dipole layer at the noble metal surface (χ_M) decreases by 0.4 V when place in aqueous electrolyte, and then increases again when its dielectric neighborhood is replaced by vacuum. This would be independent of potential, and perhaps even of some surface oxidation. The reason for such a "jellium" shift may be simply the changing of the dielectric constant of the phase in contact with the metal surface. Using this "jellium" concept, the emersion of an TiO_2-coated Au electrode is fundamentally different from that of bare Au in that the oxide has no "jellium" dipole surface layer and is not as sensitive to its neighboring phase. From this point of view, our concept that $\psi_M = \phi_S$ and that χ_S can be measured using the emersing TiO_2-coated Au electrode should be no surprise.

The results of Figure 2 (and other results (16,19,23) require rethinking as to the role of water in the double layer. It is still not known how much water is present in metal double layers emersed into gas ambient. But we have found in previous work that the amount of water in the emersed double layer on Au does not change with potential (23), and Koetz and Neff found that water was essentially absent in UHV (16) at all measured potentials. Results presented here do not decide the question of how much water is in the double layer in UHV, but they may severely restrict its role. Again, if water dipoles play an important role in double layer formation it is hard to see how their removal could cause a shift in work function which is constant with emersion potential.

Conclusions

Double layer emersion continues to allow new ways of studying the electrochemical interphase. In some cases at least, the outer potential of the emersed electrode is nearly equal to the inner potential of the electrolyte. There is an intimate relation between the work function of emersed electrodes and absolute half-cell potentials. Emersion into UHV offers special insight into the emersion process and into double layer structure, partly because absolute work functions can be determined and are found to track the emersion potential with at most a constant shift. The data clearly call for answers to questions involving the most basic aspects of double layer theory, such as the role water plays in the structure and the change in χ_M of the electrode surface as the electrode goes from vacuum or air to solution.

Acknowledgments

This work was supported by the National Science Foundation under grant number CHE-8508727.

We wish to acknowledge Kendall B. Johnson for his help in the laboratory.

Literature Cited

1. Hansen, W.N.; Wang, C.L.; Humpherys, T.W. J. Electroanal. Chem. 1978, 90, 137.
2. Hansen, W.N.; Kolb, D.M. J. Electroanal. Chem. 1979, 100, 493.
3. Kolb, D.M.; Hansen, W.N. Surf. Sci. 1979, 79, 205.
4. Hansen, W.N. J. Electroanal. Chem. 1983, 150, 133.
5. Kolb, D.M. Zeit. Phys. Chem. Neue Folge 1987, 154, 179.
6. Guggenheim, E.A. J. Phys. Chem. 1929, 33, 842.
7. Goodisman, J.G. Electrochemistry: Theoretical Foundations; John Wiley and Sons: New York, 1987; p. 10.
8. Siegbahn, H. J. Phys. Chem. 1985, 89, 897.
9. Hansen, G.J.; Hansen, W.N. J. Electroanal. Chem. 1983, 150, 193.
10. Hansen, G.J, Ph.D. Dissertation, Utah State University, Utah, 1986.
11. Randles, J.E.B. Phys. Chem. Liq. 1977, 7, 107.
12. Gomer, R.; Tryson, G. J. Chem. Phys. 1977, 66, 4413.
13. Farrell, J.R.; McTigue, P. J. Electroanal. Chem. 1982, 139, 37.
14. Hansen, W.N.; Kolb, D.M.; Rath, D.; Willie, R. J. Electroanal. Chem. 1980, 110, 369.
15. D'Agostino, A.T.; Hansen, W.N. Surf. Sci. 1986, 165, 268.
16. Neff, H.; Koetz, R. J. Electroanal. Chem. 1983, 151, 305.
17. Peuckert, M.; Coenen, F.P.; Bonzel, H.P. Surf. Sci. 1984, 141, 515.
18. Haupt, S.; Collisi, U.; Speckmann, H.D.; Strehblow, H.-H. J. Electroanal. Chem. 1985, 194, 179.
19. Koetz, E.R.; Neff, H.; Mueller, K. J. Electroanal. Chem. 1986, 215, 331.
20. Trasatti, S Pure Appl. Chem. 1986, 58, 955.
21. Hansen, W.N.; Hansen, G.J. Phys. Rev. A 1987, 36, 1396.
22. Reiss, H. J. Phys. Chem. 1985, 89, 4207.
23. Hansen, G.J.; Hansen, W.N. Ber. Bunsenges. Phys. Chem. 1987, 317, 1987.

RECEIVED May 17, 1988

Chapter 13

Applications of Scanning Tunneling Microscopy to Electrochemistry

Moris M. Dovek[1], Michael J. Heben[2], Nathan S. Lewis[2,3], Reginald M. Penner[2], and Calvin F. Quate[1,3]

[1]Department of Applied Physics, Stanford University, Stanford, CA 94305

[2]Department of Chemistry, Stanford University, Stanford, CA 94305

The Scanning Tunneling Microscope has demonstrated unique capabilities for the examination of electrode topography, the vibrational spectroscopic imaging of surface adsorbed species, and the high resolution electrochemical modification of conductive surfaces. Here we discuss recent progress in electrochemical STM. Included are a comparison of STM with other *ex situ* and *in situ* surface analytic techniques, a discussion of relevant STM design considerations, and a semi-quantitative examination of faradaic current contributions for STM at solution-covered surfaces. Applications of STM to the *ex situ* and *in situ* study of electrode surfaces are presented.

Since its introduction by Binnig *et al.* in 1982 (1,2), the scanning tunneling microscope has proven to be a powerful and unique tool in the study of surfaces. The technique is conceptually simple but technologically demanding: a conducting filament of 1-10 Å width at its point is placed within several angstroms of a conducting surface, and the tunneling current is monitored while the tip is rastered across the sample surface. Surface structure has been obtained with atomic resolution in a variety of ambients, including ultra-high vacuum (3-8), cryogenic fluids (9-10), atmospheric pressure of air (11-14), and liquid solutions (*vide infra*). Additionally, electronic information about surfaces (15-18), and vibrational information about surface adsorbates (10) has been obtained on atomic dimensions. Several review articles that describe in depth the accomplishments and promise of STM have been published (19-23). Clearly, these advances in surface characterization have the potential to impact electrochemistry in diverse areas, including small scale lithography, *in situ* characterization of electrodes and of double layer structure, spectroscopy of adsorbed intermediates, and electronic properties of electrode materials.

[3]Address correspondence to these authors.

0097–6156/88/0378–0174$08.00/0

To date, most STM studies have focused on demonstrating the breadth of the possible applications; consequently, there are relatively few studies of a single electrode/electrolyte interface as compared to, for example, surface enhanced Raman spectroscopy on the Ag/pyridine system (24-26). Application of STM to electrochemistry will require modification of current microscope designs, and will present challenges in the integration of electrochemical potential control as well as in techniques for data collection and workup. In this paper, we attempt to review the progress made in this relatively young field, and attempt to develop some of the theory and applications germane to use of the STM in an electrochemical environment. We first compare STM with other *ex situ* and *in situ* electrode surface characterization techniques. We then discuss modifications in microscope design that are necessary for use in electrochemical cells, and proceed to develop a semiquantitative discussion of the theory of STM imaging in an electrochemical environment. We then will review the uses of STM as an *ex situ* tool for the characterization of electrochemical surfaces, and finally will discuss experiments that have taken advantage of the STM for *in situ* characterization and modification of electrode interfaces.

Comparison of Scanning Tunneling Microscopy with Other Surface Characterization Methods

At present, the STM has been demonstrated to yield spectroscopic information regarding the spatial, vibrational, and electronic structure of surfaces. In all of these areas, the STM is complementary to other existing techniques. Also, and in many potential applications, STM possesses unique features and capabilities. These capabilities are discussed in the section below.

Structural Information. To a first approximation, the structural information from an STM image results from variations in the electronic wavefunctions of a surface due to the positions of the surface atoms (20,27,28). Structural information can be obtained by rastering across the surface of interest while employing negative feedback to maintain a constant tunneling current. By monitoring the movement of the tip relative to the surface an image can be obtained (2,29). If the time constant of the feedback loop is increased so that a constant height, rather than a constant current is maintained, then the spatial variation of the tunneling current will contain structural information (30). Manipulation of the tip on an angstrom level is accomplished by use of piezoelectric materials, and further details regarding this process can be found in the section on microscope design.

In STM, the localized nature of the tip-surface interaction results in an inherently sensitive probe of surface topography. Resolution in the direction perpendicular to the surface (the z-direction) can be as high as 0.1 Å (31), while resolution in the lateral direction can approach < 2 Å (20,32). The image frame area varies somewhat, depending upon the details of the microscope, but typically lies in the range of 1 μm. Due to the necessity of producing a tunneling current, STM experiments generally require a conducting or semiconducting surface, although the related atomic force microscope (AFM) can yield similar topographical information on insulating surfaces (33-35).

As an *ex situ* technique for structural information on surfaces, STM is an excellent complement to the standard electron and ion diffraction probes of surface order. The STM method can identify both short range order and long range periodicity, as well as disordered surface layers (*e.g.*, images of sorbic acid on Highly Ordered Pyrolitic Graphite (HOPG), *vida infra*). In contrast, LEED requires a coherence length of ordered domains on the order of at least 100 Å to be useful (36). SEM images are typically limited to resolution of 50 Å, but are useful due to their depth of field and wide field of view (37). High resolution TEM can yield images of solids to a resolution of 1-2 Å, but requires extensive sample preparation and also requires that the surface atoms be in registry with the remainder of the bulk sample under study (14,38). Clearly, each technique provides valuable information which complements the others, and ideally a combination of the probes would be used to obtain a complete characterization of the surface under study.

For *in situ* surface studies under modest pressures of ambient gases, or for *in situ* studies of surfaces in contact with liquids, the ion and electron diffraction techniques are not available. In these systems, STM becomes one of the few techniques capable of yielding surface topographic information. Under certain conditions, surface EXAFS and X-ray standing wave techniques can yield information concerning the periodicity and lattice properties of a surface layer (39-41), however, these techniques yield average values of the desired signal over macroscopically large (mm^2) beam areas (39-41). Ellipsometry can yield information concerning the morphology of surface layers provided that suitable models for the interface have been developed and confirmed by complementary techniques (42,43). Clearly, the overall lack of *in situ* structural probes of surfaces is a major driving force behind application of STM to electrochemical interfaces.

Vibrational and Electronic Spectroscopy. STM has recently been applied as a surface vibrational spectroscopic tool. To date, the applications have been for surfaces prepared under ambients, and examined subsequently *ex situ*. For example, Smith *et. al.* (10) have obtained vibrational information from adsorbed sorbic acid molecules on highly ordered pyrolytic graphite substrates. In this application, STM competes most closely with EELS (44), IETS (45), FTIR (46) and Raman spectroscopy (47). Each probe has particular advantages with respect to spectral range, resolution and the types of systems that may be studied. The selection rules that apply to vibrational spectroscopy with the STM have not yet been elucidated. If the interaction is dipolar in nature, then it might be expected that there must be some component of the vibrational mode's electric dipole moment which is parallel to the exciting tunneling beam in order to obtain a signal. A potential advantage of the STM technique is that molecular vibrational modes may be probed on an individual molecule-by-molecule basis (10). Although the limits of applicability of STM to vibrational spectroscopy have only recently begun to be explored, a distinguishing feature is the inherently small "effective beam size" of the technique and possible geometric information afforded by the vibrational spectra (10).

STM has also been shown to provide surface electronic information (1). The most common application to date in this area is use of STM to probe the density of states in metals and semiconductors. Materials of interest to electrochemists that have been investigated include Si (17,48), GaAs (16,49), graphite (18), Pd (15), and Au (15). Once again, STM's unique contribution in these applications arises from the small areas (ca. 2-4 $Å^2$) that are involved in the measurement process. These capabilities have enabled the mapping of the spatial distribution of electronic states. For example, cation and anion sites on GaAs surfaces have been discriminated (Stroscio, J.A.; Feenstra, R.M.; Newns, D.M.; Fein, A.P. J. Vac. Sci. Technol. A, in press)(3). Electronic structure measurements of occupied states are typically made with UPS, while unoccupied states are probed by IPS (49). EELS probes both filled and unfilled states simultaneously, and is therefore used in conjunction with either UPS or IPS to complete a band structure determination (44,49). A new electronic spectroscopy technique, Field Emission Scanning Auger Microscopy (50), utilizes STM-like technology to effect highly localized (c.a. 1 μm) Auger electron spectroscopy. The local electronic information afforded by STM is a valuable complement to these other techniques, and STM is the only one of these methods that may be applied to *in situ* investigations in condensed media.

Microscope Design for Electrochemical Applications

The critical aspects of STM design deal with the formidable task of separating detected spatial variations of electronic wavefunctions, and signals derived therein, from spurious mechanical, acoustical, and thermally derived noise (51). Stated in another way: the task is to isolate and control the vertical and lateral confines of an electron tunneling current. Important developments facilitating this goal include the introduction of magnetic levitation (1), eddy current damping (2,21), and spring supported staging (21,30) for vibration isolation. Simplicity of design was gained through the realization that stacked plates separated by lossy elastomers could achieve similar ends (52,53). High speed imaging techniques, employable on relatively smooth surfaces, were found to yield an improved signal-to-noise ratio (51,54) and, consequently, opened the real time imaging domain. The invention of the piezo tube scanner by Binnig and Smith (56) further reduced the complexity of microscopes and, since mechanical resonances of the tube occur at higher frequencies (*c.a.* 8 kHz), the tube scanner allows still higher tip speeds. Throughout these design evolutions the need for compactness and rigidity in the microscope body has been respected (55).

The design criteria for an *in situ* electrochemical STM include the above outlined considerations as well as several needs peculiar to an electrochemical environment. Sonnenfeld and Hansma (57) constructed the first STM to operate under solution. Their work highlights two important design considerations. Firstly, the tip and sample should be the only electrically active parts of the microscope exposed to solution. This first solution microscope was designed such that the piezoelectric scanning elements could remain above the solution level. In taking this design consideration further, we suggest that the tip and sample should be the only chemically reactive parts (*e.g.*, metal) of the microscope exposed to solution.

The importance of tip insulation is the second significant solution microscope design feature recognized by Sonnenfeld and Hansma (57). Except in ideally pure solutions faradaic currents will always be present between the tip and sample. If these currents are of the same magnitude as the tunneling currents, then feedback control will be difficult to maintain. Commercially available Pt-Ir tips, which are glass coated except for about 50 μm at the tip end, were employed in an effort to reduce faradaic currents. The most recent efforts by Sonnenfeld *et. al.* involved the use of Pt-Ir tips with all but 5 μm insulated from the electrolyte (58). SiO was then deposited on the uninsulated section of the tip by evaporation, and this dielectric was subsequently removed from the very end of the tip by approaching into tunneling range with a 10 V bias between the tip and sample (Schneir, J.; Hansma, P.K.; Elings, V.; Gurley, J.; Wickramasinghe, K.; Sonnenfeld, R. SPIE'88 Conference Proceedings, in press). Such tips have also been employed in a newer STM, shown in Figure 1, that employs a single tube scanner, and allows for fluid delivery and removal via a fluid transfer line. This approach, discussed in greater detail below, was successful in that Au could be plated from, and imaged with, these tips (58).

Other workers who have reported STM under solution include Itaya *et. al.* (59) and Fan *et. al.* (Fan, F-R.F.; Bard, A.J. Anal. Chem., submitted). Their instruments each employ three orthogonal piezoelectric elements as tip translators and both use glass tip insulation, though their tip preparation techniques differ. In the former case, a 10 μm Pt wire was sealed into a capillary and subsequently etched, while the latter workers used a 65 μm Pt wire which was sealed in soft glass, turned on a lathe, and then sonicated in concentrated H_2SO_4.

Morita and co-workers (60) have constructed a STM that employs a unique 3D scanner and 3D positioner that is constructed from several piezoelectric cubes. This microscope was subsequently equipped with an electrochemical cell that allows disconnection of the tip and conventional 3 electrode voltammetry to be performed (61). Itaya *et. al.* have gained similar capabilities by modifying their aforementioned STM (Itaya, K.; Higaki, K.; Sugawara, S. Chem. Lett., in press).

A noteworthy advance in the design of solution STMs was achieved by Lev *et. al.* (Lev, O.; Fan, F-R.F.; Bard, A.J. J. Electroanal. Chem., submitted) by including a Pt "flag" electrode in the STM of Fan *et. al.* (Fan, F-R.F.; Bard, A.J. Anal. Chem., submitted). A battery between the sample and this flag electrode, which remains poised at the rest potential of the solution, enables the sample to be biased away from the rest potential independently of the tip to sample bias.

We have also recently constructed a STM suitable for work under solution (see Fig. 2). The salient features of our design include the fact that only tip, sample, pyrex, and teflon are exposed to solution. An automatic approach mechanism allows for the

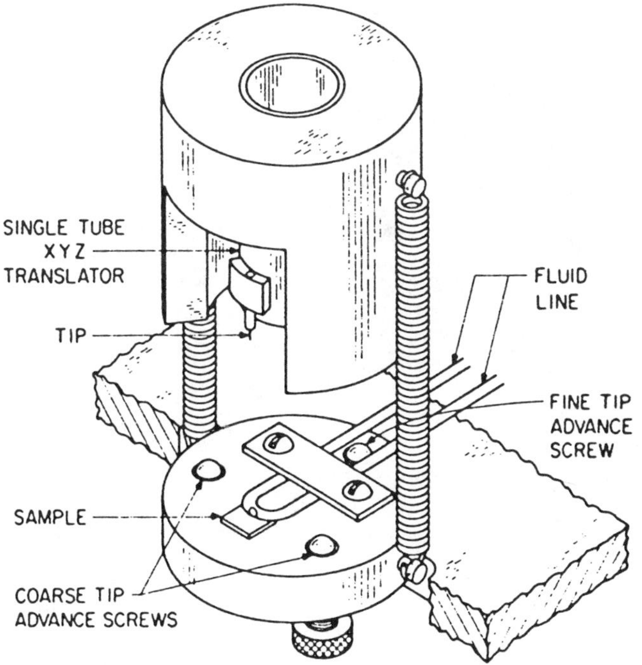

Figure 1. Single tube piezo, solution STM of Sonnenfeld and Hansma with fluid transfer line. Reproduced with permission of Ref. 58. Copyright 1987 American Institute of Physics.

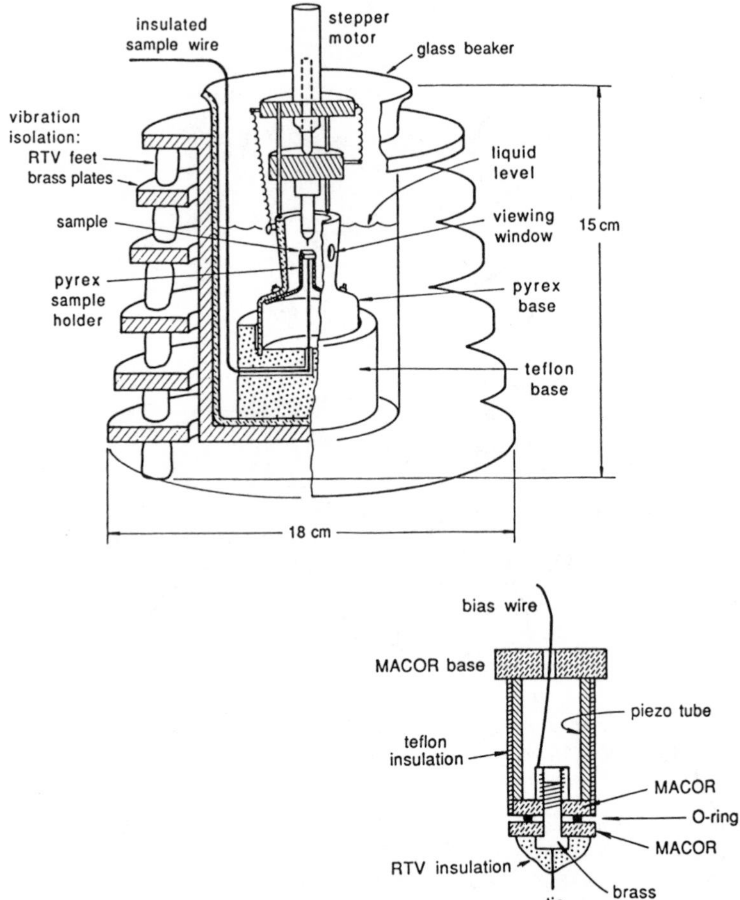

Figure 2. Solution STM design with single tube piezo, remote approach and fast retract capabilities, large solution capacity, and solution-insulated tip (tip design inset) (Reprinted with permission from ref. 98. Copyright 1988 American Institute of Physics.)

remote operation of the microscope in an N_2 drybox and facilitates the preparation of glass coated tips via a field emmision process that is similar to Sonnenfeld's. The microscope can be operated with a reference electrode, counter electrode, and bi-potentiostat to provide the above mentioned capabilities used by Lev and co-workers. A more complete discussion of this microscope design can be found elsewhere (Dovek, M.M.; Heben, M.J.; Lang, C.A.; Lewis, N.S.; Quate, C.F. Rev. Sci. Instr., submitted).

Currents in an Electrochemical STM Experiment

The key behind the imaging capabilities of the STM is the ability to control and monitor the tunneling current between the tip and the sample. To a first approximation, the tunneling current is related to the interelectrode spacing, d, by the expression (1,30):

$$i_t \propto \Delta E_t \exp(-2\kappa d) \quad [1]$$

where κ is the decay constant for the wave function in the tunneling barrier, and ΔE_t is the tunneling bias. For a typical value of κ = 1.0 $Å^{-1}$ and a constant bias, equation [1] predicts a decrease in the tunneling current of an order of magnitude with a 1 Å decrease in the tip-substrate distance. Thus, equation [1] requires that tip-sample spacings of less than 10 Å be employed for imaging at moderate biases ($\Delta E_t < 1$ V) in STM (20).

In any imaging and spectroscopic mode of the STM, a bias is required between the sample and the tip. In an electrochemical solvent, faradaic current between the tip and sample can interfere with, and sometimes completely obscure, the tunneling current. This undesirable situation makes it very difficult to control the feedback and to maintain a constant tunneling gap between the tip and the sample. For example, in our laboratory, we have found that feedback control is lost on our present microscope if the faradaic current, i_F, assumes a value greater than one-half that of the tunneling current, i_t. Use of partially insulated tips alleviates this condition, but unfortunately, does not completely eliminate the problem (57).

For the operation of an STM in a conventional two-electrode configuration, the presence or absence of significant faradaic current in the tip-sample circuit depends on three factors: 1) the redox potential(s) of the solution species, 2) the reversibility of the electron transfer events for the dissolved redox couple(s), and, 3) the extent to which solution species are permitted access to the tunneling gap. We have identified four limiting cases of electrochemical interest, and discuss each separately below.

Case I: Pure Liquids and Inert Electrolytes. In the absence of significant impurity currents, no faradaic current will flow if the applied bias between the tip and substrate, ΔE_t, is less than the total potential difference, $\Delta E_{F,rev}$, required to drive faradaic reactions at the STM tip and at the substrate. This condition can be easily calculated from the electrochemical potential data for the solvent/electrolyte system under study. This situation is most likely to exist in pure liquids or in solutions of nonelectroactive electrolytes where the faradaic reactions at both electrodes are associated with solvent or electrolyte electrolysis. The bias window available for imaging under these circumstances can be 1 V or more. In pure water at room temperature, for example, $\Delta E_{F,rev} = 1.2$ V (62). In such cases, it may be unnecessary to employ insulated STM tips. From a practical standpoint, however, such tips are desirable since the magnitude of the residual (impurity) faradaic current present at any tunneling bias is reduced.

Case II: Reversible or Quasi-Reversible Redox Species. If the tip-sample bias is sufficient to cause the electrolysis of solution species to occur, *i.e.*, $\Delta E_t > \Delta E_{F,rev}$, the proximity of the STM tip to the substrate surface (d < 10 Å) implies that the behavior of an insulated STM tip-substrate system may mimic that of a two-electrode thin-layer cell (TLC)(63). At the small interelectrode distances required for tunneling, a steady-state concentration gradient with respect to the oxidized (Ox) and and reduced (Red) electroactive species should be established between the tip and the substrate, and the resulting steady-state current will augment that present as a result of the convection of electroactive species from the bulk solution. In many cases, this steady state current is predicted to overwhelm the convective currents, so this situation is of concern when STM imaging under electrochemical conditions (64).

Davis *et. al.* (64) have calculated the steady-state thin-layer current component for a series of electrode geometries. In their derivation, these authors have assumed that the flux between the electrodes is one-dimensional (perpendicular to the plane). Particularly relevant to the STM geometry are the equations for the current in a conical electrode/planar electrode TLC, I_{con}, and those for a hemispherical electrode/planar electrode TLC, I_{hsph} (64):

$$I_{con} = (1 + \alpha^2)^{1/2}(0.5 - \gamma/4 \ln (1 + 2/\gamma)) \qquad [2]$$

$$I_{hsph} = \ln (1 + \gamma\text{-}1) \qquad [3]$$

In the above equations, α is the conical aspect ratio, r/h; γ is the ratio of the cone or hemisphere radius to the interelectrode distance, r/d; and I, the dimensionless faradaic current (either I_{con} or I_{hsph}), is the ratio between the one-dimensional current contribution, i_{TLC}, and the limiting current for an isolated hemispherical electrode, i_{hsph} (see Eq. 5)(64):

$$I = \frac{i_{TLC}}{i_{hsph}} \qquad [4]$$

A plot of I for both conical and hemispherical geometries is shown in Figure 3. The thin-layer current component for these two geometries becomes significant for interelectrode distances on the order of the cone/hemisphere radius; *i.e.*, that of the steady state diffusion layer thickness for microelectrodes with these geometries. At smaller distances, d < r, the values of I for these two cases diverge. The value of I for the hemisphere/plane system approaches infinity at small r, in analogy to twin planar electrode TLC's. In contrast the cone/plane TLC reaches a limiting value (64) at small interelectrode separations. Digital simulations performed by the

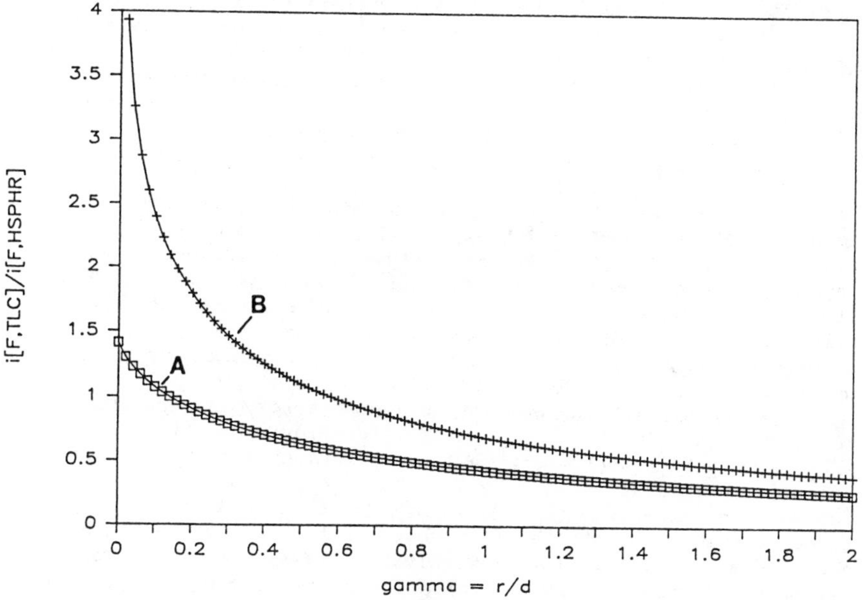

Figure 3. $i_{F,TLC}/i_{F,hsph}$ vs. γ for conical-planar two electrode cell (A) and hemispherical-planar two electrode cell (B). Calculated using Eqs. 2 and 3 (64).

authors indicate that lateral (non-one-dimensional) diffusion can be expected to contribute less than 10% of the total current for the hemisphere/plane case and for an interelectrode distance, $d < 0.01$ r (64). It should be noted that the accuracy of Equations 2 & 3 has not yet been demonstrated experimentally.

Equations 3 & 4 permit the faradaic current to be estimated for a hemispherical tip geometry and small values of γ; *i.e.*, at the limit of small interelectrode distances. At a typical tunneling distance of 10 Å, assuming the tip radius, $r = 1\ \mu m$ ($\gamma = 10^{-3}$), and $D = 10^{-5}\ cm^2\ sec^{-1}$, the concentration of electroactive species corresponding to $i_{TLC} = 1$ nA is 0.24 μM. At the limit of large interelectrode spacings, (*i.e.*, $\gamma \gg 1$) the steady-state diffusion-layer thickness $r_{ss} < d$, and Figure 3 shows that the one-dimensional TLC current, i_{TLC}, component will be small. Assuming the current, i_{hsph}, under these circumstances is limited by diffusion to a hemispherical microelectrode at the STM tip, its magnitude will be given by (65):

$$i_{hsph} = 2\pi nFRDC_{ox/red} \qquad [5]$$

Thus, for $R = 1\ \mu m$ and $D = 10^{-5}\ cm^2\ sec^{-1}$, a 1 nA limiting current is obtained for concentrations of electroactive species, $C_{ox/red} = 1.6\ \mu$M. These calculations suggest that in the presence of a reversible redox couple at micro-molar concentrations, even STM tip-sample biases of $\Delta E_t < 10$ mV will drive a faradaic current that is comparable to that of the tip-sample tunneling current. STM imaging under such circumstances is likely to be experimentally demanding using conventional feedback methodology.

This constraint may be relaxed somewhat for quasi-reversible redox couples. In this case, a significant overpotential, η, ($\eta = E - E_{rev}$), may be associated with the generation of faradaic current at one or both electrodes. Thus, it may be possible to exceed the faradaic potential window ($\Delta E_t > \Delta E_{rev}$) without drawing significant faradaic current due to the fact that reactions at one or both electrodes are kinetically slow. In this case, the effective faradaic potential window is given by, $\Delta E_F = \eta_a + \eta_c + \Delta E_{rev}$ ($i_F < 0.5\ i_t$). At smaller biases, feedback control can be maintained, and STM imaging ought to be possible. This case, however, has not as yet been demonstrated and all STM images of solution covered surfaces to date have been obtained with biases conforming to Case I conditions.

It should be noted here that the ultra thin-layer cells (UTLC) which result from the close approach of an STM tip to a conducting substrate may have important electroanalytical applications in studies other than STM imaging (64). This is because extremely large current densities should be attainable in such cells, and also because of the fast transit times (*e.g.*, 50 nsec for $d = 10$ nm) for reactants across the cell. Thus, such UTLC's might facilitate the determination of fast heterogeneous rate constants or the study of reactive electrochemical intermediates (64).

Case III: Irreversible Electron Transfer or Insoluble Products. Steady-state currents due to one-dimensional flux between the STM tip and substrate will not exist if either, 1) the products at both electrodes are insoluble, or, 2) the electron transfer reactions at both electrodes are irreversible. In both of these cases, the flux to both electrodes will only be that supplied from bulk diffusion into the gap.

Although from a diffusional standpoint, these two cases are indistinguishable, it should be noted that STM imaging during the deposition of insoluble products on either the substrate or tip is likely to be experimentally challenging. The accumulation of more than ca. 10 Å of nonelectronically conductive material on the substrate or tip, for example, will prevent the feedback from establishing the distances required for tunneling. Depositions of conductive materials on the STM tip (*e.g.*, metals) will probably result in detrimental changes in its geometry which reduce resolution. One experimentally accessible case may prove to be the deposition of electronically conductive materials on the substrate, *e.g.*, metal plating. In this case, control of the tunneling current is possible providing the deposition rate does not exceed the response time of the feedback circuit.

If the total current can be assumed to be limited by diffusion to the STM tip, Case III is similar to diffusion to a microdisk electrode (one electrode) thin-layer cell (63). Murray and coworkers (66) have shown that for long electrolysis times, diffusion to a planar microdisk electrode TLC can be treated as purely cylindrical diffusion, provided that the layer thickness is much smaller than the disk diameter (66). In contrast to the reversible case discussed above (Case I), the currents in this scenario should decrease gradually with time at a rate that is dependent on the tip radius and the thickness of the interelectrode gap. Thus, for sufficiently narrow tip/sample spacings, diffusion may be constrained sufficiently (i_F decayed) at long electrolysis times to permit the imaging of surfaces with STM.

Case IV: Spatial Exclusion From the Tunneling Gap. Throughout the discussion of diffusion above, we have assumed that the STM tip behaves as a microelectrode, *i.e.*, possesses a small metal area that is exposed to solution. This is typically the case with commercially available, glass insulated STM tips where the exposed metal surface area is typically 10 - 50 μm^2 (57,58). Moreover, the discussion has also assumed that diffusional processes between the STM tip and substrate are unperturbed by steric effects due to the presence of the narrow gap which separates these electrodes during tunneling. However, at typical tunneling distances, $d < 10$ Å, it is conceivable that electroactive species will be squeezed out of the tip-sample region and precluded from reacting at either the STM tip or the substrate. Alternatively, it may be possible to prepare insulated STM tips with geometries which prevent electroactive species from interacting with exposed metal at the tip. We have preliminary experimental results (*vide infra*) which suggest that these conditions can be met in practical STM tips. Thus, the spatial exclusion of electroactive species from the tunneling gap is considered here to be a fourth distinct classification for the imaging of solution-covered surfaces.

Ex Situ Electrochemical Applications of STM

Predictably, the first STM studies of electrode surfaces were *ex-situ* investigations of electrodes that were prepared in solution and subsequently imaged either *in vacuo* or in air. Several representative examples, illustrating the uses of STM to date, are discussed in this section.

Electrode Surface Topography. STM images of electrochemically pretreated platinum surfaces were first obtained in UHV by Baro and coworkers (67). The images reported by these workers were the first structural images of electrode surfaces that had been exposed to solution under potential control. In an initial series of experiments, Pt(111) single crystal electrodes were subjected to a large amplitude, square-wave potential perturbation (1M H_2SO_4, E_1 = 1.4 V, E_2 = 0.05 V vs. NHE, f = 2.5 kHz), and were imaged (at nanometer resolution) before and after the electrochemical treatment. The cyclic voltammograms in the H adsorption region, recorded as a function of time after the application of the square wave program, revealed an increase in the amount of "strongly" adsorbed H (E = 0.28 V)) as compared to "weakly" adsorbed H (E = 0.12V). This was consistent with the development of micro-crystalline regions of Pt(100) on the Pt(111) surface (67). STM images of the electrochemically treated surfaces revealed the development of atomically smooth facets on some regions of the surface, which was consistent with the proposed development of Pt(100) micro-crystallites. On the basis of both STM and electrochemical examination of Pt surfaces that were initially either polycrystalline Pt or Pt(111) oriented crystals, the authors concluded that preferential development of Pt(100) microcrystallites occurs with application of the square-wave, electrochemical perturbation (67).

Subsequent STM examinations of electrochemically pretreated platinum surfaces in air by Baro and coworkers (68) and by Fan and Bard (Fan, F-R.F.; Bard, A.J. Anal. Chem., submitted) have focused on identifying changes in the surface topography effected by electrochemical activation. In related experiments, Fan and Bard have imaged Pt single crystals and annealed polycrystalline Pt surfaces at nanometer resolution before and after the exposure of these surfaces to adsorbates such as I_2 and ethyl acetate. Images of the adsorbate-covered surfaces exhibited a qualitatively rougher topography as compared to clean Pt surfaces. However, the resolution in these images was insufficient to permit identification of these surface features.

Morita *et. al.* (69) have obtained STM images of electrically insulating, electrochemically generated, Al_2O_3 barrier layers. The Al_2O_3 surfaces to be imaged were coated with thin (40-300 Å) conductive metal films in order to make them suitable for STM analysis. Images such as that shown in Figure 4 possess resolution in excess of that routinely available from SEM images of these surfaces. This work demonstrated the applicability of STM to the examination of electrochemically passivated surfaces (69).

It is interesting to note that none of the STM work on metal surfaces (*i.e.*, Pt and Al) discussed above yielded images with atomic resolution. In fact, although atomic resolution images of

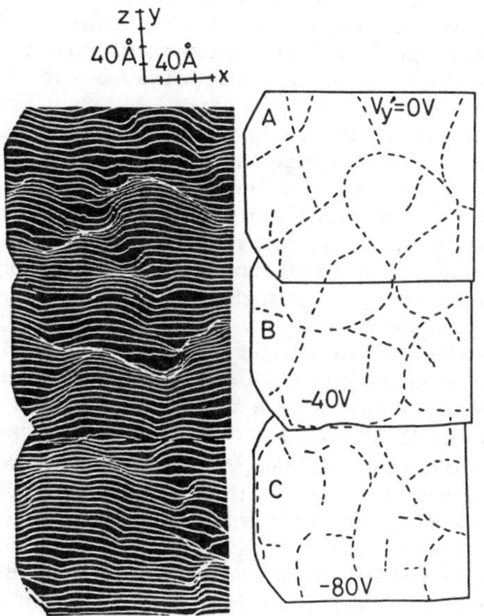

Figure 4. STM images of Al_2O_3 barrier oxide formed anodically at an applied potential of 20 V in Aq. 20 w/v% H_2SO_4. Surface was coated with 300 Å of Pt-Pd alloy prior to imaging. The image at left is a composite of the six STM images at right which were obtained with y-piezo offset voltages as shown. Reproduced with permission of Ref. 69. Copyright 1986 Japanese Journal of Applied Physics.

semi-metals such as graphite are routinely obtained in *ex situ* STM of these surfaces, the free electron nature of metals has previously been thought to be an impediment to resolving close-packed metal surfaces at atomic resolution. Notably, the close-packed surface of gold (Au(111)) has recently been imaged in both air and UHV at atomic resolution by Hallmark *et. al.* (8). This implies that atomic resolution imaging of metallic electrode surfaces may be possible in the near future. Curiously, the studies of the Au(111) surface reported an observed corrugation of 0.3 Å, which is substantially greater than that expected from theoretical calculations (70). The reasons for this discrepancy are not clear as of this date.

Adsorbate Properties. Much recent STM activity has focused on the examination of adsorbate covered surfaces. Although in several instances nonelectrochemical techniques, such as gas phase adsorption methods, have been employed to prepare these surfaces, several of these systems warrant mention here because similar adsorbates are routinely encountered in an electrochemical context.

Gimzewski *et. al.* (71) adsorbed copper phthalocyanine (CuPc) at submonolayer coverages on polycrystalline silver via sublimation *in vacuo*. These workers then imaged these surfaces in UHV with the STM. Individual copper phthalocyanine molecules were resolved at a series of tip-sample biases from 250 mV - 700 mV (tip +) at a constant current of 0.35 nA (71). At the lowest tunneling biases on this interval (narrowest tip-sample spacings), forms with disc-like geometries and diameters of ca. 10 Å were observed, as expected if CuPc were sitting flat on the surface. As the bias was increased to values > 600 mV (resulting in a greater tip-sample spacing), the shape of the observed images narrowed until a majority assumed a cone-like geometry. Significantly, stable images of individual CuPc molecules were usually observed adjacent to surface roughness. In the absence of observable roughness, stable, stationary CuPc molecules were only infrequently encountered. Instead, diffusion of CuPc molecules was observed across the STM image window at rates of ca. 2 Å/min (71). These observations demonstrate the potential for obtaining surface diffusion coefficients for adsorbed species with STM. The studies with phthalocyanines are also important from the view of electrocatalysis.

Smith *et. al.* (10) have obtained images of adsorbed sorbic acid (SA) molecules on highly ordered pyrolytic graphite substrates (HOPG) in liquid helium. Sub-monolayer coverages of sorbic acid (C_5H_7COOH) were obtained by spin-casting films from SA/benzene solutions. Images of elongated structures (Fig. 5) were associated with SA partially covering the surface. These images appeared to be composed of several SA molecules, each of which possesses dimensions 2 Å x 8 Å . At the liquid He temperatures employed for imaging in this study, gap conductivity (dI/dV) vs. tip bias spectra obtained with the STM tip positioned over SA molecules revealed peaks at energies corresponding to vibrational modes of adsorbed SA molecules as determined by IETS (72). Such spectra were typically dominated either by C=C and C-H modes, or by C=O and C-O modes, suggesting that for a given position of the STM tip over the SA molecule, tunneling current was originating primarily from the alkene end of SA molecules, or from the carboxyl moiety (10). This example is the first to demonstrate the potential of STM for extracting

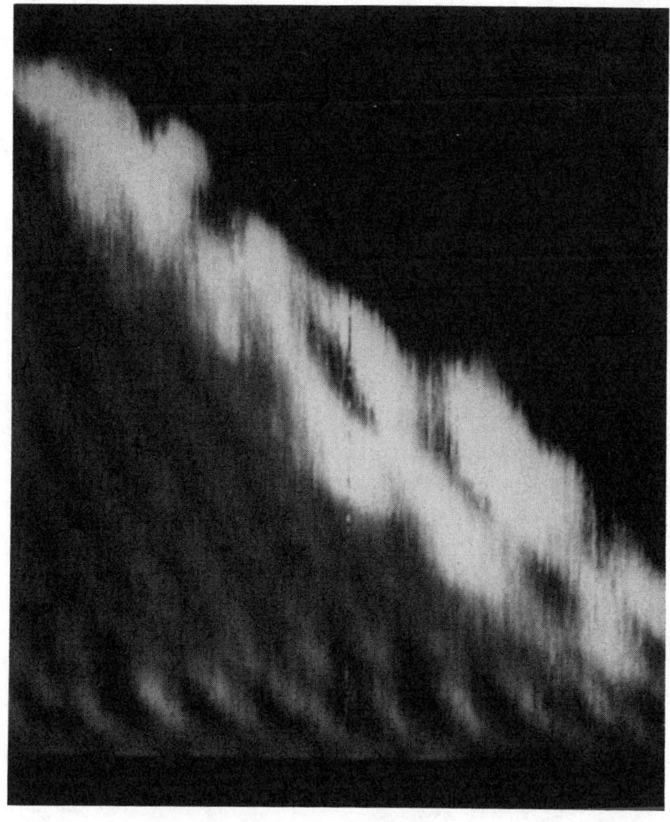

Figure 5. Sorbic acid molecules on HOPG as reported by Smith *et. al.* (10). This surface was prepared by spin-coating a dilute sorbic acid-benzene solution onto a freshly cleaved HOPG substrate. Images of the surface were obtained in liquid helium. The elongated structure shown was representative of those present on surfaces prepared with this procedure.

spectroscopic information selectively for adsorbed molecules on a molecule by molecule basis.

In addition to the two cases discussed above, images of surfaces coated with other organic adsorbates such as cadmium arachidate Langmuir-Blodgett films, (73), DNA (74), di-methyl and di-2-ethylhexyl phthalate (75), and K-24 liquid crystals (Foster, J.S.; Frommer, J.E. Nature, submitted) have been obtained in recent publications.

Surfaces modified with polymer films (*c.f.*,76-78) and self-organizing molecular assemblies (79,80) have been of particular interest to electrochemists. Albrecht *et. al.* (Albrecht, T.R.; Dovek, M.M.; Lang, C.A.; Quate, C.F.; Kuan, W.J.; Frank, C.W.; Pease, R.F.W. J. Appl. Phys., submitted) have employed both STM and atomic force microscopy (AFM) to image, and STM to modify, poly(octadecyl acrylate) (PODA) films on HOPG. The films were prepared using Langmuir-Blodgett techniques. In this work, single PODA chains were observed on surfaces containing submonolayer coverages of the polymer. The length of the polymer chain images obtained was consistent with the known molecular weight of the PODA used in this study. Crystalline-like domains of parallel polymer bundles were also present on such surfaces. Modification of the polymer-covered surface was effected by applying a 100 nsec voltage pulse of ca. 4.1 V to the tunneling bias. In crystalline regions of the polymer coating, the application of the voltage pulse resulted in the disordering of the polymer chains and in the apparent cleaving of polymer bonds resulting in the production of shorter PODA oligomers. The mechanism for polymer bond cleavage in the STM experiment may be similar to that observed previously in electron-beam irradiation of PMMA resists.

Lithography With the STM: Nonelectrochemical Methods. The prospect of atomic density information storage has spurred applications of the STM as a surface modification tool. In this application, the anisotropic current density distribution generated by an STM tip is exploited to "write" on a substrate surface. Features with critical dimensions < 5 nm have been written in UHV, in air, and under liquids.

Although the mechanism by which modification of surfaces in UHV occurs is not clear for all cases, local heating effects appear to have effected the observed modification of glassy materials such as $Pd_{81}Si_{19}$ (81) and $Rh_{25}Zr_{75}$ (82). The fluence of electrons from an STM tip has been used to accomplish nanometer scale electron beam lithography of CaF_2 coated substrates (83). A somewhat different strategy has been employed by Silver *et. al.* (84). These workers used the electric fields of $>10^7$ V/cm present in the tunneling gap to generate a micro-plasma in the presence of an organo-metallic gas (dimethyl cadmium). The organometallic compound was reduced by the plasma, and metal deposited in 20 - 50 nm features on the cathode (substrate) surface (84). A number of recent papers (85-87) report the introduction of Angstrom scale modifications to metal surfaces by the application of large tip-sample voltages ($\Delta E_t > 2V$). Notably, these modifications have occurred both in nonpolar liquids and in air, although the mechanism by which such features are produced is not as yet clear. Finally, the STM tip itself has been used to "micro machine" (88) surfaces by operating the STM with the tip in contact with a substrate.

Lithography With the STM: Electrochemical Techniques. The nonuniform current density distribution generated by an STM tip has also been exploited for electrochemical surface modification schemes. These applications are treated in this paper as distinct from true *in situ* STM imaging because the electrochemical modification of a substrate does not *a priori* necessitate subsequent imaging with the STM. To date, all electrochemical modification experiments in which the tip has served as the counter electrode, the STM has been operated in a two-electrode mode, with the substrate surface acting as the working electrode. The tip-sample bias is typically adjusted to drive electrochemical reactions at both the sample surface and the STM tip. Because it has as yet been impossible to maintain feedback control of the z-piezo (tip-substrate distance) in the presence of significant faradaic current (*vide infra*), all electrochemical STM modification experiments to date have been performed in the absence of such feedback control.

Lin *et. al.* (89) reported the first electrochemical modification of an electrode surface that was implemented with an STM. These workers electrochemically etched an illuminated GaAs surface with a 4 V bias (tip -) in the presence of 5 m<u>M</u> NaOH, 1 m<u>M</u> EDTA in either aqueous or acetonitrile solutions. Figure 6 shows an SEM image of a line etched on a GaAs surface using this procedure. Gas evolution at the tip cathode was prevented by adding a depolarizer such as nitrobenzene. In order to maintain a tip-substrate distance of ca. 1 μm during the etching of lines, the surface of the semiconductor was imaged with the STM in air, the resultant topographical data was stored on a computer disk, and the tip path was retraced during the etching process. Etched line widths of 2.0 μm and 0.3 μm were achieved in this seminal study (89).

Subsequent attempts at electrochemical lithography have focused principally on electrochemical metal deposition. The electrochemical deposition of metal features on the surface of a dissimilar metal is complicated by the fact that nucleation events always precede bulk metal plating (90). Thus, although the goal is generally to deposit a metal feature at a location directly under the STM tip, nucleation often occurs preferentially at surface defect sites that can be located microns away from this location (Dovek, M.M.; Heben, M.J. unpublished results, Stanford University; August, 1987). This problem is most pronounced for deposition on atomically smooth and unreactive surfaces such as basal plane-oriented HOPG, and is manifested in large nucleation overpotentials for the deposition of metals such as Ag and Cu (91). Thus, well controlled electrochemical modifications of these surfaces have not been demonstrated.

Schneir *et. al.* have avoided the nucleation problem by using the same metal (Au) for both deposition and substrate (Schneir, J.; Hansma, P.K., Elings, V.; Gurley, J.; Wickramasinghe, K.; Sonnenfeld, R. SPIE'88 Conference Proceedings, in press). These workers employed glass insulated STM tips to write 300 - 500 nm Au lines at a bias of 3.0 V and a tip-sample spacing of 1 μm. Craston *et. al.* (Craston, D.H.; Lin, C.W.; Bard, A.J. <u>J. Electrochem. Soc.</u>,

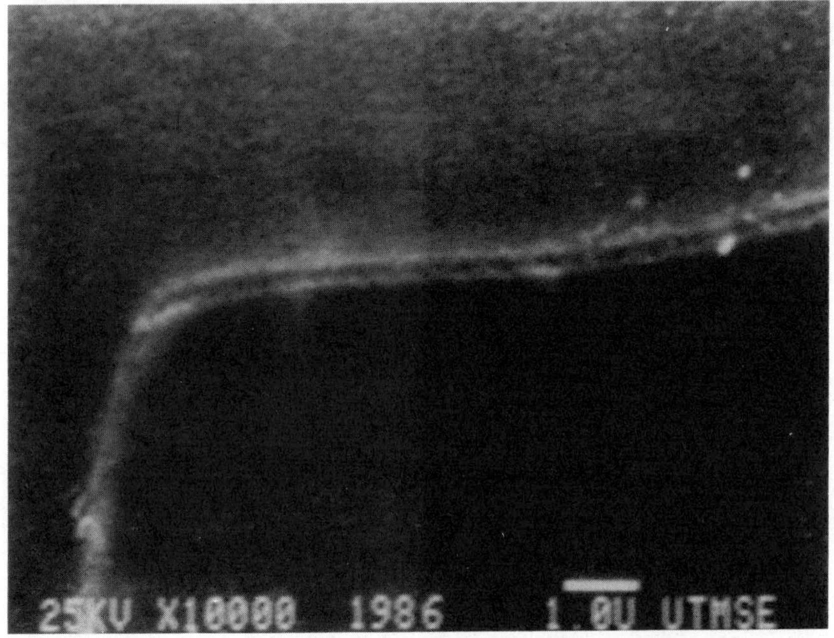

Figure 6. Scanning electron micrograph (SEM) of n-GaAs surface electrochemically etched with a scanning electrochemical and tunneling microscope (SETM). Etching was accomplished in Aq. 5 m$\underline{M}$ NaOH, 1 m$\underline{M}$ EDTA. Photoelectric current = 0.7 μA, Scan rate = 0.1 μm/sec, bias voltage = 4 V. Tip was moved in an "L" pattern. Reproduced with permission of Ref. 89. Copyright 1987 The Electrochemical Society Inc.

submitted) have achieved similar resolution by electrochemically reducing metal ions (Cu^{2+} and Ag^{+}) exchanged into Nafion-coated substrates. In this case, deposition can occur only at positions on the membrane physically contacted by the STM tip.

Clearly, the resolution attained by electrochemical methods has not yet approached that available with the other techniques discussed above. Decreasing the tip-sample distance from the 1 μm employed in previous experiments ought to localize the current density distribution and improve resolution. However, as noted above, maintaining a smaller tip-sample spacing is likely to be difficult in the absence of tunneling current feedback control. For this reason, significant improvements in electrochemical plating resolution may depend on the development of techniques for maintaining feedback controlled tunneling distances on the order of 1 - 10 nm.

In Situ STM Studies of Liquid and Solution Covered Electrode Surfaces

One of the most significant applications of STM to electrochemistry would involve the application of the full spectroscopic and imaging powers of the STM for electrode surfaces in contact with electrolytes. Such operation should enable the electrochemist to access, for the first time, a host of analytical techniques in a relatively simple and straightforward manner. It seems reasonable to expect at this time that atomic resolution images, I-V spectra, and work function maps should all be obtainable in aqueous and nonaqueous electrochemical environments. Moreover, the evolution of such information as a function of time will yield new knowledge about key electrochemical processes. The current state of STM applications to electrochemistry is discussed below.

Atomic resolution STM images of HOPG in water, and images at lower resolution of Au in aqueous 2 m$\underline{M}$ NaCl solutions, were first reported by Sonnenfeld and Hansma (57). HOPG images were obtained using total tip-sample biases of 100 mV (tip -) and total currents (faradaic + tunneling) of 50 nA. The fact that individual solvent molecules did not appear in these STM images was attributed to the fact that their high velocity relative to the tip speed resulted in time averaging (57). Itaya and Sugawara (59) subsequently reported images of HOPG in aqueous 0.05 $\underline{M}$ H_2SO_4 in the presence of residual faradaic currents of 0.2 nA (ΔE_t = 100 mV, tip -). Liu *et. al.* (92) obtained one dimensional traces with a resolution of ca. 30 nm of solution-covered, metal-coated, integrated circuits. In this case, the authors used a total current of 5 - 25 nA, which included a substantial distance-dependent faradaic current component, to control the interelectrode distance with a conventional feedback arrangement (92).

Sonnenfeld *et. al.* (58) have used STM to determine the topography of etched (0.05 v/v% Br_2/MeOH) and polished GaAs surfaces. STM images of these surfaces were obtained both in aqueous 0.01 $\underline{M}$ KOH and in concentrated aqueous NH_4OH. GaAs surfaces exposed to the chemical etchant and polished with lens paper exhibited average roughnesses of 1 nm over intervals of 5 to 1000 nm. Occasional 100 nm scale defects were also present on such surfaces. STM images of etched GaAs surfaces in 0.01 $\underline{M}$ aqueous KOH remained unchanged over periods of several minutes (58). Interestingly, the GaAs lattice could not be resolved at these solution-covered surfaces, despite the fact that previous atomic resolution images of GaAs in UHV have been reported (3,16).

Work by Schneir, Hansma and others (86,93) has extended the list of pure liquids suitable for STM studies by demonstrating imaging capabilities with a series of nonpolar organic liquids, such as paraffin oil and fluorocarbon grease. These liquids have the advantage of possessing greater viscosity than aqueous solutions, and hence exhibit inherently smaller residual currents. These workers were able to attain tip-substrate biases (ca. 3 V) greater than those routinely possible in air or in aqueous solutions. In addition, these liquids were found to improve the signal-to-noise of the STM atomic resolution images, possibly due to improved vibrational damping characteristics as a result of the liquid layer (86,93). The authors suggest that nonpolar, organic liquids may serve to protect air sensitive or otherwise reactive surfaces, and hence facilitate the imaging of such surfaces under ambient conditions.

As noted above, Schneir, Hansma and coworkers have also used the greater biases available in nonpolar organic liquids (~ 3 V) to facilitate the surface modification of gold surfaces (87). The features introduced with this procedure were imaged *in situ* under a fluorocarbon grease. Gold mounds, initially ca. 10 nm in diameter, decreased in height and broadened over a period of 24 h as a result of diffusion. Preliminary self-diffusion coefficients for gold atoms at these surfaces of $D = 10^{-20} - 10^{-16}$ cm^2 sec^{-1} were estimated from the observed rate of change of height with time. This is the first instance in which solid-state-like diffusion coefficients have been estimated from STM images. Thus, it seems likely that room temperature diffusion coefficients might be obtained with STM for well characterized systems in the future (87).

Itaya *et. al.* (Itaya, K.; Higaki, K.; Sugawara, S. Chem. Lett., in press) have examined the nano-topography of polycrystalline Pt electrode surfaces immersed in aqueous, 0.1 M H_2SO_4 in analogy to the *ex situ* work with electrochemically activated Pt electrodes discussed above. Increases in the surface area of Pt electrodes accompanying cycling in this electrolyte are associated with electrofaceting of the electrode surface as previously reported by Baro and coworkers. The crystallographic orientation of these facets, however, is not apparent in the images reported by these authors.

Using the unique four-electrode STM described above, Bard and coworkers (Lev, O.; Fan, F-R.F.; Bard, A.J. J. Electroanal. Chem., submitted) have obtained the first images of electrode surfaces under potentiostatic control. The current-bias relationships obtained for reduced and anodically passivated nickel surfaces revealed that the exponential current-distance relationship expected for a tunneling-dominated current was not observed at the oxide-covered surfaces. On this basis, the authors concluded that the nickel oxide layer was electrically insulating, and was greater than ca. 10 Å in thickness. Because accurate potential control of the substrate surface is difficult in a conventional, two-electrode STM configuration, the ability to decouple the tip-substrate bias from the substrate potential ought to facilitate the future study of electrode processes by STM.

Sonnenfeld and Schardt (94), and Schneir *et. al.* (Schneir, J., Hansma, P.K.; Elings, V.; Gurley, J.; Wickramasinghe, K.; Sonnenfeld, R. SPIE'88 Conference Proceedings, in press) have both reported *in situ* STM images of electrode surfaces on which metals have been electrochemically deposited. Morita *et. al.* (61) have examined Ag surfaces *in situ* in the presence of chloride ion. Figure 7 shows an STM image obtained by Sonnenfeld and Schardt (94) of a Ag island on the atomically smooth surface of HOPG in 0.05 M $AgClO_4$. This work provides insight into the mechanism by which nucleation processes occur on smooth surfaces. Because in each of the three above cases, images such as that shown in Figure 7 have been obtained in the presence of a reversible redox couple (i.e. either M^+/M^o or $AgCl/Cl^-$), they represent the most demanding conditions under which published STM images have been obtained under solution. As such, it is important to understand why STM imaging under these conditions is possible. In fact, if the correct bias polarity is selected, both of the above cases represent an example of Case I STM imaging.

The silver deposition experiments of Sonnenfeld and Schardt (94) provide a representative example. After the deposition of silver on HOPG, the freshly plated surface was imaged in the presence of aqueous 0.05 M $AgClO_4$ (Fig. 7)(94). Assuming a positive tip polarity is used, the STM tip will function as an anode and its potential will be that necessary to oxidize water, $E_{a,H_2O/O_2} = \sim+0.95$ V (pH = 7). The substrate cathode will drive the reduction of silver ion at the silver plated substrate at a formal potential of $E_{c,Ag^+/Ag} = +0.72$ V. Thus, an imaging window of $\Delta E_F = 230$ mV is available, inside of which neither electrochemical reaction will occur. Very different consequences result with the opposite polarity (tip -). Silver is anodically dissolved from the substrate at a reversible potential of +0.72 V, and silver ion plates at the tip cathode at the same reversible potential. Moreover, two-electrode TLC current enhancements are expected since the product at the substrate (Ag^+) is the reactant at the STM tip. In this configuration, the Ag^+/Ag system represents an example of Case II imaging, which, as noted above, is likely to be difficult with conventional, glass-coated STM tips. For this reason, Sonnenfeld and Schardt succeed in imaging Ag-plated HOPG in the presence of 50 mM Ag^+ at substantial (e.g. 100 mV) biases and positive tip polarities (94).

Thus, all of the above studies have involved Case I imaging conditions (*vide supra*). That is, in each instance, the applied biases used for imaging were less than that required to drive faradaic reactions at both tip and sample. Figure 8 shows a HOPG image obtained under 1 M NaCl at $\Delta E_t = 550$ mV (tip +) which is representative of images achieved in 1 M NaCl at biases of $\pm$ 1.5 V and in solutions of 1 M NaCl, 0.1 M $Fe(CN)_6^{3-}$/0.1 M $Fe(CN)_6^{4-}$ solutions at biases of $\pm$ 0.8 V (Dovek, M.M.; Heben, M.J.; Lewis, N.S.; Penner, R.M.; Quate, C.F. manuscript in preparation). Since the tunneling current employed in both cases was 1 nA, we believe that these images represent the first examples of Case IV imaging. This result indicates that imaging may in fact be possible under the experimentally demanding Case IV conditions, and implies that STM

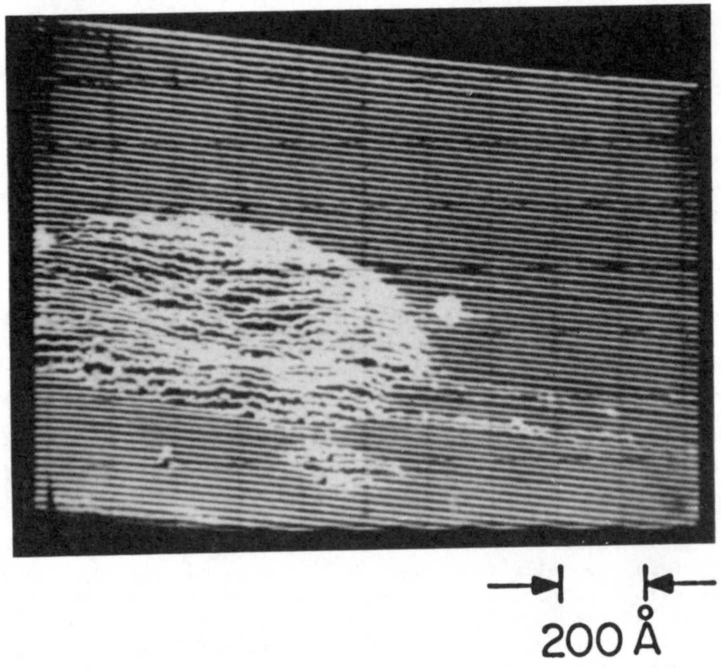

Figure 7. STM images of a silver island on HOPG in aqueous 0.05 M $AgClO_4$. Image was obtained with a glass-coated Pt-Ir tip at a bias of 100 mV (tip +) and i_t = 16 nA without removing the freshly Ag plated surface from the plating solution. Reproduced with permission of Ref. 94. Copyright 1986 American Institute of Physics.

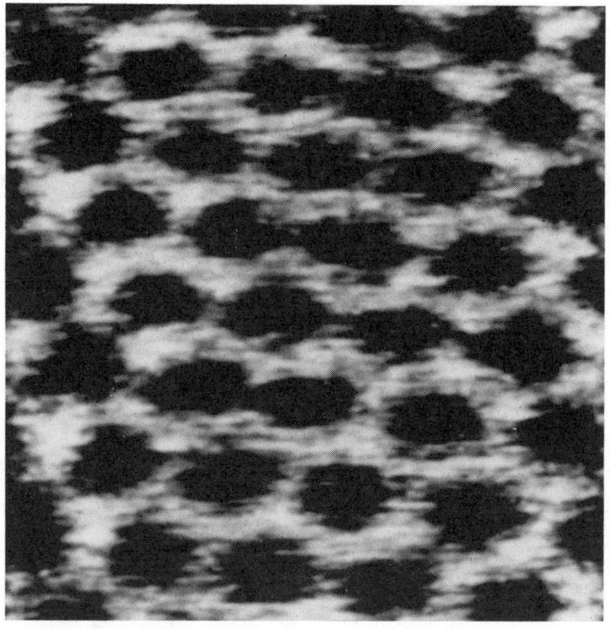

Figure 8. STM image of the basal-plane surface of HOPG obtained in 1 M NaCl. Bias = 550 mV (tip +), i_t = 1 nA. Image was obtained in the fast scan mode. Dark spots, separated by 2.5 Å, correspond to large tunneling currents.

operation may be completely decoupled from electrochemical enviroments.

Summary

It should be apparent from the discussion above that STM possesses tremendous potential for the elucidation of processes at the electrode-electrolyte interface. Particularly promising are the prospects for *in situ* studies of electrode surfaces. Vibrational, electronic, and structural information is obtainable on an atomic scale for electrodes of importance to basic electrochemical studies. Although relatively few electrochemical applications have been demonstrated to date, the availability of commercial instrumentation (*c.f.*,95-97) ought to increase the accessibility of STM to electrochemists and widespread use of the technique is expected in the near future.

Acknowledgments

We acknowledge the financial assistance of the Office of Naval Research, Grant # N00014-85-K-0805. We also acknowledge the assistance of Bruce J. Tufts. NSL also acknowledges support as a Dreyfus Teacher-Scholar and as a A.P. Sloan Fellow.

Literature Cited

1. Binnig, G.; Rohrer, H.; Gerber, Ch.; Wiebel, E. App. Phys. Lett. 1982, 40, 178-80.
2. Binnig, G.; Rohrer, H.; Gerber, Ch.; Weibel, E. Phys. Rev. Lett., 1982, 49, 57-61.
3. Feenstra, R.M.; Stroscio, J.A.; Tersoff, J.; Fein, A.P. Phys.Rev. Lett. 1987, 58, 1192-95.
4. Jaklevic, R.C.; Elie, L. Phys. Rev. Lett. 1988, 60, 120-23.
5. Becker, R.S.; Golovchenko, J.A.; Swartzentruber, B.S. Phys.Rev. Lett. 1985, 54, 2678-80.
6. Binnig, G.; Rohrer, H.; Gerber, Ch.; Weibel, E. Phys.Rev. Lett. 1983, 50, 120-23.
7. Hamers, R.J.; Avouris, Ph.; Bozso, F. Phys. Rev. Lett. 1987, 59, 2071-74.
8. Hallmark, V.M.; Chiang, S.; Rabolt, J.F.; Swalen,J.D.; Wilson, R.J. Phys. Rev. Lett. 1987, 59, 2879-82.
9. Slough, C.G.; McNairy, W.W.; Coleman, R.V.; Drake, B.; Hansma, P.K. Phys. Rev. B 1986 34, 994-1005.
10. Smith, D.P.E.; Kirk, M.D.; Quate, C.F. J. Chem. Phys. 1987, 86, 6034-38.
11. Miranda, R.; Garcia, N.; Baro, A.M.; Garcia, R.; Pena, J.L.; Rohrer, H., Appl. Phys. Lett. 1985, 47,367-69.
12. Park, S.; Quate, C.F. Appl. Phys. Lett. 1986 48, 112-14.
13. Abraham, D.W.; Sattler, K.; Ganz, E.; Mamin, H.J.;Thomson, R.E.; Clarke, J. Appl. Phys. Lett. 1986, 49, 853-55.
14. Carim, A.H.; Dovek, M.M.; Quate, C.F.; Sinclair, R.;Vorst, C. Science 1987, 237, 630-33.
15. Kaiser, W.J.; Jaklevic, R.C. IBM J. Res. Develop. 1986, 30, 411-16.

16. Stroscio, J.A.; Feenstra, R.M.; Fein, A.P. Phys. Rev. Lett. 1987 58, 1668-71.
17. Hamers, R.J.; Tromp, R.M.; Demuth, J.E. Phys. Rev.Lett. 1986, 56, 1972-75.
18. Binnig, G.; Fuchs, H.; Gerber, Ch.; Rohrer, H.; Stoll, E.; Tosatti, E. Europhys. Lett. 1986 31, 1-36.
19. Quate, C.F. Phys. Today 1986, August, 26-33.
20. Hansma, P.K; Tersoff, J. J. Appl. Phys. 1987, 61, R1-R23.
21. Binnig, G.; Rohrer, H. Sci. Am. 1985, August, 50-56.
22. Feuchtwang, T.E.; Cutler, P.H. Phys. Scr. 1986, 35,132-40.
23. Arvia, A.J. Surf. Sci. 1987 181, 78-91.
24. Fleischmann, M.; Hendra, P.J.; McQuillan, P.J. Chem. Phys. 1964 26, 163-67.
25. Jeanmaire, D.L.; Van Duyne, R.P. J. Electroanal.Chem. 1974, 84, 1-20.
26. Albrecht, M.G.; Creighton, J.A. J. Am. Chem. Soc. 1977, 99, 5215-17.
27. Tersoff, J.; Hamann, D.R. Phys. Rev. B 1985, 31, 805-13.
28. Lang, N.D. Phys. Rev. Lett. 1986, 56, 1164-67.
29. Binnig, G.; Rohrer, H. Surf. Sci. 1983, 126, 236-44.
30. Bryant, A.; Smith, D.P.E.; Quate, C.F. Appl. Phys. Lett. 1986, 48, 832-34.
31. Chiang, S.; Wilson, R.J. Anal. Chem. 1987, 59, 1267A-70A.
32. Binnig, G.; Rohrer, H. Physica B&C 1984 127B,37-45.
33. Binnig, G.; Gerber, Ch.; Stoll, E.; Albrecht, T.R.; Quate, C.F. Europhys. Lett. 1987 3, 1281-86.
34. Heinzelmann, H.; Grutter, P.; Meyer, E.; Hidber, H.; Rosenthaler, L.;Ringger, M.; Guntherodt, H.-J., Surf. Sci. 1987 189/190, 29-35.
35. Albrecht, T.R.; Quate, C.F. J. Appl. Phys. 1987, 62, 2599-602.
36. Heinz, K.; Muller, K. In Structural Studies of Solids Springer Tracts in Modern Physics: New York, 1982; vol. 91, pp 1-17.
37. Broers, A.N. Rev. Sci. Instr. 1969 40, 1040-45.
38. Allen, F.M.; Smith, B.K.; Buseck, P.R. Science 1987 238, 1695-97.
39. Blum, L.; Abruna, H.D.; White, J.; Gordon II, J.G.; Borges, G.L.; Samant, M.G.; Melroy, O.R. J. Chem. Phys. 1986, 85, 732-38.
40. Samant, M.G.; Borges, G.L.; Gordon II, J.G.; Melroy, O.R.; Blum, L.; J. Am. Chem. Soc. 1987, 109, 5970-74.
41. Materlik, G.; Schmah, M.; Zegenhagen, J.; Uelhoff, W. Ber. Bunsenges. J. Phys. Chem. 1987, 91, 292-96.
42. Kruger, J. In Advances in Electrochemistry and Electrochemical Engineering; Muller, R.H., Ed.; Wiley and Sons: New York, 1973; vol. 9, pp 227-280.
43. Muller, R.H. In Advances in Electrochemistry and Electrochemical Engineering; Muller, R.H., Ed.; Wiley and Sons: New York, 1973; vol. 9, pp 167-226.
44. Lehwald, S. and Ibach, H. In Vibrations at Surfaces; Caudano, R., Gilles, J.-M. and Lucus, A.A. Eds.; Plenum Press: New York, 1982; pp 137-152.
45. Wolf, E.L. Principles of Electron Tunneling Spectroscopy; Oxford University Press: New York, 1985, Chapter 10.
46. Pons, S., Davidson, T.; Bewick, A. J. Electroanal. Chem. 1984, 160, 63-71.

47. Birke, R.L., Lombardi, J.R., and Sanchez, L.A. In Electrochemical and Spectroscopic Studies of Biological Redox Components; Kadish, K. Ed.; ACS Symposium Series No.210; American Chemical Society: Washington, DC, 1981; p 69.
48. Feenstra, R.M.; Stroscio, J.A.; Fein, A.P. Surf. Sci. 1987, 181, 295-306.
49. Demuth, J.; Avouris, P. Phys. Today 1983, November, p 62.
50. Reihl, B.; Gimzewski, J.K. Surf. Sci. 1987, 189/190, 36-43.
51. Bryant, A. Ph.D. dissertation, Stanford University, June, 1986.
52. Binnig, G.; Gerber, C.; Marti, O. IBM Tech. Discl. Bull. 1984 27, 3137.
53. Gerber, Ch.; Binnig, G.; Fuchs, H.; Marti, O.; Rohrer, H. Rev. Sci. Instr. 1986 57, 221-24.
54. Bryant, A.; Smith, D.P.E.; Quate, C.F. Appl. Phys. Lett. 1986 48, 832-834.
55. Pohl, D.W. IBM J. Res. Develop. 1986, 30, 417-27.
56. Binnig, G.; Smith, D.P.E. Rev. Sci. Instru. 1986 57, 1688-89.
57. Sonnenfeld, R.; Hansma, P.K. Science 1986 232, 211-13.
58. Sonnenfeld, R.; Schneir, J.; Drake, B.; Hansma, P.K.; Aspnes, D.E. Appl. Phys. Lett. 1987 50, 1742-44.
59. Itaya, K.; Sugawara, S. Chem. Lett. 1987, 1927-30.
60. Morita, S.; Okada, T.; Ishigame, Y.; Mikoshiba, N. Surf. Sci. 1987 181, 119-25.
61. Morita, S.; Otsuka, I.; Okada, T.; Yokoyama, H.; Iwasaki, T.; Mikoshiba, N. Jap. J. Appl. Phys. 1987 26, L1853-55.
62. Bard, A.J.; Faulkner, L.R. Electrochemical Methods: Fundamentals and Applications; John Wiley & Sons: New York, 1980, p 699.
63. Hubbard, A.T.; Anson, F.C. In Electroanalytical Chemistry; Bard, A.J., Ed.; Marcel Dekker: New York, 1970; vol. 4, p 129.
64. Davis, J.M.; Fan, F-R.F.; Bard, A.J. J. Electroanal. Chem. 1988, 238, 9-31.
65. Wightman, R.M. Anal. Chem. 1981, 53, 1126A-31A.
66. Geng, L.; Reed, R.A.; Longmire, M.; Murray, R.W. J. Phys. Chem. 1987, 91, 2908-14.
67. Gomez, J.; Vasquez, L.; Baro, A.M.; Garcia, N.; Perdriel, C.L.; Triaca, W.E.; Avria, A.J. Nature 1986 323, 612-14.
68. Vasquez, L.; Gomez, J.; Baro, A.M.; Barcia, N.; Marcos, M.L.; Velasco, J.G.; Vara, J.M.; Arvia, A.J.; Presa, J.; Garcia, A.; Aguilar, M. J. Am. Chem. Soc. 1987, 109, 1730-33.
69. Morita, S.; Itaya, K.; Mikoshiba, N. Jap. J. Appl. Phys. 1986 25, L743-45.
70. Batra, I.P.; Barker, J.A.; Auerbach, D.J. J. Vac. Sci. Technol. A 1984, 2, 943-47.
71. Gimzewshi, J.K.; Stoll, E.; Schlittler, R.R. Surf. Sci. 1987, 181, 267-77.
72. Hall, J.T.; Hansma, P.K. Surf. Sci. 1978, 76, 61-76.
73. Smith, D.P.E.; Bryant, A.; Quate, C.F.; Rabe, J.P.; Gerber, Ch.; Swalen, J.D. Proc. Natl. Acad. Sci. USA 1987, 84, 969-72.
74. Binnig, G. Bull. Am. Phys. Soc. 1986, 31, 217.
75. Foster, J.S.; Frommer, J.E.; Arnett, P.C. Nature 331, 324-26.
76. Wrighton, M.S. Science 1986, 231, 32-37.
77. Murray, R.W. Acc. Chem. Res. 1980, 13, 135-41.
78. Buttry, D.A.; Anson, F.C. J. Am. Chem. Soc. 1983, 105, 685-89.
79. Miller, C.J.; Majda, M. J. Am. Chem. Soc. 1986, 108, 3118-20.

80. Finklea, H.O.; Robinson, L.R.; Blackburn, A.; Richter, B.; Allara, D.; Bright, T. Langmuir 1986, 2, 239-44.
81. Ringger, M.; Hidber, H.R.; Schlog, R.; Oelhafen, P.; Guntherodt, H.J. Appl. Phys. Lett. 1985, 46, 832-34.
82. Staufer, U.; Wiesendanger, R.; Eng, L.; Rosenthaler, L.; Hidber, J.R.; Guntherodt, H.-J. Appl. Phys. Lett. 1987, 51, 244-46.
83. McCord, M.A.; Pease, R.F.W. J. Vac. Sci. Technol. B 1987, 5, 430-33.
84. Silver, R.M.; Ehrichs, E.E.; de Lozanne, A.L. Appl. Phys. Lett. 1987, 51, 247-49.
85. Becker, R.S.; Golovchanko, J.A.; Swartzentruber, B.S. Nature 1987, 325, 419-21.
86. Schneir, J.; Hansma, P.K. Langmuir 1987, 3, 1025-27.
87. Schneir, J.; Sonnenfeld, R.; Marti, O.; Hansma, P.K.; Demuth, J.E.; Hamers, R.J. J. Appl. Phys. 1988, 63, 717-21.
88. McCord, M.A.; Pease, R.F.W. Appl. Phys. Lett. 1987, 50, 569-70.
89. Lin, C.W.; Fan, F-R.F.; Bard, A.J. J. Electrochem. Soc. 1987, 134, 1038-39.
90. Kolb, D.M. In Advances in Electrochemistry and Electrochemical Engineering; Gerischer, H. and Tobias,C.W., Eds.; John Wiley & Sons: New York, 1978; vol. 2, p 125.
91. Morcos, I. J. Electroanal. Chem. 1975, 66, 250-57.
92. Liu, H-Y.; Fan, F-R.F.; Lin, C.W.; Bard, A.J. J. Am. Chem. Soc. 1986, 108, 3838-39.
93. Giambattista, B.; McNairy, W.W.; Slough, C.G.; Johnson, A.; Bell, L.D.; Coleman, R.V.; Schneir, J.; Sonnenfeld, R.; Drake, B.; Hansma, P.K. Proc. Nat. Acad. Sci. USA 1987, 84, 4671-74.
94. Sonnenfeld, R.; Schardt, B.C. Appl. Phys. Lett. 1986, 49, 1172-74.
95. Nanoscope, Digital Instruments, Inc., 5901 Encina Rd., Goleta, CA 93117.
96. WA Technology Inc., 41 Accord Park Drive, Norwell, MA 02061.
97. McAllister Technical Services, Inc., 2414 Sixth Street, Berkeley, CA 94710.
98. Dovek, M.M.; Heben, M.J.; Lang, C.; Lewis, N.S.; Quate, C.F. Rev. Sci. Instr., in press.

RECEIVED August 26, 1988

Chapter 14

Electrochemical Surface Characterization of Platinum Electrodes Using Elementary Electrosorption Processes at Basal and Stepped Surfaces

J. Clavilier

Laboratoire d'Electrochimie Interfaciale du Centre National de la Recherche Scientifique, 1, Place A. Briand, 92195 Meudon Principal Cedex, France

An attempt has been made at using voltammetry of hydrogen adsorption-desorption on platinum as a tool for in-situ checking the crystalline surface structure of platinum at the atomic level. The behaviour of Pt(111), Pt(100) and stepped surfaces is discussed in connection with the rôle of the long-range order on the unusual adsorption states. Application is made to the investigation of surface changes as the effect of annealing and characterization of platinum black crystallites.

Electrochemistry of surface processes at solid metal electrodes is undergoing a profound change due to the mastery of the use of metal single crystals as electrodes in parallel with the other techniques of surface science.

By comparing the electrochemical properties of selected metal single crystals with basal orientations, and then introducing surface defects in a well-defined manner like periodic arrays of monoatomic steps, the electrochemical response of the electrode surface gains a physical meaning which approaches the surface processes at the atomic level of the surface. This becomes the major way for the renewal of surface electrochemistry which is now a promising branch of surface science.

Among the metal electrodes with surface structure sensitive properties,the first ones to be considered are those with electrocatalytic activity, such as platinum group metals. Thus platinum is the most attractive material because of its exceptional spectrum of catalytic properties and its good stability in various electrolytic media.

A few years ago platinum was thought to be a well known solid electrode and its adsorption and electrocatalytic properties widely investigated, particularly its adsorption properties for hydrogen and oxygen. Thus smooth polycrystalline surfaces, finely divided particles, were characterized both by the existence of two main electrochemical hydrogen adsorption states i.e. weakly and strongly bonded hydrogen more easily observable with the latter electrodes than with the former. Extremely reproducible hydrogen adsorption

0097–6156/88/0378–0202$06.00/0

states are observed after application of the "activation" procedure, first applied to smooth platinum samples by Hammett (1). A review of work on platinum activation was published in 1967 by James (2) which gives a complete discussion of the question at that time.

This procedure was applied to platinum single crystal electrodes by Will (3) in his pionnering work to assign the weakly and strongly bonded hydrogen to surface crystalline heterogeneity. Will's experiments have shown that the weakly bonded hydrogen was the major species on the Pt(110) activated electrode while the strongly bonded hydrogen was the major one on the activated Pt(100) surface. He misinterpreted the species adsorbed on the (111) orientation for the reasons we will see below. In this work the activation was first considered as a procedure for the electrochemical cleaning of the surface. Nevertheless the author concluded that a change in crystalline surface structure was the consequence of activation by repeated oxidation and reduction of the oriented platinum surfaces. Until recently the resulting surface damage has been underestimated as well as its exact nature and effect.

Clean and Well-Ordered Platinum Surfaces for Electrochemical Purpose

a) Initially clean surfaces. In order to avoid the difficulties related to the activation procedure for obtaining efficient adsorbing platinum surfaces, platinum single crystals with oriented surfaces were cleaned and characterized by using ultra high vacuum technique (UHV) then transferred into an electrochemical cell (4). The use of a thin layer cell reduced the amount of adsorbable solution impurities (5) interfering with hydrogen adsorption. Experiments carried out in this way confirmed Will's conclusions on the origin of the two hydrogen adsorption states and showed definitely that electrosorbed hydrogen could be used as a surface sensitive probe.Nevertheless if we consider the electrochemical results obtained with clean platinum oriented surfaces prepared during UHV experiments, studied in the range of potential of the reversible adsorption-desorption of hydrogen, the amount of adsorbed hydrogen was not the same with the various experimental devices used. This quantity was remarkably different for the (111) orientation, varying from 0.03 to 0.4 of a monolayer (6) ,(7). When such an electrode was activated, after few cycles of electrochemical adsorption and desorption of oxygen, the amount of hydrogen adsorbed was increased and the voltammetric profile of hydrogen adsorption desorption was changed (8). The general behaviour of the (111) orientation of platinum was indicative that the technique used did not succeed in preparing clean surfaces after the contact with the electrolyte was established. Another route for clean surface preparation, clean transfer of the sample and clean contact with the electrolyte was required.

b) Clean platinum surfaces in contact with an electrolytic solution. The technique which has been developed since 1980 consists in using the catalytic properties of platinum for oxidation of carbonaceous and sulphur surface compounds which are the major contaminants of the surface of platinum samples exposed to the atmospheric

environment and which readsorb at the surface after application of cleaning procedure if no special care is taken. The removal of impurities was achieved by heating the platinum in a gas-oxygen flame and quenching the sample into a stream of ultra pure water during cooling in air. A droplet of water stayed attached to the oriented surface of the crystal protecting it against back contamination by decreasing to a great extent the rate of transfer of impurities from atmosphere to the surface (9).
The platinum single crystals treated in this way are small spherical platinum beads with a diameter ranging from 1.5 to 1.9 mm obtained by melting of a wire. They are oriented, cut and polished according to the technique described in (10) with an accuracy within 3 minutes of the nominal orientation. After polishing, the samples are annealed at 1300°-1500°C to eliminate the perturbed surface layer. In the initial experiments reported in ref.(9) the thermal decontamination treatment was carried out at high temperature ($t > 900°C$). This technique has been improved in order to maintain the quality of the crystal for repeated treatments. With the new procedure the single crystal is not put in direct contact with the flame but heated indirectly through the thermal conduction of the platinum wire. The temperature of annealing is nearly 500°C for few seconds, then the crystal is cooled in air and quenched when its temperature is nearly 200°C i.e. when catalytic oxidation of impurities coming into contact with the surface ceases to be efficient. This thermal treatment applied for surface decontamination must be distinguished from the treatment for surface re-ordering which requires higher annealing temperatures as an example will demonstrate below.
Platinum surfaces with (111) and (100) orientations treated in this way have been checked by using LEED characterization on "as received" samples, both showed the characteristic LEED pattern with their respective (1x1) surface symmetry. The non observation of the (5x20) symmetry for the (100) orientation was due to the presence of residual adsorbed impurities at the surface of "as received" samples. Simply this confirms the crystalline surface quality of the platinum samples prepared according to this technique (10).

Voltammetry for Surface Analysis and Structural Characterization

It may be seen in fig.1 an example of surface analysis by using voltammetry. In these experiments the thermally treated samples, have been put in contact with the solution at a controlled potential chosen in such a way that no or negligible current transients were detected when the samples contacted the solution. This allows the analysis of initial surface conditions (11).The oxygen adsorbed during the thermal treatment is reduced in the first negative sweep in the potential range 0.9-0.6 V/R.H.E. The amount of electric charge for this reduction is a characteristic quantity and corresponds to the maximum coverage which is formed on each orientation by the thermal treatment in air.
In the subsequent process, which is hydrogen adsorption, each orientation has its maximum adsorption activity. Hydrogen is adsorbed in an amount corresponding to one H adatom per surface Pt atom of the (111) and (100) orientations and significantly more for

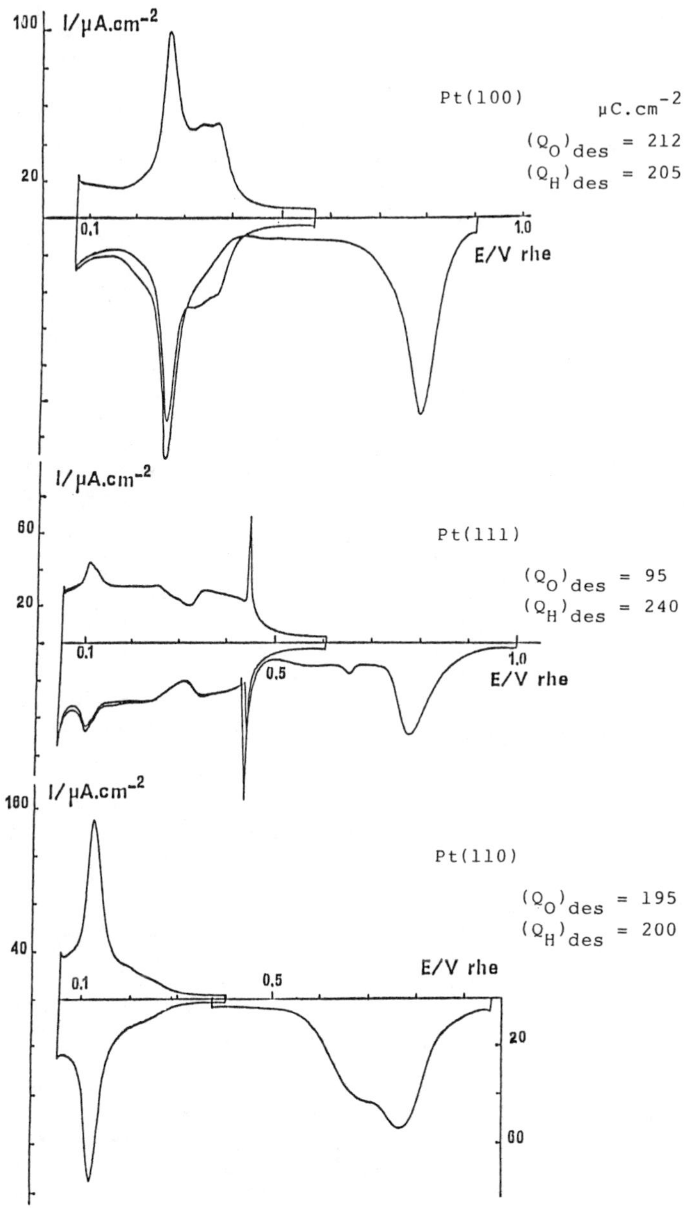

Figure 1. First voltammograms of flame-treated platinum, basal orientations. Standard experimental conditions for all figures:electrolyte 0.5 M H_2SO_4; sweep rate : 50 $mV.s^{-1}$.

Pt(110) due to a tendency for this orientation to undergo the (1x2) surface reconstruction. Evidently both thermally adsorbed oxygen and electrochemically adsorbed hydrogen are related through the surface density of platinum atoms. This set of initial data is indicative of the cleanliness of samples prepared and transferred by this method. The availability of clean platinum surfaces with defined orientation in contact with the electrolytic solution allows the separation of the various processes occurring during the electrochemical activation apart from the oxidation of impurities.

However the more important question which can be solved with such samples is related to characterization of surface sites because each basal orientation shows well-characterized hydrogen adsorption-desorption peaks which might be helpfully used for this purpose.

The main feature of electrosorbed hydrogen at basal orientation is the presence of several peaks or adsorption states for surfaces containing a single type of site like Pt(111) and Pt(100). This means that the information a voltammogram contains, concerns more than the simple interaction between adsorbed hydrogen and the surface site (short range interaction). Other surface information must be found particularly by using stepped surfaces which allows a modulation of long range interactions when they are suspected to be at the origin of some adsorption states.

This possible effect of long range interactions on the hydrogen adsorption states is suggested by the fact that platinum electrodes with (111) and (100) orientations, after thermal treatment, showed adsorption states at higher potential than those observed previously with all types of platinum electrodes which were subjected to activation or were insufficiently clean.

This puzzling difference since has been studied extensively either by using pure electrochemical technique (10),(12,13) or UHV surface science methods (14-17).

Independently of the technique used to investigate the prime cause of the existence of the adsorption states at the higher potentials on Pt(111) and Pt(100), the conclusion is they result from the existence of atomically flat extended surface domains with the respective orientation, i.e. existence of long range surface order. These adsorption states appear as an intriguing and a unique property of well-ordered Pt(111) and Pt(100) surfaces in contact with various electrolyte solutions because no equivalent effects are known with the same orientations in gas phase experiments. In this sense, these properties may be considered as a specific aspect of the electrochemistry of surface processes.

The question of the species involved in these unusual adsorption states is still controversial. No direct information (such as spectroscopic data) on the nature of the species is available. Only electrochemical data have been obtained and these require several cross determinations and careful discussion taking into account the whole experimental data set and the parallel behaviour of the (100) and (111) orientations.

With this in mind the species which fits the best with the experimental data for the anomalous adsorption states would be a fully reversibly discharged species including transfer of one electron. It could be an adsorbed hydrogen-like species under full control of the specific adsorption of anions of the base electrolyte

(18). This could account for the absence of any pH dependence of the unusual adsorption states in presence of specific adsorption of sulphuric anion (13). Recent results (Krauskopf, E. K.; Rice, L. M.; Wieckowski, A. J. Electroanal. Chem., in press) obtained with labeled sulphuric anions are in agrement with the hydrogen interpretation. The major difficulty with the hydrogen hypothesis is the potential of the unusual adsorbed hydrogen in absence of anion specific adsorption cannot be correlated with energy for hydrogen adsorbed from gas phase at the same orientation. There is too large a difference in the hydrogen binding energy values (15).In this respect it was found in gas phase for Pt m(111)x(111) that hydrogen is more strongly bonded on step than on terrace (19). It will be shown below that the potential sequence for the respective hydrogen states gives the inverse order. This conclusion is valid for stepped surfaces with (100) terraces. Thus a direct comparison between the gas phase binding energy and the adsorption potential does not seem possible. Other hypotheses introduce difficulties too, the most apparent being either a large difference in the surface chemistry of adsorbed anions at Pt(111) and Pt(100), or the change of species in the presence or absence of specific adsorption. Whatever the species involved these anomalous states oblige us to re-examine carefully, classical interfacial concepts. An example is given with the new concept of the rôle of surface long range order in adsorption properties which is now clearly established.

As an illustration of the above comments, results obtained with (111) and (100) orientations are discussed in parallel.

Surface processes related to electrochemical activation

The understanding of activation effects on the electrochemical behaviour of ordered platinum surfaces will allow the possibility of correlating the classical knowledge of platinum adsorption behaviour with the new observations.

a/ High oxygen coverage. Let us consider first the effect of a high oxygen coverage by electrosorbed oxygen at a clean Pt(111)/0.5M H_2SO_4 interface. The initial voltammogram is shown in fig.1.At such a surface oxygen adsorption occurs above 1.25 V/R.H.E. as shown in fig.2A curve 1. It amounts to nearly 500 $\mu C.cm^{-2}$ and the balance of charge is exact for desorption. This is a requirement in agreement with surface cleanliness. Subsequent, hydrogen adsorption-desorption looks similar to that at a polycrystalline surface and the unusual bonded states have decreased and shifted positively.This effect is reinforced in the second cycle where oxygen adsorption potential is is decreased. After thirty cycles between 0.05-1.45 V the reversible peaks at 0.110 V/R.H.E. are well developed, and the unusual states are spread in a wide range of potential with a low amplitude.

The same type of voltammogram has been obtained with a Pt(111) electrode after its ordered surface was subjected to argon ion bombardment, introducing structural defects like randomly distributed steps (14). The similar effects of oxygen electrosorption and ion bombardment show clearly that the former perturbs the surface order.

The first consequence of both treatments is the considerable decrease of the anomalous adsorption states. This is one of the

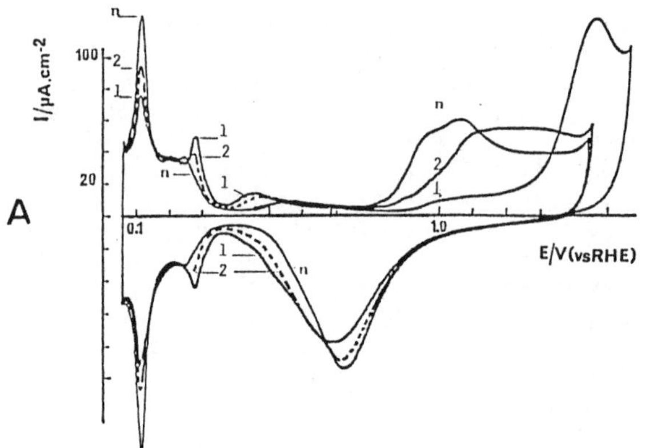

Figure 2 A. Effect of high oxygen coverage on Pt(111) : first (1),second (2) and thirtieth sweeps (n).

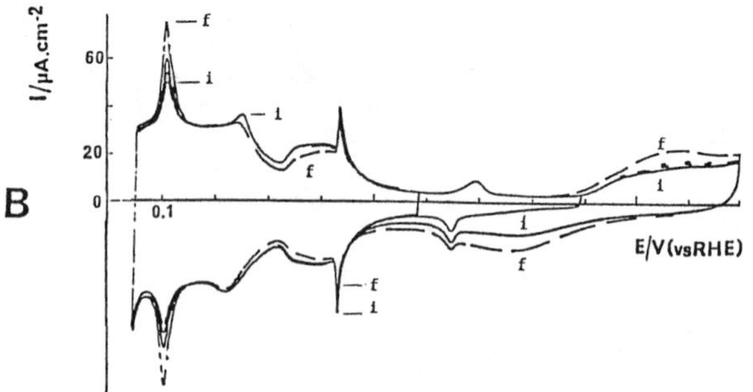

Figure 2 B. Low oxygen coverage on Pt(111), (i) initial, (f) last voltammograms.

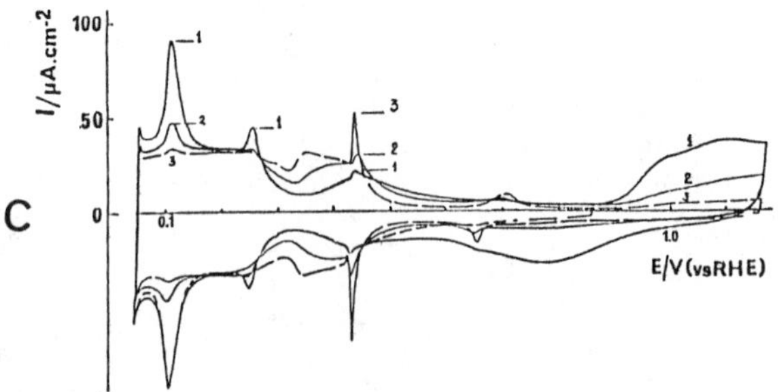

Figure 2 C. Effect of annealing on the reordering of polished Pt(111), annealing time: (1) no annealing; (2) 1.5 mn ; (3) 6.5mn

pieces of evidence for the responsibility of long-range order in the existence of these states.

A second consequence of both treatments is that a majority of new sites appears giving rise to peaks at 0.110 V the exact potential of the main peaks observed with Pt(110) fig.1 consequently they are assumed of the same nature. The hydrogen adsorbed at this potential corresponds to the weakly bonded hydrogen on a polyoriented surface. Reversible peaks corresponding to the classical strongly bonded hydrogen grow at the first cycle and disappear progressively with cycling meaning that (100) sites reach a rather low surface density.

The third consequence is that the transformation of the voltammogram in fig.2A occurs with a constant amount of electric charge when the charge is summed over all the reversible adsorption states including the unusual ones. The decrease in charge for unusual states is compensated by the increase in charge for the normal states (12)(20). Such behaviour is coherent with the assumption of hydrogen as the adsorbed species for all the reversible adsorption states.
The same behaviour is observed in perchloric or hydrofluoric acid solutions. Unfortunately a quantitative analysis cannot be carried out because the unusual adsorption state current is merging into the current of oxygen adsorption at the surface defects, the former decreases while the latter increases with surface defects density (17).
In view of the large contribution of (110) sites to the hydrogen adsorption at an electrochemically perturbed (111) orientation, stepped surfaces with (110) steps and (111) terraces have been used for the modeling of the main voltammetric features of perturbed Pt(111). Fig.5 shows the voltammogram of a flame-treated Pt(332) electrode otherwise noted 6(111)x(111) or 5(111)x(110) because the step sites have the (110) configuration. As expected the voltammogram shows the reversible peaks at 0.110 V characteristic of hydrogen adsorption-desorption at Pt(110). The other states are due to the hydrogen adsorption-desorption at terraces. This voltammogram compared to that of Pt(111) fig.1 shows clearly a smaller amount of the unusual adsorption state at a more positive potential much like after the first oxygen adsorption-desorption at high coverage. On the other hand, the voltammogram of Pt(332) in fig.5 is another confirmation of the effect of long-range order on the unusual adsorption states.
On this voltammogram the hydrogen adsorbed at step sites corresponds to a state near the lower limit of the potential range i.e. to the weakly bonded state, a conclusion at variance with gas phase results (19).

b/ Low oxygen coverage. Let us now consider the effect of low oxygen coverage for the observation of the first steps in the Pt(111) surface transformation. The progressive transformation observed in fig.2B is obtained by applying successively fast repetitive sequences of cycling between 0.05 - 1.2 V. It can be seen that the following changes are simultaneous : increase of the peaks at 0.110 V, decrease of the unusual adsorption states, increase of the amount of adsorbed oxygen. These effects allow the definition of criteria for the evaluation of the quality of a Pt(111) surface. It is evident that these changes are due to an increase in structural surface defect density, and oxygen adsorption occurs particularly at these defects while simultaneously (110) sites may be detected by hydrogen adsorption.

Thermal reordering of Pt(111) after polishing

These criteria are shown to be useful to investigate the reorganisation of the surface layers after the polishing of a (111) platinum surface. Fig.2C shows the voltammograms of a freshly polished Pt(111) sample. The first curve was recorded with the sample treated few seconds in the range 400-500°C in presence of air for surface cleaning. The first voltammogram shows the surface cleanliness. It shows common features with the curve 1, fig.2A, well developed peaks for the hydrogen adsorption at (110) sites at 0.110 V small reversible peaks at 0.26 V corresponding to strongly bonded hydrogen on (100) sites and a non negligible contribution of the unusual adsorption states. This is correlated with oxygen adsorption at surface structural defects. The presence of a non negligible amount of unusual adsorption states means that wide domains with (111) orientation are present at the surface after polishing. After that the electrode was annealed for 1.5 mn at 1300-1400°C.Then the subsequent voltammogram was recorded, curve 2, fig.2C. A substantial decrease of the peaks at 0.110 V is observed, due to the elimination of the (110) sites while the unusual adsorption states are enhanced. At the same time oxygen is adsorbed to a lesser extent. After new annealing in the same temperature range for 5 mn the voltammogram for a well ordered (111) Pt surface is practically achieved. Peaks at 0.110 V have disappeared, the unusual adsorption states are well developed with the sharp spike at 0.440 V, adsorption of oxygen has practically disappeared too. Fig.3 shows the voltammogram of a Pt(111) electrode with reordered surface by annealing at high temperature with the minimum amount of surface defects.

Simple model for Pt(111) perturbed by oxygen electrosorption.

It may be assumed that oxygen adsorption as a result of place exchange mechanism leaves after desorption some platinum atoms in a position of adatoms at the (111) ordered surface.

If we consider in a hard sphere model of fcc crystals, a dense row of adatoms on the (111) surface, on each side of the row new sites are created at the surface in same amount, (110) on one side (fig. 4a) and (100) (fig.4b) on the other side. The experimental results show that the majority of the new sites have the (110) symmetry this implies that such a row is not a stable configuration for the adatoms. More stable small triangular islands whose edge sites have the (110) symmetry seem to be formed, fig.4c possibly (100) kinks may appear, fig.4d. This type of surface arrangement is not unusual at a (111) orientation and probably is at the origin of the shape of corrosion pits observed with on the (111) face of many fcc metals. The mean widths of (111) ordered domains may be estimated from the shape and position of the unusual adsorption state observed with m(111)x(111) surfaces.

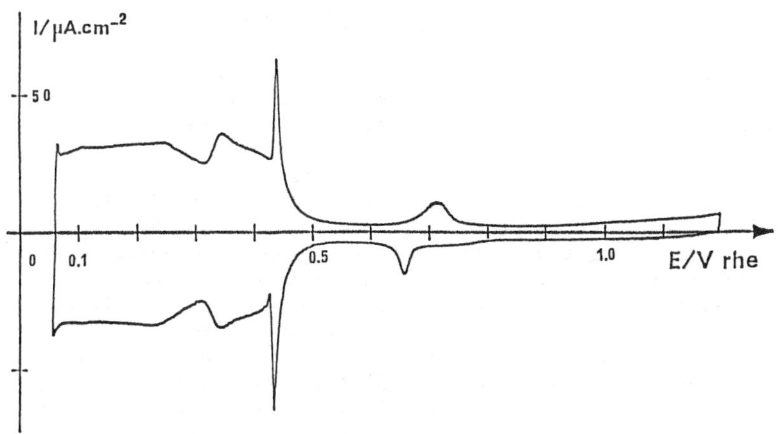

Figure 3. Voltammogram of Pt(111) with the lowest surface defect contribution to hydrogen and oxygen adsorption.

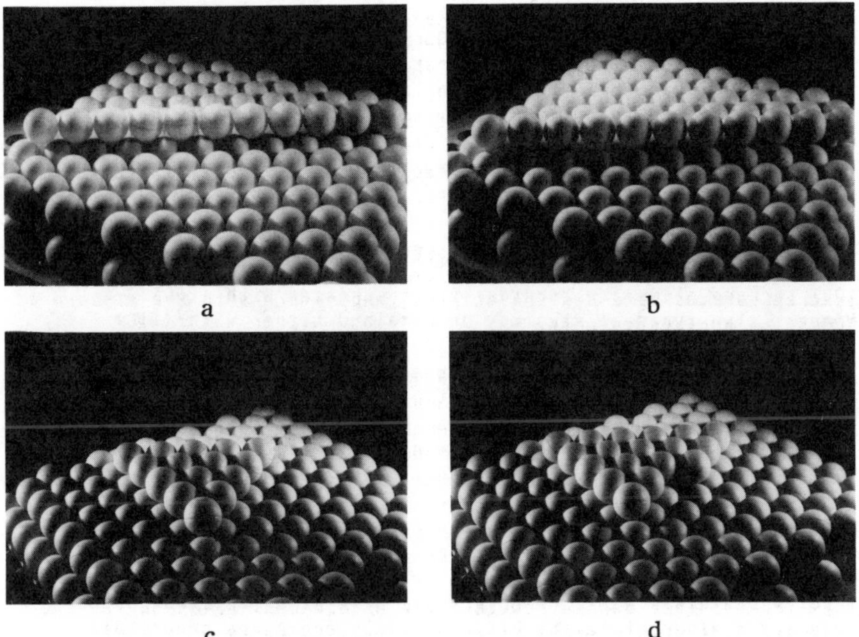

Figure 4. Hard sphere model for perturbed Pt(111). (a) dense row of Pt adatoms forming (110) sites, front view, (b) rear view showing (100) sites.(c) island forming (110) edge sites, (d) a (100) kink site.

Pt(100) orientation and (100) sites at stepped surfaces.

The Pt(100) orientation has a more complex electrochemical behaviour. The voltammogram on fig.1 corresponds to a surface cooled in air. Cooling in the presence of hydrogen the same adsorption states are observed but with a different distribution, the most positive being the highest one (11),(21). A similar distribution is obtained if the electrode is subjected to few cycles in a 1M HCl solution prior to the voltammogram recording in a 0.5M H_2SO_4 solution (22). Finally the enhancement of the reversible peaks at highest potential may be obtained by a fast cycling in a 0.5M H_2SO_4 solution as studied in (12). It is shown in fig.5 the two limiting voltammetric profiles for a (100) orientation. Experimental results indicate that (100) surface with a well- developed peak at high potential is probably that with the lowest surface defect content in view of the lowest current in the region 0.05-0.2 V as a conclusion of the results obtained with stepped surfaces (see below). Because of its absence in normal voltammetry of activated platinum the reversible state at 0.365 V is an unusual state too. When the Pt(100) orientation is subjected to oxygen adsorption-desorption at high coverage, this state disappears progressively less rapidly than in the case of Pt(111) while the adsorption-desorption current in the range 0.05 - 0.2 V increases (fig.5) without appearance of defined adsorption-desorption states like on Pt(111).Contrary to the latter orientation, even after a large perturbation of the (100) surface, the unusual state may be restored by fast cycling (12). The disappearance of the unusual state with extensive adsorption-desorption of oxygen suggests that surface long-range order is responsible for its presence too.

Stepped surfaces with (100) sites

(100) terrace sites. We consider first surfaces with (100) terraces. Two types of step may be obtained either with (110) sites for surfaces chosen in the [001] zone or with (111) sites for surfaces chosen in the [011] zone.
In fig.5 the voltammogram of the Pt(310) and Pt(511) orientations both containing (100) terraces with three atomic rows and (110) or (111) steps respectively are represented. Their similar adsorption states above 0.2 V may be unambiguously ascribed to the terrace site contribution, while characteristic states at lower potentials which are not present on the voltammogram of an ordered Pt(100) surface may be ascribed to step contribution on both surfaces. Considering the states related to the terrace contribution it may be noted the absence of the reversible peak at 0.365 V which confirms its assignment to the bidimensional long-range order. All the step contributions are located in the low potential range below 0.2 V. Both (110) and (111) step sites have their contribution to hydrogen adsorption in the same range of potential,this remark will be useful in the interpretation of the weakly bonded hydrogen particularly on activated platinum surfaces. Here the step contributions correspond to the states in the lower potential range i.e. to the weakly bonded hydrogen.

<u>(100) step sites</u>. The adsorption of hydrogen on (100) step sites may be checked by using surfaces chosen in the [011] zone. This is the case for the Pt(211) orientation which contains (111) terraces with three atomic rows. Fig.5 shows that the (100) sites contribution is limited to a single reversible peak at the potential of the least strongly bonded hydrogen on Pt(100), Pt(310) and Pt(511), while the terrace contribution gives rise to the broad peak in the low potential region. Consequently it may be concluded from the results obtained with the three orientations that (111) sites included into surface domains with no bidimensional long-range order (steps, terraces) give a contribution to hydrogen adsorption in the low potential region which is superposed with the (110) site narrow contribution at 0.110 V. Nevertheless the presence of (111) terraces may be detected by the presence of residual unusual states above 0.4 V.

Voltammetric surface analysis of platinized platinum

Recently in this laboratory we have re-investigated on platinized platinum surfaces the conditions for the observation of the various adsorption states as a consequence of the existence of ordered domains with basal orientations at the surface of the crystallites. When we consider classical results with platinized platinum electrode a voltammogram similar to that reported on Fig. 6a is observed. This type of voltammogram was obtained after cycling the electrode between 0.02 - 1.6 V, in other words for perturbed crystallites. It is interesting to investigate these electrodes from the point of view of the unavoidable presence of small ordered domains at the surface of the unperturbed crystallites. If such domains exist they have to yield adsorption states which were not observe on the conventional voltammogram of platinized platinum. Their observation requires the preparation of clean samples avoiding the use of cycling up to 1.5 V. Deposits of platinum black with a surface expansion coefficient of about 110 were prepared and studied in a narrow range of potential. A typical voltammogram is given in fig.6b. This voltammogram shows the presence of adsorption states in the range 0.3 - 0.4 V corresponding to hydrogen adsorbed at (100) terraces with various widths and the low and wide state from 0.4 to 0.6 V which is characteristic of (111) narrow terraces.

Contrary to the conventional distribution of hydrogen at activated platinum the peak at 0.26 V is higher than the peak at 0.110 V. The region of the weakly bonded hydrogen shows the narrow peak corresponding to (110) sites superposed to a broader adsorption state as expected from (111) steps and terraces contributions. After applying cycles between 0.02 and 1.6 V the voltammogram is changed to that reported in fig.6a, the surface expansion decreases down to 86 and the unusual states disappear with a dramatic decrease of the current in the double layer region above 0.4 V. These results with platinum black fit well with data from stepped surfaces and confirm that voltammetry of adsorbed hydrogen can be used as a tool leading to an interesting level of description of the crystalline surface structure of a catalyst like platinum.

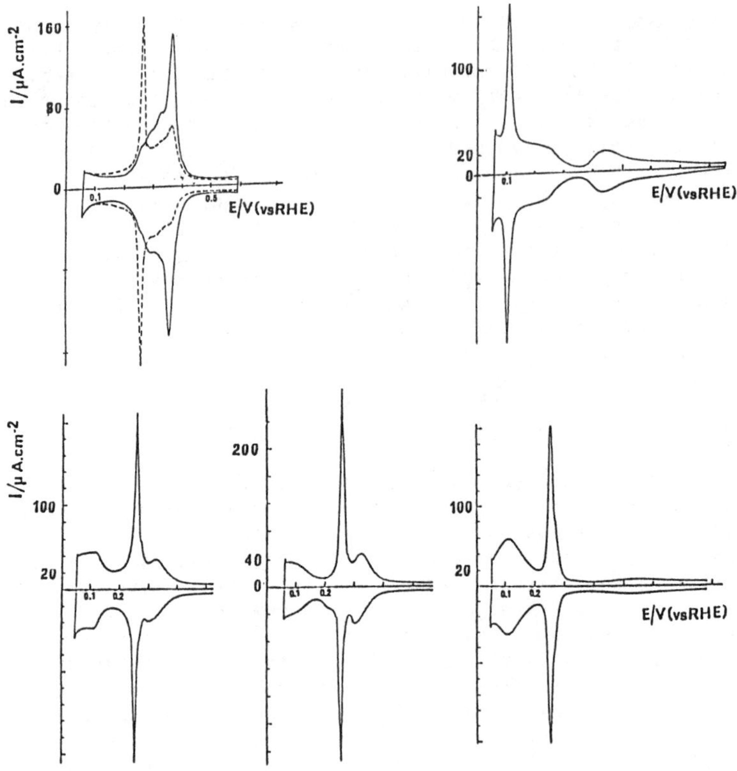

Figure 5. Voltammograms of Pt(100) (two profiles, S.G. Sun, Thesis) and stepped surfaces.

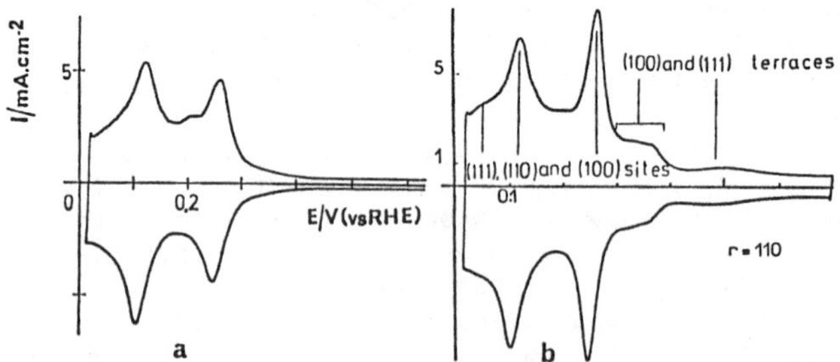

Figure 6. Voltammograms of platinum black (a) after cycles between 0.02 - 1.6 V (b) freshly electrodeposited platinum black cycled without oxygen adsorption (after Sun and Clavilier).

The use of a numerical analysis of voltammograms with a precise deconvolution technique of the adsorption states (23,24) and the comparison of resulting data with hard sphere models of the stepped surfaces in number of cases allows an understanding of the interfacial processes at the atomic level of the working surface, in the presence of the electrolyte.

Acknowledgments

The author is grateful to Drs D.Armand, M.Petit, S.G.Sun, T. Vandernoot and Mmes A.Plaza and Y. Rodier for their assistance in various ways during the preparation of this manuscript.

Literature Cited

1. Hammett, V. W. J.Am.Chem.Soc 1924, 46, 7.
2. James, S. D. J.Electrochem.Soc. 1967, 114, 1113.
3. Will, F. G. J.Electrochem.Soc. 1965, 112, 451.
4. Ishikawa, R. M.; Hubbard, A. T. J.Electroanal.Chem. 1976, 69, 317.
5. O'Grady, W. E.; Woo, M. Y. C.; Hagans, P. L.; Yeager, E. J.Vac. Sci.Technol. 1977, 14, 365.
6. Yeager, E; O'Grady, W. E.; Woo, M. Y. C.; Hagans, P. L. J. Electrochem.Soc. 1978, 125, 348.
7. Yamamoto, K.; Kolb, D.M.; Kotz, R; Lehmpfuhl, G. J.Electroanal. Chem. 1979, 96, 233.
8. Hubbard, A. T.; Ishikawa, R. M.; Katekaru, J. J. Electroanal. Chem.1978, 86, 271.
9. Clavilier, J.; Faure, R.; Guinet G.; Durand, R. J.Electroanal. Chem. 1980, 107, 205.
10. Clavilier, J.; Armand, D.; Sun, S. G.; Petit, M. J. Electroanal. Chem. 1986, 205, 267.
11. Clavilier, J.; Armand, D.; Wu, B. L. J.Electroanal.Chem. 1982, 135,159.
12. Clavilier, J.; Armand, D. J.Electroanal.Chem. 1986, 199, 187.
13. Al-Jaaf-Golze, K.; Kolb, D.; Scherson, D. J.Electroanal. Chem. 1986, 200, 353.
14. Aberdam, D.; Corotte, C.; Dufayard, D.; Durand, R.; Faure R.; Guinet, G. Proceedings of the 4th Conference on Solid Surfaces, September 22-26(1980), Cannes, France, Supplément à "Le Vide, Les Couches Minces" 1980, 201, 622.
15. Wagner, F. T.; Ross, P. N. Jr. J.Electroanal.Chem. 1985, 150, 141.
16. Wagner, F. T.; Ross, P. N. Jr. Surface Science 1985, 160, 305.
17. Aberdam, D.; Durand, R.; Faure, R.; El-Omar, F. Surface Science 1986, 171, 303.
18. Clavilier, J. J.Electroanal.Chem. 1980, 107, 211.
19. Christmann, K.; Ertl, G. Surface Science 1976, 60, 365.
20. Furuya, N.; Motoo, S.; Kunimatsu, K. J.Electroanal.Chem. 1988, 239, 347.
21. Motoo, S.; Furuya, N. J.Electroanal.Chem. 1984, 167, 309.
22. Scortichini, C. L.; Reilley, C. N. J.Electroanal.Chem. 1982, 139, 233.
23. Armand, D.,Clavilier, J. J.Electroanal.Chem. 1987, 225, 205.
24. Armand, D.,Clavilier, J. J.Electroanal.Chem. 1987, 233, 251.

RECEIVED May 17, 1988

Chapter 15

In Situ Surface Extended X-ray Absorption Fine Structure at Chemically Modified Electrodes

M. J. Albarelli, J. H. White, M. McMillan, G. M. Bommarito, and H. D. Abruña

Department of Chemistry, Baker Laboratory, Cornell University, Ithaca, NY 14853

The surface EXAFS and near edge structure of electropolymerized films of $[M(v\text{-}bpy)_3]^{+2}$ (v-bpy is 4-vinyl-4'-methyl,-2,2'-bipyrdine and M = Ru, Os) and $[Os(v\text{-}bpy)_2(phen)]^{2+}$ (phen is 4-methyl-1,10-phenanthroline) on platinum electrodes have been studied as a function of surface coverage and applied potential by measuring the characteristic RuK_a and OsL_a fluorescence intensity. For $[Ru(v\text{-}bpy)_3]^{+2}$ spectra for electrodes modified with 1, 5, 12, 25 and 50 monolayers were obtained. Analysis and comparison of EXAFS and XANES features for these films showed similar morphological and electronic characteristics in agreement with previous electrochemical studies showing that solution redox properties of monomer complexes can be transferred to electrode surfaces by electrodeposition as polymer films of varying thickness. Electrochemical oxidation of the films resulted in a shift of 2.0 eV in the edge position towards higher energy, consistent with the higher charge on the ruthenium centers. Other spectral features, however, remained essentially unchanged. Similar studies were performed on the osmium complexes but, in addition, changes in the edge position and near edge features could be correlated with the coordination environment. The applicability of these studies to the in-situ investigation of electrocatalytic systems is discussed.

0097–6156/88/0378–0216$06.00/0

Among the main goals of electrochemical research are the design, characterization and understanding of electrocatalytic systems, (1-2) both in solution and on electrode surfaces. (3) Of particular importance are the nature and structure of reactive intermediates involved in the electrocatalytic reactions. (4) The nature of an electrocatalytic system can be quite varied and can include activation of the electrode surface by specific pretreatments (5-9) to generate active sites, deposition or adsorption of metallic adlayers (10-11) or transition metal complexes. (12-16) In addition the electrode can act as a simple electron shuttle to an active species in solution such as a metallo-porphyrin or phthalocyanine.

Over the years, many systems have been investigated and a variety of experimental probes, both electrochemical and spectroscopic, have been used in their study and characterization. Thus, electrochemical techniques such as rotated-ring disk electrodes (17) and spectroscopic techniques such as Raman (18-20) (in its numerous variants) have provided much insight into the mechanisms of many of these processes, however, in-situ structural studies of such systems have, to date, proved very elusive to direct experimental probing.

The application of ultra-high vacuum surface spectroscopic methods coupled to electrochemical techniques (21-24) have provided valuable information on surface structure/reactivity correlations. These determinations, however, are performed ex-situ and thus raise important concerns as to their applicability to electrocatalytic systems, especially when very active intermediates are involved.

An added difficulty that arises in the in-situ spectroscopic study of electrocatalytic systems in solution is that the active species will be located in the vicinity of the electrode so that the material in solution will generally represent a large background signal making the detection and identification of related species difficult. Thus, it would be ideal to be able to probe only that region proximal to the electrode surface and furthermore to be able to obtain structural information of the species involved.

A way to circumvent the first problem is to ensure that all of the active material is present at the electrode surface. That is, employ a chemically modified electrode where a precursor to the active electrocatalyst is incorporated. The field of chemically modified electrodes (3) is approaching a more mature state and there are now numerous methodologies for the incorporation of materials that exhibit electrocatalytic activity. Furthermore, some of these synthetic procedures allow for the precise control of the coverage so that electrodes modified with a few monolayers of redox active material can be reproducibly prepared. (3)

The second problem of concern is being able to perform an in-situ structural characterization of the redox active catalyst incorporated on the electrode surface. Surface EXAFS and X-ray Standing Waves (XSW) are perhaps the only techniques capable of yielding in-situ (the electrode in contact with an electrolyte solution and under potentiostatic control) structural information on electrodeposited layers.

We have previously employed such techniques in the study of iodide adsorption onto Pt(111) electrodes (25) as well as in the in-situ structural characterization of underpotentially deposited copper and silver on Au(111) electrodes. (26)

With the exception of Elder et. al., who used EXAFS to look at redox properties of copper complexes diffused in Nafion film modified electrodes (27), no in-situ studies have been reported on modified electrodes using EXAFS.

We now present an in-situ surface EXAFS study of electropolymerized films of $[M(v\text{-}bpy)_3]^{2+}$ (v-bpy is 4-vinyl, 4'-methyl, 2,2'-bipyridine and M = Ru, Os) and $[Os(phen)(v\text{-}bpy)_2]^{2+}$ (phen is 4-methyl phenanthroline) and on the applicability of this technique to follow the course of redox transformations.

These systems have been the object of considerable study since the monomer complexes and their analogues have been widely studied (28) and electropolymerization is readily accomplished at the vinyl substituent on the bipyridinyl ligand. (29-31) There are various aspects that we wish to address and these are the dependence of the structure of electrodeposited layers on coverage and the ability to follow redox transformations. The dependence of structure on surface coverage should indicate 1) the nature and strength of interaction of the electrodeposited film with the electrode surface and 2) the applicable range of surface coverages over which polymer local structure is similar to that of the monomeric parent complex in solution (and by inference, the range over which redox behavior is similar). In addition, the ability to follow redox transformations will be of great utility in investigating electrocatalytic systems. We are also interested in correlating changes in spectral features with coordination environment.

The information obtained from the previously described studies may be used to establish guideposts for the rational design and synthesis of electrochemical interfaces (in particular, of polymer modified electrodes) with high catalytic activity.

Experimental

Reagents. $[M(v\text{-}bpy)_3]^{2+}$ (M = Ru, Os) were prepared as previously described.(32-33) $[Os(phen)(v\text{-}bpy)_2]^{2+}$ was prepared following procedures previously described for analogous complexes.(34) Acetonitrile (Burdick and Jackson distilled in glass) was dried over 4Å molecular sieves. Tetra n-butyl ammonium perchlorate (TBAP) (G.F. Smith) was recrystallized three times from ethyl acetate and dried under vacuum at 70 °C for 72 hours.

Electrochemical Instrumentation. For the Ru complexes, a 1 cm diameter platinum disk brazed onto a brass holder was used as a working electrode. It was masked with ChemGrip (a teflon based epoxy) except for the upper face. Prior to use, it was polished with 1 micron diamond paste (Buehler) and rinsed with water, acetone and methanol. The working electrode for each Os complex was the uppermost platinum layer of a platinum/carbon layered synthetic microstructure (LSM) (Energy Conversion Devices). The LSM consisted of 200 layer pairs of carbon and platinum whose thicknesses were 24.4 and 17.0 Å, respectively and where platinum was the outermost layer. The LSM was placed in 1.0 M H_2SO_4 and cleaned electrochemicaly by holding the potential at +1.7 volts for 5 minutes, then cycling the potential from +1.2 to -0.4 volts for 10 min. Afterwards, the LSM was rinsed with water and dried. The counter and reference electrodes were a platinum coil and silver wire, respectively.

Electropolymerization was carried out in acetonitrile/0.1M TBAP solution in a conventional three compartment electrochemical cell according to previously described procedures. (29-31)

The electrochemical cell for the EXAFS experiments (Figure 1) was machined from a teflon cylinder (6 cm diameter x 6 cm high) and was provided with contacts for all electrodes as well as teflon fittings for the injection of electrolyte. The electrode was placed flat on the top of the cell and covered with a thin Tefzel (E. I. DuPont de Nemours Inc.) film (12μm) which was held in place with a viton o-ring placed 1 cm below the top and around the circumference of the cell. The surface of the platinum electrode was raised about 1 mm above the edge of the cell so as to allow near grazing incidence of the x-ray beam. The tefzel cover made effectively a thin layer cell whose thickness we estimate to be of the order of 20 μm. The cell was mounted inside a plexiglas box with Kapton (polyimide) windows (not shown in Figure 1). The top of the box was provided with entrance and exit ports so that it could be constantly flushed with He to eliminate contamination from oxygen and minimize air scattering. The box assembly was bolted on a Huber 410 goniometer stage providing very fine and reproducible rotation and translation. (Typically 4,000 steps/degree of rotation and 100 steps/μm translation.) All experiments were performed at grazing incidence in order to enhance the surface signal and decrease background scatter.

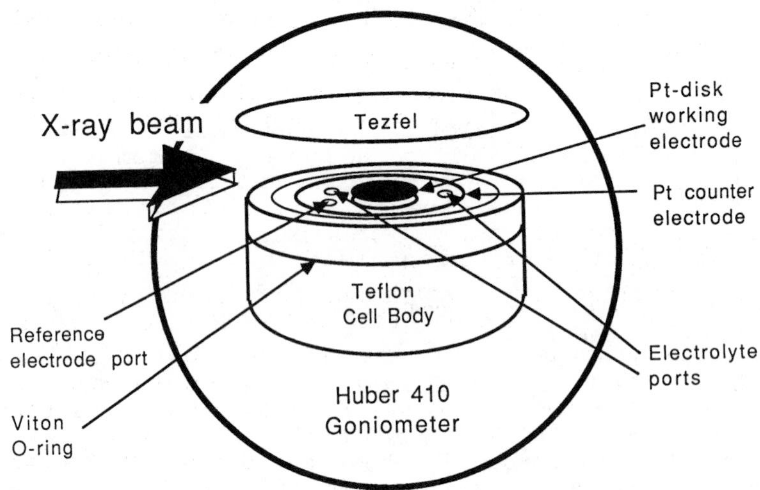

Figure 1. Electrochemical cell for in situ surface EXAFS at chemically modified electrodes.

X-ray Instrumentation. All experiments were performed at the Cornell High Energy Synchrotron Source (CHESS) operated at 5.8 GeV (Stations A-3, B and C-2). Monochromatic radiation was obtained with a Si (220) double crystal monochromator. In order to eliminate higher harmonics, 50% detunning was typically employed .

A Princeton Gamma Tech Si(Li) solid state detector in conjunction with an EG&G Ortec Model 673 Spectroscopy Amplifier and a Tennelec Model 450 Single Channel analyzer were employed to detect the characteristic Ru K_a and Os L_a fluorescence at 19.2 keV and 8.9 keV respectively. Data were analyzed using a modified version of the EXAFS analysis program by B.M. Kincaid (AT&T Bell Labs).

Bulk samples of $[M(v\text{-}bpy)_3]^{2+}$ (M = Ru, Os), and $[Os(phen)(v\text{-}bpy)_2]^{2+}$ (employed as model compounds) were evaporated from acetonitrile solutions and the EXAFS signal obtained. In this case the complexes were not in solution but in a He atmosphere.

Procedure

The platinum electrode and LSM's were modified with Ru and Os polymers respectively as described in the experimental section. Each sample was rinsed with acetone and placed in an acetonitrile/0.1M TBAP solution and a cyclic voltammogram of the Ru (or Os) $^{2+/3+}$ wave obtained. From the charge consumed (i.e. by integration of the area under the voltammetric wave) the surface coverage (in equivalent number of monolayers) was determined. Samples with 1, 5, 12, 25, and 50 monolayers were prepared for the $[Ru(v\text{-}bpy)3]^{2+}$, 1 and 3 monolayers for $[Os(v\text{-}bpy)_3]^{2+}$, and 2 monolayers for $[Os(phen)(v\text{-}bpy)_2]^{2+}$. A monolayer of complex (diameter of 14.2 Å) represents about $1X10^{-10}$ moles/cm^2. Since detection is based on the characteristic fluorescence arising from the metal centers, a more relevant figure is the amount of metal present ($5X10^{13}$ atoms/cm^2) which represents about 5% of a metal monolayer.

Each electrode in turn was mounted on the EXAFS cell and the spectrum of the reduced form of the polymeric film was performed while the potential was held at 0.0 V. Afterwards, the potential was scanned to 300 mV positive of the oxidation wave of the complex (+1.6 V for Ru and +1.1 V for Os) and held for 5 minutes to ensure complete oxidation prior to obtaining the spectrum of the oxidized form. After spectroscopic investigations, the voltammetric behavior of each modified electrode in acetonitrile/0.1M TBAP was performed to confirm there was no loss of material.

Results and Discussion

$[Ru(v\text{-}bpy)_3]^{2+}$. Figure 2 presents EXAFS spectra for bulk $[Ru(v\text{-}bpy)_3]^{2+}$ (A) as well as for electrodes modified with 5 (B) and 1 (C) monolayers of poly-$[Ru(v\text{-}bpy)_3]^{2+}$. The spectrum for bulk $[Ru(v\text{-}bpy)_3]^{2+}$ shows a well defined edge at 22.18 keV with three well defined oscillations beyond the edge. Qualitatively similar spectra are obtained for electrodes modified with 1 and 5 monolayers of the polymer. Again, a well defined edge and the first oscillation are clearly defined in both cases, but the second and especially the third oscillations are difficult to discern in the spectrum for the electrode modified with a single monolayer whereas, for the electrode modified with 5 monolayers of polymer, these oscillations are well defined. The appearance of oscillations at higher energies is more difficult to detect because of the rapid decrease in the backscattering amplitude for low Z scatterers (nitrogen in this case), in addition to the small amounts of material present on the modified electrode. (Recall that a monolayer of complex represents about 5% of a metal monolayer.) Thus, even for rather low coverages of ruthenium, clear signals can be obtained. For samples containing 25 or more monolayers, the spectrum obtained was indistinguishable from that of the bulk material.

Figure 3 shows the phase uncorrected radial distribution functions (solid line) for bulk $[Ru(v\text{-}bpy)_3]^{2+}$ (A) and an electrode modified with 50 monolayers of poly-$[Ru(v\text{-}bpy)_3]^{2+}$ (B) and both show a very prominent peak at a distance of about 1.5 Å. These peaks were fourier filtered in the indicated regions (dashed line) and back transformed into k space where they were fitted for amplitude and phase. Phase corrections were made employing the Ru-N distance reported by Rillema et al (34) for an x-ray diffraction study of $[Ru(bpy)_3]^{2+}$. Figure 4 shows the k weighted experimental data (solid line), filtered data (dashed line) and fit (squares) for an electrode modified with 50 monolayers of poly-$[Ru(v\text{-}bpy)_3]^{2+}$. A very good fit is obtained for the first three oscillations although at higher wave vector the fit degrades somewhat due to the lower signal to noise ratio in this region of the spectrum.

Virtually identical results were obtained for bulk

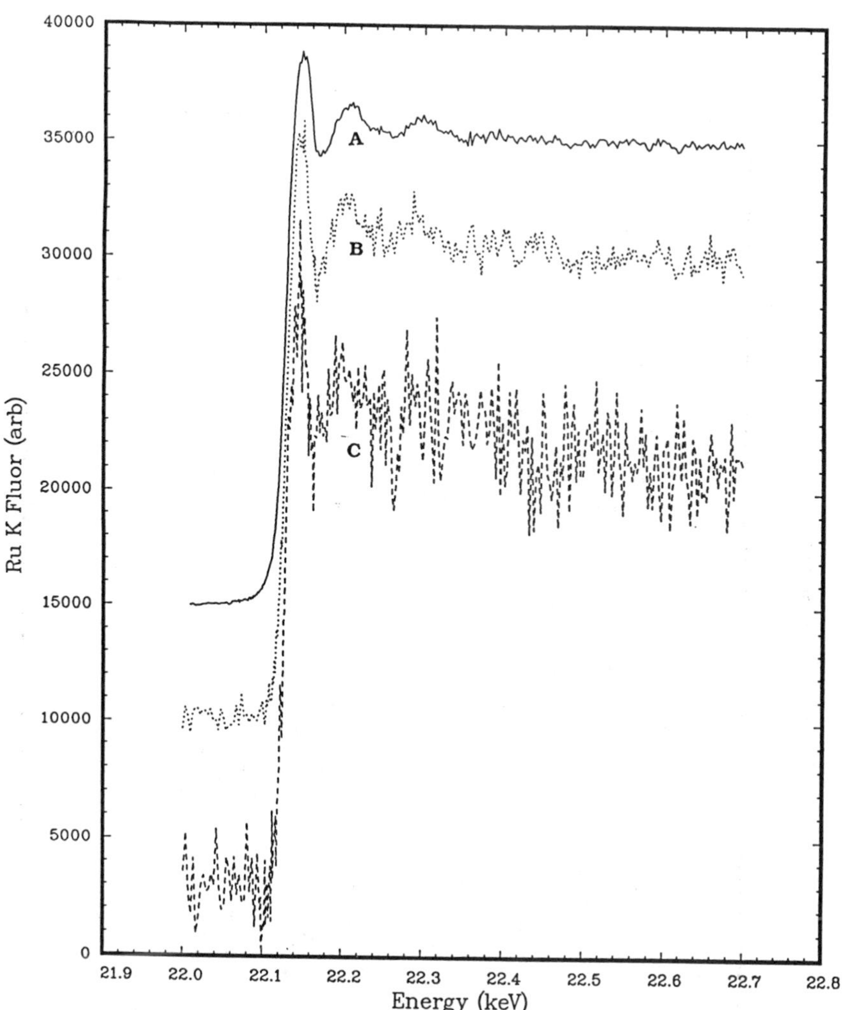

Figure 2. Normalized EXAFS spectra for (A) bulk $[Ru(v\text{-}bpy)_3]^{+2}$, (B) platinum electrode modified with 5 monolayers of poly-$[Ru(v\text{-}bpy)_3]^{+2}$, (C) same as B except the coverage is one monolayer. Spectra have been shifted vertically for clarity.

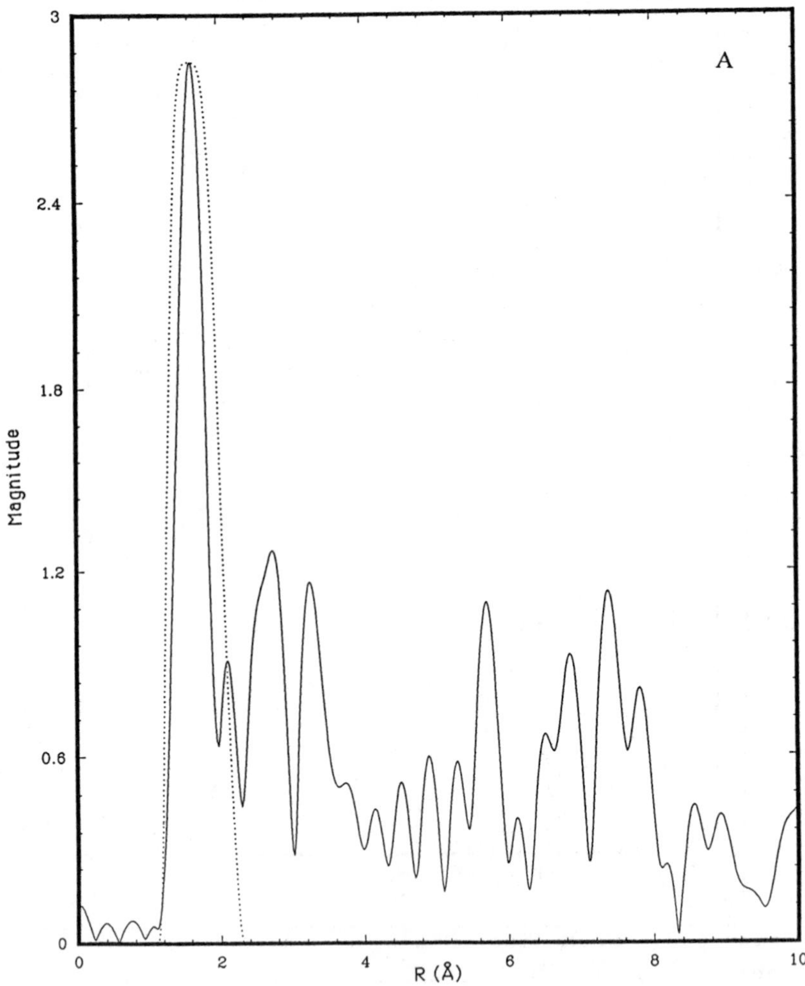

Figure 3A. Phase uncorrected radial distribution functions (solid line) and fourier filter window (dashed line) for bulk $[Ru(v\text{-}bpy)_3]^{+2}$.

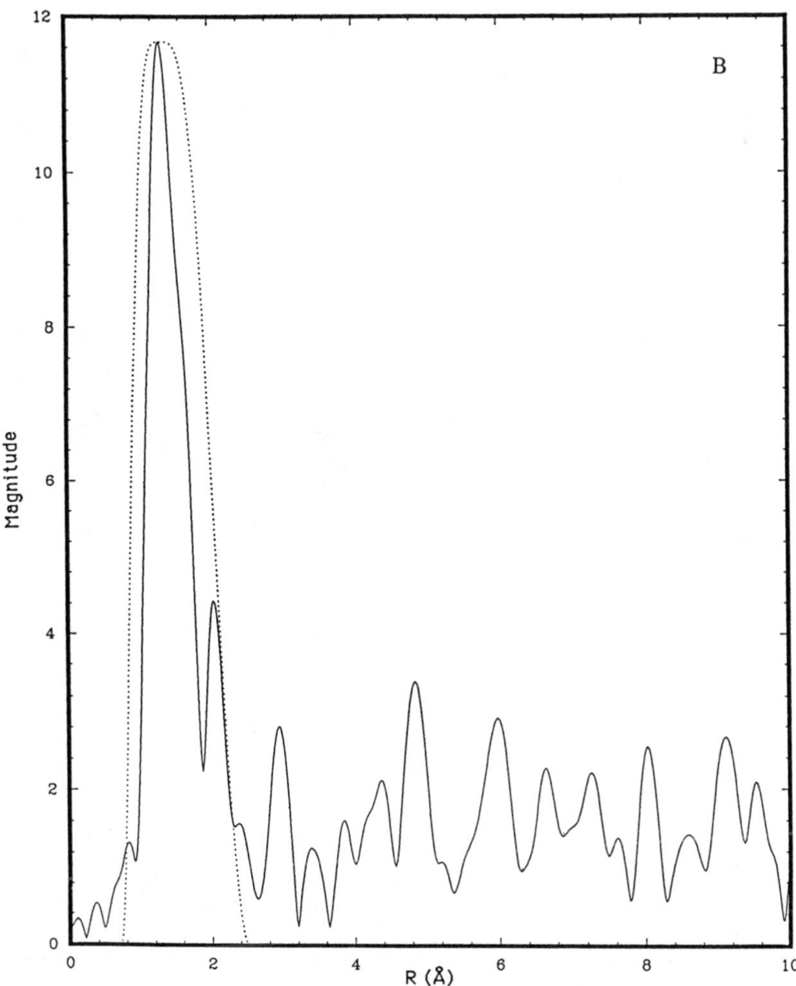

Figure 3B. Phase uncorrected radial distribution functions (solid line) and fourier filter window (dashed line) for electrode modified with 50 monolayers of poly-$[Ru(v\text{-}bpy)_3]^{+2}$.

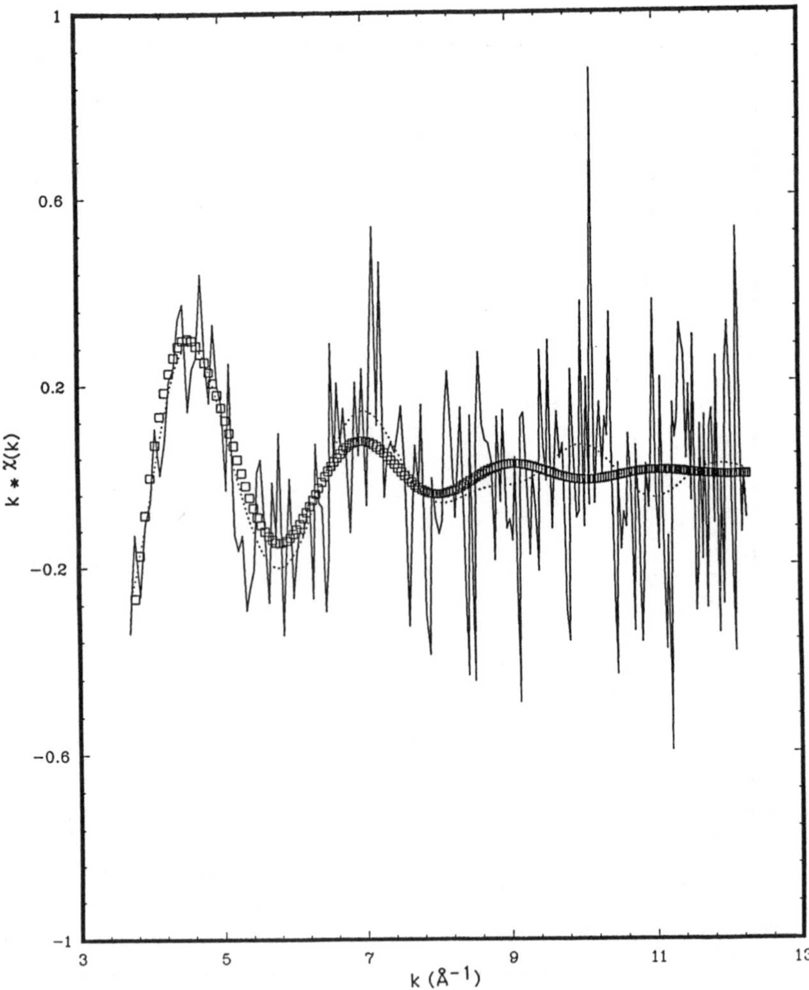

Figure 4. EXAFS as a function of wave vector for an electrode modified with 50 monolayers of poly-[Ru(v-bpy)$_3$]$^{+2}$. experimental data (solid line), fourier filtered data (dashed line), theoretical fit (squares).

$[Ru(v\text{-}bpy)_3]^{2+}$. Upon fitting the data to theoretical amplitudes and phase shifts, bond lengths of 1.9, 2.0, and 2.1 angstroms and coordination numbers of about 6 were found for 1, 5, and 50 monolayers respectively. There is doubtful statistical significance in the trend observed in the bondlengths. However, it appears that the similarity in the data for the samples is indicative of similar local structure at all stages of polymer deposition. These values correlate very well with the known coordination number of six and a Ru-N distance of 2.056 Å. (34)

In addition to the similarities aforementioned, it is also clear that there is little difference between the electrodeposited polymer and bulk compound $[Ru(v\text{-}bpy)_3]^{2+}$ in terms of the near edge spectral features pointing to a similar geometric disposition of scatterers.

The results presented here seem to indicate that 1) the local order about ruthenium centers in the polymers is essentially unchanged from that in the monomer complex and 2) that the interaction with the electrode surface occurs without appreciable electronic and structural change. This spectroscopic information corroborates previous electrochemical results which showed that redox properties (e.g. as measured by formal potentials) of dissolved species could be transferred from solution to the electrode surface by electrodepositions as polymer films on the electrode. Furthermore, it is apparent that the initiation of polymerization at these surfaces (i.e. growth of up to one monolayer of polymer) involves no gross structural change.

The results presented to this point were for films where the ruthenium was present in the 2+ oxidation state. Identical results were obtained for electrodes potentiostated at 0.0 V or at open circuit.

Having established that there were no significant structural perturbations in the coordination spheres of the ruthenium centers in the polymer films we investigated the effect of oxidation of the ruthenium to the 3+ state. This was performed in acetonitrile/0.1M TBAP by holding the potential at +1.6 V for 5 minutes to ensure oxidation of the film. A change in the color of the film from orange (typical of $[Ru(v\text{-}bpy)_3]^{2+}$) to green (typical of $[Ru(v\text{-}bpy)_3]^{3+}$) was a clear indication of complete oxidation.

Upon oxidation from Ru^{2+} to Ru^{3+} there was a well defined shift towards higher energies in the edge position. Figure 5 shows the edge regions for the films present in the reduced (A) and oxidized (B) forms. The shift, taken at half the edge jump, is 2 eV and consistent with the change in oxidation state of the metal centers. Consequent reduction of the film resulted

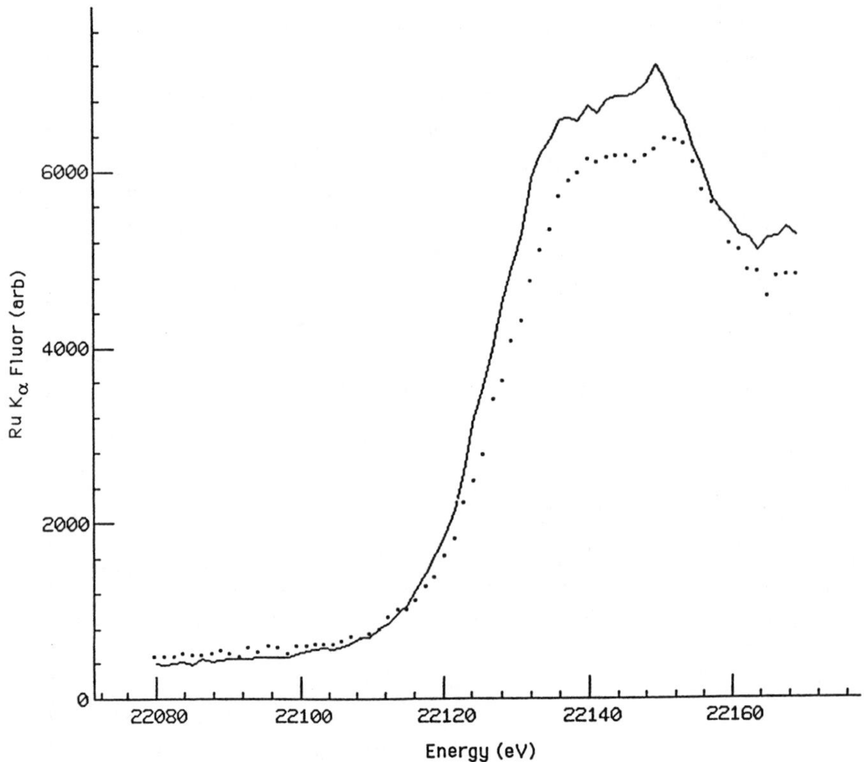

Figure 5. X-ray absorption edge of poly-$[Ru(v\text{-}bpy)_3]^{+2}$ at 0.0 V (solid line) and +1.6 V (dotted line) vs. Ag/AgCl.

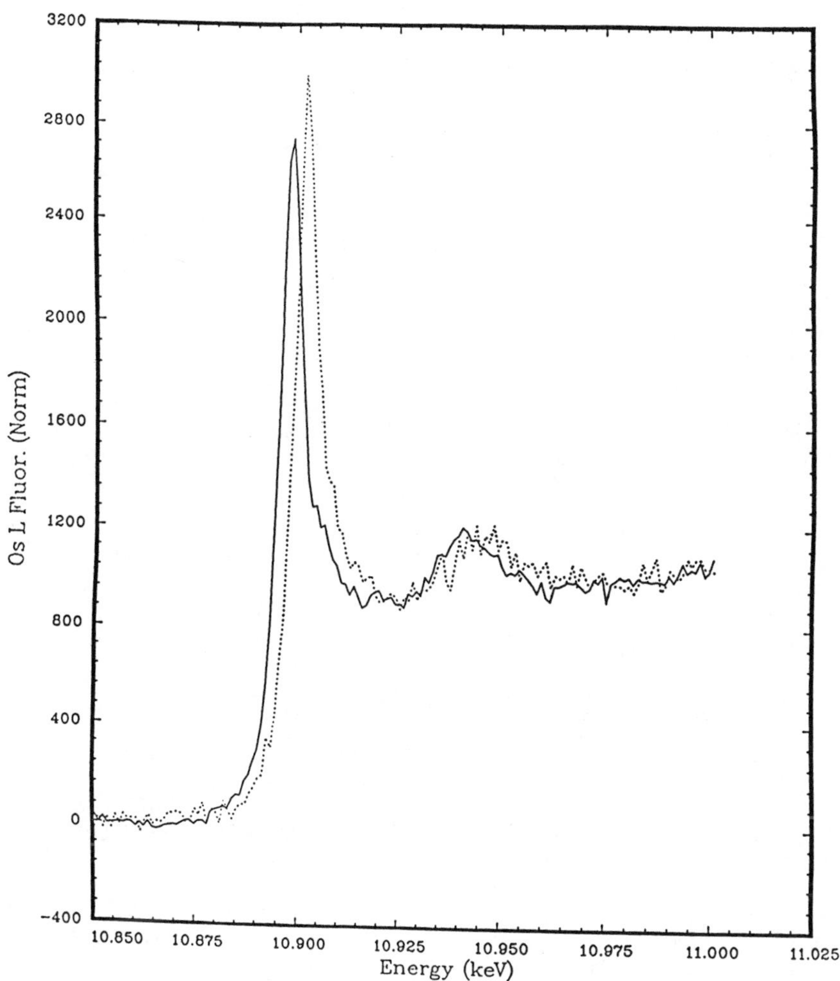

Figure 6. EXAFS spectra for electrodes modified with $[Os(v\text{-}bpy)_3]^{2+}$ (solid line) and $[Os(phen)(v\text{-}bpy)_2]^{2+}$ (dotted line).

in a reversible shift of the edge back to lower energy. The spectral features obtained in this case were essentially identical to those obtained for the film in its reduced form. This is not surprising, since the difference in metal/ligand (Ru/N) bond distance is estimated to be less than 0.01 Å.

$[Os(v\text{-}bpy)_3]^{2+}$ and $[Os(phen)(v\text{-}bpy)_2]^{2+}$. Similar results were obtained for the Os complexes in terms of the local structure around the metal center and changes in the position of the edge as a function of oxidation state. In addition, we were able to correlate changes in the position of the edge and near edge features with the coordination environment around the metal center. Figure 6 presents spectra for $[Os(v\text{-}bpy)_3]^{2+}$,and $[Os(phen)(v\text{-}bpy)_2]^{2+}$ and the difference in the edge position (3.5 eV) and area under the first peak are immediately apparent. These changes can be ascribed to the presence of the phenanthroline ligand whose higher pi acidity relative to bipyridine results in a higher degree of pi backbonding. Since a higher degree of backbonding effectively removes electron density away from the metal center the shift in the edge position to higher energies is consistent with this effect. This is also consistent with the change in the area under the sharp transition near the edge as has been previously demonstrated for Pt, Rh and other metals.(35)

There are some points from this study that need to be emphasized. First, it is possible to obtain good quality in-situ surface EXAFS spectra for electrodes modified with one to five monolayers of a transition metal complex (depending on the metal of interest) and changes in oxidation state can also be monitored. In addition, the near edge spectral features and edge position changes can also be correlated with the coordination environment around a metal center. These aspects should aid significantly in the study and identification of electrocatalytic intermediates in-situ and we are currently pursuing such studies. In addition, we are currently involved in a x-ray standing wave study of the same LSM samples (with the deposited osmium polymers) to determine the degree of order in the electropolymerized layer.

Acknowledgments

This work was supported by the Materials Science Center at Cornell University, the National Science Foundation, the Army Research Office, the Office of Naval Research and the Dow Chemical Co. HDA acknowledges support by the Presidential Young Investigator Award Program of the National Science Foundation and the Alfred P. Sloan Foundation.

Literature Cited

1. O'Grady, W. E.; Ross, P. N.; Wiull, F. G., Eds.; Proc. Symp. Electrocat.; The Electrochemical Society, Inc.: New Jersey, 1982; Vol. 82-2.
2. McIntyre, J. D. E.; Weaver, M. J.; Yeager, E. B., Eds.;Symposium of the Chemistry and Physics of Electrocatalysis; The Electrochemical Society, Inc.: New Jersey, 1983.
3. Murray, R. W. In Electroanalytical Chemistry; Bard, A. J., Ed.; Marcel Dekker: New York, 1983; Vol. 13, pp. 191-368.
4. Meyer, T. J. J. Electrochem. Soc. 1984, 131, 221C.
5. Hance, G. W.;Kuwana, T. Anal. Chem. 1987, 59, 131.
6. Poon, M.; McCreery, R. L. Anal. Chem 1986, 58, 2745.
7. Kak, J.; Kuwana, T. J. Am. Chem. Soc. 1982,104, 5515.
8. Ergstrom, R. C. Anal. Chem. 1982, 54, 2310.
9. Cabaniss, G. E.; Diamantis, A. A.; Murphy, W. R.; Linton, R. W.;Meyer, T. J. J. Am. Chem. Soc. 1985, 107, 1845.
10. Kolb, D. M. In Advances in Electrochemistry and Electrochemical Engineering; Gerischer, H.; Tobias, C., Eds.; Pergamon Press: New York, 1978, Vol. 11, p. 125.
11. Adzic, R. R. In Advances in Electrochemistry and Electrochemical Engineering; Gerischer, H.; Tobias, C., Eds.; Wiley Interscience: New York, 1984, Vol. 13, p. 159.
12. Collman, J. P.;Marrocco, M.;Denisevich, P.; Koval, C.; Anson, F. C. J. Electroanal. Chem. 1979, 101, 117.
13. Collman, J. P.; Kim, K. J. Am. Chem. Soc. 1986, 108, 7847.
14. Lieber, C. M.; Lewis, N. S. J. Am. Chem. Soc. 1984, 106, 5033.
15. Abruña, H. D.; Walsh, J. L.; Meyer, T. J.; Murray, R. W. J. Am. Chem. Soc. 1980, 102, 3272.
16. Abruña, H. D.; Calvert, J. M.; Denisevich, P.; Ellis, C. D.; Meyer, T. J.; Murphy, W. R.; Murray, R. W.; Sullivan, B. P.; Walsh, J. L. In Chemically Modified Electrodes in Catalysis and Electrocatalysis; Miller, J. S., Ed.; ACS Symposium Series No. 192, ACS Press: Washington, D. C., 1982.
17. Bard, A., J.; Faulkner, L. R. Electrochemical Methods; Wiley & Sons: New York, 1980.
18. Fleischmann, M.; Hendra, P. J.; McQuillan, A. J. Chem. Phys. Lett. 1974, 26, 173.
19. Fleischmann, M.; Hendra, P. J.; McQuillan, A. J. J. Electroanal. Chem. 1975, 65, 933.
20. Jeanmarie, D. J.; Van Duyne, R. P. J. Electroanal. Chem. 1977, 84, 1.
21. Hubbard, A. T. Accts. Chem. Res. 1980, 13, 177.
22. Yeager, E. J. Electroanal. Chem. 1981, 128, 1600.

23. Ross, P. N. Surf. Sci. 1981, 102, 463.
24. Bange, K.; Grider, D. E.; Madey, T. E.; Sass, J. K. Surf. Sci. 1984, 136, 381.
25. Gordon, J. G.; Melroy, O. R.; Borges, G. L.; Reisner, D. L.; Abruña, H. D.; Chandrasekhar, P.; Albarelli, M. J.; Blum, L. J. Electroanal. Chem. 1986, 210, 311.
26. Blum, L.; Abruña, H. D.; White, J. H.; Albarelli, M. J.; Gordon, J. G.;Borges, G. L.; Samant, M. G.; Melroy, O. R. J. Chem. Phys. 1986, 85, 6732.
27. Elder, R. C.; Lunte, C. E.; Rahman, A. F. M. M.; Kirchhoff, J. R.; Dewald, H. D.; Heineman, W. R. J. Electroanal. Chem., 1988, 240., 361.
28. Meyer, T. J. Accts. Chem. Res. 1978, 11, 94.
29. Abruña, H. D.; Denisevich, P.; Umaña, M.; Meyer, T. J.; Murray, R. W. J. Am. Chem. Soc. 1981, 103, 1.
30. Denisevich, P.; Abruña, H. D.; Leidner, C. R.; Meyer, T. J.; Murray, R. W. Inorg. Chem. 1982, 21, 2153.
31. Ghosh, P. K.; Spiro, T. G. J. Electrochem. Soc. 1981, 128, 1281.
32. Abruña, H. D. J. Electroanal. Chem. 1984, 175, 321.
33. Abruña, H. D. J. Electrochem. Soc. 1985, 132, 842.
34. Rillema,D. P.; Jones, D. S.; Levy, H. A. J. Chem. Soc. Chem. Commun. 1979, 849.
35. Bart, J. C. J. Adv. in Cat. 1986, 34, 203.

RECEIVED May 17, 1988

Chapter 16

Charge-Transfer Reaction Inverse Photoemission at Gold(111) and Polycrystalline Silver Electrodes

R. McIntyre, D. K. Roe, J. K. Sass, and H. Gerischer

Fritz-Haber-Institut der Max-Planck-Gesellschaft, Faradayweg 4–6, D–1000 Berlin 33, Federal Republic of Germany

Emission spectra have been recorded for electron injection into Au and Ag spherical electrodes and hole injection into Au(111) planar electrodes. These processes were brought about in solutions of acetonitrile containing tetrabutylammonium hexafluorophosphate (TBAHP), using the trans-stilbene radical anion as the electron injector and the thianthrene radical cation as hole injector. The spectrum for the hole injection process into planar Au(111) electrodes has been resolved into the P & S-polarised components of the emitted light. A comparison of the spectral distribution of emitted light for the above electron injection process, occurring at both Au and Ag electrodes, with that obtained for the emission process occurring at the equivalent Al/Al_2O_3/metal tunnel junction is reported.

The emission of light from metal and semi-metal surfaces in contact with a condensed phase containing charge acceptors or donors has recently been the subject of several papers /1-5/. In this work, we are going to deal with three new aspects which have recently emerged. First, an experimental modification to the original cell geometry has been made, to avoid an unusual edge effect caused by a lower cell resistance at the edge of the electrode than at the centre. The consequence is a non-uniformity of potential which is enhanced during transient changes of potential with electronic compensation for the IR drop in solution. Second, with this new modified cell geometry and hence better defined conditions, we have resolved the emitted light from the Au(111) planar surface into P and S-polarised components. The enhanced emission of P-polarised light has been attributed to direct sp to d-band transitions occurring in the Γ-L direction for planar Au(111) electrodes. Third, by comparing the spectral distribution of the emitted light for the electron injection process from electronic donor states in solution into polycrystalline gold and silver electrodes with the spectrum for the analogous process occurring at

0097–6156/88/0378–0233$06.00/0

the Al/Al_2O_3/metal tunnel junction /6/, we propose that information concerning the shape of the distribution curve which describes the electronic states in solution can be obtained.

EXPERIMENTAL

Experiments were performed with a conventional three electrode electrochemical system controlled by a PG-potentiostat (HEKA Electronik) modified to give +25V output with a maximum current of 2A. The single crystal Au(111) electrode was mechanically polished, orientated and finally electrochemically polished using a procedure described in detail elsewhere /7/. The gold and silver spherical electrodes were produced by heating the respective wires in a reducing flame. To avoid exposing the wire to the charge injection process, it was sealed into glass at the point of contact with the sphere. The Au(111) crystal was sealed into a Kel-F holder using a silicon rubber gasket material. The platinised Pt mesh counter electrode was then held firmly in place, against the Kel-F holder, directly opposite the Au(111) working electrode such that the volume of electrolyte between the working and counter electrodes could be described as a cone with the counter electrode as base ($A = 0.64cm^2$) and the working electrode as top ($A = 0.125cm^2$). The separation between the two electrodes was ~3mm. This holder substantially decreases the edge effect. The Kel-F holder was located in the glass cell such that the Au(111) electrode was in line with the rotation axis. The Luggin capillary was positioned between the working and counter electrodes by means of a 0.5mm diameter hole in the Kel-F holder.

The solvent acetonitrile, the supporting electrolyte, TBAHP, and the reactant thianthrene were purified by well-known procedures described in detail elsewhere /8/. The reactant t-stilbene (Fluka Gmbh) was recrystallised twice from a methanol water mixture. The optical arrangement consisted of focusing lenses, a high efficiency Bausch and Lomb monochromator and a polarising filter. The electrochemical cell was mounted on an X, Y, Z manipulator with calibrated rotation facilities (Fritz-Haber-Institut). The detection equipment has been described in detail elsewhere /1,2/.

RESULTS

It has previously been shown that the trans-stilbene

radical anion is a very suitable species for injecting electrons into Pt electrodes, in acetonitrile solutions, with excess energies of up to 2.2eV /5/. Fig 1 shows the voltammogram for the reversible reduction of t-stilbene (10^{-3}M) at a Au surface (the reversible potential U_r = -2.5V vs Ag/Ag^+). The potential range positive of the wave is shown to be almost perfectly polarisable to +1.0V.

To create the excited states we employed square-wave potential modulation from -2.6V where the radical anion is produced to various positive potentials within the polarisable region. The oscilloscope trace of the current-time and photon-time data are shown in Fig 2, using electronic compensation for about 90% of the cell resistance and a modulation amplitude of 3.1V (ie. -2.6V to 0.5V). We see clearly that light is produced only during the half-cycle when the radical anion is oxidised. During the negative half-cycle production of the t-stilbene radical anion occurs, on stepping positive there is initially no light since the time constant of the cell is about 20μs due to the residual uncompensated resistance. After 30-50 s, the desired potential change is sufficient and we observe emission throughout the remaining period of the positive cycle. On returning to the negative half-cycle, the excitation energy is reduced and the photon pulses disappear. The spikes observed on the current-time curves are an unfortunate aberration due to the high impedance of the junction between reference and working electrode compartments. The curvature of the photon spikes is simply due to the finite time constant associated with the inherent capacitance of the coax cable and the large load resistor (10k Ω) necessary to register the photomultiplier current on the oscilloscope.

The spectra recorded for the electron injection process by the t-stilbene radical anion into both Au and Ag spherical electrodes are shown in Fig 3. The excitation energy E_{ex} is defined as $(U_p - U_r)/e$ where U_p is the positive potential limit. For the silver electrodes U_p was limited to -0.1V to avoid appreciable dissolution of the electrode. Two observations can be made, both of which have previously been identified with the CTRIPS process /1-4/. First, spectral peak values (E_{sp}) and the high energy threshold values (E_{th}) are observed to shift to lower energies with decreasing excitation energy. Second, the emission intensity is observed to fall with decreasing values of E_{ex}. In this case, however, we no longer observe the high energy light (ie. photons with energies>0.5eV higher than expected from

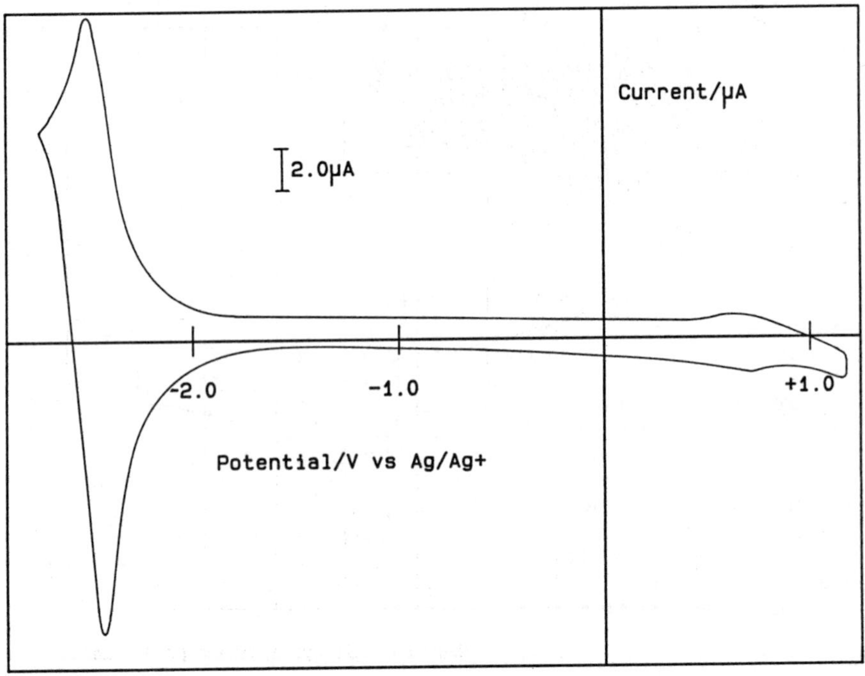

Figure 1. The current-voltage curve for the reversible reduction of t-stilbene (10^{-3}M) at a Au sphere electrode (area = 0.125cm^2) in acetonitrile containing TBAHP (0.2M) for a sweep rate = 0.1Vs^{-1}.

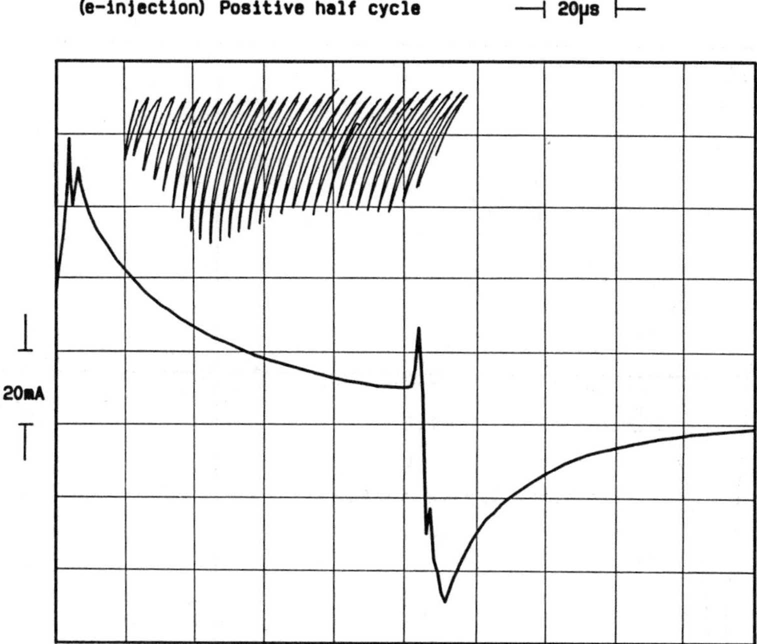

Figure 2. Current-time and photon-time data for the electron injection process by t-stilbene into a Au sphere electrode. These data were obtained using electronic compensation for the residual IR drop in solution, with a modulation amplitude of 3.1V (ie. -2.6V to 0.5V).

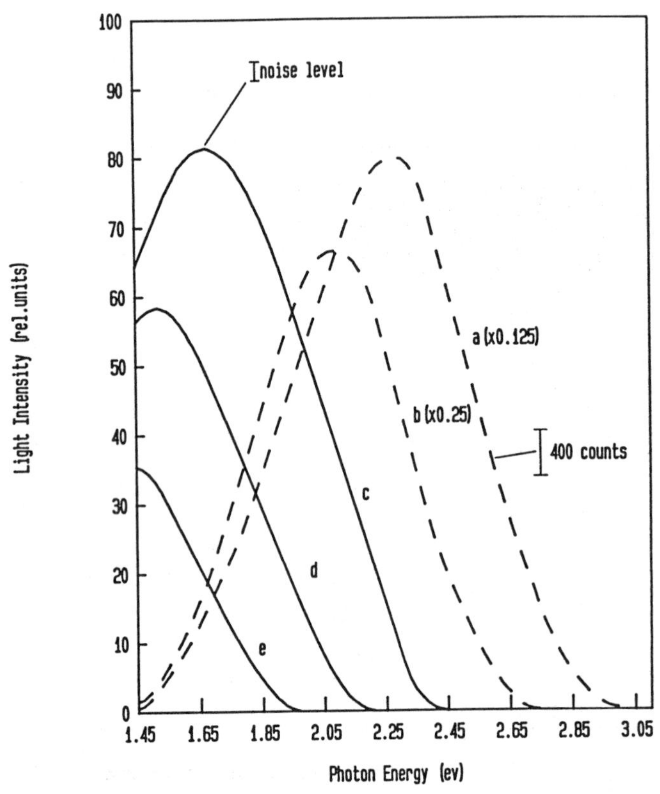

Figure 3. CTRIPS spectra recorded for the system, metal/MeCN, TBAHP (0.2M) t-stilbene (0.07M) for the electron injection process by the radical anion into a spherical Au electrode (dashed line), area = 0.22cm^2, a) E_{ex} = 2.6eV; b) E_{ex} =2.6eV; c) E_{ex} = 2.4eV and a spherical Ag electrode area = 0.125cm^2; d) E_{ex} = 2.2eV and e) E_{ex} = 2.0eV.

the maximum excitation energy) when compensating electronically for the IR drop in solution, as previously reported /3,4/. Furthermore, with this particular geometry, the precision with which E_{th} and E_{sp} move as a function of E_{ex} is much closer to that expected from simple energetic considerations, than previously observed /1-4/.

As a consequence of having performed the above experiments with spherical electrodes we suspected that the origin of the high energy light, previously observed with planar electrodes, was a result of a non-uniform potential distribution at the planar electrode due to the lower solution resistance at the unshielded edge compared to the central area of the electrode. We tested this hypothesis with a planar gold disc electrode evaporated onto a mica substrate, similar to that used previously /2-4/. By having two redox couples benzophenone (U_r = -2.0V vs Ag/Ag^+) and thianthrene (U_r = +0.95V vs Ag^-/Ag^+) present in solution and then setting the modulation limits to -1.5V and +0.4V (ie. 500mV away from the reversible potential of either couple) we could not expect to observe light from either the electrogenerated chemiluminescence or inverse photoemission processes. However, by applying electronic compensation we were able to observe orange light originating from the edges of the gold electrode. This observation can be explained if we consider that for a given volume element at the edge of the electrode, the resistance between working and counter electrodes is slightly less than for the same volume element at the centre, simply because more current paths exist at the edge than in the centre. Therefore, during the positive feedback process, used to compensate for the IR drop, it is possible to sustain higher potentials at the edge than in the centre, thus exceeding the set potential limits at the edge and hence giving rise to electrogenerated chemiluminescence at the edges only. A quantitative analysis which attempts to explain the high energy light reported earlier /3,4/ will be given elsewhere /9/. To avoid this situation with planar electrodes, it is necessary to eliminate the extra current paths available at the edges of the electrodes. This was achieved for the Au(111) electrode using the Kel-F holder described in Section 2.

The spectra recorded for the hole injection process by the thianthrene radical cation into the Au(111) electrode are shown in Fig 4, for three different excitation energies. Once again, square wave modulation was employed with a fixed positive value of +1.0V where the cation radical is produced /4/ to negative values

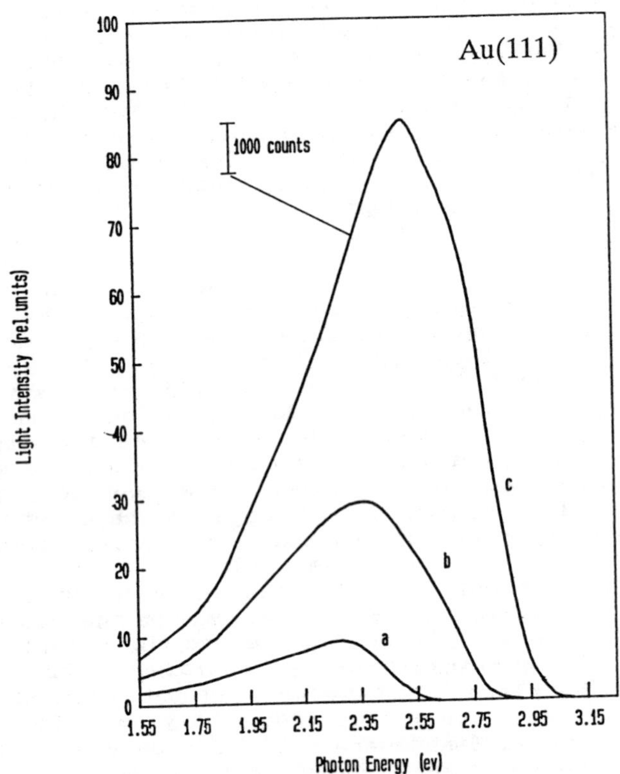

Figure 4. CTRIPS spectra recorded for the system Au(111)/ MeCN, TBAHP (0.2M), thianthrene (10^{-2}M) for the hole injection process with excitation energies, a) E_{ex} = 2.6eV; b) E_{ex} = 2.8eV and c) E_{ex} = 3.0eV for an electrode area = $0.125cm^2$.

chosen to give the required excitation energy. The characteristic shifts of E_{sp} and E_{th} to lower energies with decreasing values of E_{ex} are observed. Also, in contrast to the electron injection process, where light pulses are observed in the positive half-cycle (see Fig 2), we observed emission only during the negative half-cycle in accordance with expectations for a hole injection process. Fig 5 shows an emission peak for P-polarised light from the Au(111) surface. This spectrum was obtained by resolving the emitted light into P and S-polarised components for an emission angle = 30°. The P and S spectra were then normalised to account for the wavelength dependent transmission of the monochromator and the photo-response of the photomultiplier tube, then subtracted to give the P-polarised emission spectrum. The emission peak is centred at about 2.1eV for an excitation energy E_{ex} = 3.0eV.

DISCUSSION

With this new cell geometry for planar electrodes the threshold energy is no longer dependent on the degree of electronic compensation for the IR drop and always coincides closely with the excitation energy. Therefore, these spectra are more likely to represent the true joint optical density of states for the system than those reported previously /1-4/. Consequently, this data does merit more rigorous interpretation with respect to the spectral distribution of the emitted light and the polarisation dependence of the emission.

The emission of P-polarised light, for hole injection into Au(111) electrodes, occurs predominantly in the energy range 1.7 to 2.5eV. If we consider that for orientation P the transmitting axis of the filter was parallel to the plane of the electrode but rotated 90° about the line of observation, then by rotating the electrode to make a finite angle (in our case 30°) with the normal, the emission of P-polarised light would be enhanced if the source were a dipole radiating perpendicular to the surface of the gold electrode. Inspection of the bulk band structure for Au /10/ shows that direct sp to d-band transitions, fulfilling the above requirements, are possible in the Γ-L direction with energies similar to those observed above.

The principle of the light-emitting tunnel junction warrants particular mention because of the similarities which exist between the CTRIPS process for electron injection and the emission process which occurs in tunnel junctions /11,12/. In the first case electrons

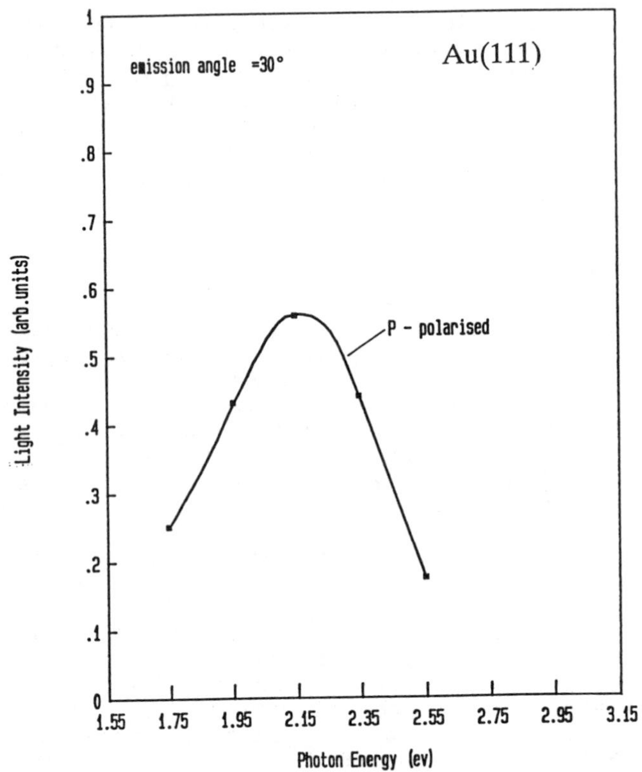

Figure 5. The P-polarised emission spectrum obtained for hole injection by the thianthrene radical cation into Au(111) for emission angle = 30° and excitation energy E_{ex} = 3.0eV. (NB. electrochemical conditions as for spectrum C, Fig 4).

are injected from a distribution of electronic states in solution into a metal electrode with typical excitation energies of 2.0-3.0eV and currents of the order of 100mA. In the second case electrons tunnel from one metal to another with excitation energies of 2.0-3.7eV and typical current values between 8-100mA. Data exists for Al-Al_2O_3 tunnel junctions with top Au and Ag films /6/. The spectra obtained from these junctions, for the top electrodes biased 3.5V positive with respect to the aluminium, show high energy threshold values at about 3.5eV, which is consistent with the maximum excitation energy. The spectrum for the Ag top electrode shows almost constant emission intensity down to 1.5eV. Similar behaviour is observed for the Au top electrode with the exception of a slight dip at 2.5eV which is reportedly due to a Au interband transition. The CTRIPS spectra for electron injection, shown in Fig 3 for both Au and Ag, have high energy threshold values which compare favourably with the maximum excitation energy, however, the emission intensity in both cases peaks and falls off quite rapidly. This is expected since the electronic states in solution are not continuous as they are for the metal injector. However, the exact distribution function which describes the states in solution has been the subject of controversy for a number of years. Unfortunately, the existing tunnel junction data is not directly comparable with existing CTRIPS data because of a miss-match of the excitation energies. However, by carrying out CTRIPS and tunnel junction experiments, experimental information could be obtained to compare with theoretical descriptions of the distribution function for the electronic states in solution.

REFERENCES

1. R McIntyre and J K Sass, J Electroanal. Chem., 196,199 (1985)

2. R McIntyre and J K Sass, Phys. Rev. Lett. 56,651 (1986)

3. R McIntyre, D K Roe, J K Sass and H Gerischer, Ber.Bunsenges. Phys. Chem. 91,488 (1987)

4. R McIntyre, D K Roe, J K Sass and W Storck, J Electroanal Chem. 228,293 (1987)

5. J Ouyang and A J Bard, J Phys. Chem. 91,4058 (1987)

6. A Adams and P K Hansma, Phys. Rev. B, 23,3597 (1981)

7. H Laucht, J K Sass, E Piltz and H Neff Surface Sci. 80,141 (1979)

8. R McIntyre, B Smandek and H Gerischer, Ber. Bunsenges. Phys. Chem. 89, 78 (1985)

9. D K Roe, R McIntyre, J K Sass and H Gerischer, to be published

10. N E Christensen and B O Seraphin, Phys. Rev. B 4,3321 (1971)

11. J Lambe and S J McCarthy, Phys. Rev. Lett. 37,923 (1976)

12. Tien-Lai Hwany, S E Schwarz and R K Jain, Phys. Rev. Lett. 36,379 (1976)

RECEIVED May 17, 1988

Chapter 17

Radioelectrochemistry at Well-Defined Electrodes

Andrzej Wieckowski

Department of Chemistry, University of Illinois, Urbana, IL 61801

Radiochemical measurements of adsorption on single crystal, well-defined electrodes are reported and the method of the measurements is described. Adsorption of acetic acid on Pt(111) surface was studied: the surface concentration data were correlated with voltammetric profiles of the Pt(111) electrode in perchloric acid electrolyte containing 0.5 mM of CH_3COOH. It is concluded that acetic acid adsorption is associative and occurs without a significant charge transfer across the interface. Instead, the recorded currents are due to adsorption/desorption processes of hydrogen, processes which are much better resolved on Pt(111) than on polycrystalline platinum. A classification of adsorption processes on catalytic electrodes and atmospheric methods of preparation of single crystal electrodes are discussed.

The methodology of surface electrochemistry is at present sufficiently broad to perform molecular-level research as required by the standards of modern surface science (1). While ultra-high vacuum electron, atom, and ion spectroscopies connect electrochemistry and the state-of-the-art gas-phase surface science most directly (1-11), their application is appropriate for systems which can be transferred from solution to the vacuum environment without desorption or rearrangement. That this usually occurs has been verified by several groups (see ref. 11 for the recent discussion of this issue). However, for the characterization of weakly interacting interfacial species, the vacuum methods may not be able to provide information directly relevant to the surface composition of electrodes in contact with the electrolyte phase. In such a case, *in situ* methods are preferred. Such techniques are also unique for the nonelectrochemical characterization of interfacial kinetics and for the measurements of surface concentrations of reagents involved in

0097–6156/88/0378–0245$06.00/0

steady-state processes. Many *in situ* techniques have recently emerged and some of them can provide valuable information as to the composition and structure of the "wet" solid-liquid interface. Vibrational analysis by SERS (12) and surface IR (12,13), *in situ* X-ray diffraction (12), Raman second harmonic generation (14,15) and scanning tunneling microscopy (16) are the best examples of probing the solid/liquid interface without system emersion.

We have recently modified (17) one of the several radiochemical methods (18) which have been used for surface electrochemistry investigations in order to characterize adsorption on well-defined, single crystal electrodes. Below, we will describe the technique and identify some challenging issues which we will be able to address. The proposed method is sensitive to a few percent of a monolayer at smooth surfaces, is nondestructive and simple to use. The radiochemical measurements can be made with all compounds which can be labelled with reasonably long-lived, preferably β^- emitting radioisotopes. We believe this technique will fulfill the quantitative function in *in situ* surface analysis as Auger spectroscopy currently does in vacuum, *ex situ* characterization of electrodes.

The Method. The sketch of the radio-electrochemical cell (17) used in our laboratory is shown in Figure 1. Figure 2 illustrates the essential element of the cell: the single crystal electrode and the detector of the nuclear radiation, i.e., the glass scintillator. Two positions of the electrode against the scintillator are depicted: in position A the electrode is situated sufficiently away from the detector to allow for a complete attenuation, in the bulk electrolyte, of the radiation emitted from the electrode surface. In this position the counting efficiency of the system is determined. In position B, the electrode is pressed against the glass scintillator and the amount of adsorption is measured. The counting contribution due to the radioactivity in the trapped film can, in some instances, be negligible, or can easily be calibrated in the "squeeze" (B) position when the adsorption is not occurring. A suitable electrode potential can very often be found to perform such calibration.

Surface concentration and counting rate data are connected by Equation 1:

$$\Gamma = \frac{N_{ads}}{N_{backg}} \frac{N_A \times 10^{-3}\, c}{\mu R f_b \exp(-\mu x)} \tag{1}$$

where: N_{ads} is the signal measured in the "squeezed" position, Figure 2-B, N_A is the Avogadro constant, c is the bulk concentration of the adsorbate (M), N_{backg} is the counting rate measured in the "open" position, Figure 2-A, μ is the absorption coefficient of the β^- radiation in solution, R is the roughness factor, f_b is the backscattering factor, and x is the thickness of the trapped solution, Figure 2-B.

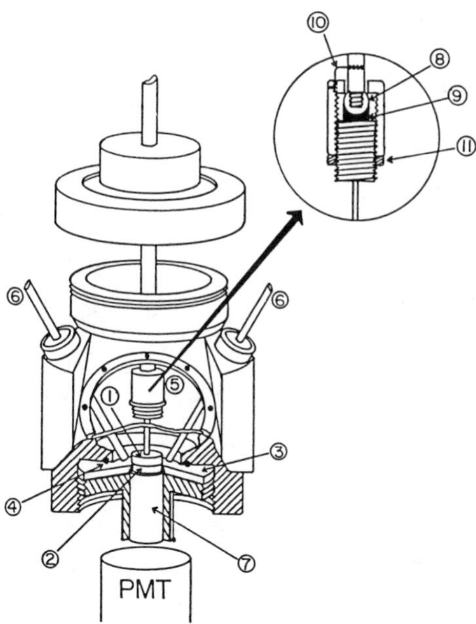

Figure 1. Diagram of Teflon cell: (1) platinum electrode; (2) glass scintillator; (3) Macor ceramic disk cell bottom; (4) Teflon O-ring; (5) flexible elbow (see insert); (6) cell ports (six of them around cell body); (7) light pipe. Inset shows the details of the flexible elbow: (8) stainless steel sphere; (9) concave Teflon spacer; (10) platinum wire for electrical connection across elbow; (11) lock nut.

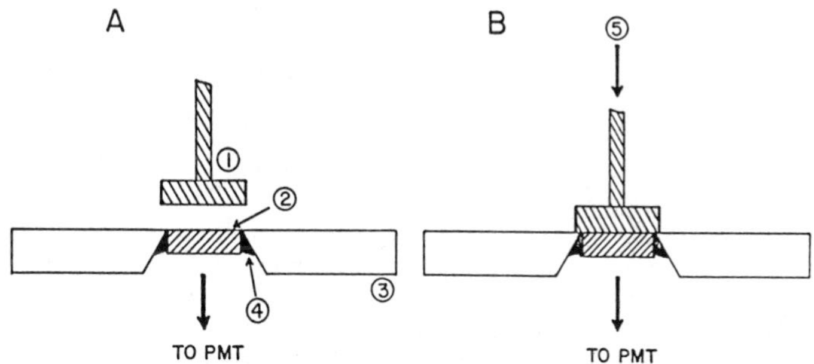

Figure 2. Diagram of electrode in adsorption position and "squeezed" position for adsorbate measurement: (1) smooth, polycrystalline platinum electrode; (2) polished glass scintillator detector; (3) Macor ceramic disk, which forms cell bottom; (4) resin; (5) electrode shaft.

In the squeezed position, the β^- radiation is trapped in the gap between the two surfaces and the electrons are backscattered from both of them (19). The larger contribution comes from the scattering from the high density material of the metal electrode (due to the Z-dependence of the backscattering power). The f_b correction which accounts for the backscattering is included in Equation 1.

A powerful characterization of electrode surfaces can be achieved by the combined radiochemistry and electrochemistry procedures; with polycrystalline materials many measurements of this type have been reported (18,20,21). A successful study of adsorption at well-defined surfaces has recently been conducted in our laboratory and it is reported below (22,23).

Electrochemical Adsorption at Catalytic Electrodes. A classification of adsorption processes at catalytic electrodes, such as platinum or rhodium, first proposed by Horanyi (24) and further developed by Wieckowski (21,25,26), categorizes adsorption processes into three fundamental groups:

1) surface complexing processes;
2) processes leading to formation of CO-type products;
3) adsorption processes of weakly interacting species.

Two first two groups can best be understood in terms of chemisorption with energetics, irreversibility and surface diffusion characteristics analogous to those known from chemisorption research in the gas phase (2). They encompass both dissociative and nondissociative adsorption processes. The discrimination between the first and the second group was possible (21,25,26) due to the identification of two separate potential ranges for the voltammetric electrooxidation of respective chemisorption products. This is, in turn, related to the differences in adsorption energies: for the first group between 30 and 40 kcal/mole and for the second group, higher than 40 kcal/mole. The high adsorption energy associated with these surface reactions makes the chemical contribution to the their overall energetics much more pronounced than the electrical contribution. Such systems, in their compositional and structural aspects, can safely be studied by the vacuum methods: any desorption (or rearrangement) from the surface during solution/vacuum transfer at room temperature can be ruled out from energetic reasons.

The processes classified in the third group are of primary importance in elucidating the significance of electric variables in electrosorption and in the double layer structure at solid electrodes. These processes encompass interactions of ionic components of supporting electrolytes with electrode surfaces and adsorption of some organic molecules such as saturated carboxylic acids and their derivatives (except for formic acid). The species that are concerned here are weakly adsorbed on platinum and rhodium electrodes and their heat of adsorption is well below 20 kcal/mole (25). Due to the reversibility and significant mobility of such weakly adsorbed ions or molecules, the application of the in situ methods for the surface concentration measurements is more appropriate than that of the vacuum techniques. The primary measurements that can be done by the use of the radio-electrochemistry regarding this group of processes are as follows:

- potential dependence of adsorption at various metals and at different crystallographic planes of a given metal;
- adsorption isotherms and isotherm-related parameters such as free energy of adsorption and interaction parameters (again related to surface crystallography);
- dynamics of the interfacial processes including kinetics of adsorption and desorption and those of molecular and atomic exchanges between the species in the surface attached and dissolved states;
- steady-state characteristics;
- molecular orientation of molecules that have more than one possible modes of surface attachment.

To implement this program of measurements on well-defined electrodes, the method of preparation of clean and well-ordered single crystal surfaces is essential. Such surfaces can be obtained either by the use of ultra-high vacuum or "atmospheric" procedures, the latter methods will briefly be described below.

Preparation of Well-Defined Electrodes for In Situ Measurements in Surface Electrochemistry. There are three methods of preparation of catalytic, single crystal electrodes that do not require an ultra-high vacuum environment: (i) crystal flame annealing followed by water quenching (27); (ii) crystal flame annealing followed by cooling in gaseous hydrogen (28) and (iii) crystal electrical annealing and cooling in an inert gas atmosphere containing iodine vapor (29,30). To create an iodine-free surface, this third method requires some additional steps developed in our laboratory: a replacement of iodine by carbon monoxide and anodic stripping of chemisorbed CO has been proposed (31). Contrary to the electrooxidation of iodine, the electrooxidation of carbon monoxide does not cause surface disorder.

A voltammetric characterization of platinum single crystal surfaces produced by these three methods shows that a very similar surface order and cleanliness are obtained in each case. The features embodied by the third method which we use for the radio-electrochemistry work, are as follows:

- up to the stage of I/CO replacement, the single crystal preparation procedure was strictly verified by LEED and Auger spectroscopy (29,30). A close correspondence between the surface electrochemistry and the gas-phase surface science of platinum has resulted due to this earlier work.
- slow cooling in the iodine atmosphere does not cause polygonization of platinum crystal reported by some authors to trouble the quenching technique (32). For the applications requiring high level of surface smoothness (as in the case of radio-electrochemistry) deformations of the electrode material must be avoided.
- the clean, iodine-free single crystal surface is created in the electrolytic solution thus minimizing possible

contaminations from the gaseous atmosphere before the electrochemical steps are executed.

- extension of the preparation of single crystal surfaces to rhodium has become possible (33). Aqueous quenching of hot rhodium causes a formation of black surface deposit and makes further well-defined work with this metal difficult, if not impossible.

Case Studies: Adsorption of Acetic Acid on Pt(111) Single Crystal Electrodes. Acetic acid is one of a very few organic compounds which is reversibly adsorbed on platinum at room temperature (20,25). We report below our radiochemical results on adsorption of this compound on Pt(111) and on polycrystalline Pt.

The Pt(111) electrode was characterized by the low electron energy diffraction in the ultra-high vacuum instrument to be described shortly. A proper hexagonal LEED pattern and its low-intensity background (Figure 3) indicate a high degree of surface order. The crystal was disconnected from the UHV chamber and transferred via air to the radiochemical cell, Figure 1 (17). The cleaning and ordering of the surface was performed according to the I/CO method described above and in ref. 31. Addition of 0.5 mM acetic acid to the perchloric acid solution produced the voltammogram shown in Figure 4. The formation of some new voltammetric features between 0.1 and 0.4 V and the suppression of the peaks between 0.3 and 0.6 V, characteristic for clean perchloric acid electrolyte should be noted. Similar voltammetric behavior of the Pt(111) electrode has previously been observed after addition to perchloric acid electrolyte of sulfuric acid in the millimolar range of its bulk concentration (27).

The radiochemical results obtained with the Pt(111) electrode in the electrolyte containing C-14 labeled acetic acid are shown in Figure 4 (circles). The adsorption is negligible at 0.15 V, small at more negative potentials and very pronounced at potentials more positive than 0.15 V. In the potential range between 0.3 and 0.85 V, the radiochemical signal from the surface, i.e., the surface concentration of acetic acid, equation 1, is potential independent. This result is interesting since very strong potential effects in this potential range were found in the case of CH_3COOH adsorption on polycrystalline platinum (see refs. 20, 25, 34 and Figure 5 of this work). If the products on both electrodes are the same, this observation would be indicative of an increased stability of adsorbed acetic acid on the well-ordered, single crystal electrode vs. the polycrystalline substrate. In general, such behavior reflects the properties of interfacial atoms and molecules which, when adsorbed on well-defined substrates, tend to create stable and distinct phases of two-dimensional periodicity and order (35).

As shown in Figure 4, the increase in adsorption of acetic acid on Pt(111) occurs in the potential range of 0.15 to 0.3 V, which roughly coincides with position of the current-potential peak of the voltammogram recorded in the same solution. An interdependence can thus be sought between acetic acid adsorption

Figure 3. LEED pattern of the Pt(111) electrode used for the reported research. The beam energy was 65 eV.

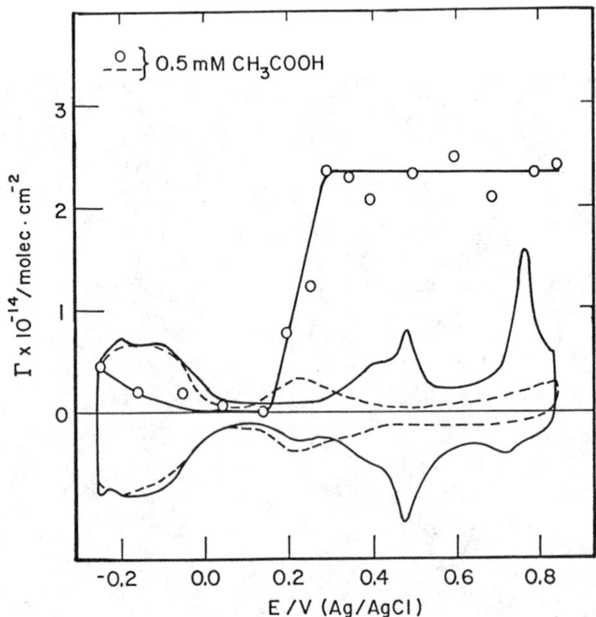

Figure 4. A combined figure representing the following relationships: cyclic voltammogram of the Pt(111) electrode in 0.1 M $HClO_4$ (solid line); cyclic voltammogram following addition of 0.5 mM CH_3COOH to the 0.1 M $HClO_4$ electrolyte (broken line); surface concentration of adsorbed acetic acid plotted as a function of the electrode potential (solid line and circles). (Reprinted with permission from ref. 23. Copyright 1988 Elsevier.)

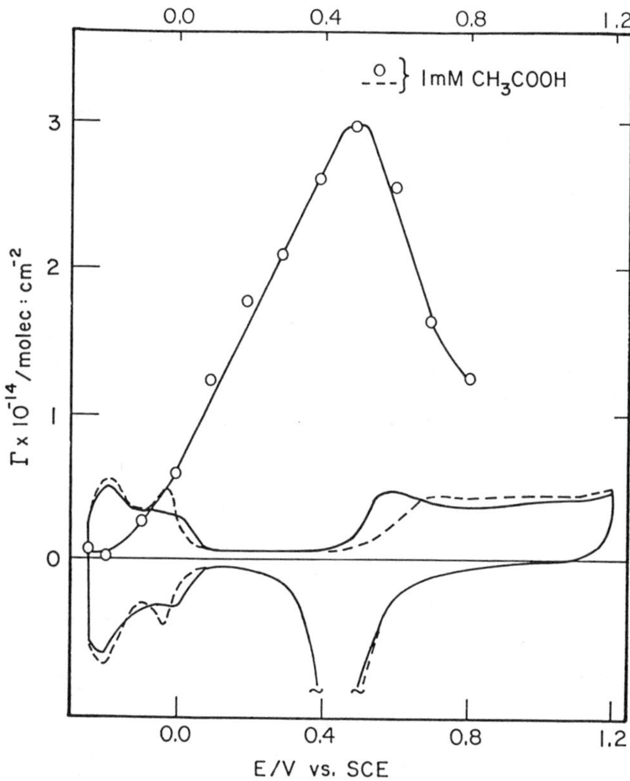

Figure 5. A combined figure representing the following relationships: cyclic voltammogram of the polycrystalline Pt electrode in 0.1 M $HClO_4$ (solid line); cyclic voltammogram following addition of 1 mM CH_3COOH to the 0.1 M $HClO_4$ electrolyte (broken line); surface concentration of adsorbed acetic acid plotted as a function of the electrode potential (solid line and circles). (Reprinted with permission from ref. 23. Copyright 1988 Elsevier.)

and the respective current-generating process. The simplest explanation is that adsorption of acetic acid is dissociative (36) and followed by the electron charge transfer process, according to equation 2:

$$CH_3COOH \dashrightarrow CH_3COO_{ads} + H^+ + e^- \quad (2)$$

Contrary to the single-crystal results discussed above, the effect of addition of acetic acid to the perchloric acid electrolyte on the voltammograms of the polycrystalline Pt (34) was insignificant (Figure 5). Only a small extra charge in the hydrogen range and an initial suppression of platinum oxidation were observed (dotted curve) versus the base voltammogram obtained in pure $HClO_4$ electrolyte (solid voltammogram). Noticeably, the current-potential curve in the double layer region (from 0.1 to 0.4 V) was unaffected by acetic acid present in the electrolyte. In the same potential range, the surface concentration of acetic acid was monotonically increasing, as shown by the radiochemical measurements (circles). We may therefore conclude that accumulation of acetic acid molecules at the electrode surface between 0.1 and 0.4 V was not associated with a measurable electron charge transfer process across the interface. In other words, adsorption of acetic acid on the polycrystalline substrate is associative.

Comparison of Results Obtained with Pt(111) and Polycrystalline Electrodes. We have shown above that potential dependence of adsorption of acetic acid on Pt(111) and on polycrystalline Pt is different. However, the central question whether adsorption of CH_3COOH can be dissociative on Pt(111) (equation 2) and associative on the polycrystalline Pt can not unambiguously be answered by the radiochemical and electrochemical studies alone (23). To address the issue of the "anomalous" single-crystal voltammetry (27), we have recently studied adsorption of sulfate (bisulfate) anions on Pt(111) (37). We found that the specific adsorption model of the anomalous behavior (38) would require an exchange of three electrons between adsorbed anion and the metal. Since this appeared unlikely, we concluded that the anomalous charge is due to adsorption of high-energy hydrogen, as originally proposed by Clavilier (27). We have also indicated the possibility that the potential range of adsorption/desorption of this high-energy hydrogen can be controlled by the extent of coadsorption with hydrogen of anions or other electron-rich molecules. Since the voltammetry of the Pt(111) electrode in solutions containing acetic acid and sulfuric acid is very similar, we consequently conclude that desorption of high-energy hydrogen is a more probable source of electric current than the acetic acid discharge represented by equation 2.

Whether such strongly-bound hydrogen is present on the Pt(111) surface has been a subject of debate for some time. By analyzing then existing data on adsorption energy of gas phase hydrogen on Pt(111), Wagner and Ross (39) concluded that electrochemical adsorption of hydrogen is not energetically allowed to occur in the high potential range closely preceding platinum electrooxidation. However, some newer data indicate that at least 10 - 15% of a monolayer of hydrogen desorbs at a temperature of 100 K higher than that where most of the monolayer hydrogen desorbs (40). We believe that this observation gives an additional support to the original interpretation of the Pt(111) voltammetry in acidic media (27). We would also like to note that more data regarding adsorption of hydrogen on single-crystal surfaces may soon be provided by the in situ spectroscopies of electrochemical surface science. Recent progress in surface IR (13,41,42) and in Raman second harmonic generation (15,43) is particularly promising here.

Acknowledgments

This work was supported by the National Science Foundation under Grant NSF DMR-86-12860 (administered by the Materials Research Laboratory of the University of Illinois) and by Dow Chemical U.S.A.

Literature Cited

1. Hubbard, A. T. Acc. Chem. Res. 1980, 13, 177.
2. Somorjai, G.A. Chemistry in Two Dimensions: Surfaces; Cornell University Press: Ithaca, 1981.
3. Ishikawa, R. M.; Hubbard, A. T.; J. Electroanal. Chem. 1976, 69, 317.
4. Ross, R. N. J. Electroanal. Chem. 1977, 78, 139.
5. Yeager, E.; O'Grady, E. W.; Woo, M. Y. C.; Hagans, P. J. Electrochem. Soc. 1978, 125, 346.
6. Hubbard, A. T. J. Vac. Sci. Technol. 1980, 17, 49.
7. Yeager, E. J. Electrochem. Soc. 1981, 128, 160c.
8. Ross, P. N.; Wagner, F. T. Adv. in Electrochem. and Electrochem. Eng; Gerisher, H., Ed.; Wiley-Interscience, New York, 1984; Vol. 31, p. 69.
9. D'Agostino, A. T.; Hansen, W. N. Surf. Sci. 1986, 165, 268.
10. Kotz, E. R.; Neff, H.; Muller, K. J. Electroanal. Chem. 1986, 215, 331.
11. Kolb, D. M. Z. Phys. Chem. 1987, 154, 179.
12. Fleischmann, M. Advances in Electrochemistry; The Robert A. Welch Foundation Conferences on Chemical Research, 1986; No. 3-5, p. 91.
13. Bewick, A.; Pons, S. Adv. in Infrared, Raman Spectr.; Clark, R. J. H.; Hester, R. E., Ed.; Wiley Heyden: New York, 1985; Vol. 12, p. 1.
14. Chen, C. K.; Heinz, T. F.; Ricard, D; Shen, Y. R. Phys. Rev. B 1983, 27, 1965.
15. Richmond, G. L. Langmuir 1986, 2, 132.
16. Sonnenfeld, R.; Hansma, P. K. Science 1986, 232, 211.
17. Krauskopf, E. K.; Chan, K.; Wieckowski, A. J. Phys. Chem. 1987, 91, 2327.
18. Kazarinov, V. E.; Andreev, V. N. Comprehensive Treatise of Electrochemistry; Yeager, E.; Bockris, J. O'M.; Conway, B. E.; Sarangapani, S., Ed.; Plenum: New York, 1984; Vol. 9, p. 393.

19. Price, W. J. McGraw-Hill Series in Nuclear Engineering; Zinn, W. H.; Luntz, J. D., Ed.; McGraw-Hill: New York, 1958; p. 131.
20. Horanyi, G. Electrochimica Acta 1980, 25, 43.
21. Krauskopf, E. K.; Wieckowski, A. Modern Aspects of Electrochemistry; submitted.
22. Wieckowski, A.; Krauskopf, E. K. Presented at the 194th National Meeting of the ACS, August 30-September 4, 1987, New Orleans, LA.
23. Rice, L.; Krauskopf, E. K.; Wieckowski, A. J. Electroanal. Chem. in press.
24. Horanyi, G. J. Electroanal. Chem. 1974, 51, 163.
25. Wieckowski, A. Electrochimica Acta 1981, 26, 1121. Wieckowski, A.; Sobkowski, J.; Zelenay, P.; Franaszczuk, K. Electrochimica Acta 1981, 26, 1111.
26. Wieckowski, A. The Analysis of Electrochemical Adsorption Processes on Platinum; P. W. U., Warsaw, Poland, 1984.
27. Clavilier, J. J. Electroanal. Chem. 1980, 107, 211.
28. Motoo, S.; Furuya, N. J. Electroanal. Chem. 1984, 172, 339.
29. Wieckowski, A.; Rosasco, S. D.; Schardt, B. C.; Stickney, J. L.; Hubbard, A. T. Inorg. Chem. 1984, 23, 565.
30. Wieckowski, A.; Schardt, B.C.; Rosasco, S. D.; Stickney, J. L.; Hubbard, A. T. Surf. Sci. 1984, 146, 115.
31. Zurawski, D.; Rice, L.; Hourani, M.; Wieckowski, A. J. Electroanal. Chem. 1987, 230, 221.
32. Aberdam, D.; Durand, R.; Faure, R.; El-Omar, F. Surf. Sci. 1986, 171, 303.
33. Hourani, M.; Wieckowski, A. J. Electroanal. Chem. 1987, 227, 259.
34. Corrigan, D. S.; Krauskopf, E. K.; Rice, L. M.; Wieckowski, A.; Weaver, M. J. J. Phys. Chem. in press.
35. Roelofs, L. D.; Estrup, P. J. Surf. Sci. 1983, 125, 51.
36. Avery, N. R. J. Vac. Sci. Technol. 1982, 20, 3.
37. Krauskopf, E. K., Rice, L. M., and Wieckowski, A. J. Electroanal. Chem., in press.
38. Al Jaaf-Golze, K.; Kolb, D. M.; Scherson, D. J. Electroanal. Chem. 1986, 200, 353.
39. Wagner, F. T.; Ross, P. N., Jr. J. Electroanal. Chem. 1983, 150, 141.
40. Zhou, X.-L.; White, J. M. Surf. Sci. 1987, 185, 450.
41. Leung, L-W. H. and Weaver, M. J. J. Electroanal. Chem., 1988, 240, 341.
42. Sun, S. G., Clavilier, J., and Bewick, A. J. Electroanal. Chem., 1988, 240, 147.
43. Campbell, D. J. and Corn, R. M. J. Phys. Chem., 1987, 91, 4668.

RECEIVED May 17, 1988

Chapter 18

Applications of In Situ Mössbauer Spectroscopy to the Study of Transition Metal Oxides

Daniel A. Scherson[1], Dennis A. Corrigan[2], Cristian Fierro[1], and Raul Carbonio[1]

[1]Case Center for Electrochemical Sciences and the Department of Chemistry, Case Western Reserve University, Cleveland, OH 44106
[2]Physical Chemistry Department, General Motors Research Laboratories, Warren, MI 48090–9055

A detailed understanding of the electrochemistry of transition metal oxides is of importance to the fields of energy storage, corrosion protection, and electrocatalysis.(1) It is thus not surprising that much research has been devoted to the **ex situ** and **in situ** spectroscopic characterization of such materials in an effort to elucidate their structure and overall physicochemical properties. Although **ex situ** techniques offer significant advantages in terms of sensitivity and specificity, considerable care must be exercised in the analysis of information derived from their use. This is due primarily to the fact that the transfer of electrodes from electrochemical to other environments often results in a loss of potential control, structural rearrangements, such as those induced by the exposure of specimens to air or vacuum, and/or actual chemical transformations associated with the irradiation of specimens with UV, X-rays and electrons, as is the case with the electron-based surface science analytical methods. It is precisely because of these uncertainties that much attention has been focused over the last ten to fifteen years towards the development of techniques capable of providing **in situ** microscopic level information regarding the structure and properties of metal electrolyte interfaces.

This work summarizes some applications of **in situ** Mossbauer spectroscopy to the study of certain aspects of the electrochemistry of iron and iron containing transition metal oxides. A number of illustrations of the use of this technique to the investigation of a wide variety of interfacial phenomena may be found in two recent monographs.(2)

THE IRON OXYHYDROXIDE SYSTEM

One of the first **in situ** Mossbauer investigations of the behavior of iron oxides in electrochemical enviroments was the result of a fortuitous incident in which a specimen containing iron phthalo-

0097–6156/88/0378–0257$06.00/0

cyanine, FePc, dispersed on a high area carbon was accidentally decomposed during a rather mild heat treatment to yield a very fine dispersion of some form of ferric oxide.(3) FePc is a highly conjugated transition metal macrocycle which has been found to exhibit high activity for the electrochemical reduction of dioxygen when supported on a host electronically conducting substrate.(4)

Fig. 1 shows the cyclic voltammetry of an FePc/XC-72 dispersion, heated at 280°C in an inert atmosphere, in the form a thin porous Teflon bonded coating electrode in a 1 M NaOH solution. A description of the methodology involved in the preparation of this type of electrode may be found in Ref. 3. As can be clearly seen, the voltammetry of this specimen exhibits two sharply defined peaks separated by about 330 mV. The potentials associated with these features are essentially identical to those found by other workers for the reduction and oxidation of films of iron oxyhydroxide formed on a number of host surfaces, including iron and carbon.(5)

A 10% w/w highly enriched ^{57}FePc/XC-72 dispersion, prepared in exactly the same fashion, was used in the Mossbauer measurements. (The isomer shifts, δ, are referred to the α-Fe standard, and δ, the quadrupole splittings Δ, and widths Γ, are all given in $mm \cdot s^{-1}$, throughout the text.)

The **in situ** Mossbauer spectra obtained at 0.0 V is given in curve A, Fig. 2. The parameters associated with this doublet (Table I) are similar to those reported by various groups for high spin ferric oxyhydroxides (Table II). Also, they appear in agreement with those observed for certain magnetically ordered oxides for which the characteristic six line spectra collapses into a doublet as the particles become smaller in size. This phenomenon known as superparamagnetism,(6) is attributed to the flipping of the magnetic moment of each microcrystal between easy directions. This occurs in a shorter time than either the Larmor precessional period of the nucleus or the lifetime of the excited 3/2 state of the ^{57}Fe nucleus or both, when the temperature is sufficiently high. Such behavior has been observed, for example, by Hassett et al.(7) for magnetite dispersed in lignosulfonate.

The quadrupole splitting of the heat treated FePc/XC-72 electrode measured **ex situ**, prior to the electrochemical experiments, was larger than that found **in situ**. Smaller values for Δ have been reported for certain ferric hydroxide gels and for small particles of FeOOH (Table II), and thus the effect associated with the immersion of the specimen in the electrolyte is most probably related to the incorporation of water into the oxide structure. For this reason, the material observed **in situ** at this potential will be referred to hereafter as FeOOH(hydrated), without implying any specific stoichiometry.

The **in situ** spectra obtained at -1.0 V, shown in curve B, Fig 2, yielded a doublet with values of δ and Δ in excellent agreement with those of crystalline $Fe(OH)_2$ (Table I). This provides rather definite evidence that the redox process associated with the voltammetric peaks is given by:

$$FeOOH\ (hydrated) + e^- + H^+ \longrightarrow Fe(OH)_2\ (crystalline) \quad (1)$$

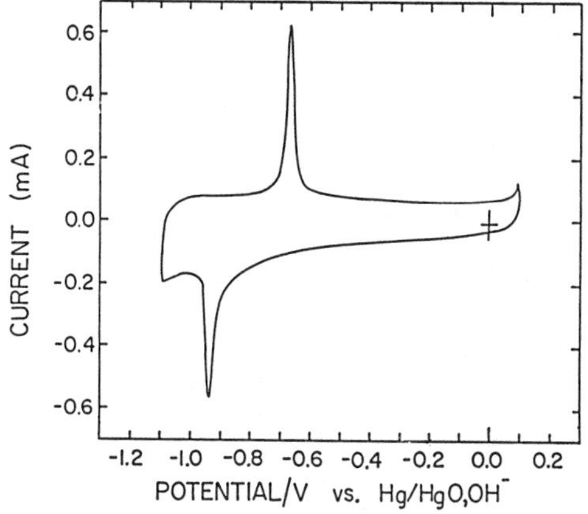

Fig. 1. Cyclic voltammetry of 7% w/w iron phthalocyanine, FePc dispersed on Vulcan XC-72 carbon, after a heat treatment at 280°C in a flowing inert atmosphere. The measurement was conducted with the material in the form a thin porous Teflon bonded coating in 1 M NaOH at 25°. Sweep rate: 5 mV/s. (Reproduced with permission from ref. 3. Copyright 1985 Elsevier.)

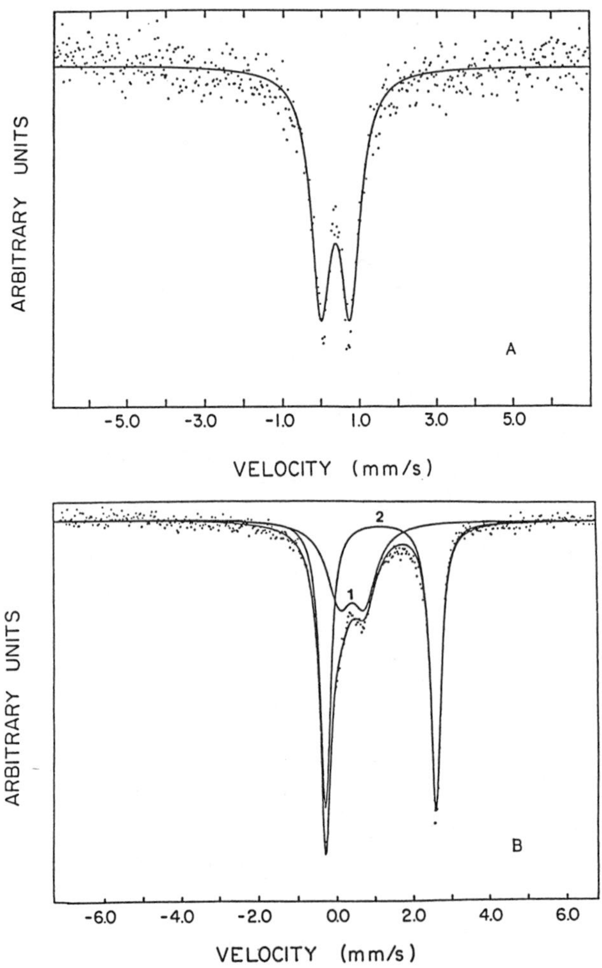

Fig. 2. **In situ** Mossbauer spectra for ^{57}FePc dispersed in Vulcan XC-72 carbon and subsequently heat treated at 300°C obtained at 0.0 V (Curve A), and -1.05 V vs. Hg/HgO,OH^- (Curve B). (Reproduced with permission from ref. 3. Copyright 1985 Elsevier.)

TABLE I

In situ Mossbauer parameters for small particles of a hydrated ferric oxyhydroxide, FeOOH(hydrated), dispersed on high area Vulcan XC-72 carbon *

Potential V vs. Hg/HgO,OH^-		Isomer Shift $\delta/mm\cdot^{-1}$ vs. α-Fe	Quadrupole Splitting $\Delta/mm\cdot s^{-1}$	Width $\Gamma/mm\cdot s^{-1}$	Figure
0.0		0.37	0.76	0.62	2A
-0.4		0.37	0.74	0.65	--
-0.75		0.37	0.66	0.57	--
-0.85		0.35	0.63	0.60	--
-1.05	1	0.41	0.65	0.77	2B
	2	1.14	2.85	0.32	

* This material was prepared by the heat treatment of iron phthalocyanine dispersed on the carbon in an inert atmosphere at 280°C.

TABLE II

Ex situ Mossbauer parameters of various iron oxides and oxyhydroxides at room temperature *

Specimen	Isomer Shift δ/mm·$^{-1}$ vs. α-Fe	Quadrupole Splitting Δ/mm·s^{-1}	Width Γ/mm·s^{-1}	H_{eff} kOe
α-FeOOH (goethite)	0.44	0.16	0.86	367
α-Fe_2O_3[a] (diam.< 10 nm)	0.32	0.98		
β-FeOOH	0.38 (61.4) 0.39 (38.6)	0.53 0.88	0.26 0.30	
γ-FeOOH (lepidocrocite)	0.38	0.59	0.27	
α-Fe_2O_3 (hematite)	0.38	0.24	0.29	523
γ-Fe_2O_3 (maghemite)	0.43	0.06	0.45	506
Fe_3O_4 non-stoichiometric	0.39 (49.6) 0.78 (50.4)	0.11 0.28	0.53 0.38	506 465
γ-Fe_3O_4 stoichiometric (magnetite)	0.37 (4.8)[o] 0.34 (34.4) 0.72 (59.8)	0.59 0.12 0.10	0.43 0.29 0.31	 491 461
γ-Fe_3O_4[b] (diam.< 5 nm)	0.37	0.89		
$Fe(OH)_2$[c]	1.18	2.92		
Ferric oxide[d] (small particles)	0.33	0.70		
Ferric oxide[e] (hydrated)	0.35	0.62		
FeOOH[f] (small particles)	0.39	0.62		

* Data from S. Music, I. Czako-Nagy, S. Popovic, A. Vertes and M. Tonkovic, Croat. Chem. Acta, 59,833(1986), except where otherwise indicated.
o This feature is due to FeOOH.

a. W. Kundig, H. Bommel, C. Constabaris and H. Lindquist, Phys. Rev.,142, 327(1966);
b. S. Aharoni and M. Litt, J. Appl. Phys.,42,352(1971);
c. A. M. Pritchard and B. T. Mould, Corros. Sci.,11,1(1971);
d. D. G. Rethwisch and J. A. Dumesic, J. Phys. Chem.,90,1863(1986);
e. P. P. Bakare, M. P. Gupta and A. P. B. Sinha, Indian J. Pure Appl. Phys.,18,473(1980);
f. P. O. Vozniuk, V. N. Dubinin, Sov. Phys.-Solid State,15,1265(1973).

It may thus be concluded that the specific methodology involved in the dispersion of FePc on high area carbon leads to the thermal decomposition of the macrocycle at temperatures much below those expected for the bulk material, generating small particles of hydrated FeOOH upon exposure to an alkaline solution.

More recently, Fierro et al.(8) have reported a series of **in situ** Mossbauer experiments aimed at investigating the electrochemical behavior of the iron oxyhydroxide system in strongly alkaline media. For these measurements, a hydrated form of a ferric oxyhydroxide precipitated by chemical means on a high area carbon was used. This material was prepared by first dissolving a mixture of highly enriched metallic ^{57}Fe and an appropriate amount of natural iron in concentrated nitric acid to achieve about one-third isotope enrichment in the final product. This solution was then added to an ultrasonically agitated water suspension of Shawinigan black, a high area carbon of about 60 $m^2 \cdot g^{-1}$, and the iron subsequently precipitated by the addition of 4 M KOH. A Teflon bonded electrode was prepared with this material following the same procedure as that described in Ref. 3, except that no heat treatment was performed to remove the Teflon emulsifier. The ^{57}Fe/XC-72 w/w ratio was in this case 50% and thus much higher than that involved in the heat treated FePc experiments described earlier. The electrochemical cell for the **in situ** Mossbauer measurements is shown in Fig. 3.

The **ex situ** Mossbauer spectrum for the partially dried electrode yielded a doublet with $\delta = 0.34$ and $\Delta = 0.70$ $mm \cdot s^{-1}$. A decrease in the value of Δ was found in the **in situ** spectra of the same electrode immersed in 4 M KOH at -0.3 V vs Hg/HgO,OH^- (see Table III, and Curve a, Fig. 4), in direct analogy with the behavior observed for the heat treated FePc. It is thus conceivable that this material is the same as that found after the thermal decomposition of FePc dispersed on carbon and that reported by other workers, and that the variations in the value of Δ are simply due to differences in the degree of hydration of the lattice.

No significant changes in the spectra were found when the electrode was polarized sequentially at -0.5 and -0.7 V, by scanning the potential to these values at 10 $mV \cdot s^{-1}$. This is not surprising since the cyclic voltammetry for an identical, although non-enriched, iron/carbon mixture, shown in the Inset, Fig. 4, indicated no significant faradaic currents over this voltage region for the sweep in the negative direction.

In a subsequent measurement at a potential of -0.9 V, the resonant absorption of the doublet underwent a marked drop, an effect that may be due to an increase in the solubility of the oxide, and thus in a loss of solid in the electrode, and/or to a modification in the recoilless fraction of the solid induced by the hydration of the lattice.

The electrode was then swept further negative to -1.1 V, a potential more negative than the onset of the faradaic current in the voltammogram, yielding after about two hours of measurement, a strong, clearly defined doublet (Curve b, Fig. 4), with parameters in excellent agreement with those of $Fe(OH)_2$ (see Tables II and III). The potential was then stepped to -0.3 V. In contrast to the

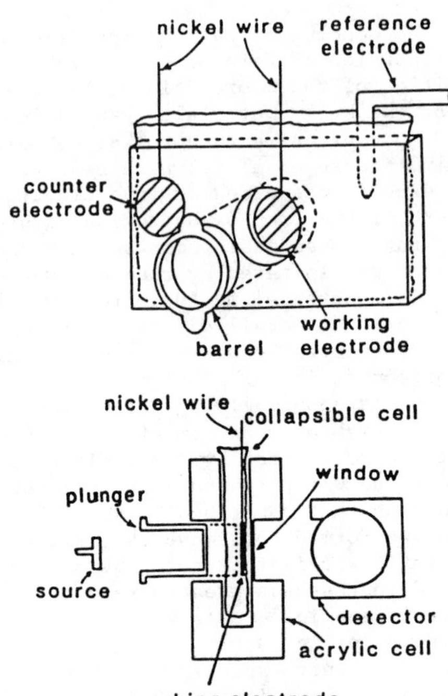

Fig. 3. Electrochemical cell for **in situ** Mossbauer spectroscopy measurements. (Reproduced with permission from ref. 3. Copyright 1985 Elsevier.)

TABLE III

In situ Mossbauer parameters for iron oxides and oxyhydroxides dispersed on high area Schawinigan Black carbon electrode

Potential V vs. Hg/HgO,OH⁻	Isomer Shift δ/mm·$^{-1}$ vs. α-Fe	Quadrupole Splitting Δ/mm·s^{-1}	H_{eff} kOe	Figure
-0.3 (initial)	0.33(0.34)*	0.58(0.70)		3A
-1.1	1.10	2.89		3B
-0.3 (step)	0.37	0.04	406	4A
-0.3 (sweep)	0.33	0.57		4B
	0.28		463	
	0.56	0.15	437	
	0.24			
-1.2	0.00		330	4C
	1.15	2.90		

* Values in parenthesis are those obtained for the same electrode dry.

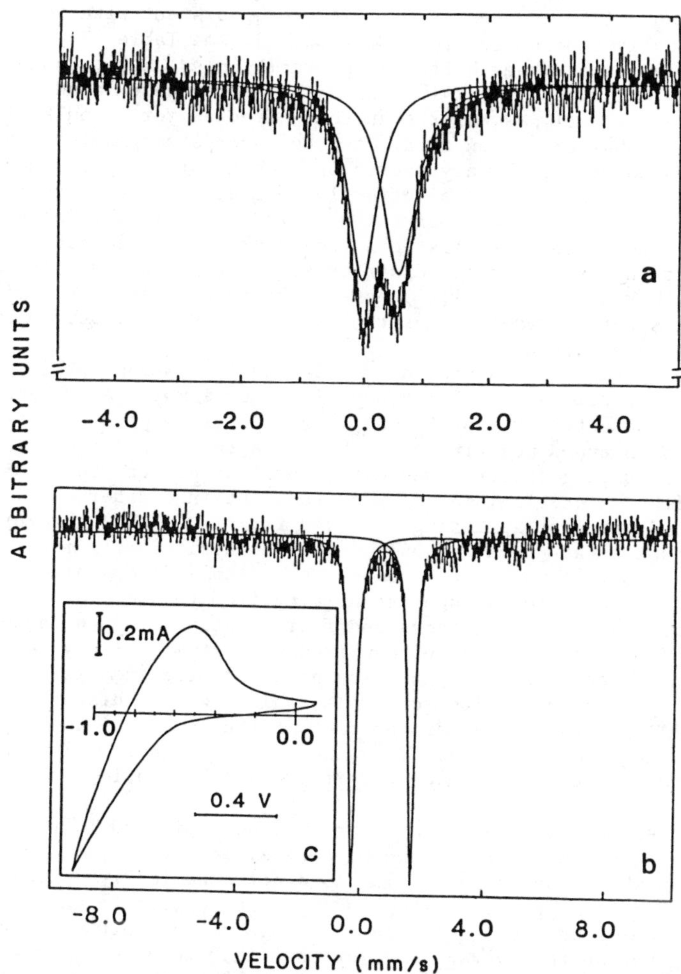

Fig. 4. **In situ** Mossbauer spectra of a Teflon bonded 50% w/w ^{57}Fe enriched hydrated ferric oxide precipitated on Schawinigan black high area carbon electrode in 4 M KOH at -0.3 V (Curve a), and -1.1 V vs Hg/HgO,OH^- (Curve b). Inset: Cyclic voltammogram of the same, although non enriched, material ferric oxide on carbon material in the form of a thin porous Teflon bonded coating electrode deposited on an ordinary pyrolytic graphite electrode, in 4 M KOH. Scan rate: 10 mV/s.

doublet obtained originally at this voltage, a magnetically split six-line spectrum was obtained in this case (Curve a, Fig. 5). The Zeeman effect and the value of δ are consistent with those of a magnetically ordered ferric oxide species (see Table II). Unfortunately, the strength of the internal field, H_{eff}, cannot be used as a definite identifying parameter as the calculated value seems significantly smaller than that expected for a bulk iron oxide, a behavior often attributed to superparamagnetism (**vide supra**). Furthermore, the asymmetric broadening of the peaks may be ascribed to a distribution of effective magnetic fields, providing evidence for the presence of an ensemble of small particles of varying sizes. From a statistical viewpoint this is accounted for in the fitting by a Gaussian distribution of Lorentzians. It may be noted that Hassett et al.(7) have reported a strikingly similar six-line spectrum for small particles of magnetite dispersed in a lignosulfonate matrix.

In a subsequent measurement, the potential was swept in the negative direction to -1.1 V, yielding once more a Mossbauer spectrum characteristic of $Fe(OH)_2$, and later swept at 2 $mV \cdot s^{-1}$ rather than stepped positive to -0.3 V. As shown in curve b, Fig. 5, the resulting spectrum was different than either that associated with the original material or that obtained after a voltage step. The apparent splitting observed for two of the absorption lines located at negative velocities is typical of magnetite (Fe_3O_4) in bulk form at room temperature (Table II). This is due to the superposition of spectra arising from ferric cations in tetrahedral sites and ferrous and ferric cations in octahedral sites. The broad background centered at 0.24 $mm \cdot s^{-1}$ may be the result of several effects including particle size and structural disorder among magnetite crystals which would distort the Mossbauer spectra. The sharp doublet in the center of the spectrum, as judged by the parameter values given in Table III, can be attributed to the same hydrated ferric oxyhydroxide observed originally.

The results of these experiments may be explained in terms of differences in the nature of the particles generated by the specific way in which the ferrous oxide is electrochemically oxidized. In particular, a potential step is expected to promote the formation of multitude of nuclei large enough to exhibit a Zeeman splitting but on the average smaller than that required to yield a spectra characteristic of bulk magnetite. When the oxidation is performed by sweeping the potential, however, a few magnetite nuclei are generated which grow to a size sufficiently large to show bulk like behavior.

At the end of these measurements, the electrode was polarized by sweeping the potential to -1.2 V, yielding a six-line spectrum corresponding to metallic iron with some contribution from $Fe(OH)_2$ (curve c, Fig. 5). The potential was then scanned up to -0.3 V and a spectrum essentially identical to that recorded at -1.2 V was observed. This result clearly indicates that the iron metal particles formed by the electrochemical reduction are large enough for the contributions arising from the passivation layer to be too small to be clearly resolved. After scanning the potential several

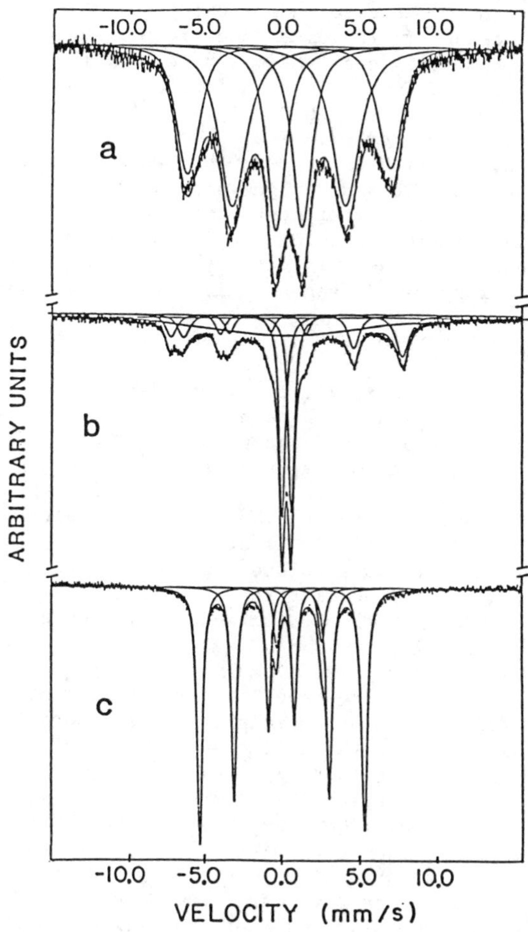

Fig. 5. **In situ** Mossbauer spectra of the same electrode as in a) Fig. 4 at -0.3 V, after a potential **step** (Curve a), and after a potential **sweep** (Curve b), from -1.1 V. Curve c was obtained at a potential of -1.2 V.

times between -0.3 and -1.2 V, however, the doublet associated with the $Fe(OH)_2$ disappeared.

It may be interesting to note that Geronov et al.(9) in much earlier studies investigated the spectral changes induced by the charge and discharge of high area iron-carbon polymer bonded electrodes in strongly alkaline media. Despite the fact that the experiments were not conducted under strict potential control, as the circuit was opened during data acquisition, these authors made a number of interesting observations regarding the behavior of iron electrodes in 5 M KOH. In particular, electrodes in the fully charged state (-0.9 V vs. $Hg/HgO,OH^-$) were found to exhibit features associated with metallic iron and $Fe(OH)_2$, whereas two additional peaks, attributed to β-FeOOH, were observed upon discharge of these electrodes under galvanostatic conditions (-0.5 V vs. $Hg/HgO,OH^-$). Similar experiments conducted in the presence of LiOH in the same solution lead to a conversion of the β-FeOOH into bulk magnetite (Fe_2O_3) as evidenced by the appeareance of the characteristic strong field Zeeman split six-line spectrum.

MIXED NICKEL-IRON OXYHYDROXIDES

The presence of iron in nickel oxyhydroxide electrodes has been found to reduce considerably the overpotential for oxygen evolution in alkaline media associated with the otherwise iron free material.(10) An **in situ** Mossbauer study of a composite Ni/Fe oxyhydroxide was undertaken in order to gain insight into the nature of the species responsible for the electrocatalytic activity.(11) This specific system appeared particularly interesting as it offered a unique opportunity for determining whether redox reactions involving the host lattice sites can alter the structural and/or electronic characteristics of other species present in the material.

Thin films of a composite nickel-iron (9:1 Ni/Fe ratio) and iron-free oxyhydroxides were deposited from metal nitrate solutions onto Ni foils by electroprecipitation at constant current density. A comparison of the cyclic voltammetry of such films in 1M KOH at room temperature (see Fig. 6) shows that the incorporation of iron in the lattice shifts the potentials associated formally with the $NiOOH/Ni(OH)_2$ redox processes towards negative potentials, and decreases considerably the onset potential for oxygen evolution. The oxidation peak, as shown in the voltammogram, is much larger than the reduction counterpart, providing evidence that within the time scale of the cyclic voltammetry, a fraction of the nickel sites remains in the oxidized state at potentials more negative than the reduction peak.

The **in situ** Mossbauer experiments were conducted with 90% ^{57}Fe enriched 9:1 Ni/Fe oxyhydroxide films which were deposited in the fashion described above onto a gold on Melinex support(12) in a conventional electrochemical cell. Prior to their transfer into the **in situ** Mossbauer cell, the electrodes were cycled twice between 0 and 0.6 V vs. $Hg/HgO,OH^-$ in 1 M KOH. Two such films were used in the actual Mossbauer measurements in order to reduce the counting time. **The in situ** Mossbauer cell involved in these experiments was previously described.

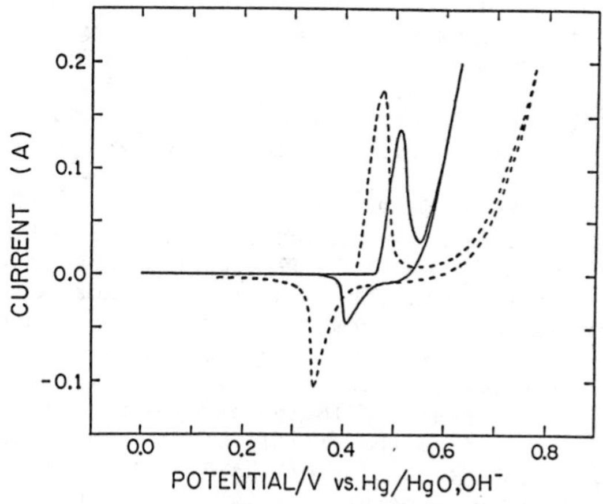

Fig. 6. Cyclic voltammograms for a composite Fe/Ni oxyhydroxide (Fe/Ni 1:9) on a Ni foil substrate in 1 M KOH (solid curve). Scan rate: 10 mV/s. The dashed curve was obtained for an iron-free Ni oxyhydroxide film under the same experimental conditions.

The **in situ** spectrum obtained at 0.5 V vs. Hg/HgO,OH^- (oxidized state) is shown in curve a, Fig. 7. Following this measurement, the potential was swept to 0.0 V (reduced state), and a new **in situ** spectrum recorded after the current had dropped to a very small value (curve b, Fig. 7). Essentially identical results were obtained when the films were examined first in the reduced and then in the oxidized state.

The spectrum of the oxidized form was successfully fitted with a singlet yielding an isomer shift of 0.22. For the spectrum in the reduced state, a satisfactory fit could be achieved with two singlets, which when regarded as the components of an asymmetric doublet yielded $\delta = 0.34$ and $\Delta = 0.43$. In view of the fact that the cyclic voltammetry indicated a slow reduction of the oxidized state, a statistical analysis of the data in curve b, Fig. 7 was attempted with a symmetric doublet and a singlet to account for a possible contribution due to the oxidized phase. This approach afforded excellent results, yielding an isomer shift for the singlet very similar to that of the oxidized species in curve a, Fig. 7. Furthermore, the values of δ and Δ for the symmetric doublet were found to be nearly the same as those in the fit involving the asymmetric doublet. These are listed in Table IV.

The isomer shift of 0.32 associated with the reduced state of the composite material is similar to that reported for various Fe(III) oxyhydroxides (see Table II) and indicates that iron is present in the ferric form. The much smaller quadrupole splitting, however, provides evidence that the crystal environment of such ferric sites is different than that in common forms of ferric oxyhydroxides. The lower value of the isomer shift observed upon oxidation of the composite film indicates a partial transfer of electron density away from the Fe(III) sites, which could result indirectly from the oxidation of the Ni(II) sites to yield a highly oxidized iron species. It is interesting to note that recent Raman measurements have provided evidence for the presence of highly symmetrical Ni sites in oxidized nickel hydroxide films.(13) Based on such information, it was postulated that the film structure could be better represented as NiO_2 rather than as NiOOH. It seems thus reasonable that the oxidized composite oxyhydroxide could contain symmetrical sites which might be occupied by Fe ions. This would be consistent with the presence of a singlet, rather than a doublet, in the spectra shown in curve a, Fig. 7. Based on these arguments, Corrigan et al.[11] concluded that the composite metal oxyhydroxide may be regarded as a single phase involving distinct iron and nickel sites as opposed to a physical mixture of $Ni(OH)_2$ and FeOOH particles. This is not surprising since the composite hydroxide is a better catalyst than either of the individual hydroxides.

In summary, the results of this investigation indicated that the formal oxidation of the nickel sites in a composite nickel-iron oxyhydroxide modifies the electronic and structural properties of the ferric sites yielding a more d-electron deficient iron species. Although it may be reasonable to suggest that the electrocatalytic activity of this composite oxide for oxygen evolution may be related to the presence of such highly oxidized iron sites,

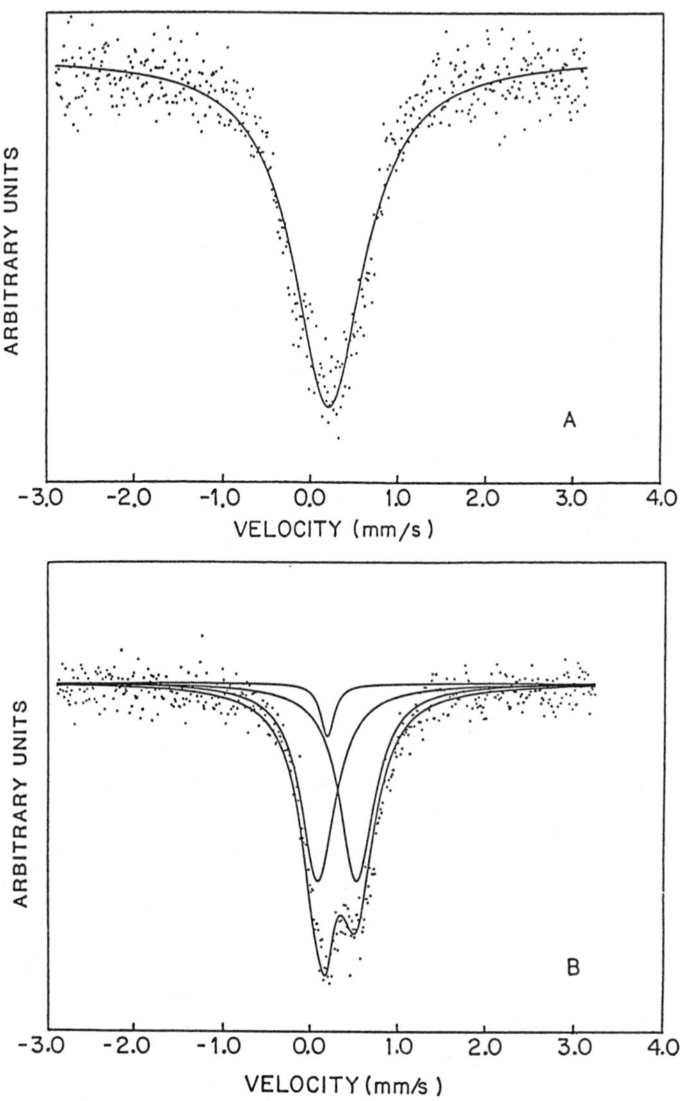

Fig. 7. **In situ** Mossbauer spectrum of a composite Fe/Ni oxyhydroxide (Fe/Ni 1:9) polarized at 0.5 V (oxidized state), curve A, and at 0.0 V vs Hg/HgO,OH^- (reduced state), curve B, in 1 M KOH.

TABLE IV

In situ Mossbauer parameters for an iron-nickel mixed oxyhydroxide in 1 M KOH

Potential V vs. Hg/HgO,OH^-	Isomer Shift δ/mm·s^{-1} vs. α-Fe	Quadrupole Splitting Δ/mm·s^{-1}	Width Γ/mm·s^{-1}	Figure
0.5	0.22		0.97	6A
0.0	0.32(95)†	0.44	0.47	6B
	0.19(5)		0.19	

† Values in parentheses represent the fraction of the total resonant absorption associated with each peak assuming a common recoilless fraction for both species.

additional **in situ** spectroscopic measurements such as **EXAFS** may be necessary in order to support this view.

CONCLUDING REMARKS

Although important information can be derived from an analysis of **in situ** data of the type described in this work, considerable insight into the nature of these electrochemically formed materials could be obtained by examining their spectral behavior as a function of temperature and the strength of an externally applied magnetic field. Unfortunately, the range of temperatures in which electrochemical measurements can be carried out is rather narrow. An approach that could to a certain extent circumvent some of these difficulties, however, currently being implemented in this laboratory, involves the fast quenching of polarized electrodes at cryogenic temperatures with the electrochemical cell placed in the cavity of the superconducting magnet. Should this quasi **in situ** technique prove successful it could provide the means of acquiring information of much more detailed character than that afforded by present **in situ** techniques.

ACKNOWLEDGMENTS

Financial support for this work was provided by NASA Lewis Research Center, and by IBM through a Faculty Development Award to one of the authors (D.A.S.).

LITERATURE CITED

1. For a recent review in this area see: E.J.M. O'Sullivan and E. J. Calvo **Reactions at Metal Oxide Electrodes**, in Comprehensive Chemical Kinetics, Vol. 27, Chapter 4.
2. a. Vertes, A. and Czako-Nagy, I. Izv. Khim., 1988, 19, 380.
   b. Scherson, D. in Spectroelectrochemistry: Theory and Practice, R. Gale, Ed. Plenum Press, New York, 1988.
3. Scherson, D.; Fierro, C.; Tryk, D.; Gupta, S. L.; Yeager, E. B.; Eldridge, J.; Hoffman, R. W. J. Electroanal. Chem. 1985 184, 419.
4. Tanaka, A.; Fierro, C.; Scherson, D.; Yeager, E. B. J. Phys. Chem. 1987, 91, 3799.
5. a. Burke, L. D.; Murphy, O. J. J. Electroanal.Chem. 1980 109, 379.
   b. Macagno, V. A.; Vilche, J. R.; Arvia, A. J. J. Appl. Electrochem. 1981, 11, 417.
6. a. Simpson, A. W. J. Appl. Phys. 1962, 33, 1203.
   b. Kunding, W.; Bommel, H.; Constabari, G.; Lindquist, R. H. Phys. Rev. 1966, 142, 327.
   c. McNab, T. K.; Fox, R. A.; Boyle, A. J. F. J. Appl. Phys. 1968 39, 5703.
7. Hassett, K. L.; Stecher, L. C.; Hendrickson, D. N. Inorg. Chem. 1980, 19, 416.
8. Fierro, C.; Carbonio, R.; Scherson, D.; Yeager, E. B. J. Phys. Chem. 1987, 91, 6597.

9. Geronov, Y.; Tomov, T.; Georgiev, S. J. Appl. Electrochem. 1975, 5, 351.
10. a. Cordoba, S.I.; Carbonio, R.; Lopez-Teijelo, M.; Macagno, V. A. Electrochim. Acta 1986, 31, 1321.
b. Corrigan, D.; J. Electrochem. Soc. 1987, 134, 377.
11. Corrigan, D.; Conell, R. S.; Fierro, C.; Scherson, D. J. Phys. Chem. 1987, 91, 5009.
12. Kordesch, M. E.; Hoffman, R. W. Thin Solid Films 1983, 107, 365.
13. Desilvestro, J.; Corrigan, D.; Weaver, M. J. J. Phys. Chem. 1986, 90, 6408.

RECEIVED May 17, 1988

Chapter 19

Use of the Frozen Electrolyte Electrochemical Technique for the Investigation of Electrochemical Behavior

Ulrich Stimming

Electrochemistry Laboratory, Department of Chemical Engineering and Applied Chemistry, Columbia University, New York, NY 10027

The background, experimental procedures and some results using the FREECE technique are described. Properties of liquid and frozen $HClO_4*5.5H_2O$) with respect to electrical conductivity and deuteron NMR are discussed. Electrochemical behavior of the metal-(liquid or frozen) electrolyte interface as observed in capacity measurements and charge transfer reactions is described. In the case of hydrogen evolution on copper, silver and gold, temperature dependent transfer coefficients are found. For copper and silver, straight Tafel lines with a slope of 85mV independent of temperature result in a transfer coefficient that is proportional to temperature. Some implications of the results for the understanding of electrochemical processes on a molecular level are discussed. The importance of UHV work on water-ion coadsorption on metal surfaces for the understanding of electrochemical processes on a molecular level is emphasized.

In studying interfacial electrochemical behavior, especially in aqueous electrolytes, a variation of the temperature is not a common means of experimentation. When a temperature dependence is investigated, the temperature range is usually limited to 0-80°C. This corresponds to a temperature variation on the absolute temperature scale of less than 30%, a value that compares poorly with other areas of interfacial studies such as surface science where the temperature can easily be changed by several hundred K. This "deficiency" in electrochemical studies is commonly believed to be compensated by the unique ability of electrochemistry to vary the electrode potential and thus, in case of a charge transfer controlled reaction, to vary the energy barrier at the interface. There exist, however, a number of examples where this situation is obviously not so.

0097–6156/88/0378–0275$06.00/0

There are currently a number of attempts to overcome the limitations of the limited temperature accessible in aqueous electrolytes:
(i) To use organic solvents that allow one to work at much lower temperatures (1-5);
(ii) To use an electrolyte with a high boiling point such as concentrated phosphoric acid (6);
(iii) To work in an aqueous electrolyte under high pressure in order to extend the available temperature range to higher temperatures (7,8);
(iv) To work in aqueous electrolytes below the freezing point of the electrolyte thus increasing the available temperature range to lower temperatures (9).

The latter approach termed **FREECE** (**FR**ozen **E**lectrolyte **E**lectro**ChE**mistry) has been pursued in our laboratory. In addition to increasing the available temperature down to approx. 120K which allows one to vary the temperature by more than a factor of two, there are other interesting aspects of these technique:

- One can study specific low temperature effects, such as possible quantum effects;
- One can study the influence of solidification, i.e. the development of a long range order or a possible glass formation of the electrolyte on electrochemical behavior;
- One is in a better position to compare results in electrochemistry with results in surface science involving water ad- or coadsorption on metal surfaces which are usually investigated in the temperature range 100-250K as done by Sass and co-workers (10,11), Wagner (12) and Madey (13).

In this paper, some of the possibilities associated with the FREECE technique will be described. Results referring to the charge distribution at the electrode-electrolyte interface and to charge transfer reactions will be presented and briefly discussed.

Experimental

Design of Experimental Set-up. In performing electrochemical measurements at cryogenic temperatures, two different approaches can be chosen: Either to adapt an electrochemical cell to a commercially available cryostat or to adapt a cooling system to an optimized electrochemical cell.

While it is convenient to use a standard cryostat, the cell has to be fitted in the sample space of the cryostat and the leads to the cell are usually fairly long - a specific draw-back for impedance measurements. Another disadvantage is the long cool-down times of these systems which may not allow one to rapidly freeze the electrolyte. However, for experiments not requiring sophisticated electrochemical experimentation this may be the most convenient experimental set-up.

Most of the work in our laboratory has been done using specially designed cells that are incorporated in a rather simple cooling device. In working with polycrystalline metal foils, a cell design as pictured in Fig.1 has been used. Metal

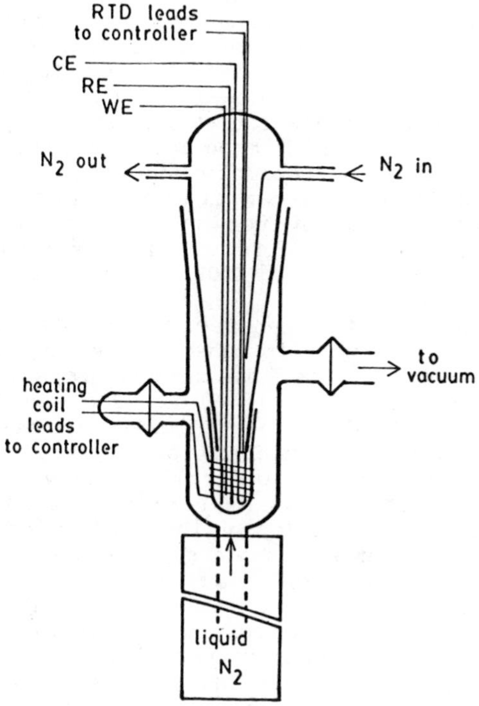

Figure 1. Electrochemical cell for cryogenic measurements.

foils for working and counter electrode are separated by approx. 1mm, with a reference electrode made of wire close to the gap between the two foils. The reference electrode can be either a hydrogen loaded Pd wire or, in case of a redox electrolyte, a platinum wire or any other metal that establishes the equilibrium potential of the investigated reaction. For the latter situation, overpotentials at all temperatures are directly given by the difference between working and reference electrode. The cell further contains two platinum resistance elements sealed in glass for the temperature control and an independent temperature measurement. The electrolyte volume is approx. $1cm^3$. A heating wire is wrapped around the cell which is connected to the temperature controller. The cell is placed in an outer jacket which is hooked up to a vacuum regulation system and, in its lower part, is connected with a liquid nitrogen reservoir. By regulating the vacuum, a variable stream of evaporating nitrogen passes the cell and cools it. The temperature controller adjusts a constant temperature by switching the heat on and off. The accuracy of the temperature regulation is about $\pm 1K$ and the cell has been operated down to 100K. Cells made either of glass or of Kel-F have been used.

In order to investigate specific crystal faces, a cell design has been chosen that is based on the idea of the hanging drop technique often used in single crystal studies. A schematic picture is shown in Fig.2. The central part is a glass block that extends into the lower part where cooling is provided by evaporating liquid nitrogen, while heating wires connected to the temperature controller and wrapped around the glass block establish a constant temperature in a comparable way as described above. On top of the glass block, a gold disk which forms the counter electrode sits in a depression. A drop of electrolyte of 50-100 mm^3 can be supplied onto the Au disk. A crystal rod with the desired crystal face pointed down can be lowered by means of a micrometer to contact the electrolyte. A Pd/H wire as reference electrode can also be positioned with a micrometer. The use of micrometers allows for an optimization of the geometry which is crucial at lower temperatures to keep the cell as low-ohmic as possible. A glass sealed thermocouple measures the actual temperature of the electrolyte. Because of the higher thermal mass of this cell, special insulation is necessary in order to reach temperatures below 170K.

Electrochemical Instrumentation. For all electrochemical experiments, regular instrumentation common in electrochemical research can be used. One has to consider, however, that the electrical characteristics of the cell change by many orders of magnitude by changing the temperature from room temperature to cryogenic conditions. The electrolyte that has been widely used in our laboratory is aqueous perchloric acid of the composition $HClO_4*5.5H_2O$ which exhibits a relatively high conductivity even a low temperatures. More details regarding this electrolyte will be given below.

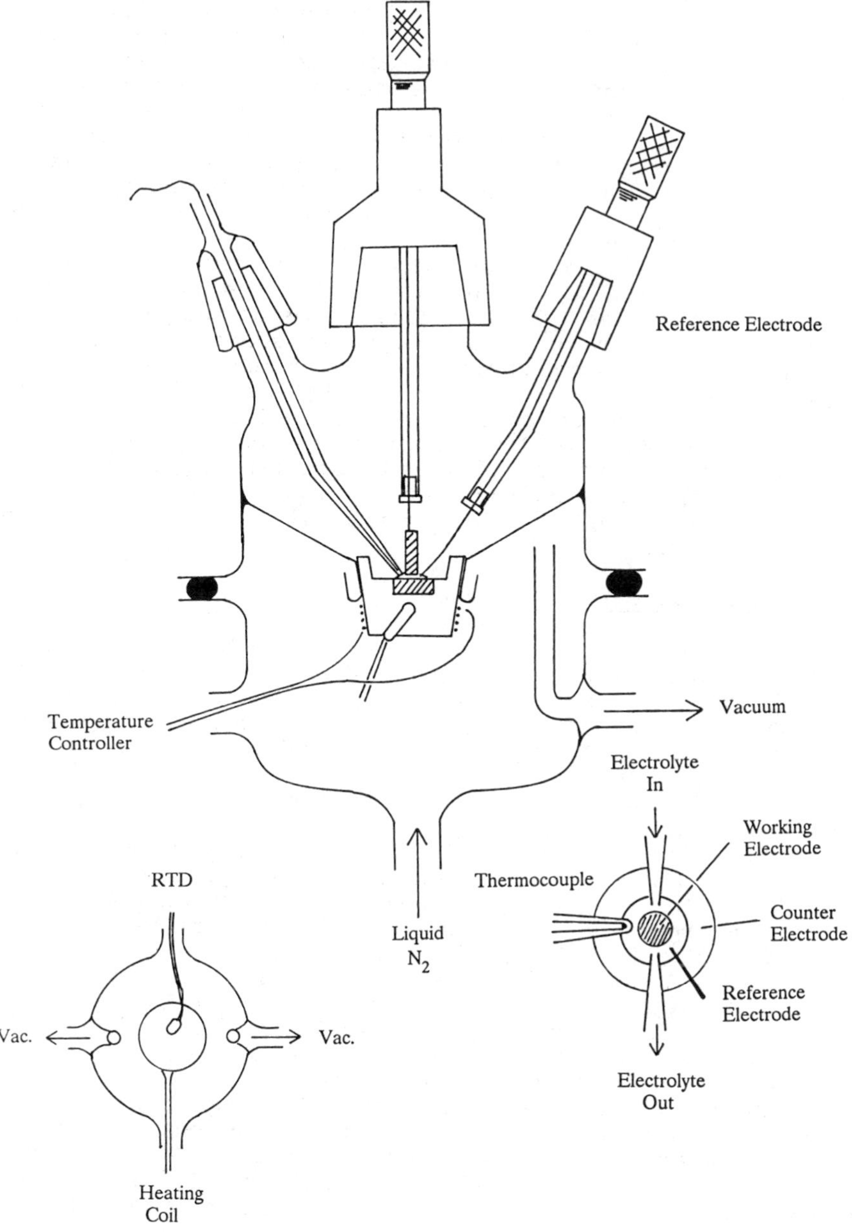

Figure 2. Electrochemical cell for measurements with single crystals at cryogenic measurements.

For capacity measurements, several techniques are applicable. Impedance spectroscopy, lock-in technique or pulse measurements can be used, and the advantages and disadvantages of the various techniques are the same as for room temperature measurements. An important factor is the temperature dependent time constant of the system which shifts e.g. the capacitive branch in an impedance-frequency diagram with decreasing temperature to lower frequencies. Comparable changes with temperature are also observed in the potential transients due to galvanostatic pulses.

For the investigation of charge tranfer processes, one has the whole arsenal of techniques commonly used at one's disposal. As long as transport limitations do not play a role, cyclic voltammetry or potentiodynamic sweeps can be used. Otherwise, impedance techniques or pulse measurements can be employed. For a mass transport limitation of the reacting species from the electrolyte, the diffusion is usually not uniform and does not follow the common assumptions made in the analysis of current or potential transients. Experimental results referring to charge distribution and charge transfer reactions at the electrode-electrolyte interface will be discussed later.

Charge distribution

The charge distribution at metal electrode-electrolyte interfaces for liquid and frozen electrolytes has been investigated through capacity measurements using the lock-in technique and impedance spectroscopy. Before we discuss some of the important results, let us briefly consider some properties of the electrolyte in its liquid and frozen state.

$HClO_4*5.5H_2O$ is an aqueous perchloric acid of stoichiometric composition that represents the highest hydrate of the perchloric acid. It crystallizes in a clathrate type of structure at 228K. The crystallographic parameters have recently been determined using X-ray diffraction (14,15). In clathrates one usually has a host lattice with different types of voids. In case of $HClO_4*5.5H_2O$ the host lattice is formed by H_2O with ClO_4^- and H^+ as guests. Conductivity measurements indicate a high conductivity in the liquid state, but upon freezing at 228K no abrupt, just a steady, decrease is observed with decreasing temperature; the mobile species are presumably the protons. In order to gain more insight into the conduction process, NMR studies were performed. The measurements were carried out in deuterated perchloric acid solutions since there are some advantages of deuteron versus proton NMR. Below the freezing point, a line broadening increasing with decreasing temperature can be observed. The line broadening can be related to the frequency of the diffusion-reorientation of the deuterons. Fig.3 shows an Arrhenius plot of the conductivity, plotted as the inverse resistance, and of the inverse halfwidth of the line broadening (Huang, T.-H.; Ang, T.T.; Frese, U.; Stimming, U. submitted for publication). It is interesting to note that below 228K, the freezing point of the electrolyte,

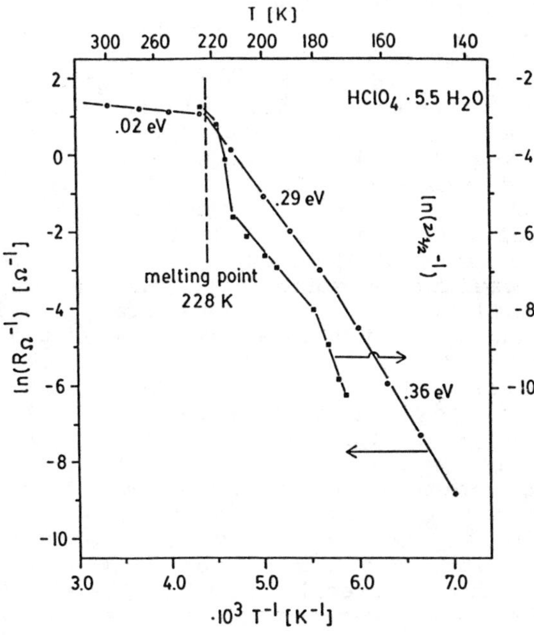

Figure 3. Arrhenius plot of electrical conductivity of $HClO_4$*$5.5H_2O$ and of the halfwidth of the ^2H-NMR signal.

the slopes for the line broadening and the conductivity are very similar, corresponding to an activation energy of approx. 0.3eV. At 170K, another break in the slope is observed. This temperature has been identified as a reversible higher order phase transformation (14,15). While the activation energy of the conductivity increases to approx. 0.38eV, the decrease in frequency in the NMR signal is much steeper with approx. 0.7eV. It has to be mentioned, however, that the shape of the NMR signal at very low temperatures suggests that the deuterons are still mobile with a frequency of about 10^3Hz.

Capacity data has been obtained for gold, silver and copper. As mentioned above, the main effect of temperature on the impedance is a shift of the spectrum to lower frequencies with decreasing temperatures. This is illustrated in Fig.4 for a polycrystalline gold electrode in $HClO_4$*5.5H_2O (Tellefsen, K.; Stimming, U. unpublished results). Comparing the two impedance spectra, the one measured at 278K in the liquid electrolyte and the other measured at 198K in the frozen electrolyte, the effect of temperature is clearly seen. The shape of the spectra is essentially not changed which indicates that the equivalent circuit underlying this interfacial behavior is preserved in the frozen electrolyte. This behavior is also reflected in the potential dependence of the capacity of polycrystalline gold (16). The characteristic peaks in capacity-potential curves observed in the liquid electrolyte are preserved in the frozen electrolyte. The curves in Fig.5 show a structure that is similar to what is observed at room temperature in the liquid electrolyte. With decreasing temperature, these features become less pronounced, and at temperatures below 150K, the curves become rather flat. Similiar results were also obtained by Iwasita et al. (17) for polycrystalline gold and by Hamelin et al. (18) for gold {210}.

In a similar way, capacity measurements for polycrystalline silver and copper indicate that "electrochemistry" stays intact below the freezing point of the electrolyte. When the minima of capacity-potential curves are plotted as a function of temperature, as shown in Fig.6, a straight line is obtained in the temperature range 200-300K (19). The freezing of the electrolyte does not seem to have any major effect on the electrode capacity of silver and copper. As can be seen from Fig.6, the capacity exhibits a positive temperature coefficient, dC/dT; this is also observed for gold. This is in contrast to the early results of Grahame on mercury (20) and also on gold by Schmid and Hackerman (21). Hamelin et al. (22), however, observed for Au {210} a positive temperature coefficient at potentials positive of the pzc and a negative temperature coefficient at potentials negative of the pzc in the temperature range 274 to 323K for dilute $HClO_4$. It would be interesting to see if e.g. the Grahame results, where the capacity increases with decreasing temperature, are also reflected in capacity behavior of mercury in the frozen electrolyte. Overall, the capacity behavior as found for gold, silver and copper suggests the possibility to use the FREECE technique to study electrochemical behavior in a wide range of

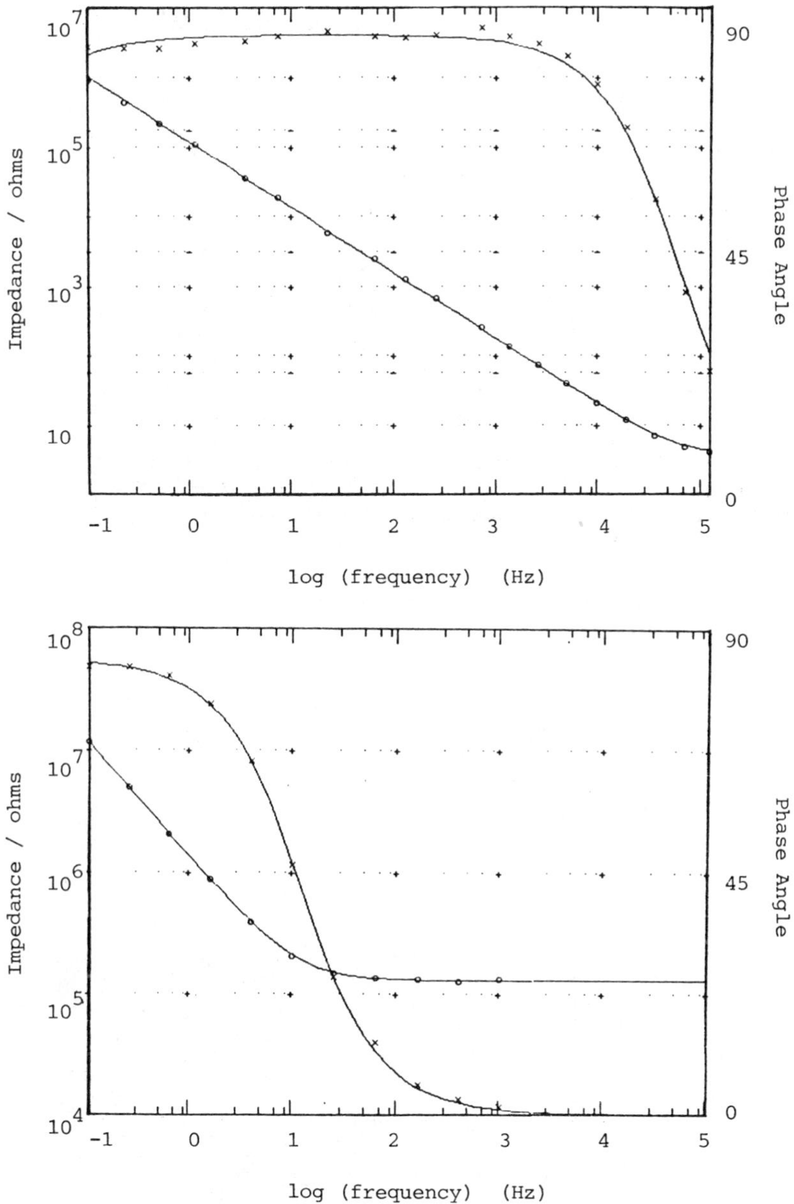

Figure 4. Impedance spectrum of gold in liquid (278K) and frozen (198K) $HClO_4*5.5H_2O$.

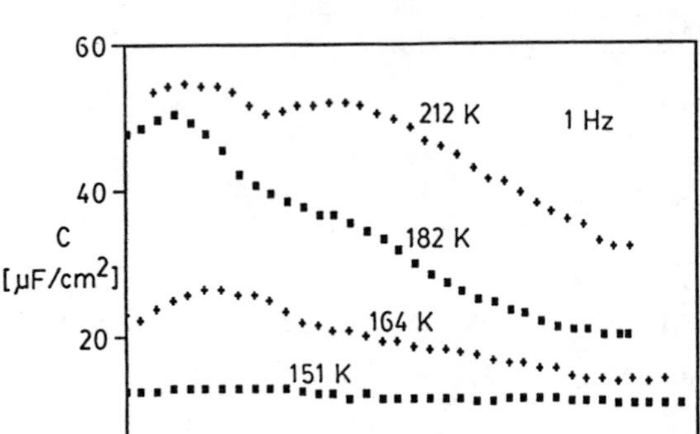

Figure 5. Capacity-potential curves of gold at various temperatures in the frozen electrolyte (taken from ref. 16, with permission of the Electrochemical Society, Pennington, NJ).

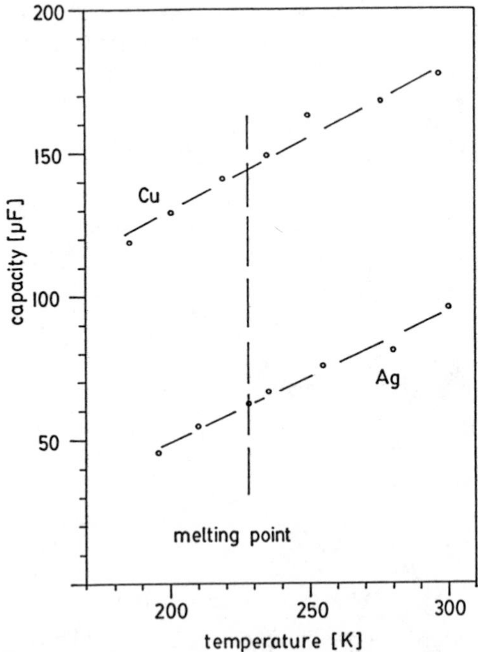

Figure 6. Temperature dependence of the electrode capacity of copper and silver (taken from ref. 16 , with permission of the Electrochemical Society, Pennington, NJ).

temperatures by passing the freezing point of the electrolyte.

It is tempting to consider, on the basis of these capacity measurements, the consequences for an understanding of the electrochemical double layer. Although it is much too early, with respect to the available data, to justify a qualified attempt, a few aspects seem to be clear enough to allow for some qualitative remarks. The results indicate that the double layer is still intact in the frozen electrolyte. The solidification of the electrolyte with its development of a long range order obviously affects the interfacial region only in a minor way. One reason certainly is that the electrolyte is highly concentrated; $HClO_4$*$5.5H_2O$ corresponds to an approx. 7M $HClO_4$ in water. This limits the potential drop to the very interface, excluding diffuse layer effects, i.e. there is no extension of the potential drop into the bulk of the electrolyte. The charging of the interface as a function of the electrode potential can still be achieved since, as found from the NMR measurements, the protons are mobile in the frozen electrolyte. Reversing the point of view, one can then ask why is the capacity behavior in the liquid electrolyte so similar to that in the frozen electrolyte. In the frozen state, the electrolyte will, ideally, terminate at the electrode in some kind of structured way, at least in the two dimensions parallel to the metal surface. Normal to the surface some differences may exist, especially at the locations of charge and countercharge where a high electric field exists. The locations of charge depends on the proton mobility not only normal but also parallel to the surface. Any anisotropy effects of the proton mobility are not known for $HClO_4$*$5.5H_2O$ and the crystal structure does not suggest any (14,15). This arrangement can stay essentially preserved upon melting, especially close to the surface where the field may stabilize advantageous orientations of the water with the perchlorate ions and the protons. In addition, it is well known that aqueous solutions and pure water form large agglomerates in the liquid state that allow e.g. X-ray diffraction to be observed. In summary, regarding the molecular processes at the electrode-electrolyte interface, they do not seem to be strongly affected by the phase change from liquid to solid of the bulk electrolyte.

Charge transfer reactions

Charge transfer reactions represent an important category of electrochemical behavior. As already pointed out above, an appropriate investigation of kinetic parameters of electrochemical reactions in aqueous electrolytes suffers from the small temperature range experimentally accessible. In the following, some preliminary results using the FREECE technique are presented for the Fe^{2+}/Fe^{3+} redox reaction and for hydrogen evolution at various metal electrodes.

The redox reaction Fe^{2+}/Fe^{3+} was the first reaction for which it could be demonstrated that the investigation of electrochemical behavior in frozen aqueous electrolytes is possible (9). The electrolyte was 1M $HClO_4$ which has the disadvantage that below 273K one has a temperature dependent composition of the liquid electrolyte in addition to ice and eventually a solid mixture of ice and the eutectic of ice and $HClO_4$*5.5H_2O. This rather complicated situation led to the choice of $HClO_4$*5.5H_2O as electrolyte which has a defined freezing point at 228K and is always a single phase except at the freezing point itself.

The exchange current density of the Fe^{2+}/Fe^{3+} reaction in $HClO_4$*5.5H_2O has been investigated using galvanostatic pulses in the potential regime of the linear current-potential relationship. Current pulses have to be chosen such that a steady state potential is reached. This procedure allows for an easy separation and also determination of the ohmic drop and the charge transfer overpotential. Results were obtained for various concentrations and are plotted in an Arrhenius plot in Fig.7 (Dinan, T.; Stimming, U. unpublished results). As expected, straight lines are found in the liquid electrolyte with a slope that corresponds to an activation energy of 0.3eV, the common value for this reaction. The concentration dependence also represents, as expected, a reaction order of one. Upon freezing, an increase of the exchange current is observed and the temperature has to be several ten degrees below the freezing point in order for the current to come back to the same value as at the freezing point. The current for the two highest concentrations, 0.03M and 0.01M, becomes almost identical upon freezing. At lower temperatures, the differences between the values for 0.01M and 0.003M also seem to diminish to eventually become of comparable value. The difference between the curves for 0.001M and 0.003M, however, stays constant and is approx. a factor of three, as expected for a rection order of one. The reason for the non-proportionality of the current at higher concentrations is probably due to a limited solubility of the redox system in the electrolyte. Precipitation, or better, crystallization, as opposed to a solid solution of the redox system in the electrolyte, may lead to a reduction of the effectively available number of Fe^{2+} and Fe^{3+} sites at the interface. If the concentration is low enough, the distribution is obviously as random as one would expect it to be similarly to the liquid. From the slope of the curves for lower concentrations an activation energy of 0.35eV can be calculated slightly higher than the value in the liquid electrolyte. This, together with the current increase at the freezing point, results in a considerably higher pre-exponential factor. It will be shown later that the current increase upon freezing of the electrolyte seems to be a rather typical phenomenon and is also observed for the hydrogen evolution reaction at various metal electrodes. Recently, Matsunaga et al. (23) have investigated the Fe^{2+}/Fe^{3+} reaction in 1M $HClO_4$ and in $HClO_4$*5.5H_2O in the liquid and frozen state of the electrolyte. While their results in 1M $HClO_4$ are very similar to our earlier results (9), the Arrhenius plot for $HClO_4$*5.5H_2O as electrolyte differs considerably from the results shown in Fig.7. In particular, Matsunaga et al. observe a drop in the exchange current density by more than three orders of magnitude between 180 and 175K. It is currently not clear what the reasons for these differences are.

Hydrogen evolution, the other reaction studied, is a classical reaction for electrochemical kinetic studies. It was this reaction that led Tafel (24) to formulate his semi-logarithmic relation between potential and current which is named for him and that later resulted in the derivation of the equation that today is called "Butler-Volmer-equation" (25,26). The influence of the electrode potential is considered to modify the activation barrier for the charge transfer step of the reaction at the interface. This results in an exponential dependence of the reaction rate on the electrode potential, the extent of which is given by the transfer coefficient, α.

The first system studied using FREECE was the hydrogen evolution on platinum (27). The reaction proceeds similarly in liquid and frozen $HClO_4$*$5.5H_2O$. It exhibits a potential dependence of the current with α=2 which is practically the same in the liquid and the frozen electrolyte. The potential dependence indicates, however, that not the charge transfer step but the recombination of hydrogen atoms and/or the diffusion of molecular hydrogen from the surface is rate determining (28).

In contrast to platinum, the hydrogen evolution on copper, on silver and on gold show a different behavior. Similarly to platinum, slow potentidynamic sweeps have been used to measure the current-potential curves (29). While the behavior of gold shows potential and temperature dependent changes of the slope, silver and copper exhibit straight lines in a Tafel plot, however, with no or only little change of the slope with temperature. This is illustrated in Fig.8 for silver in the temperature range of 140-300K. As can be seen from Fig.8, the lines are straight over about three order of magnitude and almost ideally parallel in the range T=160-300K with a Tafel slope of b=85mV. The results for copper are very similar. Calculating the transfer coefficient according to $b=RT/\alpha nF$ consequently results in a temperature dependent transfer coefficient as shown in Fig.9 for silver and copper (ref.19). In the temperature range from 160 to 300K a linear dependence of a vs. T is found with a neglible intercept at T=0K.

This result is quite in contrast to the common expectation that the electrode potential changes the activation barrier at the interface which would result in a temperature independent transfer coefficient α. Following Agar's discussion (30), such a behavior indicates a potential dependence of the entropy of activation rather than the enthalpy of activation. Such "anomalous" behavior in which the transfer coefficient depends on the temperature seems to be rather common as recently reviewed by Conway (31).

Analyzing the data in terms of the Arrhenius equation leads to a complementary result (ref.19). An Arrhenius plot of the data obtained on copper is shown in Fig.10. Straight lines are obtained for different potentials in the liquid and in the frozen electrolyte, again with an increase of the current at the freezing point. The lines are nearly parallel for the

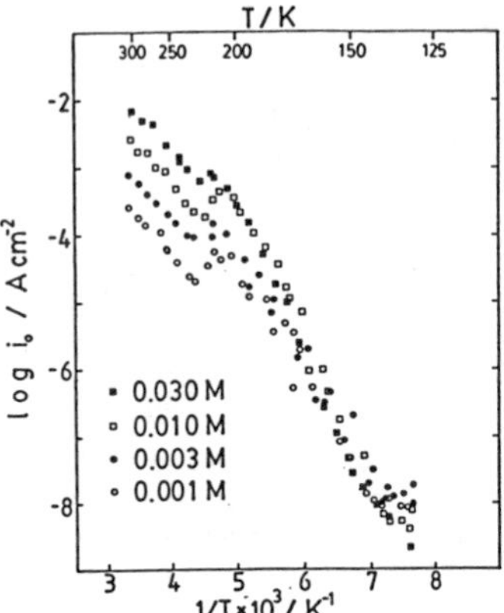

Figure 7. Arrhenius plot of the exchange current density of the Fe^{2+}/Fe^{3+} reaction for various concentrations.

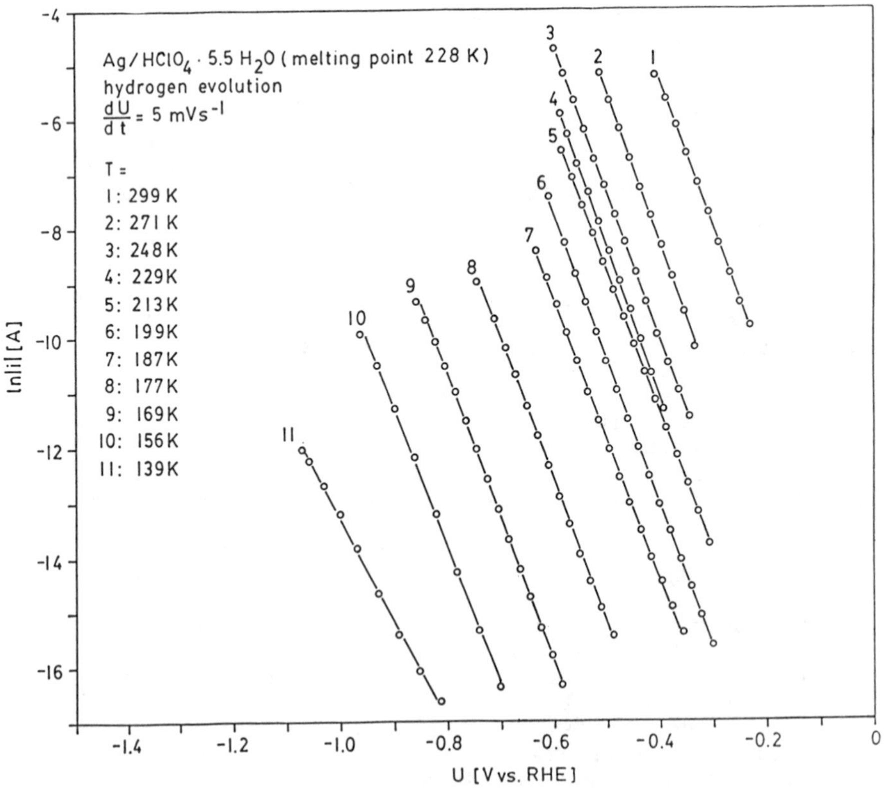

Figure 8. Tafel plot of the hydrogen evolution on silver at various temperatures.

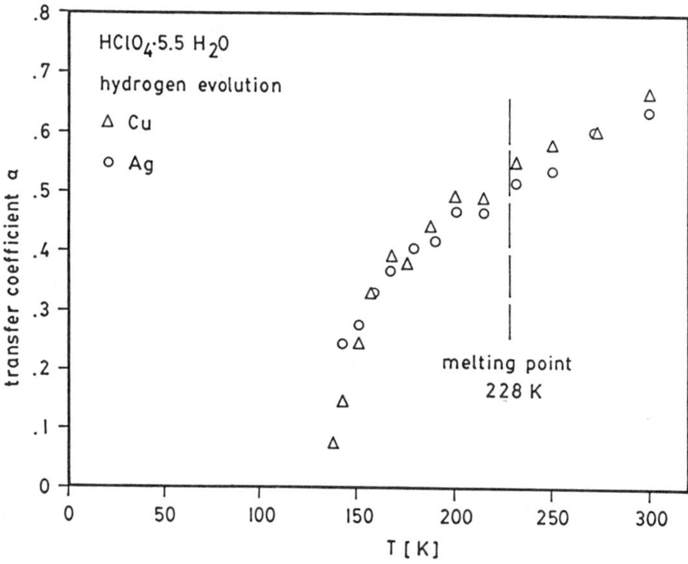

Figure 9. Temperature dependence of the transfer coefficient of the hydrogen evolution on silver and copper.

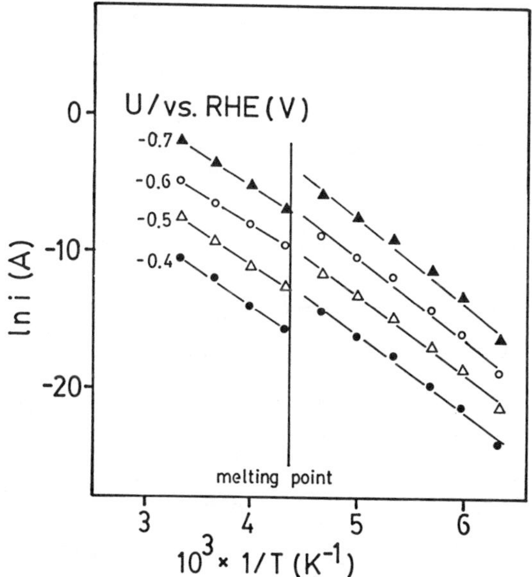

Figure 10. Arrhenius plot of the hydrogen evolution on copper at various potentials.

various potentials. This results in almost potential independent activation energies between 0.4 and 0.5eV (Fig.11a). The pre-exponential factor, on the other hand, as obtained from the extrapolation to 1/T=0, is strongly potential dependent (Fig.11b). It is higher in the frozen electrolyte and changes by approx. five orders of magnitude over a range of just 300mV in overpotential!

The latter discussion confirms the results of the potential dependence of the current in that the activation barrier for the hydrogen evolution reaction is, at least on copper and silver, not affected by the electrode potential. This behavior is, on the other hand, connected with the observation of straight lines in a Tafel plot. It would be premature to come up with a comprehensive model that would explain this behavior; more experimental work is necessary to substantiate and quantify the effects for a larger variety of systems and reactions. A few aspects, however, should be pointed out.

(i) It is difficult to conceive that impurity adsorption or the like could account for such radical effects, i.e. that the potential dependence of the activation energy is completely shifted to a potential dependence of the pre-exponential factor.

(ii) A potential dependent coverage with hydrogen may account for some of the observed effects. However, surface science studies indicate that hydrogen is not adsorbed on silver or copper (10), but electrochemical systems may be different since the change of the electrode potential, i.e. the Fermi level, may alter this situation.

(iii) Tunneling effects may play a role which would express themselves in the pre-exponential factor; it is not clear, however, if the tunneling probability would exhibit such a strong potential and temperature dependence.

For an understanding of these processes more information from in-situ surface science experiments is desirable. The study of water adsorption on silver is a good example for providing molecular information that may be relevant for the understanding of electrochemistry. As reviewed by Thiel and Madey (13), water, even in amounts of fractions of a monolayer, does not specifically interact with the silver surface. No differences are found in the thermodesorption spectra are found from submonolayer to multilayer coverages: they all show a single desorption peak typical of bulk ice. The bond structure within the phases is obviously much stronger than between the phases, i.e. in the interfacial region. The orientation of the water molecules is thus determined by the water phase rather than by the influence of the substrate.

Applying this to our electrochemical system could qualitatively explain some of the observed effects. Assuming that there is only a weak interaction between metal and the perchloric acid hydrate, protons being part of the clathrate structure may not be in a favorable position for a charge transfer reaction at the interface. This could result in small pre-exponential factors. The electrode potential, however, may

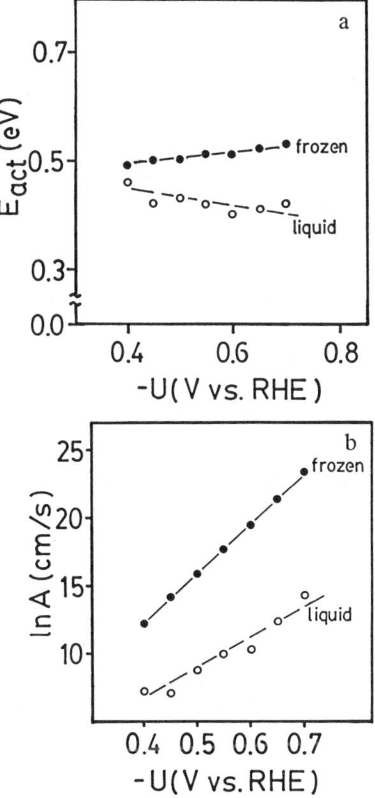

Figure 11. Potential dependence of the activation energy, E_a, (a) and pre-exponential factor, A, (b) as obtained from Fig.10.

have a large influence on it since the protons are also the charge carrier that allow the potential difference to be established at the interface. Thus by charging the interface, the protons with the water structure surrounding them, will also be changed which, in turn, influences the hydrogen evolution reaction. Such an effect would be expressed in a potential dependent pre-exponential factor.

Conclusions

It is possible to investigate electrochemical behavior in aqueous electrolytes in a wide range of temperatures by using the FREECE technique. This way, using $HClO_4*5.5H_2O$ as electrolyte, temperatures down to approximately -140°C can be reached. The results show that double charging and faradaic reactions occur in the frozen electrolyte in a comparable way as is known for the liquid electrolyte. The electrode capacities for gold, silver and copper decrease with decreasing temperature. The results for the hydrogen evolution on these metals indicate an anomalous behavior in that the transfer coefficient shows a temperature dependence. In the case of copper and silver the transfer coefficient appears to be proportional with temperature. It seems that current models have little ability to explain this behavior. It is expected that microscopic models of the electrochemical interface are better suited. In this respect, the concept of studying the co-adsorption of water and ions on defined surfaces under u.h.v. conditions developed by Sass et al. (10,11) and recently also applied by Wagner et al. (12) can be very helpful for a better understanding of these questions.

Acknowledgments

The discussions with Prof. J.K. Sass are very much appreciated. Thanks are also due to Drs. A. Pinkowski and K. Tellefsen for their help in preparing this manuscript. This work has been supported in part by the National Science Foundation and the Office of Naval Research.

Literature Cited

(1) VanDyne, R.P.; Reilley, C.N. Anal.Chem. 1972 44 142.
(2) VanDyne, R.P.; Reilley, C.N. Anal.Chem. 1972 44 153.
(3) VanDyne, R.P.; Reilley, C.N. Anal.Chem. 1972 44 158.
(4) Conway, B.E.; Salomon, M. J.Chem.Phys. 1964 41 3169.
(5) Conway, B.E.; McKinnon, D.J.; Tilak, B.V. Trans.Faraday Soc. 1970 66 1203.
(6) Yeager, E.; Scherson, D.; Simic-Glavaski, B. 163rd Meeting of the Electrochemical Society, San Francisco, 1983, Ext.Abstract #706.

(7) McDonald, A.C.; Fan, F.F.; Bard, A.J. J.Phys.Chem. 1986 90 196.
(8) Flarsheim, W.M.; Tsou, Y.; Trachtenberg, I.; Johnston, K.P.; Bard, A.J. J.Phys.Chem. 1986 90 3857.
(9) Stimming, U.; Schmickler, W. J.Electroanal.Chem. 1983 150 125.
(10) Sass, J.K. Vacuum 1983 (1983) 741.
(11) Stuve,E.M.; Bange, K.; Sass, J.K. In Trends in Interfacial Electrochemistry; Silva, A.F., Ed.; D.Reidel Publishing Company: Dordrecht, 1986; p.255.
(12) Wagner, F.T.; Moylan, T.E. Surf.Sci. 1987 182 125.
(13) Thiel, P.A.; Madey, T.E. Surf.Sci.Rep. 1987 7 211; and references of T.E.Madey's work therein.
(14) Wiebke, M.; Mootz, D. Z.Kristallogr. 1985 170 194.
(15) Mootz, D.; Oellers, E.-F.; Wiebke, M. J.Am.Chem.Soc. 1987 109 1200.
(16) Dinan, T.; Stimming, U. J.Electrochem.Soc. 1986 133 2662.
(17) Iwasita, T.; Roettgermann, S.; Schmickler, W. J.Electroanal.Chem. 1985 196 203.
(18) Hamelin, A.; Roettgermann, S.; Schmickler, W. J.Electroanal.Chem. 1987 230 281.
(19) Frese, U.; Stimming, U. In Principles of Electrochemical Processes; Schultze, J.W., Ed.; VCH-Verlagsgesellschaft: Weinheim, 1986; DECHEMA Monographs Vol. 102, p.247.
(20) Grahame, D.C. J.Am.Chem.Soc. 1957 79 2093.
(21) Schmid, G.M.; Hackerman, N. J.Electrochem.Soc. 1963 110 440.
(22) Hamelin, A.; Stoicoviciu, L.; Silva, F. J.Electroanal.Chem. 1987 229 107.
(23) Matsunaga, A.; Itoh, K.; Fujishima, A.; Honda, K. J.Electroanal.Chem. 1986 205 343.
(24) Tafel, J. Z.Phys.Chem. 1905 50 641.
(25) Butler, J.A.V. Trans.Faraday Soc. 1924 19 729.
(26) Erdey-Gruz, T.; Volmer, M. Z.Phys.Chem. 1930 150 203.
(27) Frese, U.; Iwasita, T.; Schmickler, W.; Stimming, U. J.Phys.Chem. 1985 89 1059.
(28) Ludwig, F.; Sen, R.K.; Yeager, E. Elektrokhimiya 1977 13 847.
(29) Frese, U.; Stimming, U. J.Electroanal.Chem. 1986 198 409.
(30) Agar, J.N. Disc.Faraday Soc. 1947 11 81.
(31) Conway, B.E. In Modern Electrochemistry; Bockris, J.O'M., Conway, B.E. and Yeager, E., Eds.; Plenum Press: New York, 1987; Vol.16.

RECEIVED August 1, 1988

Chapter 20

Second Harmonic Generation Studies of Chemisorption at Electrode Surfaces

Deborah J. Campbell and Robert M. Corn

Department of Chemistry, University of Wisconsin—Madison, Madison, WI 53706

Optical second harmonic generation from electrode surfaces is employed *in situ* to study the electrochemical processes of lithium deposition and monatomic hydrogen chemisorption at silver electrodes in acetonitrile, and the reversible deposition of hydrogen at platinum electrodes in aqueous perchloric acid solutions. The changes observed in the second harmonic intensity are ascribed to modifications in the electronic structure of the metal surface upon adsorption. These studies demonstrate how the nonlinear optical response from the surface can provide both a qualitative and quantitative probe of chemisorption.

Optical second harmonic generation (SHG) is sensitive to the interface of two centrosymmetric media. At metal/electrolyte interfaces, the surface nonlinear susceptibility arising from the sharp gradients in the static electric fields have been shown to dominate the bulk magnetic dipole sources for SHG when using p-polarized light in a reflection geometry (1-2). This sensitivity of the SHG to the surface has been exploited at electrodes for measurements of charge density (3-4), anionic adsorption (5), oxide formation (6) and chemisorption (7-8). In the majority of the chemisorption studies to date, changes in the nonlinear optical response of the metal surface upon reaction have been used to indirectly follow the adsorption process. For example, the process of underpotential deposition (upd) of a monolayer of Pb or Tl has been studied through its effect on the SHG from silver surfaces (3-4,7).

In this paper we extend our SHG studies to silver electrodes in acetonitrile solutions and to platinum electrodes in aqueous solutions. Three different examples are chosen to demonstrate how SHG can be used both qualitatively and quantitatively to study the adsorption of chemical species onto

0097–6156/88/0378–0294$06.00/0

an electrode surface. The systems examined are the deposition of lithium onto a silver surface in acetonitrile, the chemisorption of monatomic hydrogen during the evolution of molecular hydrogen in nonaqueous acidic media, and the reversible deposition of hydrogen on platinum electrodes prior to the hydrogen evolution reaction in aqueous perchloric acid solutions.

The SHG studies at silver electrodes employed a p-polarized fundamental beam at 1064 nm from a Q-switched Nd:YAG laser (10 Hz repetition rate, 8 ns pulse width) with an incident angle of 30° relative to the surface normal, and the platinum electrode experiments used p-polarized light at 578 nm from a pulsed dye laser (10 Hz repetition rate, 6 ns pulse width) with an incident angle of 60°. The current and second harmonic signal observed during the cycling of the electrode potential were recorded simultaneously and stored on computer. In the two nonaqueous studies, potentials are reported versus a Ag|0.100M $AgNO_3$ reference electrode, and in the aqueous study the potentials are reported versus a saturated calomel electrode (SCE). Further experimental details are published elsewhere (8).

Lithium Deposition on Silver Electrodes in Acetonitrile

The cyclic voltammogram for a silver electrode in 0.1M $LiClO_4$ acetonitrile solution is shown in Figure 1 (curve a). At a potential of -1.5 V, cathodic current due to the reduction of Li^+ ions commences. The upd of lithium has been reported previously by Kolb et al. for positive potential sweeps after substantial lithium reduction (9); however, due to the reactivity of the metallic lithium with impurities in solution, the adsorbed layer formed on the negative potential sweep is not as stable as other upd monolayers (9). An additional cathodic wave due to the reduction of lithium is observed at approximately -2.5V, and on the return sweep the lack of an anodic wave is indicative of the reactivity of the chemisorbed atoms.

The second harmonic signal at 532 nm obtained from the electrode during this cyclic voltammogram is also shown in Figure 1 (curve b). The second harmonic signal steadily increases as the potential is swept negatively through the lithium reduction region, and decreases back to its original level on the positive scan. The independence of the second harmonic signal on the scan direction suggests that the surface undergoes a change that is solely a function of potential. We attribute the increase in second harmonic signal from the surface to the presence of a potential-dependent surface coverage of chemisorbed lithium. The amount of adsorbed lithium at a given potential is determined by the rates of formation and destruction of the reactive metal overlayer. The nonlinear susceptibility of the electrode surface increases upon lithium deposition due to the increase in free electron density on the surface through the delocalization of the lithium 2s electron.

Similar increases of the surface second harmonic intensity have been observed during the formation of alkali metal monolayers on Ge and Rh surfaces *in vacuo* (10-11).

Hydrogen Adsorption and Evolution at Silver Electrodes in Acetonitrile

The cyclic voltammogram and the second harmonic signal at 532 nm for a silver electrode in a 0.1M $LiClO_4$ + 3.5mM $HClO_4$ acetonitrile solution is shown in Figure 2. At a potential of -0.83V there is a cathodic current peak due to the irreversible evolution of molecular hydrogen. A reversible decrease is observed in the second harmonic signal that starts prior to the cathodic current wave. We attribute this decrease in the optical signal to the modification of the nonlinear susceptibility of the silver surface by the steady state formation of a monatomic hydrogen intermediate on the electrode during the hydrogen evolution reaction.

The cyclic voltammogram and the second harmonic signal at 532 nm for a silver electrode in a 0.1M $TBABF_4$ (tetrabutylammonium tetrafluoroborate) + 35mM acetic acid acetonitrile solution is shown in Figure 3. A reversible loss of second harmonic signal and an irreversible current wave from hydrogen evolution are present just as in the perchloric acid solution. However, the potentials for these processes have been shifted by approximately -1V. These shifts are directly related to the pH of the solution; millimolar solutions of other acids show shifts of varying magnitude in accordance with their acidity (8).

We can quantitate the relative surface coverage of adsorbed hydrogen, θ, from the second harmonic signal, $I(2\omega)$:

$$I(2\omega) = I(2\omega)_0(1 + c\theta)^2 \tag{1}$$

where $I(2\omega)_0$ is the second harmonic signal from the surface in the absence of chemisorption, and c is a constant that can be determined by the second harmonic signal at maximum surface coverage ($\theta = 1$). Equation 1 assumes that there is no change in phase of the nonlinear susceptibility from the surface. The relative surface coverage calculated from the second harmonic signal using Equation 1 for the case of acetic acid is shown as the solid circles in Figure 4. The potential at which the relative surface coverage is one half, $E_{1/2}(\theta)$, is -1.51V. For the perchloric acid solution a similar curve can be constructed, and $E_{1/2}(\theta)$ is found to be -0.45V.

These two facts, (i) the shape of the relative surface coverage-potential curve and (ii) the pH dependence of $E_{1/2}(\theta)$ can both be accounted for by the following reaction mechanism:

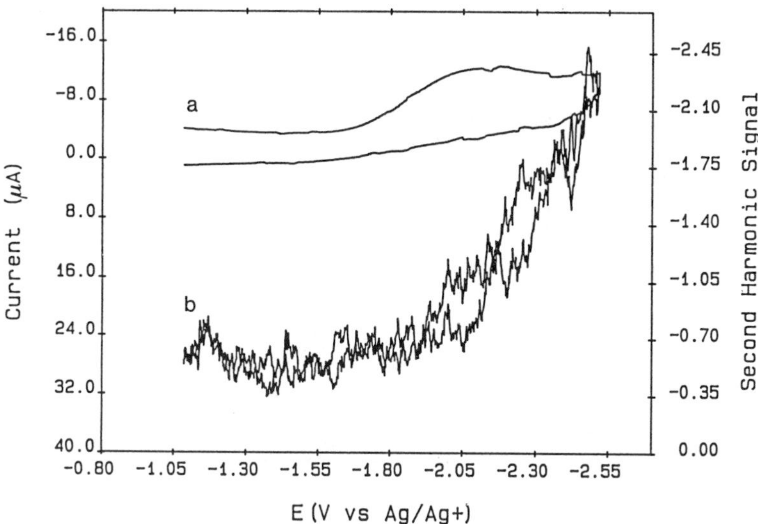

Figure 1. The current (a) and the second harmonic signal (b) obtained during the potential cycling (10 mV s^{-1} scan rate) from a polycrystalline silver electrode in a 0.1M $LiClO_4$ acetonitrile solution.

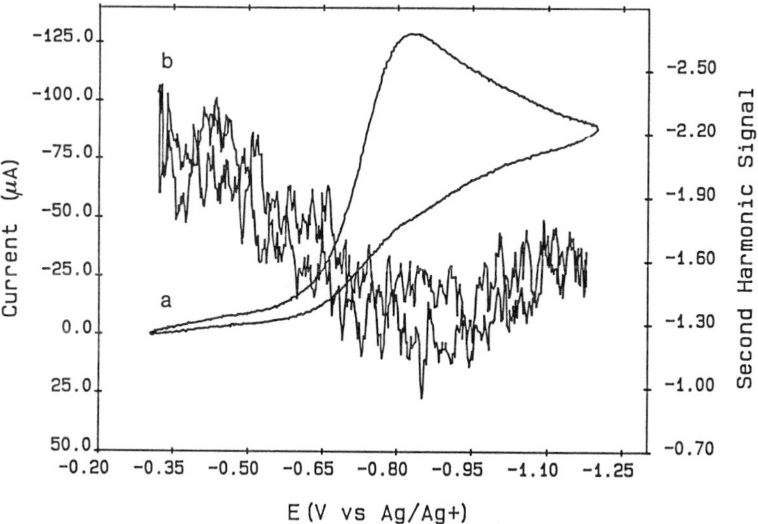

Figure 2. The current (a) and the second harmonic signal (b) obtained during the potential cycling (20 mV s^{-1} scan rate) from a polycrystalline silver electrode in a 0.1M $LiClO_4$ + 3.5mM $HClO_4$ acetonitrile solution.

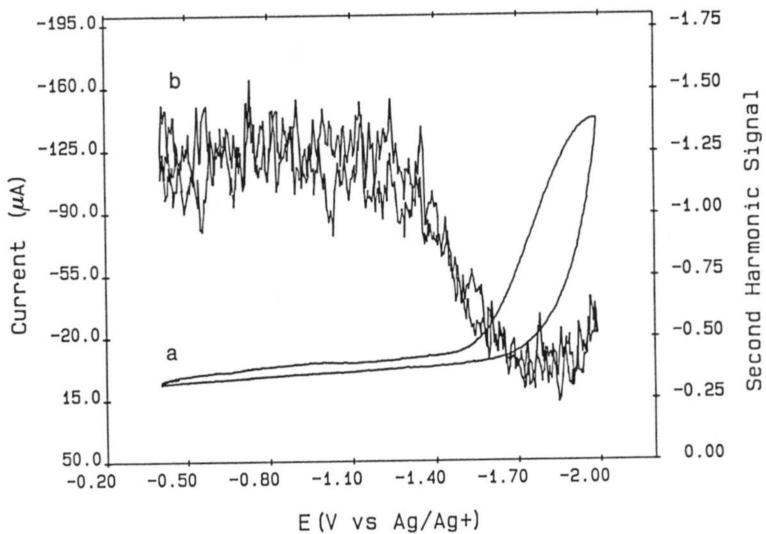

Figure 3. The current (a) and the second harmonic signal (b) obtained during the potential cycling (10 mV s^{-1} scan rate) from a polycrystalline silver electrode in a 0.1M $TBABF_4$ + 35mM acetic acid acetonitrile solution.

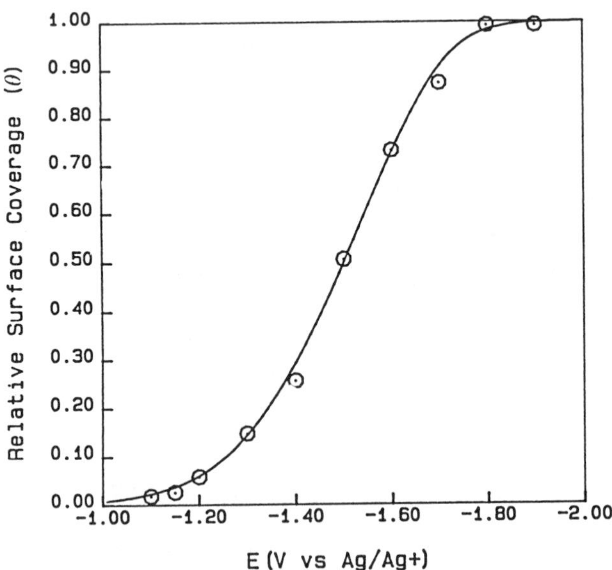

Figure 4. The relative surface coverage of adsorbed hydrogen as a function of electrode potential for acetic acid. The circles are calculated from the second harmonic signal using Equation 1, and the solid line is the theoretical curve predicted from the reaction mechanism (Equations 2 and 3).

$$H^+ + e \underset{k_{-1}}{\overset{k_1}{\rightleftharpoons}} H_{ads} \qquad (2)$$

$$2H_{ads} \xrightarrow{k_2} H_2 \qquad (3)$$

The solid line in Figure 4 is the potential dependence of the relative surface coverage of H_{ads} by this reaction scheme in the steady state approximation. Details of the calculation and reaction scheme determination have been presented elsewhere (8).

Hydrogen adsorption on platinum electrodes in perchloric acid solutions

As a final example of the study of chemisorption by the changes in the SHG from metal surfaces, Figure 5 depicts the corrected current and the second harmonic signal at 289 nm from a platinum electrode in a 0.35M $HClO_4$ aqueous solution during a cathodic potential sweep in the hydrogen adsorption region. The current waveform has been corrected for double layer and hydrogen evolution contributions by subtraction of the current observed in the presence of 1 millimolar potassium iodide (which is known to block hydrogen upd on the platinum surface) (12). The two peaks in the current waveform at -0.045V and -0.200V correspond to the formation of strongly (H_s) and weakly (H_w) adsorbed hydrogen on the platinum surface (Equation 2). Concomitant with the chemisorption of hydrogen, we observe in Figure 5 a large increase in the second harmonic signal (curve b). In contrast, the second harmonic signal is virtually unchanged in the double layer and oxide regions of the potential sweep (not shown). The increase in signal is exactly the opposite of the decrease that was observed for the silver electrodes in acetonitrile. It is unclear from this data at a single wavelength whether the increase is due to changes in the platinum surface's nonlinear susceptibility, or whether it is due to the intrinsic nonlinear susceptibility of the chemisorbed hydride species.

In spite of this uncertainty, we can quantitate this increase in the second harmonic signal due to chemisorption through the use of Equation 1. Since, in this case, the hydrogen is chemisorbed prior to the formation of molecular hydrogen, we can also monitor the surface coverage of hydride by the charge passed during the deposition. A plot of "$c\theta$" $\{ = [I(2\omega)/I(2\omega)_0]^{1/2} - 1 \}$ vs. q, the charge passed due the hydrogen deposition, is shown in Figure 6. A full monolayer corresponds to approximately 251 $\mu C/cm^2$ (12). As predicted by Equation 1, the nonlinear susceptibility of the surface depends linearly upon the amount of chemisorbed hydrogen, and

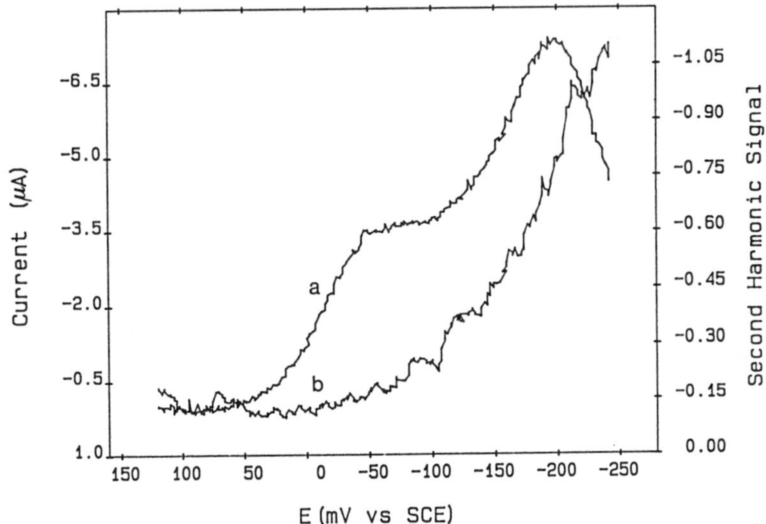

Figure 5. The corrected current (a) and the second harmonic signal (b) obtained during cathodic sweep (10 mV s^{-1} scan rate) from a polycrystalline platinum electrode in a 0.35M $HClO_4$ aqueous solution.

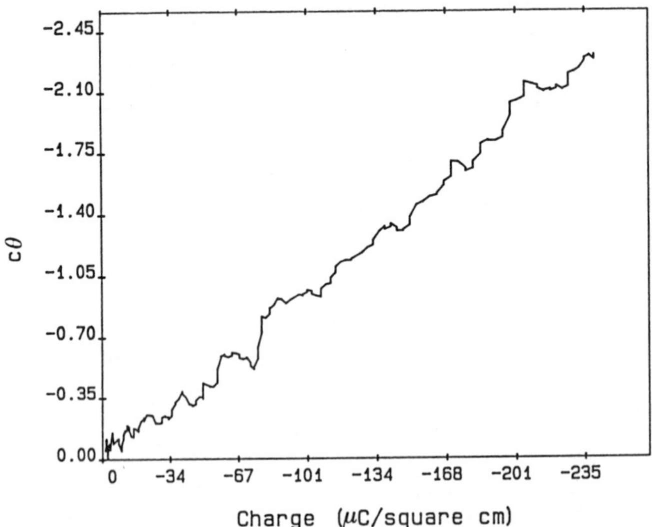

Figure 6. The function cθ (= $[I(2\omega)/I(2\omega)_0]^{1/2} - 1$) determined from the second harmonic signal plotted versus the charge passed due to hydrogen chemisorption on the platinum electrode.

surprisingly, is independent of the chemical nature of the hydrogen species (H_S vs. H_W). We have found this latter fact not to be the case in sulfuric acid solutions (Campbell, D. J.; Corn, R. M. "Second Harmonic Generation Studies from Platinum Electrodes in Sulfuric Acid Solutions", submitted to J. Phys. Chem.).

Summary

In this paper we have utilized the changes in the SHG from metal surfaces to monitor chemisorption processes at electrode surfaces. In particular, we have seen that:

i) During the deposition of lithium onto silver electrodes in acetonitrile there is a potential dependent increase in the second harmonic signal from the silver surface due to a steady state coverage of unreacted metallic lithium. This increase in the second harmonic signal is thought to arise from the delocalization of the electrons associated with the metallic lithium overlayer.

ii) The chemisorption of a monatomic hydrogen species onto silver electrodes in acetonitrile is found to decrease the SHG from the surface during the irreversible process of molecular hydrogen evolution. The quantitation of the second harmonic signal for this process yields the relative surface coverage of chemisorbed hydrogen as a function of electrode potential, and the shape and pH dependence of this relative surface coverage-potential curve can be used to ascertain the reaction mechanism for hydrogen evolution at silver electrodes in acetonitrile.

iii) On platinum electrodes in perchloric acid solutions the second harmonic signal quantitatively increases during hydrogen atom underpotential deposition, and this increase is independent of the type of chemisorbed hydrogen.

Although use of the second harmonic signal from metal surfaces as an indirect probe of chemisorption lacks structural specificity, the utility of the technique is clearly demonstrated by the experiments reported here. The next step in our studies will be to demonstrate how one can directly monitor chemisorption with some degree of molecular specificity via the resonant second harmonic signal from the nonlinear susceptibility of adsorbed species.

Acknowledgments

The authors gratefully acknowledge the assistance of Matt Lynch and Eric Miller in the construction of the experimental apparatus for these experiments. This work was supported by the National Science Foundation.

Literature Cited

1. Bloembergen, N.; Chang, R. K.; Jha, S. S.; Lee, C. H. Phys. Rev. 1968, 174 , 813.
2. Shen, Y. R. The Principles of Nonlinear Optics; Wiley: New York, 1984.
3. Corn, R. M.; Romagnoli, M.; Levenson, M. D.; Philpott, M. R. Chem. Phys. Lett. 1984, 106, 30.
4. Corn, R. M.; Romagnoli, M.; Levenson, M. D.; Philpott, M. R. J. Chem. Phys. 1984, 81, 4127.
5. Richmond, G. L. Langmuir 1986, 2, 132 and references therein.
6. Biwer, B. M. Surf. Sci. 1986, 176, 377 .
7. Furtak, T. E. ; Miragliotta, J.; Korenowksi, G. M. Phys. Rev. 1987, B35, 2569.
8. Campbell, D. J.; Corn, R. M. J. Phys. Chem., 1987, 91, 5668.
9. Kolb, D. M.; Przasnyski, M.; Gerischer, H. J. Electroanal. Chem. 1974, 54, 25.
10. Chen, J. M.; Bower, J. R.; Wang, C. S.; Lee, C. H. Optics Comm. 1973, 9, 132.
11. Tom, H. W. K.; Mate, C. M.; Zhu, X. D.; Crowell, J. E.; Heinz, T. F.; Somorjai, G. A.; Shen, Y. R. Phys. Rev. Lett. 1984, 52, 348.
12. Hubbard, A. T.; Ishikawa, R. M.; Katekaru, J. J. Electroanal. Chem. 1978, 86, 271.

RECEIVED May 17, 1988

Chapter 21

Vibrational Molecular Probes of Electrochemical Interfaces

Comparisons and Chemical Applications

M. J. Weaver, D. S. Corrigan, P. Gao, D. Gosztola, and L.-W. H. Leung

Department of Chemistry, Purdue University, West Lafayette, IN 47907

Some characteristics of, and comparisons between, surface-enhanced Raman spectroscopy (SERS) and infrared reflection-absorption spectroscopy (IRRAS) for examining reactive as well as stable electrochemical adsorbates are illustrated by means of selected recent results from our laboratory. The differences in vibrational selection rules for surface Raman and infrared spectroscopy are discussed for the case of azide adsorbed on silver, and used to distinguish between "flat" and "end-on" surface orientations. Vibrational band intensity-coverage relationships are briefly considered for some other systems that are unlikely to involve coverage-induced reorientation. Two examples of the application of SERS and potential-difference IRRAS methods to the identification of adsorbed intermediates and reaction mechanism elucidation are also described, involving the catalytic electrooxidation of carbon monoxide and small organic molecules on transition-metal surfaces.

A major emerging area of research activity in interfacial electrochemistry concerns the development of in-situ surface spectroscopic methods, especially those applicable in conventional electrochemical circumstances. One central objective is to obtain detailed molecular structural information for species within the double layer to complement the inherently macroscopic information that is extracted from conventional electrochemical techniques. Vibrational spectroscopic methods are particularly valuable for this purpose in view of their sensitivity to the nature of intermolecular interactions and surface bonding as well as to molecular structure. Two such techniques have been demonstrated to be useful in electrochemical systems: surface-enhanced Raman spectroscopy (SERS) (1) and several variants of infrared reflection-absorption spectroscopy (IRRAS) (2).

We have recently become interested in developing and applying

0097–6156/88/0378–0303$06.00/0

both of these techniques to the characterization of fundamental electrochemical phenomena. In some respects, the SERS and IRRAS approaches can be regarded as providing complementary techniques for the vibrational characterization of electrode surfaces. Thus SERS has the important virtue of enabling absolute spectra to be obtained for a variety of adsorbates, usually without significant bulk-phase interferences. On the other hand, infrared spectroscopy is applicable to a much wider range of electrode surfaces, including oriented single-crystal faces.

A substantial portion of our efforts in this area can be characterized on the basis of three general themes. Firstly, we have been exploring methods to extend the applicability of SERS to surfaces beyond the usual "SERS active" coinage metals, silver, copper, and gold (3). This has been accomplished for a range of materials (3a-d), including transition metals (3c,d), and also for metal oxides (3f,g) by depositing them as thin (ca. 1-4 monolayer) films on a SERS-active gold substrate. Secondly, we have been exploring the utilization of SERS and IRRAS for following redox-induced molecular transformations on electrode surfaces, especially for the elucidation of electrochemical reaction mechanisms (3c-e,4-6). Thirdly, comparisons between corresponding surface Raman and infrared spectra have been pursued in order to examine differences in surface selection rules as well as a means of ascertaining to what extent the adsorbate sensed by the former probe may differ from the preponderant species presumably sensed by the latter technique (3c,d,7). This last question is of some importance given the likelihood that SERS may sense only a minority of the adsorbate, located at or close to the surface morphologies responsible for the surface enhancement (1).

The present conference paper provides a discussion of some representative findings from our recent studies on these topics, with the aim of comparing and contrasting some of the distinctive properties of SERS and IRRAS as applied to fundamental interfacial electrochemistry. We limit the presentation here to a brief overview; further details can be found in the references cited. All electrode potentials quoted here are with respect to the saturated calomel electrode (SCE).

Comparisons Between Corresponding Surface Raman and Infrared Spectra: Band Intensity-Coverage Relationships

Although by now a large number of electrochemical systems have been examined using both SERS and IRRAS, including some common to both techniques (2b), the conditions employed are usually sufficiently different (e.g. disparate surface state, adsorbate concentrations) so to preclude a quantitative comparison of the spectral responses. One further hindrance to such comparisons is that it usually is difficult to remove entirely the contribution to the infrared spectra from solution-phase species. Two types of approaches are commonly used in IRRAS with this objective in mind. Firstly, modulating the infrared beam between s- and p-polarization can achieve a measure of demarcation between surface and bulk-phase components since considerably greater infrared absorption will occur for the former, but not the latter, species for p- versus s-polarized light (2,8). However, a complication is that the "surface region" in this context extends a distance of about 0.5 λ, where λ is the infrared wavelength, away from the reflecting surface (8,9). Since the thickness of the thin-layer cavity conventionally used for such measurements is of the order of λ (about 5 microns), the species in the solution as well as in the double layer will exhibit a polarization-sensitive infrared absorbance. Consequently, in practice this polarization-modulation approach will not usually achieve a complete cancellation of the spectral contribution from solution-phase species.

The second, and most widely used, approach to the separation of surface and solution-phase contributions to IRRAS involves potential-difference methods, taking advantage of the commonly anticipated dependence of the surface infrared spectra features to the electrode potential. We will refer to this general approach as "potential-difference infrared spectroscopy" (PDIRS). Although the details of the experiment are different depending if a dispersive or Fourier transform (FT) instrument is employed, in both cases PDIR spectra are obtained by subtracting (or ratioing) individual spectra acquired at a suitable pair of electrode potentials (2). [If an FT instrument is employed, this approach is often referred to as "subtractively normalized Fourier transform infrared spectroscopy" (SNIFTIRS) (2)]. Bipolar bands are therefore usually obtained in PDIR spectra, reflecting the difference in the potential-sensitive component at the particular pair of "base" and "sample" potentials chosen (2).

Provided that the surface species of interest does not undergo adsorption/desorption between these two potentials, the PDIR spectrum thus obtained will arise only from the surface species. However, if the species undergoes significant adsorption/desorption (or redox transformations) under these conditions, then the PDIR spectra may well reflect the consequent changes in the thin-layer solution composition as well as in the surface region (10). This circumstance constitutes a solution-phase interference upon the PDIR spectrum which will obviously be detrimental if the frequencies of the vibrational bands for the adsorbate in the surface and bulk-phase environments overlap substantially. On the other hand, as noted below the synergic appearance of such band partners can yield valuable surface compositional information.

A simple example, taken from reference 7, that illustrates these points in the context of infrared-Raman comparisons is provided in Figures 1 and 2. Figure 1 consists of a series of PDIR spectra obtained for azide adsorbed on silver. Each spectrum employed a base potential of -0.97 V and the series of sample potentials indicated, the latter being ratioed against the former. Under the conditions chosen, the azide coverage could be altered from very small values (at the most negative potentials) to approaching a monolayer (at the most positive potentials). At the most negative potentials, only a single positive-going band at 2048 cm^{-1} appears, the intensity increasing as the sample potential is made less negative. This band is identified as arising from loss of solution azide caused by corresponding increases in the amount of adsorbed azide at the more positive potentials (10). The constant frequency, 2048 cm^{-1}, of the positive-going band is consistent with the asymmetric N-N-N stretch, ν_{as}, for aqueous azide. (An essentially constant total amount of azide will be trapped in the thin-layer

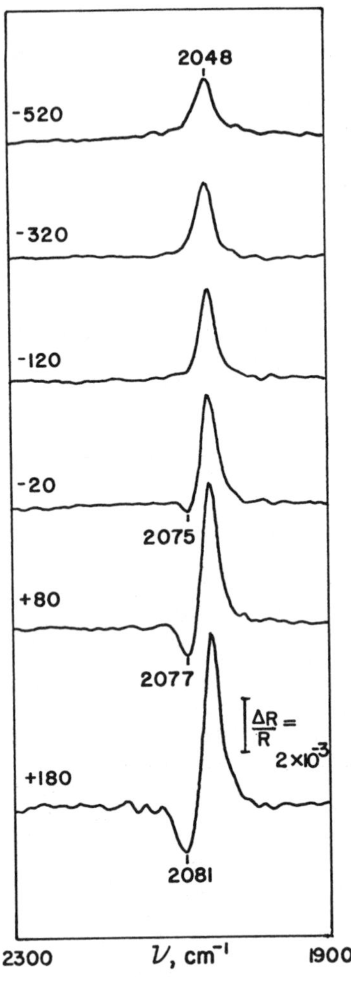

Figure 1
Potential-difference infrared (PDIR) spectra for adsorbed azide at an electrochemically roughened silver-aqueous interface in the asymmetric N-N-N, ν_{as}, stretch region. Electrolyte: 0.01 M NaN_3 + 0.1 M $NaClO_4$. Base (reference) potential was -0.97 V vs. SCE; sample potentials (mV vs. SCE) are as indicated. Spectra were obtained by acquiring 1024 interferometer scans at the base and sample potentials, the potential being altered after every 32 scans (see references 7 and 10 for further details).

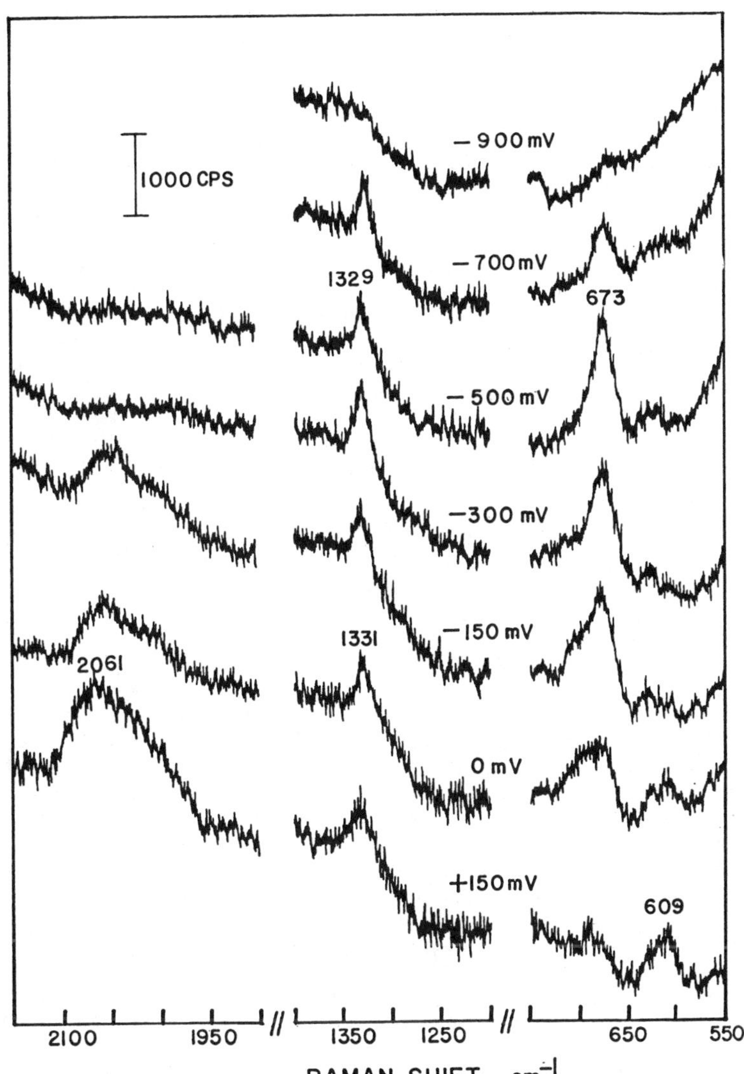

Figure 2
Surface-enhanced Raman (SER) spectra for adsorbed azide at an electrochemically roughened silver-aqueous interface at electrode potentials indicated. Frequency regions shown refer to asymmetric N-N-N stretch (ν_{as}), symmetric N-N-N stretch (ν_s), N-N-N bending modes. Electrolyte: 0.01 M NaN_3 + 0.1 M $NaClO_4$. Monochromator scan rate was 0.5 cm^{-1} s^{-1}, with 50 mW laser excitation at 514.5 nm (see reference 7 for further details).

cavity under the conditions of these experiments (10), where the use of excess supporting electrolyte essentially eliminates the contribution of N_3^- to the ion migration to and from the thin layer which occurs upon potential alteration (32).)

Interestingly, however, the corresponding negative-going ν_{as} band expected for adsorbed azide is not observed until the most positive potentials, E $\geq$ -0.05 V, whereupon a feature at 2075-2085 cm^{-1} appears and grows markedly in intensity as the sample potential, and hence the azide coverage, is increased further. From a knowledge of the effective integrated absorptivity, A_i, of the azide band in the thin-layer solution, the positive-going band intensity can yield a semiquantitative measure of the potential-dependent azide coverage since, unlike the corresponding surface infrared band, the solution feature will accurately obey Beer's Law. In favorable cases, then, plots of the integrated intensities of the negative-going (surface) versus the positive-going (solution) bands, I_{sol} and I_{sur}, once corrected for band overlap therefore provide a reliable means of assessing the coverage-dependent alteration in the absorptivity of a given infrared band brought about by surface coordination.

In the azide-polycrystalline silver case illustrated in Figure 1, $I_{sur} \ll I_{sol}$ for fractional azide coverages below ca. 0.3-0.5, whereas for higher coverages, $\Delta I_{sur} \sim \Delta I_{sol}$ (7,10). These results have been interpreted as indicating the presence of predominantly "flat" adsorbed azide giving way to a more "vertical" orientation of the N_3^- ion at higher coverages (10). This interpretation follows from the well-known surface dipole selection rule, which decrees that only vibrations having a dipole component normal to the surface can interact with the incident p-polarized light (13). Similar results were obtained with electropolished and electrochemically roughened (i.e. SERS-active) silver surfaces (7,10). Corresponding measurements for azide adsorbed at low-index single-crystal silver faces yielded comparable findings for the Ag(111) and Ag(100) planes as obtained for the polycrystalline surfaces (14). Significantly, however, the Ag(110) surfaces yielded no detectable surface ν_{as} band over the entire range of azide coverages. This was attributed to the preference of a linear N_3^- anion to be oriented flat along the parallel atomic-scale "furrows" which are characteristic of the Ag(110) surface (14).

A set of SER spectra for adsorbed azide on silver, obtained for the same surface and solution conditions and for a similar sequence of electrode potentials as for the PDIR spectra in Figure 1, is shown in Figure 2. (See the figure caption and reference 7 for experimental details.) Inspection of these SER spectra in comparison with the PDIR results illustrate some characteristic differences in the information provided by the two techniques. Most prominently, in addition to the N_3^- ν_{as} band around 2060 cm^{-1}, the former spectra exhibit three other features at lower frequencies attributable to adsorbed azide vibrations. By analogy with bulk-phase spectra for free and coordinated azide (15), the 1330 cm^{-1} SERS band is attributed to the N-N-N symmetric stretch, ν_s (7). The observation of both ν_s and ν_{as} features in the SER spectra differs from the surface infrared results in that only the ν_{as} band is obtained in the latter (7). The appearance of the ν_{as} band in SERS is of interest since this feature is symmetry forbidden in the solution azide Raman spectrum.

An interesting difference, however, is seen in the potential dependence of the SERS ν_s and ν_{as} features, in that the latter is only observed at the least negative potentials (E > -0.2 V), where the surface infrared ν_{as} band also appears, whereas the former band survives until the most negative potentials (E < -0.8 V), corresponding to virtually complete azide desorption (10) (Figure 2). The appearance of the SERS ν_{as} feature is indicative of azide coordinated "end on", i.e. bound by a single nitrogen, since the reduction in symmetry required by the vibrational selection rules is thereby obtained. No such symmetry lowering would occur, however, for azide bound in a flat surface configuration, which can account for the loss of the SERS ν_{as} feature at more negative potentials. The retention of the ν_s band under these conditions (Figure 2) is consistent with such an adsorbate orientation since, unlike surface infrared spectra, the occurrence of surface Raman transitions does not require the presence of a vibrational tensor normal to the surface plane provided that they are allowed on the basis of *bulk-phase* selection rules (1c,13a,b).

Examination of the azide bending-mode region (600-700 cm^{-1}) in the SER spectra (Figure 2) is also instructive with regard to adsorbate orientation. Thus, the pair of bands (at ca. 610 and 670 cm^{-1}) seen at the least negative potentials are characteristic of end-on coordinated azide (15); the loss of the lower-frequency partner for E < -0.15 V is therefore also indicative of the removal of azide bound in this adsorbate geometry, again in harmony with the interpretation of the infrared spectra (7).

Overall, then, this simple example illustrates how the form of surface Raman as well as infrared spectra can yield useful complementary (and even consistent!) information on surface orientation, at least for small symmetric adsorbates. Even though the surface vibrational selection rules are markedly less clearcut for the former technique, the wide frequency range and correspondingly rich spectral features commonly obtained with SERS can provide major benefits. Another variant of such applications of SERS that we have pursued recently involves the elucidation of the adsorbate-surface binding geometries for various monosubstituted benzenes on gold (16). In this case a distinction between surface binding via the aromatic ring and the substituent group could be made by examining the relative frequency shifts and intensities of some characteristic internal modes in SERS relative to the bulk-phase Raman spectra (16).

Quite apart from the structural information that can be obtained by inspecting the dependence of the surface vibrational band intensities, I, upon adsorbate coverage, θ, a knowledge of these relationships are obviously of importance when employing such spectra for surface analytical purposes. A particular case of interest to us concerns utilizing potential- and/or time-dependent band intensities as a means of monitoring the progress of electrochemical reactions involving adsorbed species (4-6,17). We have made several quantitative comparisons between the potential-dependent SERS intensities for electroactive adsorbates engaged in reversible one-electron redox equilibria with corresponding surface concentration-potential data obtained from faradaic electrochemical measurements (6a,17). Generally speaking, the corresponding I_{SER}-E and θ-E plots exhibit reasonable agreement, as would be expected if the SERS band intensity I_{SER} was at least roughly proportional to θ under the appropriate conditions.

Figure 3 shows plots of the relative I_{SER} values against the relative adsorbate coverage, θ_{rel}, for two such adsorbates at silver, $Ru(NH_3)_6^{3+}$ and $Os(NH_3)_5py^{3+}$ (py = pyridine), extracted from the corresponding potential-dependent data in references 6a and 17, respectively. Both these species are adsorbed by electrostatic attraction to the chloride monolayer present under the experimental conditions employed (18). The I_{SER} values for $Ru(NH_3)_6^{3+}$ and $Os(NH_3)_5py^{3+}$ adsorbates refer to the Ru^{III}-NH_3 stretching mode at 500 cm^{-1} and the pyridine ring "breathing" mode at 1020 cm^{-1}, respectively. For $Ru(NH_3)_6^{3+}$, the I_{SER}-θ_{rel} relation is approximately linear, whereas for $Os(NH_3)_5py^{3+}$ the plot exhibits significant deviations from linearity (Figure 3). In both cases, however, the maximum adsorbate coverage is well below that corresponding to a close-packed monolayer, the surface concentration being around 3 to 5 x 10^{-11} mol cm^{-2} (6a,17).

Figure 3 also contains an example of an I_{SER}-θ_{rel} plot for a simple specifically adsorbed species, bromide on silver (solid curve). This plot was extracted from bromide coverage-potential data, obtained from differential capacitance measurements, along with the corresponding potential-dependent intensity of the SERS bromide-surface stretching mode at ca. 160 cm^{-1} (19). In this case, the maximum (i.e. unity) value of θ_{rel} corresponds to a close-packed bromide monolayer, ca. 1.4 x 10^{-9} mol cm^{-2}. Again, the I_{SER}-θ_{rel} plot is decidedly nonlinear. Such behavior is unsurprising, not only in view of the likely influence of adsorbate-adsorbate interactions, but also given the inevitable nonuniform nature of the roughened surfaces necessary for SERS activity. Similar I_{SER}-θ_{rel} comparisons using potential-dependent data, but in a less quantitative vein, have been made for other halide and pseudohalide adsorbates on silver and gold electrodes (19,20). Generally speaking, although there is a rough correspondence between the corresponding I_{SER}-E and θ_{rel}-E plots, substantial nonlinearities in the I_{SER}-θ_{rel} relations can be inferred from these data. Recent detailed examinations of I_{SER}-θ_{rel} relations for resonantly enhanced adsorbates also indicate the occurrence of marked nonlinearities (21).

Aside from coverage-induced adsorbate structural changes such as are noted above for the azide-silver system, the form and interpretation of infrared band intensity-coverage relationships has a reasonable fundamental footing (22). Substantial nonlinearities in infrared I_{sur}-θ plots are both predicted and observed to occur in ultra-high vacuum (uhv) for adsorbed carbon monoxide at both single-crystal and polycrystalline surfaces as a result of dipole-dipole coupling effects (22). For CO adsorbed at polycrystalline platinum electrodes, however, approximately $I_{sur} \propto \theta$ over a wide range of coverages (5b,23), although significant deviations from linearity have been reported (24). Coverage measurements can be obtained in this case from the charge required for electrooxidation of irreversibly adsorbed CO (vide infra) (5b). A linear I_{sur}-θ relation has also been obtained for cyanate (OCN^-) adsorbed on a polycrystalline silver electrode, again by utilizing the relative intensities of I_{sur} and I_{sol} bands in PDIR spectra arising from potential-dependent adsorption-desorption equilibria (10).

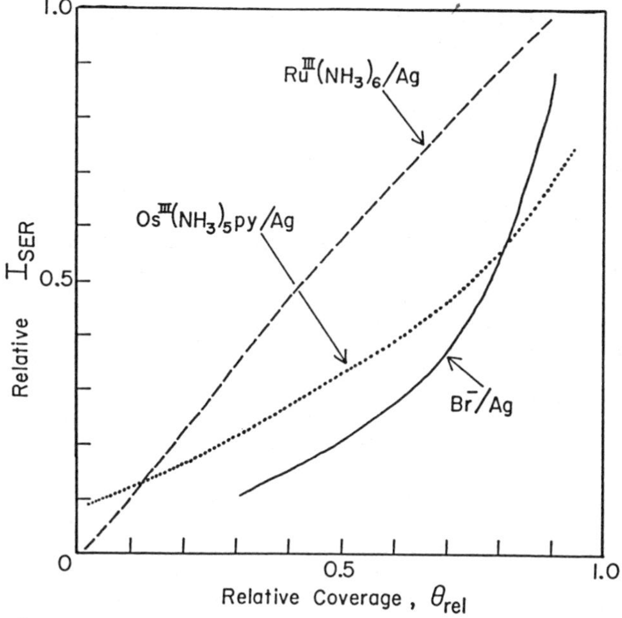

Figure 3
Plots of relative SERS band intensity, I_{SER}, against relative adsorbate coverage, θ_{rel}, for $Ru(NH_3)_6^{3+}$, $Os(NH_3)_5py^{3+}$ (py = pyridine), and Br^- adsorbed on silver, extracted from corresponding I_{SER}-potential and θ-potential data. Experimental data taken from references 6a, 17, and 19, respectively (see text and the references for further details).

Another method of probing dipole-dipole adsorbate coupling effects involves examining band intensity and frequency shifts for adsorbate isotopic mixtures (22). Similarly to results obtained for uhv systems, Severson et al. have observed strong coupling effects in the infrared spectra for $^{12}CO/^{13}CO$ mixtures adsorbed on polycrystalline platinum, manifested in substantially higher intensities of the higher-frequency forms ("intensity borrowing") and large frequency shifts for the mixed isotope layers (24).

We have recently examined such dipole-dipole coupling effects for cyanide adsorbed at gold electrodes using both surface infrared and SER spectroscopies, with the objective of probing any differences in the nature of the surface environment for the infrared and SERS-active adsorbate (25). Such differences might be anticipated given the likelihood that SERS senses only a minority of the adsorbate, probably in the vicinity of particular metal microstructures (1). Somewhat surprisingly, only relatively small isotope dipole-dipole coupling effects were observed for cyanide at gold despite the similarity in structure and mode of surface coordination for CN^- and CO (via the carbon atom (26)). Figure 4 shows a plot of the peak frequencies for high- and low-frequency bands in the isotopic mixture plotted against the percentage of CN^- present as $^{12}CN^-$. Although both bands shift to progressively lower frequencies as a given isotopic species is diluted increasingly in the $^{12}CN^-/^{13}CN^-$ mixture, the shifts are small ($\lesssim$ 8 cm^{-1}). Similar infrared results have been reported earlier by Kunimatsu et al. (27). In addition, the degree of "intensity borrowing" by the high-frequency band is only moderate in comparison with the corresponding effects for CO. Interestingly, the magnitude of these coupling effects is very similar for SER and infrared spectroscopies, both the frequencies and relative band intensities in the isotopic mixtures being virtually identical in both cases.

Irrespective of the detailed interpretation of these results (25), they provide further evidence against the oft-popular notion that SERS for cyanide and other strongly coordinating adsorbates arises from specific "surface complexes" [such as $Au(CN)_2^-$] (28) as distinct from the preponderant adsorbate species presumably sensed by infrared spectroscopy (also see reference 7).

Applications to Mechanistic Surface Electrochemistry

As noted in the Introduction, a central focus of our current interests in both surface Raman and infrared spectroscopies is directed towards their utilization in reactive electrochemical systems (3-6). An important virtue of in-situ vibrational spectroscopies for this purpose is that they can yield information on the molecular identity (as well as detailed physical state) of electrogenerated species, and hence can provide considerable mechanistic information for multistep electrode processes. We have recently discussed several facets of such applications in a conference paper for "Surface Vibrations V" (6b).

Although both SER and infrared spectroscopies are well suited to providing information of this type, several factors conspire to slant their applications in noticeably different directions. At least with suitably SERS-active surfaces, the former technique can

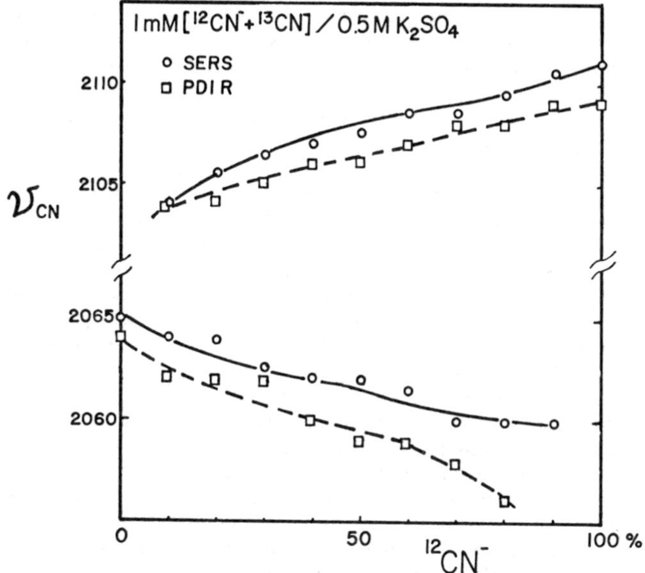

Figure 4
Plots of the C-N peak frequencies, ν_{CN}, for various $^{12}CN^-/^{13}CN^-$ mixtures adsorbed on electrochemically roughened gold electrode, observed with SERS (circles) and potential-difference infrared (PDIR, squares) methods. Total CN^- concentration ($^{12}CN^-$ + $^{13}CN^-$) held constant at 1 mM, with 0.5 M K_2SO_4 supporting electrolyte. Electrode potential was -0.8 V vs. SCE. PDIR spectra obtained by stepping from -0.8 V to 0.8 V so to remove solution interference (25).

provide absolute vibrational spectra for adsorbed species in the presence of high concentrations of solution species in conventional electrochemical cells. Consequently, SERS can yield sensitive, albeit qualitative, information on the changes in interfacial composition and structure accompanying electrode reactions for a variety of electrochemical techniques and cell configurations. For example, SERS can readily be coupled with steady-state potential-dependent techniques, either with rotating-disk voltammetry (4a,c) or in quiescent solution (3c,d,4b-h). Such tactics enable the experimental timescale to be sufficiently long to enable conventional scanning monochromators to be employed, even though the markedly shorter measurement times afforded by the use of optical multichannel analyzer (OMA) detection offer considerable benefits. The latter detection scheme also enables sequences of SER spectra to be obtained during cyclic voltammetric and other potential-time excursions (4i,j,6b).

The examination of surface compositional changes accompanying electrochemical processes under steady-state conditions is less amenable to study by IRRAS, partly because of the need to employ a thin-layer cell configuration in order to minimize solution-phase interferences. The use of potential-difference infrared techniques to further nullify solution interferences is advantageous for the examination of electrochemical reactions since these processes are inherently potential sensitive. However, a difficulty with the use of potential-*modulation* infrared methods (i.e. employing repeated potential alteration between base and sample values) is that they are sensitive only to the occurrence of strictly reversible processes, while most multi-step processes of mechanistic interest are irreversible in nature. This difficulty can be surmounted, at least in favorable cases, by employing a Fourier transform spectrometer and applying only a single potential alteration during the spectral acquisition, sequentially at the base and sample potentials (5a,b). This "single potential-alteration infrared" (SPAIR) procedure, like other PDIR techniques, has the virtue of enabling potential-induced compositional changes in the thin-layer solution as well as the electrode surface to be detected. It therefore can enable the identity of particular solution intermediates and/or products, trapped in the thin layer cavity, to be identified and their formation kinetics established. We now describe briefly two examples, chosen with the aim of illustrating some of these points.

Figure 5 shows SER spectra in the C-O stretching, ν_{CO}, region obtained at the positive-going sequence of electrode potentials indicated for carbon monoxide adsorbed on a rhodium-coated gold electrode from a CO-saturated 0.1 M $HClO_4$ solution (3d). This surface was formed by electrodepositing a film (ca. 3 equivalent monolayers) of rhodium on to the SERS-active gold substrate. This procedure enables the SERS effect to be imparted to suitable adsorbates, such as CO, bound to the transition-metal overlayer (3d). At the initial, least positive potentials ($\leq$ 0.4 V), two ν_{CO} bands are observed, a strong feature centered at 2025-2055 cm^{-1} and a weaker band at 1900 cm^{-1} (Figure 5). These features are characteristic of CO bound to rhodium in a terminal ("linear") and bridged configuration, respectively, by analogy with infrared data gathered at rhodium-gas interfaces (29).

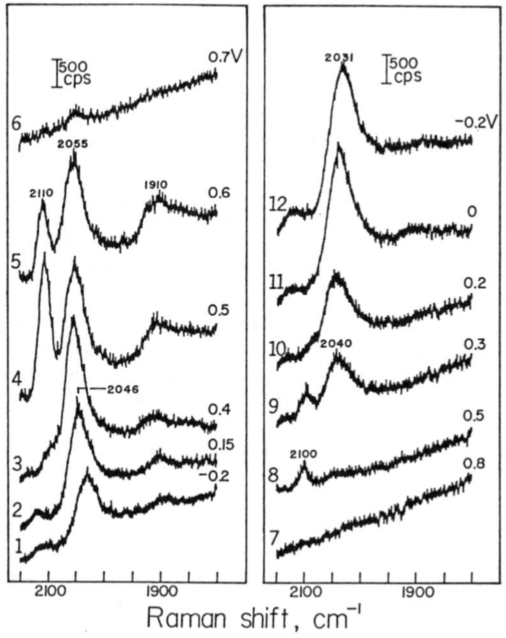

Figure 5

Representative sequence of SER spectra in C-O stretching, ν_{CO}, region for CO adsorbed on rhodium-coated gold electrode. Electrolyte was CO-saturated 0.1 M $HClO_4$. Spectra were obtained in numbered sequence for potentials (V vs. SCE) as indicated. Laser excitation was 70 mW at 647.1 nm; monochromator scan rate was 0.5 cm^{-1} s^{-1}. Rhodium electrodeposition employed a faradaic charge of 2.0 mC cm^{-2} (corresponding to ca. 3 equivalent monolayers) (3d).

As the potential is altered to sequentially more positive values in the region where CO electrooxidation commences, 0.5-0.6 V, an additional pronounced ν_{CO} feature at 2110 cm^{-1} appears (Figure 5). Each of these bands eventually disappear at potentials, > 0.7 V, corresponding to the passivation of the rhodium surface by oxide film growth, although they sequentially reappear when the potential is readjusted in the negative direction (spectra 8-12 in Figure 5). Corresponding PDIR spectra obtained under comparable conditions did not exhibit the 2110 cm^{-1} band, although the other two ν_{CO} features were observed.

One possible assignment of the 2110 cm^{-1} feature is to vibrations involving a dicarbonyl, $Rh^{I}(CO)_2$, surface species (30); however, this seems unlikely since the anticipated band partner (30) is absent. Most likely, the 2110 cm^{-1} band signals the appearance of CO bound to "oxidized" rhodium sites, where rhodium has undergone formal oxidation to Rh(I) or Rh(III), and/or adjacent to sites where surface oxide has formed (3d). A similar high-frequency ν_{CO} feature has been observed at supported rhodium exposed to gaseous CO_2 and O_2 (31).

These results, as well as similar findings obtained for CO adsorbed at ruthenium-coated gold (3d), are of relevance to the mechanism of CO electrooxidation since they indicate that a significant fraction of the CO adsorbed at potentials where reaction proceeds is modified substantially by the early stages of metal oxidation. This suggests that adsorbed O or OH may be responsible for CO electrooxidation on these surfaces, although such sites admittedly may not be electrocatalytic in nature.

The second example taken from recent work in our laboratory involves the application of SPAIRS under potential-sweep conditions for examining the electrooxidation mechanisms of organic molecules on platinum (5b,c,12b). Figure 6 shows a partial sequence of infrared spectra obtained during a positive-going potential sweep at 2 mV s^{-1} from -0.25 V, for a solution of 50 m*M* benzaldehyde at polycrystalline platinum. Each spectrum shown involved acquiring 25 interferometer scans (consuming ca. 15 s) and ratioing these to a corresponding set of scans obtained at the initial potential prior to starting the potential sweep. (The potentials indicated refer to the average values during each spectral acquisition.)

Several features of these SPAIR spectra provide information regarding the electrooxidation mechanism of benzaldehyde. Most prominently, a negative-going band at 2343 cm^{-1} is observed at potentials following the commencement of faradaic oxidation at ca. 0.6 V. This is identified as gaseous CO_2 formed in the thin layer by benzaldehyde electrooxidation. (The negative-going form of this band in the "transmittance" spectra shown denotes that the species is being formed at the sample potential.) Furthermore, the quantity of CO_2 formed can be deduced from a knowledge of the band absorptivity, obtained in separate experiments involving the formation of CO_2 from the electrooxidation of an irreversibly adsorbed CO monolayer on platinum (5b). By this means, we deduce that about 3 x 10^{-9} mol cm^{-2} of CO_2 is formed by the end (1.26 V) of the positive potential sweep.

The origin of at least a portion of this CO_2 is identified by the band appearing around 2050-2070 cm^{-1} (Figure 6). The frequency and potential-dependent characteristics of this feature are

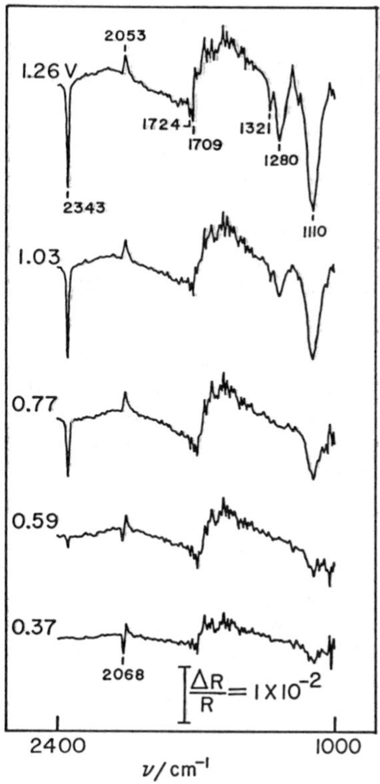

Figure 6
Sequence of single potential alteration infrared (SPAIR) spectra in 1000-2400 cm^{-1} region obtained during voltammetric oxidation of benzaldehyde at platinum-aqueous interface. Solution contained 50 mM benzaldehyde in 0.1 M $HClO_4$. Voltammetric sweep rate was 2 mV s^{-1}; initial (base) potential was -0.25 V vs. SCE. Each spectrum shown, referring to average potentials indicated, was obtained by acquiring 25 interferometer scans (consuming ca. 15 s), and ratioing against spectrum obtained similarly at the initial potential just prior to the potential sweep (see references 5b and c for further details).

diagnostic of linearly bound carbon monoxide, formed by dissociative chemisorption of benzaldehyde. We have noted recently that high coverages of CO are readily formed on platinum from a variety of aldehydes, alcohols, and amides, providing that the molecule contains at least one hydrogen bound to the α carbon atom (5c). At potentials prior to irreversible CO oxidation (ca. 0.55 to 0.75 V) a bipolar band is observed, arising from the potential-dependent shift in the ν_{CO} peak frequency. Beyond this point, a unipolar positive-going band is obtained, reflecting only the ν_{CO} feature at the initial potential (Figure 6). From the relation between infrared absorbance and CO coverage noted above (5b), the CO surface concentration is estimated to be about 1×10^{-9} mol cm^{-2}.

These results therefore indicate that a significant portion (ca. two-thirds) of the CO_2 formed during the positive-going sweep arises from electrooxidation of benzaldehyde (or possibly other interfacial species) rather than from adsorbed CO. In addition, the SPAIR spectra indicate that the formation of CO_2 commences only upon CO electrooxidation, and essentially no CO_2 from a source *other* than previously adsorbed CO appears until the CO coverage becomes very small ($\theta \leq 0.1$). Although these results do not rule out the possible role of adsorbed CO as an intermediate in the electrooxidation of solution benzaldehyde to CO_2, they suggest that the majority of the adsorbate acts as a poison for this process, removal of which is required for initiation of the electrocatalytic pathway.

We have recently performed a variety of these and related SPAIRS-voltammetric measurements on platinum and palladium (5c,12b), and have concluded that the adsorbed CO formed in most cases acts predominantly as a poison for organic electrooxidation. Interestingly, the potential at which the CO undergoes electrooxidation, and hence where the electrocatalysis commences, can be strongly dependent on the structure of the solution species involved. Thus for acetaldehyde, for example, this process occurs at about 0.3 V lower overpotentials than for benzaldehyde under comparable conditions (5c).

The third mechanistically instructive feature of the spectra in Figure 6 concerns the appearance of several additional negative-going bands at relatively positive potentials. While the intense band at 1110 cm^{-1} is due to potential-induced migration (and possibly adsorption) of perchlorate anion (32), the appearance of other bands at 1280, 1321, and ca. 1700-1720 cm^{-1} (Figure 6) results from the electrooxidation of benzaldehyde. On the basis of their frequencies and relative intensities, these features can be assigned with confidence to C-O, C-C, and C=O stretching modes for solution benzoic acid (5d,33). (The relatively weak and distorted form of the C=O band is probably due to partial overlap with the corresponding positive-going band associated with the loss of solution benzaldehyde.) Although benzoic acid preadsorbed on platinum undergoes exhaustive oxidation of all seven carbons to CO_2, solution benzoic acid is apparently electroinactive (5d), thus accounting for its appearance as a benzaldehyde oxidation product even at far positive potentials.

Similar experiments performed with other aldehydes and primary alcohols indicate that their electrooxidation on platinum produce primarily the corresponding carboxylic acid in addition to CO_2, whereas secondary alcohols yield ketones. Moreover, the product distribution can be established quantitatively as a function of potential and/or time from the measured SPAIR band intensities combined with a knowledge of the effective molar absorptivities (12b).

While such measurements utilize a relatively conventional spectroelectrochemical procedure, they have the substantial virtue for electrocatalytic purposes of providing at least partial information on the involvement of *both* surface and solution-phase intermediates and/or products.

Concluding Remarks

At the present time, both surface-enhanced Raman and infrared spectroscopies have reached an interesting level of maturity and demonstrated applicability to problems of genuine electrochemical significance. The frenzy that surrounded the early development of SERS, especially regarding the enhancement mechanism(s), has now largely subsided. Nevertheless, the technique still retains promise as a truly sensitive, as well as molecular selective, surface analytical probe. Although hampered by the limited realm of substrates displaying suitable surface enhancement effects, this difficulty is increasingly becoming circumvented so that the electrochemical (and chemical) impact of SERS should grow steadily in the future.

Surface infrared spectroscopy, on the other hand, appears to provide opportunities and challenges having a noticeably different flavor. A major emerging area is in its application to examining oriented single-crystal and other surfaces having so-called "well-defined" structures, for which SERS is likely to remain inapplicable. Within the greater limitations of sensitivity and frequency response faced by IRRAS in comparison with SERS, one might anticipate that infrared techniques will play a significant role in exploring the reactive, as well as purely structural, properties of such surfaces. The examination of chemically modified electrodes consisting of organized microstructures is another obvious area of application for IRRAS where the demands on sensitivity will often be less stringent.

As a consequence, the development paths of surface Raman and infrared techniques will, given their different strengths and weaknesses, probably continue to proceed along divergent avenues of chemical application. Nevertheless, it seems inevitable that both will contribute in an increasingly important fashion to the detailed molecular-level elucidation of reactive as well as stable electrochemical systems.

Acknowledgments

This work is supported by the National Science Foundation and the Office of Naval Research.

Literature Cited

1. Recent reviews include (a) Chang, R. K.; Laube, B. L., CRC Crit. Rev. Solid State Mat. Sci., 1984, 12, 1; (b) Weitz, D. A.;

Moskovits, M.; Creighton, J. A., in "Chemistry and Structure at Interfaces - New Laser and Optical Techniques", Hall, R. B.; Ellis, A. B., eds., VCH Publishers, Deerfield Beach, FL, 1986, p. 197; (c) Moskovits, M., Rev. Mod. Phys., **1985**, 59, 783.

2. Recent reviews include (a) Bewick, A.; Pons, S., in "Advances in Infrared and Raman Spectroscopy", Clark, R. J. H.; Hester, R. E., eds., Vol. 12, Wiley Heyden, New York, 1985, Chapter 1; (b) Foley, J. K.; Korzeniewski, C.; Dashbach, J. L.; Pons, S., in "Electroanalytical Chemistry - A Series of Advances", Bard, A. J., ed., Vol. 14, Marcel Dekker, New York, 1986, p. 309.
3. (a) Leung, L.-W. H.; Weaver, M. J., J. Electroanal. Chem., **1987**, 217, 367; (b) Leung, L.-W. H.; Gosztola, D., Weaver, M. J., Langmuir, **1987**, 3, 45; (c) Leung, L.-W. H., Weaver, J. Am. Chem. Soc., **1987**, 109, 5113; (d) Leung, L.-W. H.; Weaver, M. J., Langmuir, in press; (e) Desilvestro, J.; Corrigan, D. A.; Weaver, M. J., J. Phys. Chem., **1986**, 90, 6408; (f) Desilvestro, J.; Corrigan, D. A.; Weaver, M. J., J. Electrochem. Soc., **1988**, 135, 885; (g) Gosztola, D.; Weaver, M. J., in preparation.
4. (a) Farquharson, S.; Milner, D.; Tadayyoni, M. A.; Weaver, M. J., J. Electroanal. Chem., **1984**, 178, 143; (b) Patterson, M. L.; Weaver, M. J., J. Phys. Chem., **1985**, 89, 1331; (c) Tadayyoni, M. A.; Gao, P.; Weaver, M. J., J. Electroanal. Chem., **1986**, 198, 125; (f) Tadayyoni, M. A.; Weaver, M. J., Langmuir, **1986**, 2, 179; (g) Desilvestro, J.; Weaver, M. J., J. Electroanal. Chem., **1986**, 209, 377; (h) Desilvestro, J.; Weaver, M. J., J. Electroanal. Chem., **1987**, 234, 237; (i) Gao, P.; Gosztola, D.; Weaver, M. J., Anal. Chim. Acta, in press; (j) Gao, P.; Gosztola, D.; Weaver, M. J., J. Phys. Chem., submitted.
5. (a) Corrigan, D. S.; Leung, L.-W. H.; Weaver, M. J., Anal. Chem., **1987**, 59, 2252; (b) Corrigan, D. S.; Weaver, M. J., J. Electroanal. Chem., **1988**, 241, 143; (c) Leung, L.-W. H.; Weaver, M. J., J. Electroanal. Chem., **1988**, 240, 341; (d) Corrigan, D. S.; Weaver, M. J., Langmuir, in press.
6. (a) Weaver, M. J.; Gao, P.; Gosztola, D.; Patterson, M. L.; Tadayyoni, M. A., ACS Symp. Ser., **1986**, 307, 135; (b) Weaver, M. J.; Corrigan, D. S.; Gao, P.; Gosztola, D.; Leung, L.-W. H., J. Electron. Spect. Related Phen., **1987**, 45, 291.
7. Corrigan, D. S.; Gao, P.; Leung, L.-W. H.; Weaver, M. J., Langmuir, **1986**, 2, 744.
8. For example, Golden, W. G., in "Fourier Transform Infrared Spectroscopy", Vol. 4, Academic Press, New York, 1985, Chapter 8.
9. Seki, H.; Kunimatsu, K.; Golden, W. G., Applied Spect., **1985**, 39, 437.
10. Corrigan, D. S.; Weaver, M. J., J. Phys. Chem., **1986**, 90, 5300.
11. A complication in determining absolute A_i values for solution species in reflection-absorption cell geometries is that these will differ somewhat from the values determined using a conventional transmittance cell arrangement as a result of the influence of the metal surface on the local electromagnetic field. (8,9) For typical conditions using electrochemical IRRAS with p-polarized light, the effective A_i values for aqueous solution species appears to be ca. 2-4 times those obtained in transmittance measurements. (12)
12. (a) Corrigan, D. S.; Krauskopf, E. K.; Rice, L. M.; Wiekowski,

A.; Weaver, M. J., J. Phys. Chem., **1988**, 92, 1596; (b) Leung, L.-W. H.; Weaver, M. J., J. Phys. Chem., submitted.
13. (a) Moskovits, M., J. Chem. Phys., **1982**, 77, 4408; (b) Moskovits, M.; Suh, J. S., J. Phys. Chem., **1984**, 88, 5526; (c) Greenler, R. G., J. Chem. Phys., **1966**, 44, 310.
14. Corrigan, D. S.; Brandt, E. S.; Weaver, M. J., J. Electroanal. Chem., **1987**, 235, 327.
15. (a) Nelson, J.; Nelson, S. M., J. Chem. Soc. A, **1969**, 1597; (b) Forster, D.; Horrocks, W. DeW., Jr., Inorg. Chem., **1966**, 5, 1510.
16. Gao, P.; Weaver, M. J., J. Phys. Chem., **1985**, 89, 5040.
17. Farquharson, S.; Guyer, K. L.; Lay, P. A.; Magnuson, R. H.; Weaver, M. J., J. Am. Chem. Soc., **1984**, 106, 5123.
18. Tadayyoni, M. A.; Farquharson, S.; Weaver, M. J., J. Chem. Phys., **1984**, 80, 1363.
19. Weaver, M. J.; Hupp, J. T.; Barz, F.; Gordon II, J. G.; Philpott, M. R., J. Electroanal. Chem., **1984**, 160, 321.
20. Gao, P.; Weaver, M. J., J. Phys. Chem., **1986**, 90, 4057.
21. (a) Zeman, E. J.; Carron, K. T.; Schatz, G. C.; Van Duyne, R. P., J. Chem. Phys., **1987**, 87, 4189. (b) Cotton, T. M.; Kim, J. H.; Uphans, R. A.; Möbius, D., to be submitted to J. Phys. Chem.
22. For a review, see Hollins, P.; Pritchard, J., Prog. Surf. Sci., **1985**, 19, 275.
23. Kunimatsu, K.; Kita, H., J. Electroanal. Chem., **1987**, 218, 155.
24. Severson, M. W.; Russell, A.; Campbell, D.; Russell, J. W., Langmuir, **1987**, 3, 202.
25. Gao, P.; Weaver, M. J., in preparation.
26. Monocyanogen (CN), however, has been observed to usually be adsorbed *parallel* to the surface in uhv on the basis of electron energy loss spectroscopy: (a) Kordesch, M. E.; Stenzel, W.; Conrad, H., Surf. Sci., **1986**, 175, L689; (b) Kordesch, M. E.; Stenzel, W.; Conrad, H.; Weaver, M. J., J. Am. Chem. Soc., **1987**, 109, 1878.
27. Kunimatsu, K.; Seki, H.; Golden, W. G.; Gordon II, J. G.; Philpott, M. R., Langmuir, in press.
28. Baltruschat, H.; Heitbaum, J., J. Electroanal. Chem., **1983**, 157, 319.
29. For a review, see: Sheppard, N.; Nguyen, T. T., in "Advances in Infrared and Raman Spectroscopy", Clark, R. J. H.; Hester, R. E., eds., Heyden, London, 1978, Vol. 5, p. 67.
30. For example, Yates, J. T.; Kolasinski, K., J. Chem. Phys., **1983**, 79, 1026.
31. Kiss, J. T.; Gonzalez, R. D., J. Phys. Chem., **1984**, 88, 898.
32. Corrigan, D. S.; Weaver, M. J., J. Electroanal. Chem., **1988**, 239, 55.
33. Gonzalez-Sanchez, F., Spectrochim Acta, **1958**, 12, 17.

RECEIVED May 17, 1988

Chapter 22

In Situ Optical Vibrational Spectroscopy of Simple Ions at the Electrode–Electrolyte Interface

H. Seki

IBM Research Division, Almaden Research Center, San Jose, CA 95120–6099

The combination of surface enhanced Raman scattering (SERS) and infrared reflection absorption spectroscopy (IRRAS) provides an effective *in-situ* approach for studying the electrode-electrolyte interface. The extreme sensitivity to surface species of SERS is well known. By using polarization modulation of the infrared beam for IRRAS, the complete band shape is obtained without modulating the electrode potential. Thus, besides being sensitive to absorbing species on the electrode surface as well as in the solution in the region very close to the surface, it is possible to obtain potential dependent behavior in fine detail. We have applied these techniques to examine the interaction of simple ions such as CN^- and N_3^- with polycrystalline electrodes of silver, gold and copper. The observed vibrational spectra can be interpreted with the help of selection rules based on symmetry and analysis of *ab-initio* SCF wavefunctions of clusters. The results of these studies will be reviewed.

The discovery of the surface enhanced Raman scattering phenomenon (SERS) (1,2) in 1977 and the demonstration of straightforward, modulated specular reflectance spectroscopy in the infrared region (3) have combined to give the electrochemists a means to investigate *in-situ* the electrode-electrolyte interface which has a relatively high degree of molecular specificity. The mechanisms underlying SERS has turned out to be more complex than had been anticipated (4). The electrode surface generally requires some pretreatment which gives it an atomic scale roughness. This is usually achieved by repeated oxidation-reduction cycles, and the potential range in which SERS spectra can be observed is limited. The enhancement is also limited to certain metals such as Ag, Au and Cu. Within these limitations, however, SERS provides very high sensitivity and exceptionally wide frequency range. The ability to go down to the 100 cm^{-1} region, where the metal ligand stretching modes are found, is especially complementary to the present limitation of infrared spectroscopy to go down to low frequency.

0097–6156/88/0378–0322$06.00/0

In this paper some of the work involving *in-situ* vibrational spectroscopy, mainly those from our laboratory, will be reviewed which illustrate the kind of understanding we have been able to achieve. It has often been our experience that considerable insight, regarding the adsorption of molecules and ions, is gained when the results obtained by vibrational spectroscopy are considered in conjunction with the results of *ab initio* SCF cluster-adsorbate calculations.

Experimental

The electrochemical cell used in our laboratory has been fully described elsewhere (5). The cell body is made of chemically inert Kel-F and the electrode is mounted on a piston so that its surface can be pushed to the optical window, to a spacing of the order of 1-3 microns, in order to minimize the signal from the bulk electrolyte. For Raman scattering spectroscopy the window is of flat fused quartz, and the exciting laser beam is incident at about 60°. The scattered light is collected off-normal, but the geometry is not critical for SERS due to the high sensitivity. Details on the SERS measurements in our laboratory have been reported previously (6,7).

Various techniques have been developed for obtaining the infrared spectra and with them a series of acronyms. These include EMIRS (electrochemically modulated infrared spectroscopy) (8), SNIFTIRS (subtractively normalized interfacial fourier transform infrared spectroscopy) (9), PDIR (potential difference infrared spectroscopy) (10) and SPAIRS (single potential-alteration IR spectroscopy) (11). They all involve taking the spectral difference between two potentials in order to improve the signal to noise ratio, and a discussion of the details has been given by Corrigan et al. (11). In our laboratory we use the IBM Instruments, Inc. IR/98 FTIR (fourier transform infrared) spectrometer and, besides PDIR, we often employ PM FTIRRAS (polarization modulated FTIR reflection absorption spectroscopy). This latter technique evolved from the work of Golden et al. (12,13) and uses a photoeleastic modulator to switch the polarized infrared beam (incident at ca. 60°) between p and s polarization (14). It takes advantage of the surface selection rule (13) to obtain essentially the absorption reflection spectra of species very close to the electrode surface which have dynamic dipole components normal to the surface. The detector signal consists of an AC component corresponding to $(I_p - I_s)$ and a DC component corresponding to $(I_p + I_s)$, where I_p and I_s are detected intensities of the p and s polarized light. By circuitry outlined elsewhere (13) we obtain $(I_p - I_s)/(I_p + I_s)$. The main advantage of this technique is that $(I_p - I_s)$, which contains the absorption spectra of the species on/near the electrode surface, is amplified directly by the lock-in amplifier before digitization, resulting in a high signal to noise ratio. Furthermore this is done at a fixed potential and the band shape is observed directly.

The optical configuration which we have been using recently for IRRAS is shown in Figure 1. This arrangement has fewer beam reflections after the sample compared to what was shown in an earlier report (5). The highest sensitivity is obtained by using an InSb detector but this limits our lowest wavenumber to about 1850 cm^{-1}. With a slight sacrifice in sensitivity, use of a HgCdTe (MCT) detector would lower the limit to 1030 cm^{-1}, the cutoff of the CaF_2 prism window which we use. However for the studies presented here this was not a limitation.

Adsorbate Structure

The cyanide ion was one of the first adsorbed ions which was found to be surface enhanced (8,9) on silver electrodes. In this first stage of SERS investigation of the cyanide ion, it was assumed to be linearly bonded to Ag, through the carbon atom, perpendicular to the metal surface (9). However faced with the pressure of explaining the SERS mechanism, alternative structures were proposed. With the conception of the 'ad-atom' model for SERS, Otto conjectured that the main SERS band was due

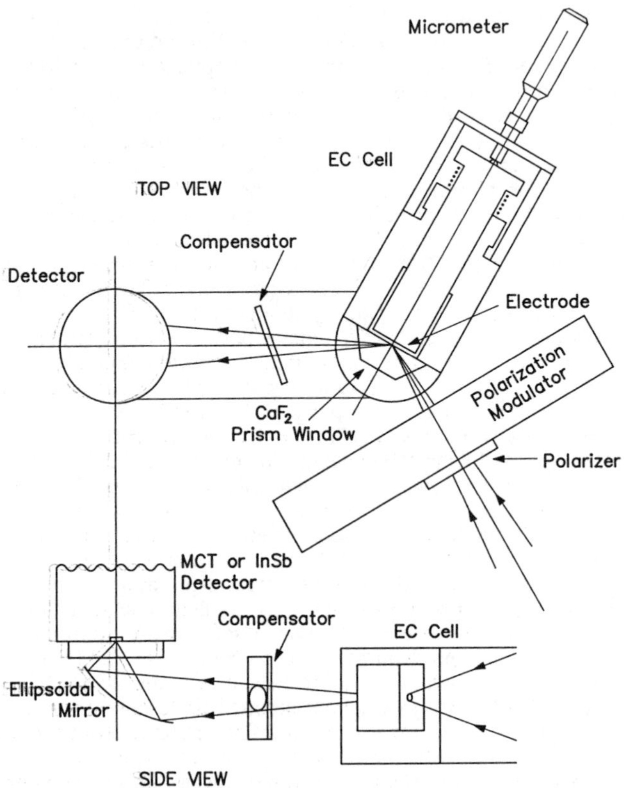

Figure 1. A schematic diagram giving the top view and side view of the optics and electrochemical cell for obtaining PM FTIRRAS.

to an $[Ag(CN)_3]^{2-}$ surface complex (10). Based on a well concieved SERS experiment involving isotopes, Fleischmann et al. (11) proposed that the main band was due to $[Ag(CN)_2]^-$. Unfortunately the resolution of the spectra was insufficient to back this proposal firmly. Baltruschat and Vielstich independently also proposed that the main adsorbed species was $[Ag(CN)_2]^-$ based on electrochemical arguements (19). This model was further reinforced by SERS work of cynide on Au electrode (20).

As we built up our PM-IRRAS capability, one of the interests was to compare the infrared and SERS spectra of the same species. Figure 2 presents spectra taken by IRRAS and SERS for cyanide on Ag (21). In the series of IRRAS spectra, there are three distinct bands (22,23). Those at 2080 cm^{-1} and 2136 cm^{-1}, which do not change their band positions are readily assigned to cyanide and the $[Ag(CN)_2]^-$ complex, respectively, in the solution in agreement with reported values (24). The third band, around 2100 cm^{-1}, shifts to lower wavenumbers as the electrode potential becomes more negative and matches the main band in the SERS spectra. If the surface species were a dicyano complex, it could be bent with a C_{2v} symmetry, and have vibrational modes corresponding to the symmetric, raman active mode and the asymmetric, infrared active mode of the solution complex with $D_{\infty h}$ symmetry. The near perfect match of the IRRAS and SERS bands strongly suggests that the adsorbed species responsible for these two bands is not likely to be due to a cyanide complex ion but due to a single cyanide ion. Similar results are obtained for cyanide on Au (23,25), i.e., the position and potential dependent shift of the IRRAS and SERS bands of the adsorbed cyanide are identical.

Another indication, that the surface species is a single cyanide ion, is seen in the reversible series of IRRAS spectra in Figure 3 (25). As the potential is stepped positive from -1.0V up to -0.7V and then reversed, the 2146 cm^{-1} band which is assigned to the $[Au(CN)_2]^-$ complex ion in solution (23,25) undergoes a reversible increase and then a decrease. The appearance of the 2146 cm^{-1} band is accompanied by the disappearance of the 2080 cm^{-1} solution cyanide band, and vice versa. When the potential drops to -0.9V all the $[Au(CN)_2]^-$ complex in solution is reduced as indicated by the disappearance of the 2146 cm^{-1} band. It is highly improbable that the complex could remain unreduced on the gold surface at potentials more negative than -0.9V. The frequency shifting band around 2100 cm^{-1} must therefore be assigned to a single cyanide, linearly bonded to the surface.

Since the band of the surface cyanide is slightly sharper for IRRAS than for SERS, isotope experiments were carried out with IRRAS, partly to reexamine the SERS isotope experiments reported by Fleischmann et al. (18). In Figure 4 a series of spectra are shown, each for a different ratio of $^{12}CN^-$ and $^{13}CN^-$. The total concentration of the cyanide was maintained at 0.01M and the spectra were taken at -1.0V where the single isotope ($^{12}CN^-$) solution gave the maximum intensity. Although a slight shift in the band frequency is seen with the change in composition (which will be discussed later), there is no noticeable line broadening which would have been expected due to the formation of some mixed isotope complexes, if the surface species were a dicyano complex. The fact that there are only two distinct frequency bands in the presence of the two isotopes regardless of their ratio supports the single cyanide surface model. This conclusion is also consistent with the cluster-adsorbate computations which are discussed later. At this point suffice it to say that calculations performed for the single linearly adsorbed cyanide could account for all the essential spectral observations.

Potential Dependent Band Shift

One of the important variables in the electrochemical system is the electrode potential. By controlling the electrode potential, very high electric fields, up to the order of 10^7 V/cm, can be applied to an adsorbed molecule or ion, which is not as easily accomplished for metal-vacuum or metal-gas interfaces. The first observation of field dependent shift of the vibrational band was reported in 1981 by SERS

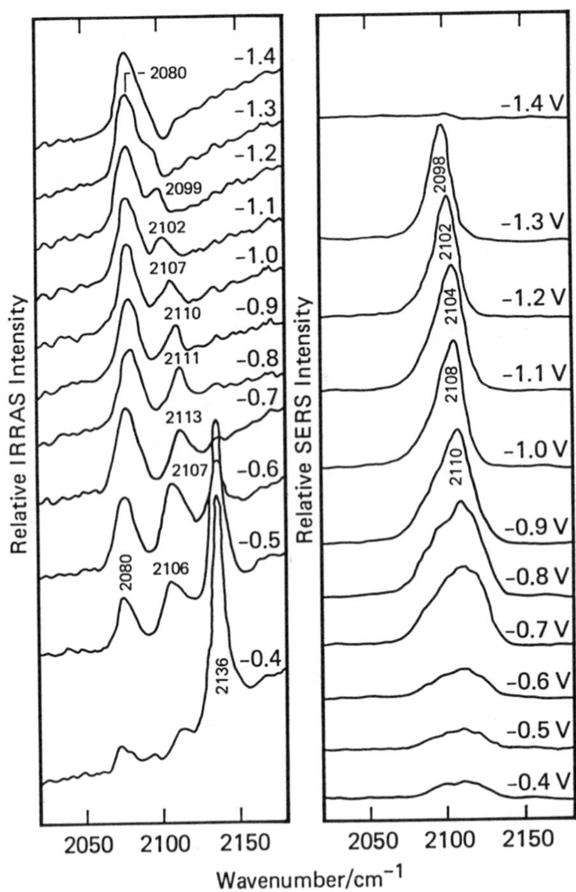

Figure 2. PM-IRRAS (left) and SERS (right) spectra for Ag electrode in 0.01M cyanide solution for various potentials, -0.4V to -1.4V (Ag/AgCl).

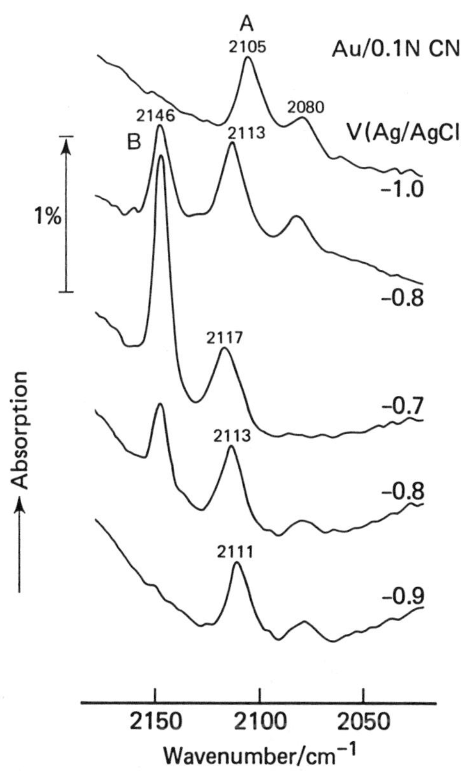

Figure 3. The reversible change of the IRRAS spectra seen at the Au/0.01M CN^- interface when the potential is changed from -1.0V to -0.7V and reversed[25].

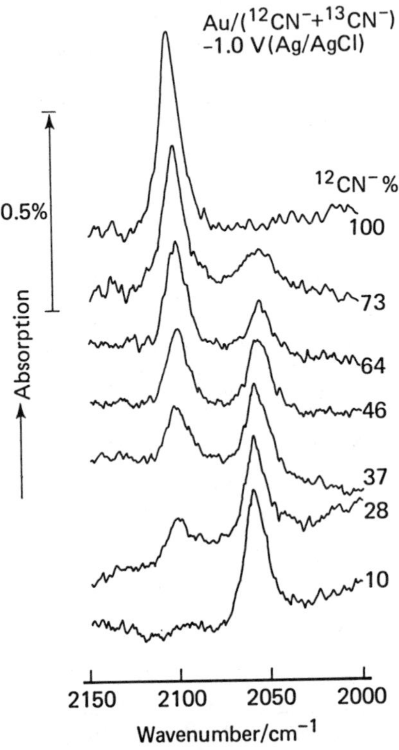

Figure 4. Change in the bands of the adsorbed CN^- with isotopic composition of $^{12}CN^-$ and $^{13}CN^-$ while the electrode potential was at −1.0 V and the total cyanide concentration in the solution is 0.01 M/1. (Reprinted from ref. 25. Copyright 1988 American Chemical Society.)

investigators (26,27), i.e., the shift with potential of the SERS band from an Ag electrode in cyanide solution. In the following 1982, potential dependent shift in the CO stretching band was observed with IRRAS (28,29). Experiments have also been carried out for CO on Ni in UHV (ultra high vacuum) (30) , but for lower applied fields.

Since the detection of infrared spectra is not dependent on special active sites as in SERS and the bands are slightly narrower, the use of PM-IRRAS is especially suitable for studying the potential dependent behavior, and careful studies have been made of CO on Pt (31,32) and CN^- on Ag (22,23), Au (23,25) and Cu (33). The infrared bands of both linearly and bridge bonded CO have been observed on Pt immersed in acid solutions (31). The potential dependent behaviour of the linearly bonded CO is summarized for three different acids in Figure 5, in which the integrated band intensity and the band positions are plotted as a function of the potential. The frequency shift irrespective of the anion is ca. 30 cm^{-1}/V. It should be noted that the intensity of the CO band is constant over most of the potential range indicating that the coverage is constant. Therefore the shift in frequency seen here is not due to change in the lateral adsorbate-adsorbate interaction which was observed as coverage dependent frequency shift in UHV (34-36).

Our investigations of the potential dependent behavior of adsorbed cyanide have shown that the potential shifts are similar to the case of CO on Pt. The frequency shift seen from CN^- on Au is shown in Figure 6 (25). It is seen that between -1.2 and -0.5V(Ag/AgCl) the band intensity is relatively constant and in this region there is a linear rate of shift of about 33 cm^{-1}/V which can be attributed mainly to the potential change. Outside the above potential range, the coverage is decreasing and the shift deviates from the linearity. This deviation from linearity can be considered to be due to decrease in frequency with surface concentration which is seen in Figure 4 and more explicitly plotted in Figure 7. Figures 6 and 7 show that the the total shift due to the field or potential is larger than that due to the coverage.

The direction of the potential shift of the surface C-N band is the same as for the C-O band on Pt and the shift rates are ca. 30 cm^{-1}/V for Ag and Au (22,23,25) and ca. 45 cm^{-1}/V for Cu (33). The measurement for copper is less accurate due to formation of ion complexes with large absorption crossections and the relatively weak signal from the surface species itself.

The most widely accepted model of CO linearly adsorbed on metal, based on studies of metal carbonyls (37), is with C towards the metal. The bond is covalent with the lone pair 5σ orbital donating charge to the metal while there is a π back donation into the empty CO $2\pi^*$ level (38). The occupation of the antibonding $2\pi^*$ orbital weakens the C-O bond and causes the lowering of the C-O vibration of the adsorbed CO. Thus it is reasonable to propose that this balance of charge exchange in the orbitals is affected by the applied field, altering the C-O force constant (39,40). Faced with the similarity of the potential dependent band shift and the fact that CO and CN^- are isoelectronic, it could be argued that the observed potential dependent shift is due to similar mechanisms for both CO and CN^-. Indeed this explanation was given for the initial observation of the CN^- frequency shifts (41).

Lambert however proposed an interesting alternative explanation, i.e., the potential dependent frequency shift can be explained by a first-order Stark effect without involving a change in the covalent bonding due to the field (42). This was demonstrated by calculating the Stark tuning rates using first order perturbation theory and experimentally measured vibrational frequencies and molecular constants of a number of molecules (including CO) in the gas phase.

SCF Cluster Calculations

The development of *ab-initio* self-consistent field (SCF) wave functions for clusters by Bagus et al. provides a powerful basis for interpreting the experimental observations related to the localized bonding of adsorbed molecules (43-47). In these

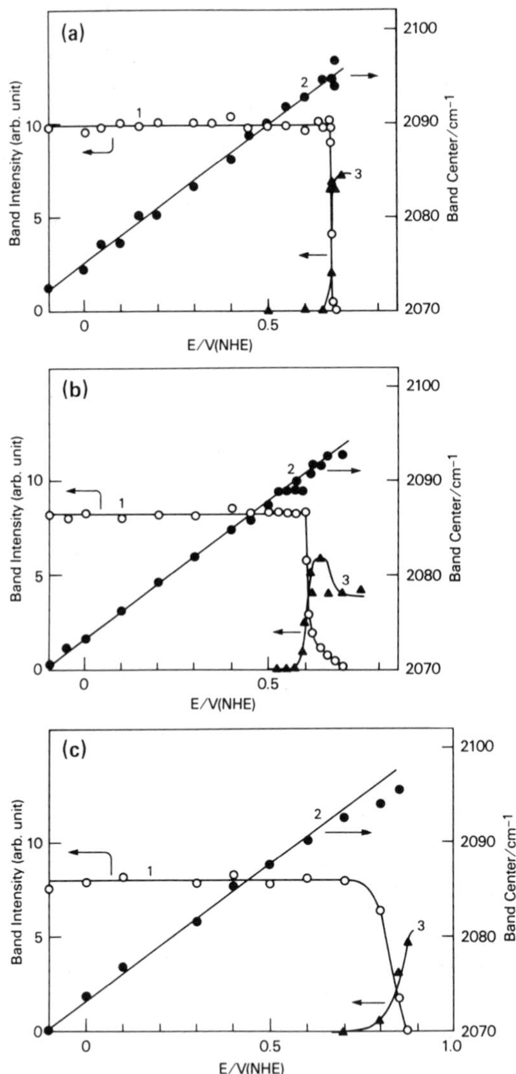

Figure 5. The integrated band intensity (1) and the band position (2) of linearly adsorbed CO are plotted as a function of potential in (a) 1 M $HClO_4$, (b) 1 M H_2SO_4, and (c) 1 M HCl. (Reprinted from ref. 31. Copyright 1985 American Chemical Society.)

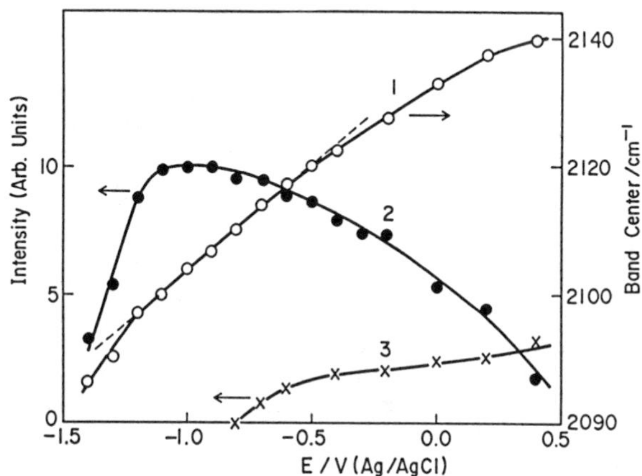

Figure 6. The potential dependence of (1) the IR band center and (2) the integrated band intensity of adsorbed CN^- on Au, and (3) the integrated band intensity of the $[Au(CN)_2]^-$ complex. (Reprinted from ref. 25. Copyright 1988 American Chemical Society.)

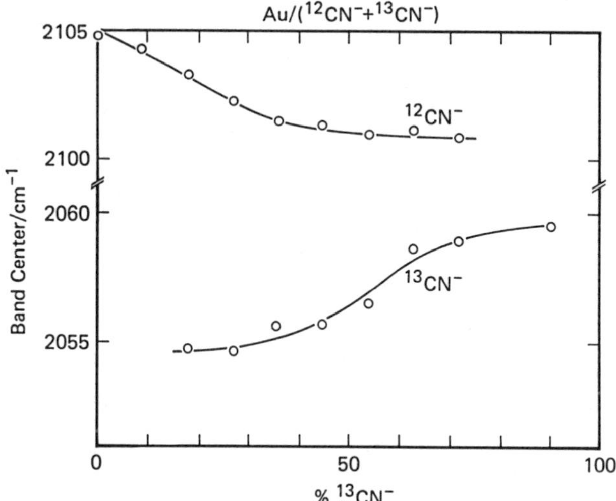

Figure 7. The dependence of the C–N bands of the adsorbed $^{12}CN^-$ and $^{13}CN^-$ on the isotopic composition based on the data of Figure 4. (Reprinted from ref. 25. Copyright 1988 American Chemical Society.)

calculations the metal surface is represented by a metal cluster which ranges in size from a single metal atom up to over 30 metal atoms, the larger cluster giving, generally, more realistic results at the cost of increased computational time. Substitution of the real surface with a metal cluster is an approximation and weakens the validity of the computational results relative to the cases of an isolated molecule. However the advantage of the *ab initio* method, that it does not involve adjustable parameters to fit the experimental results, is still quite persuasive.

Very briefly what this means is the following: The electrons in the system (cluster) are described by molecular orbitals which generally consist of a sum of a finite set of basis functions (e.g., atomic orbitals), each basis function having a coefficient. The wavefunction of the system is a Slater determinant of these molecular orbitals. The SCF wavefunction, for a given fixed position of the atoms, is determined by carrying out variational SCF computations of the total energy of the system, i.e., the coefficients of the basis functions are varied in order to select a set of coefficients which gives the minimum energy. There are no empirical parameters which can be adjusted to match the results to experiment. The investigator chooses the model for the calculation, i.e., the basis set, its size, the level of sophistication of the computation (e.g., configuration interaction) cluster size, the use of pseudo potentials to approximate the core electrons of some of the metal atoms, etc. The validity of the computation is determined by these choices. Physical insight and computational experience are needed to make the appropriate selection of the computational model depending on the information being sought. This usually involves trade off with the computational power and time available. (A more complete discussion of the *ab-initio* methods can be found e.g., in reference (48)).

The degree of confidence we can have in the computational result is ultimately determined by the ability of the model to produce results which agree with experiments. As an example, the equilibrium interatomic distance of a molecule is obtained by computing the SCF energy for various atomic separations and finding the location for the minimum energy. The models used in the work here have been successful in arriving at the proper adsorption bond distances (44) and in interpreting photoemission spectra (43).

Most of the calculations have been done for Cu since it has the least number of electrons of the metals of interest. The clusters represent the Cu(100) surface and the positions of the metal atoms are fixed by bulk fcc geometry. The adsorption site metal atom is usually treated with all its electrons while the rest are treated with one 4s electron and a pseudopotential for the core electrons. Higher z metals can be studied by using pseudopotentials for all the metals in the cluster. The adsorbed molecule is treated with all its electrons and the equilibrium positions are determined by minimizing the SCF energy. The positions of the adsorbate atoms are varied around the equilibrium position and SCF energies at several points are fitted to a potential surface to obtain the interatomic force constants and the vibrational frequency.

The *ab initio* SCF cluster wavefunction has been used to investigate the bonding of CO and CN^- on Cu_{10} (5,4,1), (5 surface layer, 4 second layer and 1 bottom layer atoms), and to calculate their field dependent vibrational frequency shifts in fields up to 5.2×10^7 V/cm(46). A schematic view of the Cu_{10} (5,4,1)CO cluster is shown in Figure 8. In order to assess the significance of Lambert's proposal, that the linear Stark effect is the dominant factor in the field dependent frequency shift, the effect of the field was calculated by three methods. One is by a fully variational approach (i.e., the adsorbate is allowed to relax under the influence of the applied field) in which the Hamiltonian for the cluster in a uniform electric field, F, is given by

$$H(F) = H(0) + F\sum_i r_i - F\sum_i R_i,$$

where r_i and R_i are the position of the electrons and the nuclei of the adsorbate. The energy obtained from this Hamiltonian is $E_{SCF}(F)$ and the resulting wavefunction includes the total effect of the field, changes in the chemical bonding and the Stark effect. The second method uses first-order perturbation theory energy, $E_p(F) = E_{SCF}(0) - \mu(0)F$ where $E_{SCF}(0)$ and $\mu(0)$ are the zero field SCF energy and dipole moment, respectively. The third method follows the formalism of Lambert (42) in which Taylor series expansions of the ligand potential and the dipole moment are used. The coefficients in the expansions are calculated from the zero field SCF wavefunctions, instead of using the measured molecular parameters as was done by Lambert. The latter two methods calculate changes based on zero field parameters and do not involve changes in the chemical bonding. (For a complete description and discussion of the SCF cluster analysis in general, references given here (43-47) and references therein should be consulted).

The frequency shifts calculated by the latter two methods agree quite well as might be expected since both are essentially variations of first order perturbation theory. More importantly, the difference between the shifts calculated by the fully SCF method and the pertubation methods are relatively small and this endorses the proposal by Lambert (42) that the frequency shifts due to fields less than the order of 10^7 V/cm arise mainly from a first order Stark effect. (A similar conclusion was claimed by Bauschlicher (49) but it was based only on results of SCF calculations on the isolated CO in an applied field parallel to the molecule). The results of these analyse hs also showed that the metal-ligand bonding for CO and CN^- is quite different. In the case of CO the bond is covalent with dominantly metal to CO $2\pi^*$ back donation and for CN^- the bonding is mainly ionic. It is also interesting that the bond energy did not change very much, when the orientation of CN^- was changed, consistent with the picture that covalency of the bonding is small. The fact that the infrared intensity of the surface C-N stretch is weak on Cu most likely is an indication that all the CN^- is not oriented perfectly normal to the surface. Since the intensity is stronger on Ag and Au electrodes, this may indicate that as the metal d shell is filled the covalency of the metal ligand bond increases and the degree of normal orientation improves.

It is quite interesting that due to this difference in the bonding, the way by which the field affects the frequency is different for the two adsorbates. In the case of CO the change is a direct change in ligand frequency but in the case of CN^- the change in the calculated ligand frequency is considerably smaller than for CO. This does not agree with the experimental observation that the frequency shifts were comparable for CO and CN^-. This discrepancy is resolved by recognizing that the ligand vibration and the metal ligand vibration are coupled. Due to the ionicity of the metal-CN^- bond, the Cu-C distance and the metal ligand frequency are changed appreciably by the applied field. Through the interaction between the two oscillators the frequency of the normal mode corresponding to the C-N stretch is also increased. This has been investigated in more detail using a Cu_{14} (5,4,5)CO cluster (47) which can be visualized by surrounding the bottom Cu atom in Figure 8 with 4 more Cu atoms. In this study it is shown more clearly how the change in the Cu-C distance also causes a change in the ligand force constant due to a "wall" effect of the metal surface electrons and this brings the C-N vibrational frequency shift within a factor of 2 of the C-O frequency shift. The shift in the metal cyanide frequency was, in fact, reported in the initial SERS observation of the potential dependent shifts (26,27). The final experimental verification of these analyses will be to see if the metal-ligand stretching frequency for CO is little affected by the field as our model predicts.

Azide on Silver

Recently the vibrational spectra of azide (N_3^-) on Ag has been investigated using PM FTIRRAS (50). In solution the azide ion has two vibrational modes, the Raman active symmetric mode at ca. 1340 cm^{-1} and the ir active asymmetric mode at ca. 2050 cm^{-1}. The surface species was identified with some difficulty due to the narrow potential range in which it could be seen and the presence of an intense band of the solution N_3^-. The spectra of the surface species can be seen in Figure 9 shifting from 2048 cm^{-1} to 2083 cm^{-1}. The band due to the surface species is seen in the potential range of about 0 to 0.25V(Ag/AgCl). Differential capacitance measurements however have indicated that azide is adsorbed on the Ag electrode over the range extending down to -0.9V (50,51). The frequency of the band shifts with potential at about 100 cm^{-1} /V, assuming a linear dependence. These spectra were extracted by subtracting the spectrum at -0.95V where differential capacitance measurements (51,52) and SERS measurements indicate there is no azide adsorption (53,54).The interpretation of the azide IRRAS spectra was also facilitated by the computations of the *ab-initio* SCF wavefunctions (50). In this case the simplest cluster, consisting of a single Ag atom, was used. The objective was more to identify trends and changes in the values relevent to the surface spectra rather than to obtain precise values of molecular parameters and vibrational frequencies. The clusters, $[AgN_3]^0$ and $[AgN_3]^-$, in two geometries were investigated. In one with $C_{\infty v}$ symmetry, the Ag atom and the azide are colinear, modelling the azide adsorbed normal to the Ag surface, and in the other with C_{2v} symmetry, an Ag atom is located beneath the center of the linear azide, modelling azide adsorbed parallel to the surface. The calculations show that azide is strongly ionic, similar to CN^- and the frequency shift is also due mainly to the first order Stark effect. When N_3^- is colinear with the Ag atom the effect of applied fields of $\pm$ 5.2 × 10^7 V/cm is to greatly alter the Ag-azide distance. As in the case of CN^- (47), this then causes the asymmetric mode frequency to shift. However the analysis, interestingly, shows that the field dependent shifts in the symmetric mode frequency for both the colinear and flat orientations are much smaller than that in the asymmetric mode.

When the experimental results are considered together with the calculations the following model can be envisioned. In the region of negative potential N_3^- is adsorbed parallel to the surface and can not be observed by IRRAS. The symmetric vibration and the bending modes are observed in this region by SERS (53) and the vibrational frequency is not seen to shift very much with potential, consistent with the SCF calculations. As the potential is made more positive the concentration increases and adsorption tends to perpendicular orientation which is detected by IRRAS.

Summary

Examples of investigation, which involve *in-situ* vibrational spectroscopy of the electrolyte interface, have been reviewed, demonstrating the kind of insight such spectral data can provide. By a combination of SERS, IRRAS and isotope effects it can been clearly seen that the structure of cyanide adsorbed on Cu, Ag and Au electrodes is linearly bonded to the surface. This simple conclusion is important since alternative, more complex structures have been discussed in the literature until quite recently. *Ab-initio* SCF cluster calculations strongly suggest that the bonding of cyanide to the metal is essentially ionic. The ability to control the potential in an electrochemical cell and hence the field applied to the adsorbate provides a powerful handle in the investigation of the adsorbate substrate interaction. The potential dependent shift of the vibrational band can serve to identify whether the source is a surface species. The case of CO and CN^- illustrate that insight into the mechanisms underlying this shift can be gained in molecular detail through calculations of *ab-initio* SCF wavefunctions of clusters. What has been presented here is only a beginning. The very rapid advance in the computor technology will insure that we will see increasingly more sophisticated computations of more complex molecules and clusters. This will greatly enhance the value of data obtained by vibrational spectroscopy. In combination with other *in-situ* techniques such as differential capacitance measurements, radio tracer measurements, X-ray scattering and quartz

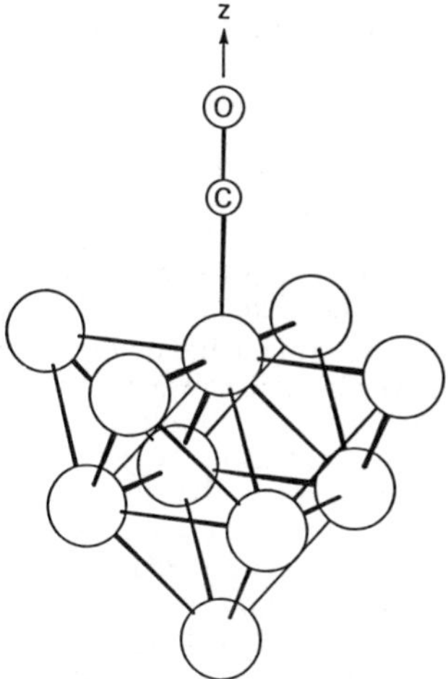

Figure 8. A schematic view of the Cu_{10} (5,4,1)CO cluster. (Reprinted with permission from ref. 45b. Copyright 1985 American Institute of Physics.)

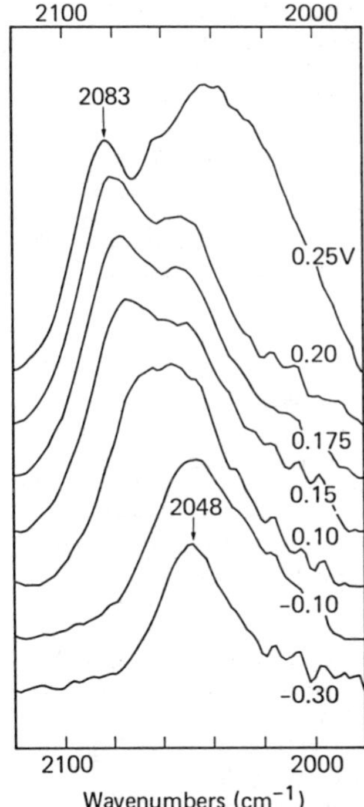

Figure 9. PM-IRRAS spectra for Ag electrode in 0.03 M azide in 0.1 M $NaClO_4$. These are obtained by taking the difference of the spectra taken at the specified potential and at −0.95 V (Ag/AgCl). (Reprinted with permission from ref. 50. Copyright 1988 American Institute of Physics.)

crystal microbalance we can expect exciting progress in our understanding of the electrode electrolyte interface.

Acknowledgment

The work has been supported in part by the Office of Naval Research. A large part of the work discussed here was done in collaboration with my colleagues, P. S. Bagus, W. G. Golden, J. G. Gordon, K. Kunimatsu, C. J. Nelin, M. R. Philpott, M. G. Samant. I have benefitted greatly from this pleasant interaction for which I am very grateful. I thank O. Melroy for constructive comments regarding the manuscript.

References

1. Jeanmaire, D. J.; Van Duyne, R. P. J. Electroanal. Chem. 1977, 84, 1.
2. Creighton, J. A. J. Am. Chem. Soc. 1977, 99, 5215.
3. Bewick, A.; Kunimatsu, K.; Pons, S. B. Electrochim. Acta 1988, 25, 465.
4. Surface Enhanced Raman Scattering Chang, R. K.; Furtak, T. F. Ed.; Plenum Press, N. Y., 1982.
5. Seki, H.; Kunimatsu, K.; Golden, W. G. Appl. Spectrosc. 1984, 39, 437.
6. Kunz, R. E.; Gordon, J. G., II; Philpott, M. R.; Girlando, A. J. Electroanal. Chem. 1980, 112, 391.
7. Pettinger, B.; Philpott, M. R.; Gordon, J. G., II, J. Chem. Phys. 1981, 74, 934.
8. Beden, B.; Lamy, C.; Bewick, A.; Kunimatsu, K. J. Electroanal. Chem. 1981, 122, 343.
9. Bewick, A.; Pons, S. Adv. Infrared Raman Spectrosc. 1985, 12, 1.
10. Corrigan, D. S.; Weaver, M. J. J. Phys. Chem. 1986, 90, 5300.
11. Corrigan, D. S.; Leung, L. H.; Weaver, M. J. Anal Chem. 1987, 59, 2252.
12. Golden, W. G.; Dunn, D. S.; Overend, J. J. Catal. 1981, 71, 395.
13. Golden, W. G.; Saperstein, D. D. J. Electron Spectrosc. 1983, 30, 43.
14. We have been using the Hinds International, PEM-80 Series II ZnSe photoelectic modulator which modulates the polarization at 74kHz.
15. Greenler, R. G. J. Chem. Phys. 1966, 44, 310.
16. Otto, A. Surface Sci. 1978, 75, L392.
17. Furtak, T. E. Solid State Commun. 1978, 28, 903.
18. Fleischmann, M.; Hill, I. R.; Pemble, M. E. J. Electroanal. Chem. 1982, 136, 362.
19. Baltruschat, H.; Vielstich, W. J. Electroanal. Chem. 1983, 154, 141.
20. Baltruschat, H.; Heitbaum, J. J. Electroanal. Chem. 1983 .us 157, 319.
21. H. Seki, K. Kunimatsu and W. G. Golden, presented at the meeting of the International Society of Electrochemistry, Berkeley, 1984.
22. Kunimatsu, K.; Seki, H.; Golden, W. G. Chem. Phys. Letters 1984, 108, 195.
23. Kunimatsu, K.; Seki, H.; Golden, W. G.; Gordon, J. G., II and Philpott, M. R. Surface Sci. 1985, 158, 596.
24. Jones, L. H.; Penneman, R. A. J. Chem. Phys. 1954, 22, 965.
25. Kunimatsu, K.; Seki, H.; Golden, W. G.; Gordon, J. G., II; Philpott, M. R. Langmuir in press, 1988.
26. Kötz, R.; Yeager, E. J. Electroanal. Chem. 1981, 123, 335.
27. Wetzel, W.; Gerischer, H.; Pettinger, B. Chem. Phys. Lett. 1981, 80, 159.
28. Kunimatsu, K. J. Electroanal. Chem. 1982, 140, 205.
29. Russell, J. W.; Overend, J.; Scanlon, K.; Severson, M.; Bewick, A. J. Phys. Chem. 1982, 86, 3066.
30. Lambert, D. K. Solid State Comm. 1984, 51, 297.
31. Kunimatsu, K.; Golden, W. G.; Seki, H.; Philpott, M. R. Langmuir 1985, 1, 245.

32. Kunimatsu, K.; Seki, H.; Golden, W. G.; Gordon, J. G., II; Philpott, M. R. Langmuir. 1986, 2, 464.
33. Lee, K. A. B.; Kunimatsu, K.; Gordon, J. G., II; Golden, W. G.; Seki, H. J. Electrochem. Soc. 1987, 134, 1676.
34. Shigeishi, R. A.; King, D. A. Surface Sci. 1976, 58, 379.
35. King, D. A., in Vibrational Specttroscopy of adsorbates, edited by Willis, R. F., Springer Series in Chem. Phys. 15, Springer-Verlag, Berlin, 1980.
36. Hayden, B. E.; Bradshaw, A. M. Surf. Sci. 1983, 125, 787.
37. Cotton, F. A.; Wilkinson, G. Advanced Inorganic Chemistry, 3rd edition, John Wiley, New York, 1972.
38. Blyholder, G. J. Phys. Chem. 1964, 68, 2773.
39. Ray, N. K.; Anderson, A. B. J. Phys. Chem. 1982, 86, 4851.
40. Holloway, S.; Norskov, J. K. J. Electroanal. Chem. 1984, 161, 193.
41. Anderson, A. B.; Kötz, R.; Yeager, E. Chem. Phys. Lett. 1981, 82, 130.
42. (a) Lambert, D. K. J. Electron. Spectrosc. 1983, 30, 59; (b) Lambert, D. K. Phys Rev. Letters 1983, 50, 2106; (c) Lambert, D. K. Solid State Commun. 1984, 51, 297.
43. Bagus, P. S.; Hermann, K.; Seel, M. J. Vac. Sci. Technol. 1981, 18, 435.
44. (a) Bagus, P. S.; Nelin, C. J.; Bauschlicher, C. W. Phys. Rev. 1983, B28, 5423; (b) Bagus, P. S.; Nelin, C. J.; Bauschlicher, C. W. J. Vac. Sci. Technol. 1984, A2, 905.
45. (a) Bagus, P. S.; Müller, W. Chem. Phys. Letter 1985, 115, 540; (b) Müller, W.; Bagus, P. S. J. Vac. Sci. Technol. 1985 A, 3, 1623; (c) Müller, W.; Bagus, P. S. J. Electron. Spectrosc. 1986, 38, 103.
46. Bagus, P. S.; Nelin, C. J.; Müller, W.; Philpott, M. R.; Seki, H. Phys. Rev. Letters 1987, 58, 559.
47. Bagus, P. S.; Nelin, C. J.; Hermann, K.; Philpott, M. R. Phys. Rev. B 1987, 36, 8169.
48. H. F. Schaefer, III, Electronic Structure of Atoms and Molecules, Addison-Wesley, Reading, Mass., 1972, and Quantum Chemistry, Clarendon Press, Oxford, 1984.
49. Bauschlicher, C. W. Chem. Phys. Letter 1985, 118, 307.
50. Samant, M. G.; Viswanathan, R.; Seki, H.; Bagus, P. S. Nelin, C. J.; Philpott, M. R. J. Chem. Phys. 58, submitted, 1988.
51. Larkin, D.; Guyer, K. L.; Hupp, J. T.; Weaver, M. J. J. Electroanal. Chem. 1982, 113, 401.
52. Hupp, J. T.; Larkin, D.; Weaver, M. J. Surf. Sci. 1983, 125, 429.
53. Kunz, R. E.; Gordon, J. G., II; Philpott, M. R.; Girlando, A. J. Electroanal. Chem. 1980, 112, 391.
54. Corrigan, D. S.; Gao, P.; Leung, L. H.; Weaver, M. J. Langmuir 1986, 2, 744.
55. Corrigan, D. S.; Weaver, M. J. J. Phys. Chem. 1986, 90, 5300.

RECEIVED May 17, 1988

Chapter 23

Infrared Spectroelectrochemistry of Surface Species

In Situ Surface Fourier Transform Infrared Study of Adsorption of Isoquinoline at a Mercury Electrode

Daniel Blackwood[1], Carol Korzeniewski[2], William McKenna[1], Jianguo Li[1], and Stanley Pons[1]

[1]Department of Chemistry, University of Utah, Salt Lake City, UT 84112
[2]Department of Chemistry, University of Michigan, Ann Arbor, MI 48109

Subtractively normalized interfacial Fourier transform infrared spectroscopy (SNIFTIRS), has been used extensively to examine interactions of species at the electrode/electrolyte interface. In the present work, the method has been extended to probe interactions at the mercury solution interface. The diminished potential dependent frequency shifts of species adsorbed at mercury electrodes are compared with shifts observed for similar species adsorbed at d-band metals.

Several spectroscopic techniques have been developed for the investigation of electrode - solution interfacial phenomena (1-7). The infrared techniques include (i) subtractively normalized interfacial Fourier transform infrared spectroscopy, (SNIFTIRS) (7) in which infrared spectra are collected at two different potentials and a difference spectrum obtained by subtraction and normalization, (ii) electrochemically modulated infrared spectroscopy, EMIRS (4), a dispersive spectrophotometric technique in which the electrode potential is modulated at a set frequency and the resulting attenuation of the reflecting infrared radiation is analyzed by phase sensitive detection, and (iii) infrared reflection absorption spectroscopy, IRRAS (6), in which the polarization of the incident infrared radiation is modulated at a high rate (ca. 74 kHz) between the s- and the p- states to distinguish between adsorbed and solution dissolved infrared absorbers. The attenuated infrared signal is again monitored with the aid of phase sensitive detection techniques. The characteristics of the two types of spectra are listed in Table I. The selection rules that determine interaction of infrared radiation with adsorbates are well known, and arise from the differences in the relative intensities of the s- and p-polarized components of the electromagnetic field vectors of the incident infrared radiation at the metal-solution interface. The s-polarized component undergoes a phase shift upon reflection from a metal surface which results in zero field strength at the surface, while the p-component field strength increases at the surface as the angle

0097–6156/88/0378–0338$06.00/0

Table I. A Comparison of the Characteristics of Infrared Vibrational Bands Arising from Bulk Solution Species and Adsorbed Species

BULK SOLUTION SPECIES	ADSORBED SPECIES
i) Bands present with either s- or p-polarized light.	i) Bands present only with p-polarized light.
ii) Band positions independent of potential.	ii) Band positions may shift with potential.
iii) Relative intensity of bands independent of potential.	iii) Relative intensity may change with potential.
iv) Normally IR inactive bands not observed.	iv) Some may be observed.

of incidence increases. As a result, only p-polarized light can interact with species which are either adsorbed onto or are very close to the electrode surface. In addition, the normal modes of an adsorbate that interact with the radiation must have a component of the derivative of the dipole moment (with respect to the normal coordinate) perpendicular to the metal surface. In addition, modes of an adsorbate activated by vibronic coupling to the surface may interact with the radiation (vide infra).

Potential dependence of the frequencies of infrared bands arise because of the change in extents of bonding of the adsorbate to the metal surface as the potential is changed. The effect has been explained by several models (vide infra). To illustrate the potential dependence of adsorbed species, the SNIFTIRS spectra of the b_{3u} ring bending mode of p-difluorobenzene at a platinum electrode is illustrated in Figure 1 (8) The upward pointing band is clearly independent of potential and is assigned to the solution species whereas the position of the downward pointing band shifts monotonically (at constant ionic strength) with potential and is therefore due to an adsorbed species.

The particular system investigated in this work was the adsorption of isoquinoline at a mercury electrode, which has been previously studied by several techniques, including electrocapillary measurements (9), ellipsometry (10), double layer capacity measurements (11), and a range of potential step techniques (12-15). The interest in this system is due in part to the fact that one observes well defined transitions in the measurements as the adsorbed molecules undergo transitions in surface orientation and packing density under certain experimental conditions. Isoquinoline has been shown to be adsorbed on mercury in four different orientations (Figure 2). The previous investigations indicate the following behavior for the isoquinoline orientation as a function of potential and concentration: at low negative potentials and low bulk concentrations, the molecules are believed to lie flat on the mercury surface (molecular plane parallel to the surface). On increasing either the potential (in the negative direction) or the bulk concentration, the isoquinoline molecules are forced up into either the 4,5 position (10) or the 5,6 position (9). This reorientation occurs gradually with the changing coordinates, and proceeds through a series of phases containing mixtures of these three orientations of isoquinoline molecules.

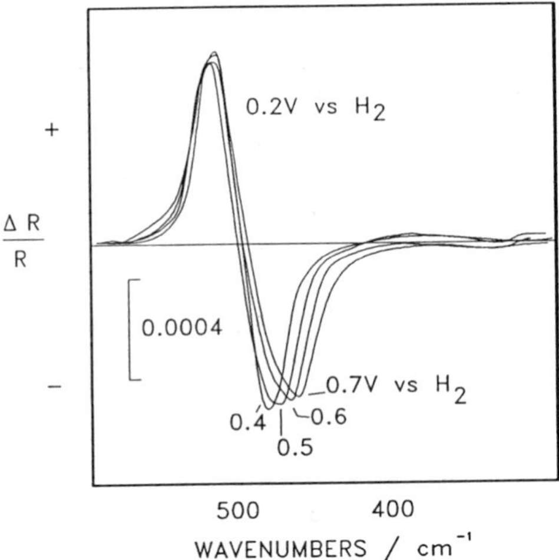

Figure 1. SNIFTIRS difference spectra of the b_{3u} ring bending mode of p-difluorobenzene at a platinum electrode as a function of modulation potential.

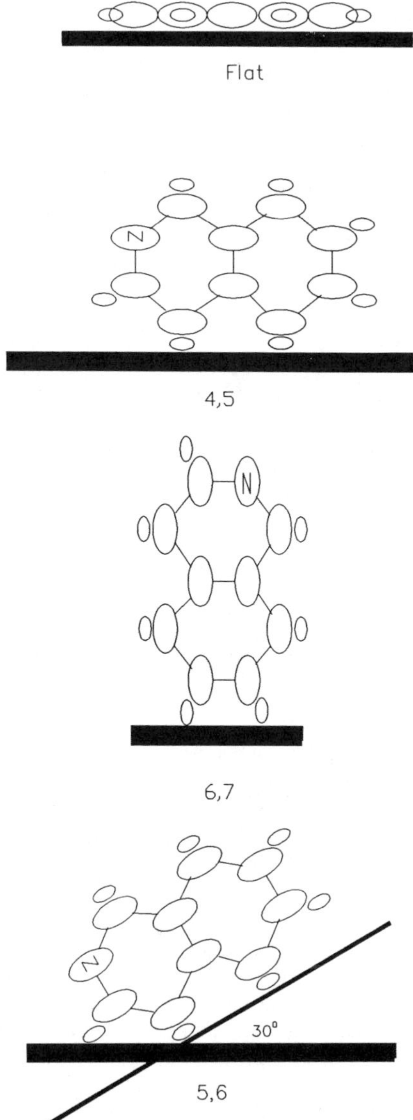

Figure 2. Likely orientations for the adsorption of isoquinoline on mercury.

Increases in the potential to more negative values (at sufficiently high concentrations) results in an abrupt reorientation to the 6,7 position. The reason that this second transition is much sharper than the first lies in the fact that mixed phases which would contain the 6,7 orientation are energetically less favorable than a complete monolayer of any of the pure standing orientational phases. Gierst et al (9) have produced a graph showing the dependence of the superficial excess on both potential and bulk concentration from their electrocapillary data; we reproduce some of their data in Figure 3.

For an isoquinoline molecule adsorbed onto the surface of mercury, the component of its total dipole moment that is perpendicular to the surface will increase as its orientation changes from:

Flat - 4,5 - 5,6 - 6,7

As the isoquinoline molecule reorients in the order listed above, the absorption of infrared radiation by the in-plane vibrational modes would be expected to increase, while that of the out-of-plane modes would be predicted to decrease (in accordance with the surface selection rule as described above). In the flat orientation there is no component of the dipole moment perpendicular to the surface for the in-plane modes, and under the surface selection rule these modes will not be able to absorb any of the incident radiation. However, as mentioned above, infrared active modes (and in some cases infrared forbidden transitions) can still be observed due to field-induced vibronic coupled infrared absorption (16-20). We have determined that this type of interaction is present in this particular system.

EXPERIMENTAL

Isoquinoline (Aldrich 97%) was purified by refluxing with BaO for 30 minutes and distilling in vacuo. The resulting white crystalline solid had a melting point of 26°C. The purified isoquinoline was stored in the dark, at $0^{o}C$ and under an argon atmosphere. Mercury was triply distilled (American Scientific) and all other chemicals were of AnalaR grade. Solutions were prepared with triply distilled water. Glassware was cleaned in a 50:50 (v:v) mixture of HNO_3 and H_2SO_4, and rinsed and steamed (triply distilled water) for half an hour. A thin layer cell was designed (Figure 4) which could be mounted vertically on top of the sample chamber of the spectrometer. The mercury was held in position by a glass tube, and electrical contact was achieved with a piece of platinum wire inserted into the mercury. Potentials are measured with respect to a saturated calomel electrode (SCE).

The technique used to acquire the data in this paper was SNIFTIRS. A schematic diagram of the required apparatus is shown in Figure 5, and has been described in detail elsewhere. The FTIR spectrometer used was a vacuum bench Bruker IBM Model IR/98, modified so that the optical beam was brought upwards through the sample compartment and made to reflect from the bottom of the horizontal mercury surface. The methods used herein are adapted from a configuration that has been used by Bewick and co-workers (21) at Southampton.

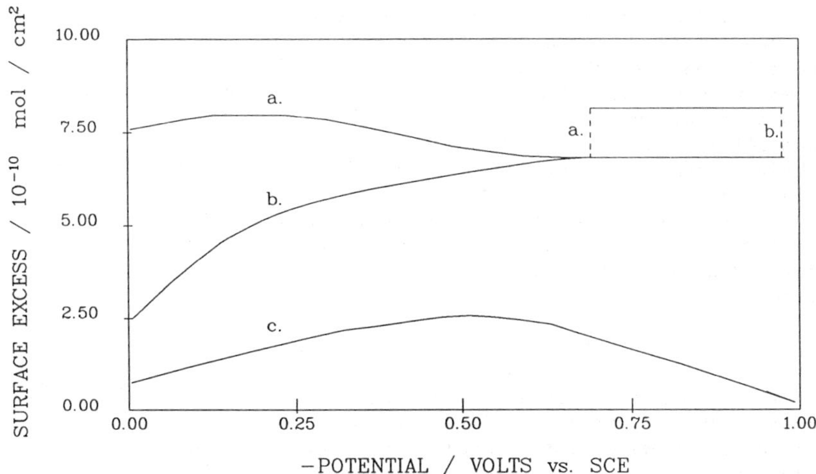

Figure 3. Superficial excess as a function of potential for a mercury electrode in 0.5 mol dm^{-3} Na_2SO_4 and the following isoquinoline concentrations: (a) 2.1 x 10^{-2} (saturated), (b) 6.3 x 10^{-3}, and (c) 2.1 x 10^{-4} mol-dm^{-3}. (Data from Gierst et al. (9))

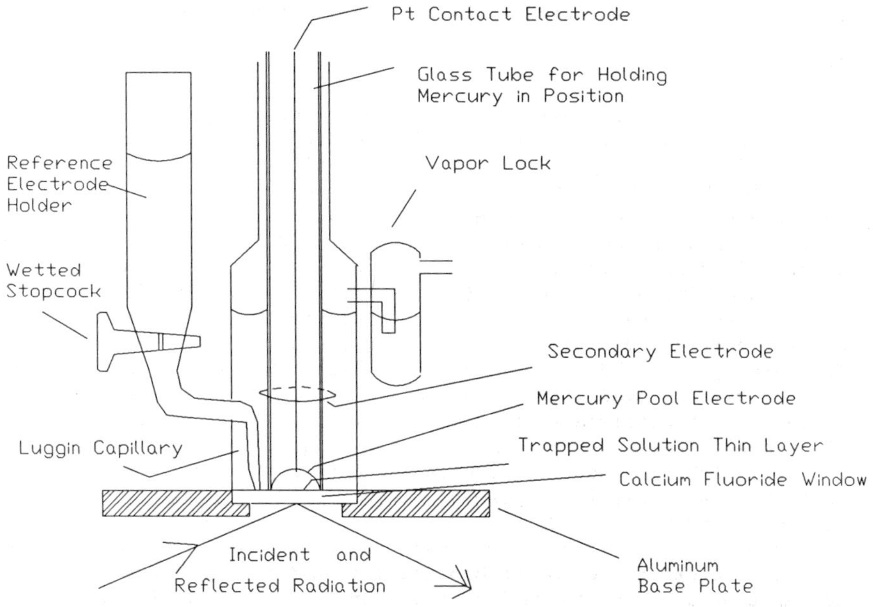

Figure 4. Thin layer mercury reflectance cell.

RESULTS AND DISCUSSION

The abrupt transition to the 6,7 orientation manifests itself in cyclic voltammetry as a sharp current spike (Figure 6). The cathodic spike was found to contain a charge of 2.9 $\mu C\ cm^{-2}$, while its anodic counterpart contained 3.3 $\mu C\ cm^{-2}$. The peak separation was 100 mV; this large value is due to the large iR losses suffered in the thin layer cell.

Figure 7 shows SNIFTIRS spectra for isoquinoline molecules adsorbed on mercury. The reference spectrum in each case was obtained at 0.0V vs. a SCE reference electrode; at this potential the molecules are believed to be oriented flat on the metal surface. The vibrational frequencies of the band structure (positive values of absorbance) are easily assigned since they are essentially the same as those reported by Wait et al. (22) for pure isoquinoline. The differences in the spectra are that the bands for the adsorbed species exhibit blue shifting of 3-4 cm^{-1} relative to those of the neat material, and the relative intensities of the bands for the adsorbed species are markedly changed.

The major vibrational modes observed for isoquinoline are listed in Table II. The assignments made by Wait et al. (22) are also included. These authors made their assignments from considerations of the higher $D_{\infty h}$ symmetry parent species, instead of the C_s symmetry group; they demonstrated that the assignments arising from this representation are reasonable.

Table II. Assignments of the Major Infrared Bands for Isoquinoline at the Mercury/water Interface

Band	Assignment	Symmetry C_s	Symmetry D_{2h}	In- or out-of-plane
1628	ν_8	A′	B_{3g}	In
1589	ν_9	A′	A_g	In
1575	$\nu_{26} + \nu_{38}$	A′+ A″	$A_g + B_{3u}$	In + Out
1500	ν_{11}	A′	B_{2u}	In
1462	ν_{12}	A′	A_g	In
1435	ν_{13}	A′	B_{3g}	In
1380	ν_{14}	A′	A_g	In
1376	ν_{15}	A′	B_{1u}	In
1273	ν_{17}	A′	B_{1u}	In
1257	ν_{18}	A′	B_{2u}	In
1215	$\nu_{36} + \nu_{42}$	A″+ A″	$A_u + B_{1g}$	Out + Out
1180	ν_{19}	A′	B_{3g}	In

The difference spectra show a complete absence of bands with negative absorbances (Figure 7). This can be explained if the vibrational frequencies of the bands do not shift with changes in the electrode potential, and if they are adsorbed over the entire potential region investigated. This is consistent with results of

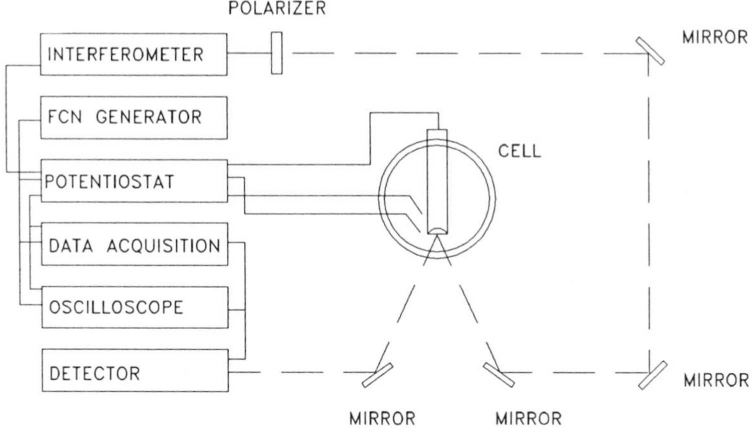

Figure 5. Schematic representation of the SNIFTIRS instrumentation.

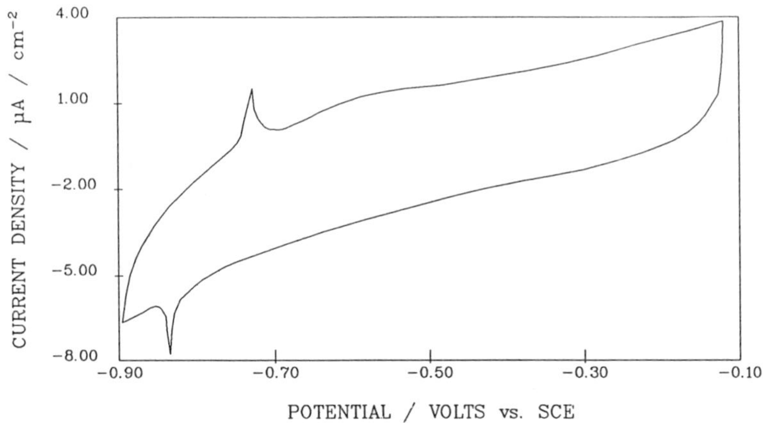

Figure 6. Cyclic voltammogram for a 2.1 x 10^{-2} mol-dm^{-3} isoquinoline / 0.5 mol-dm^{-3} Na_2SO_4 solution at a mercury electrode at 10 mV-s^{-1}.

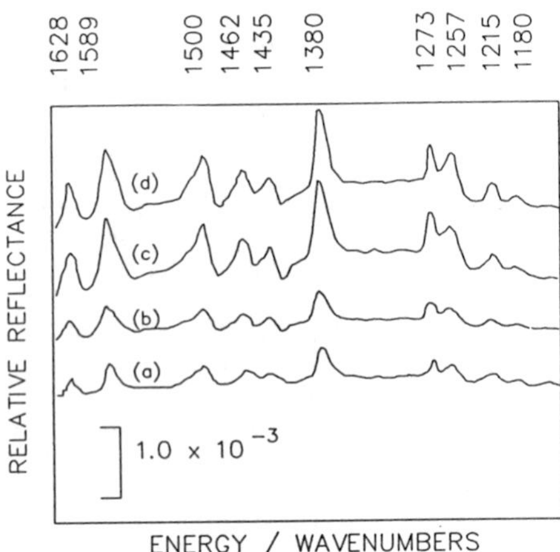

Figure 7. SNIFTIRS difference spectra for 1.3 x 10^{-2} mol-dm^{-3} isoquinoline solution at a mercury electrode. Reference potential was 0.0 V vs. SCE, sample potential (a) -0.60V, (b) -0.75V, (c) -0.80V, and (d) -0.9V vs. SCE.

electrochemical double layer experiments. The spectra can now be interpreted as relatively simple changes in the absorption of infrared radiation. The intensities of the bands are markedly potential dependent; an especially large change in the intensities is observed at potentials where the orientation changes to the vertical 6,7 configuration.

Two types of mechanisms have been proposed to explain the potential dependent shift of vibrational frequencies. The first involves molecular orbital arguments; the second is based on arguments for interactions between the electric field across the double layer and the polarizable electrons of the adsorbed molecule (an electrochemical Stark effect). In the molecular orbital mechanism, electrons can be donated to empty metal orbitals through σ-type overlap with filled ligand orbitals of the appropriate symmetry. The metal can "back" donate electrons from filled d-orbitals to empty π^* antibonding orbitals on the adsorbate. When a molecule is adsorbed on a clean uncharged metal surface, its vibrational frequency may either increase or decrease from the frequency of the unadsorbed molecule depending upon the relative contributions of the σ- and π-bonding interactions. If the π-bonding interaction is dominant the frequency will decrease; conversely, the frequency will increase if the σ-bonding interaction is dominant. When the charge on the electrode is made negative, the bond is weakened due to donation of charge from the metal into adsorbate π^* orbitals and the band frequency shifts to lower wavenumber. When the charge on the metal is made positive a shift to higher frequency occurs. At a mercury electrode, however, there are no p- or d-electrons available to participate in a back-bonding interaction. The observation of potential dependent frequency shifts are therefore not expected according to this model.

The electric field mechanism involves coupling of the electric field across the double layer with highly polarizable electrons of the adsorbate. According to the Gouy-Chapman-Stern model, for high concentrations of supporting electrolytes, most of the potential drop will occur within the first 5-10Å of the electrode surface, and the drop will be approximately linear with distance. When a layer of adsorbed species is present, it can act as a dielectric across which the greatest portion of the potential drop will occur. Electric fields on the order of 10^9 V m^{-1} can exist in this region. Interaction of this electric field with the dipole moment of the molecule leads to changes in the vibrational frequency of the molecule.

The absorbances in Figure 7 have positive values. This indicates that the interaction with infrared radiation is strongest when the isoquinoline molecules are lying flat on the electrode surface (in these difference spectra, positive values of absorbance denote stronger absorption at the positive potential, i.e. potentials where the isoquinoline is adsorbed in the flat configuration). This is an opposite result than that expected from the surface selection rule, and suggests that there is a strong field-induced absorption for the in-plane modes in this configuration, similar to that observed in previous work for pyrene adsorbed on platinum (18).

A closer examination of the SNIFTIRS difference spectra shows that there are marked differences in the changes in intensity of the in-plane and out-of-plane vibrational modes of the adsorbed

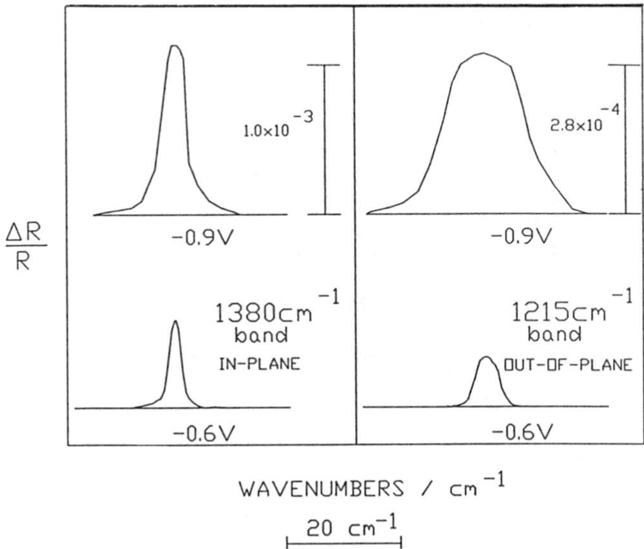

Figure 8. Expanded section of Figure 5 showing examples of differences in the magnitude of integrated area for two absorption bands with the same changes in electrode potential.

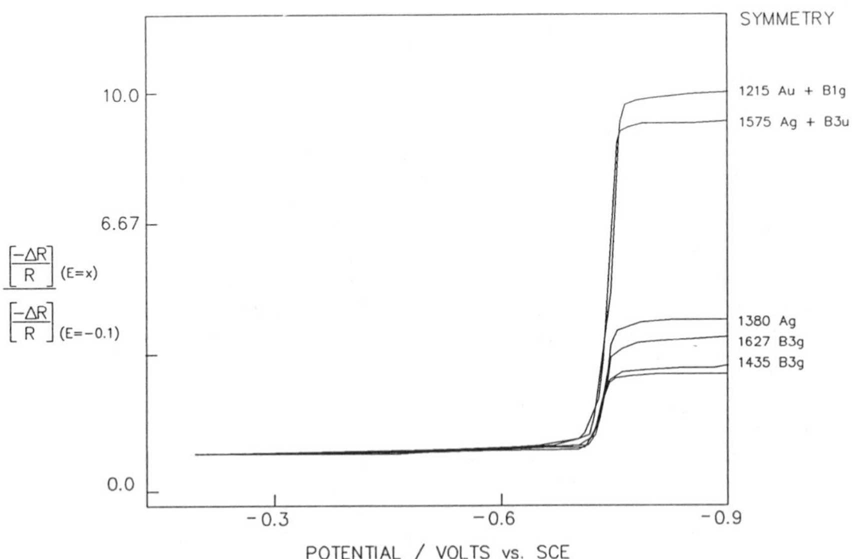

Figure 9. Plot of the normalized intensities of the bands observed in the SNIFTIRS difference spectra at a mercury electrode in 1.3 x 10^{-2} mol-dm^{-3} isoquinoline solution vs. those at the sample potential (see text). Reference potential = 0.0V vs SCE.

isoquinoline with potential (Figure 8). In the figure, the potentials for the difference spectra are -0.6 and -0.9V. The change in area for the 1380cm^{-1} in- plane mode has increased by a factor of ≈4 , whereas that of the 1215cm^{-1} out-of-plane mode has increased by a factor of ≈10. The normalized intensities (against their intensity at -0.1V vs. SCE) of the vibrational bands shown in Figure 7 were plotted against potential (Figure 9). Since the bands in Figure 7 have positive values of absorbance, the positive vertical axis in Figure 9 represents a decrease in the amount of infrared radiation absorbed. The Figure clearly shows that the amount of radiation absorbed by the out-of-plane vibrational modes decreases up to a factor of 10 as the potential is changed from -0.1 to -1.0V, whereas the in-plane vibrational modes only change by a factor of 3-4 over the same potential range. (We point out at this point that solution soluble isoquinoline would not exhibit this effect).

The explanation of the trends seen in Figure 9 is that the out-of-plane vibrational modes have dipole derivative changes perpendicular to the metal surface when the molecules are lying flat on the surface. Absorption of radiation by the surface selection rule is thus allowed. When the orientation changes to the 6,7 configuration, absorption by these modes, which are now parallel to the surface is forbidden, as is any field induced interaction since most of the molecule lies outside of most of the field gradient. A large decrease in absorption (large increase in positive absorbance in the difference spectra) is thus expected and observed. In the case of the in-plane modes, the high electric field in the double layer leads to strong field induced absorption for the flatly adsorbed isoquinoline as expected. When reorienting to the 6,7 configuration, the modes become allowed by the surface selection rule. A smaller decrease in absorption is then observed.

CONCLUSIONS

Subtractively normalized interfacial Fourier transform infrared spectroscopy has been used to follow the reorientations of isoquinoline molecules adsorbed at a mercury electrode. Field induced infrared absorption is a major contribution to the intensities of the vibrational band structure of aromatic organic molecules adsorbed on mercury. Adsorbed isoquinoline was observed to go through an abrupt reorientation at potentials more negative than about -0.73 V vs SCE (the actual transition potential being dependent on the bulk solution concentration) to the vertical 6,7 position.

There was a lack of any potential dependence in the frequencies of the vibrational modes of the adsorbed isoquinoline. The explanation of this may be the fact that mercury has no available vacant p- or d-orbitals into which back-bonding (which is observed for adsorbates on d metals) can occur.

ACKNOWLEDGMENTS
We thank the Office of Naval Research for support of this work.

LITERARURE CITED

1. Bewick, A.; Robinson, J. J. Electroanal. Chem. 1975, 60 ,163.
2. Bewick, A.; Robinson, J. Surf. Sci. 1976, 55, 349.
3. Adzic, R.; Cahan, B.; Yeager, E J. Chem. Phys. 1973, 58, 1780.

4. Pons, S. Ph.D. thesis, University of Southampton, England, 1979.
5. Bewick, A.; Mellor, J. M.; Pons, S. Electrochim. Acta 1980, 25, 931.
6. Golden, W. G.; Dunn, D. S.; Overend, J. J. Catal. 71 (1981) 395.
7. Pons, S. J. Electroanal. Chem. 1983, 150, 495.
8. Pons, S.; Bewick, A. Langmuir 1985, 1, 141.
9. Buess-Herman, C.; Vanlaethem-Meuree, N,; Quarin, G.; Gierst, L. J. Electroanal. Chem. 123 (1981) 21.
10. Humphreys, M. W.; Parsons, R. J.; Electroanal. Chem. 1977, 82, 369.
11. Greef, R. In ref. 9.
12. Quarin, G.; Buess-Herman, C.; Gierst, L.; J. Electroanal. Chem. 1981, 123, 35.
13. Buess-Herman, C. J. Electroanal. Chem. 1985, 186, 27.
14. Buess-Herman, C. J. Electroanal. Chem. 1985, 186, 41.
15. Buess-Herman, C.; Franck, C.; Gierst, L. Electrochim. Acta 1986, 31, 965.
16. Crawford, M.F.; MacDonald, R. E. Can. J. Phys. 1958, 36, 1022.
17. Korzeniewski, C.; Shirts, R.B.; Pons, S. J. Phys. Chem. 1985, 89, 2297.
18. Korzeniewski, C.; Pons, S. Langmuir 2 (1986) 468.
19. Korzeniewski, C.; University of Utah Ph.D Thesis (1987).
20. Foley, J. K.; Korzeniewski, C.; Daschbach, J.; Pons, S. In Electroanalytical Chemistry; Bard, A. J., Ed.; Marcel Dekker: New York, 1986; Vol.14, p 309.
21. A. Bewick, private communication.
22. Wait, Jr, S.C.; McNerney, J. C. J. Mol. Spectrosc. 1970, 34, 56.

RECEIVED May 17, 1988

Chapter 24

Fourier Transform Infrared Spectroscopic Investigation of Adsorbed Intermediates in Electrochemical Reactions

J. O'M. Bockris and K. Chandrasekaran[1]

Surface Electrochemistry Laboratory, Texas A&M University, College Station, TX 77843–3255

The adsorption of electrochemical intermediates during the hydrogen evolution reaction, the reduction of CO_2 and the oxidation of methanol has been investigated using Fourier Transform Infrared Spectroscopy. Two Fe-H vibrations at 2060 and $980cm^{-1}$ were identified. A slope of F/6RT is obtained for a plot of θ vs. η, if one assumes that the area under the peak at $2060cm^{-1}$ is proportional to the hydrogen coverage, θ_H. The mechanism of the electrochemical reduction of CO_2 in acetonitrile on a platinum electrode has been shown to involve adsorbed CO_2 and adsorbed $\dot{CO_2}^-$ concentrations of the adsorbed species. A new mechanism is proposed for the oxidation of methanol on platinum in aqueous sulfuric acid accounting for the observed structure of the electrochemical intermediates.

The infrared spectroscopy of electrodes was first introduced by Hansen (1), Kuwana and Osteryoung (2). However, the early methods were internal reflectance methods. Bewick et al. developed direct infrared methods (3-5) and Pons et al. developed the Fourier transform version thereof (6,7). The use of polarization modulation in the electrochemical situation was introduced by Habib and Bockris in 1985 (8).

These several techniques for the solid-solution interface give different kinds of information. However, the one which gives most information about the nature of entities on the surface, and potentially near the surface, is fourier transform IR spectroscopy, which is not restricted to a particular metal, or, indeed, to the type of substrate (except that this must be reflecting).

ASPECTS OF THE FTIR TECHNIQUE IN SOLUTION

One of the most important parts of the FTIR technique is the movable electrode in order to allow proper diffusion of species to and fro

[1]Current address: E. I. du Pont de Nemours and Company, P.O. Box 505, Towanda, PA 18848–0505

0097–6156/88/0378–0351$06.00/0

from the electrode surface and the ability to reestablish the electrode to this position. Another aspect of the positioning of the electrode near the cell arises because of the danger that current lines will not fall uniformly onto the electrode.

The basic advantage of FTIR is the overcoming of noise. In order of magnitude terms, the solution before the electrode is about 1μm in thickness in a 0.1 M solution of a solute, the adsorption of which it is desired to measure, and this layer will thus contain about 10^{-8} moles per cm^{-2}. At 10% coverage, the solution would contain about 10^{-10} moles, and therefore the signal-to-noise ratio would be expected to be on the order of 0.01. There are circumstances when the signal-to-noise ratio may be as little as 10^{-4}. This results due to the use of signal averaging. Such low signal-to-noise ratios are commonly obtained with non-aqueous electrolytes which eliminate the introduction of noise from the broad water bands.

To get over this difficulty, FTIR uses the basic principle of an interferogram. A source of IR light (containing a broad band of IR frequencies) is incident upon a fixed mirror, but passes through an optical device, the beam splitter at which about half the light is reflected and half allowed to pass through shown in Figure 1a (9).

After striking the fixed mirror, the light is reflected back to the beam splitter, and part of this light then interferes with light returning from the moving mirror. When the light reflected from the fixed mirror meets that from the moving mirror, a number of extinctions and augmentations of light occur, and the detailed nature of these depend on the position of the mirror at a given instance.

The degree to which augmentation or destruction occurs when the two light beams meet depends on the wavelengths in the beam and the distance of displacement of the mirror at a given instance. As the mirror sweeps to and fro, certain (changing) frequencies are augmented and transferred to the electrode where, after absorptive interaction with the adsorbed material, and reflection, the light is analyzed in respect to intensity.

When the signal-to-noise ratio is poor, repetition of the signal means that minor peaks which are merely due to fluctuations and errors are not repeated and therefore not enhanced, only those peaks which are repeated with every sweep of the mirror (i.e., peaks representing absorbed entirely), are finally left in the information.

The final step in obtaining the spectrum by the FTIR method is turning back the data obtained as a result of the repetitive interference action of the moving mirror into an intensity wavelength line. It is here that Fourier Transform mathematics is utilized. It is the signal intensity that is stored in a digital representation of the interferogram. This information is then Fourier transformed by the computer into the frequency spectrum.

There are two light sources involved, a white light and a laser source. The white light uses the same moving mirror and therefore makes up a second interferometric system within the spectrometer. When the moving mirror and the fixed mirror of this secondary interferometer are equidistant, a centerburst is produced which is used to trigger the electronic collection system so that each scan begins when the moving mirror is at the same position, known as zero path difference. The laser source being monochromatic, has a known frequency and therefore can be used to digitize the collected data.

The spectrum obtained in the way described so far is, however, not sufficiently sensitive, and two methods are used, then, to obtain results which may be utilized for the investigation of surface phenomena in electrochemistry.

The SNIFTIRS approach. The acronym SNIFTIRS means Subtractively Normalized Interfacial Fourier Transform Infrared Spectroscopy. The basic concept of this method involves the fact that the raw data obtained directly from the Fourier Transform process contain components which are undesirable. Firstly, there is material in the solution which may have affected the spectrum. Secondly, unwanted information on certain material on the electrode (adsorbed water, for example) is best eliminated.

In an ideal case, it is possible to find (perhaps from other methods, e.g., a voltammogram) the potential region of adsorption. If it is then possible to obtain a spectrum (using parallel polarized light) this spectrum is subtracted from the spectrum obtained in the potential region in which the adsorption is known to occur. The resulting intensity-wavelength line (free from intensities of material in solution, or from bonds on the electrode which do not change with potential) is the SNIFTIRS spectrum.

The SNIFTIRS equations used for working out the results, also shown below. The SNIFTIRS spectrum is noisy, and suffers from the fact that it is sometimes not possible to designate a region in which there is no adsorption.

$$A \approx \frac{\Delta R}{R_o} = \left(\frac{I - I_{ref}}{I_{ref}}\right) \tag{1}$$

$$A \approx \left(\frac{(I_p)_V - (I_p)_R}{(I_p)_R}\right) \tag{2}$$

Polarization modulation. Figure 1b (9) shows relevant properties of the interaction of the two principal modes of polarization of light with the surface. This is different (as was first shown by Greenler (10).) depending on whether one is utilizing the parallel or the vertical plane of polarization. In respect to the vertical plane of polarization (the left-hand diagram in Figure 1b), the electrostatic components of light which arise when the light interacts with the surface are 180° out of phase, so the reflected light contains no information from the surface, only from the solution.

On the other hand, with the parallel plane of polarization the result of the light interacting with the surface is to produce components which add to each other; thus, the reflected beam contains information from the surface as well as that from the solution.

The statements made above are only true when the angle of incidence of this light is nearer to 90°. An instant angle of about 85° is used.

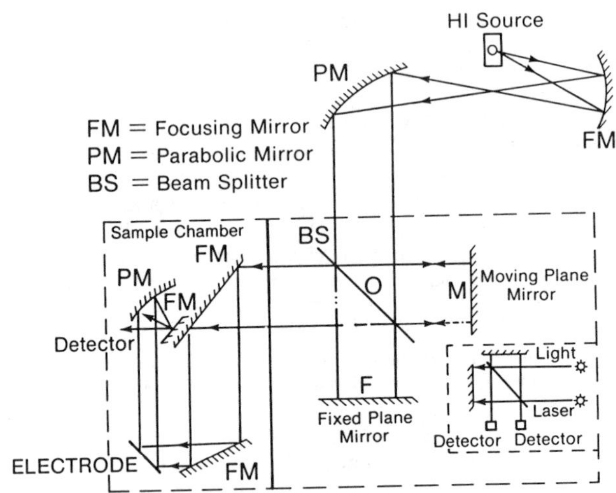

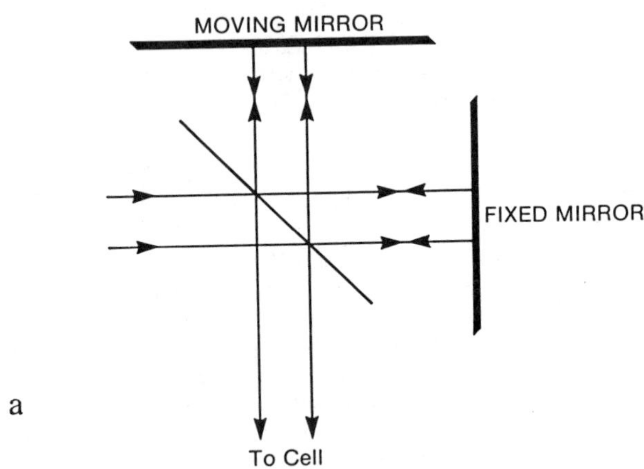

Figure 1a. The optical arrangements of an FT-IR Spectrophotometer with reflectance attachments in the sample chamber for electrochemical experiments. (Reproduced with permission from Ref. 9. Copyright 1984 Elsevier.)

b **REFLECTED LIGHT**

PHASE CHANGE OF REFLECTED LIGHT DEPENDS UPON

1. ANGLE OF INCIDENCE
2. STATE OF POLARIZATION

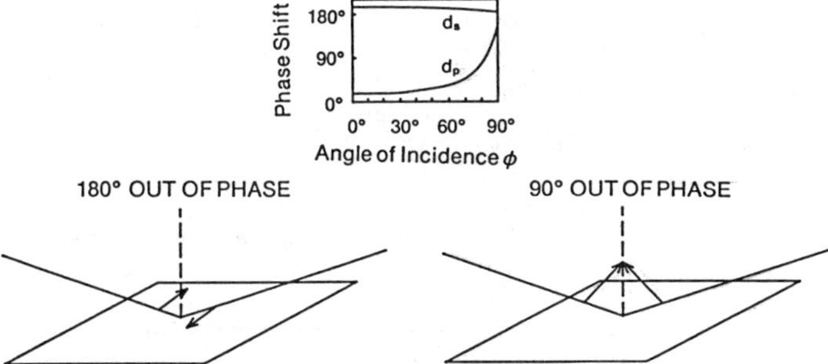

Figure 1b. Schematic diagram of polarized light used with the polarization modulation technique. (Reproduced with permission from Ref. 9. Copyright 1984 Elsevier.)

It can be readily seen from the above discussion that a simplistic version of the polarization modulation method for using Fourier Transform Infrared method would be to say that if one subtracted the message obtained from the vertical light from that obtained from the parallel light, the result will be information from the surface only. The relevant equation is shown below.

$$\text{Absorbance} \approx -\frac{\Delta R}{R} \approx 2\left(\frac{I_p - I_s}{I_p + I_s}\right) \qquad (3)$$

The statements made hitherto are all based upon Greenler's paper. If the parallel light interacts with surface and the solution, but the vertical light only with the solution. In the case of adsorption from the gas phase, the adsorbed phase is sharp and consists essentially only of molecules actually in contact with the surface. In electrochemical situations, however, substantial amounts of "absorbed" solute are in the layer near the electrode. A careful examination of the Greenler paper shows that the net signal from the parallel and vertical components of the light does carry information from the solution phase as well as from the electrical phase.

For this reason, Chandrasekaran and Bockris (11) carried out an examination of the extent to which this fact introduced an error into the electrochemical Fourier Transform measurements of surface occupancy, an error which perhaps would mean that the information contained contributions from material in the solution.

A calculation which can be made to show essentially ΔR/R for the parallel and vertical components as a function of distance from the electrode. The calculations involve separating the parallel and perpendicular components for each phase utilizing Fresnel coefficients.

Figure 2a shows the result of these calculations. The values obtained are dependent upon the assumptions made, in particular the concentration of the solution for which the calculations are carried out, and the coverage on the electrode. The situation is worse (maximum interference from the solution) when the coverage is small and the concentration of the solution high, and then 60% of the information can come from the solution. In more usual circumstances, however, when the solution concentration is medium or low, and the electrode concentration is medium or high, more than 90% of the information does come from the electrode surface itself (Figure 2b). The details are given elsewhere (11).

The conclusions apply to the information before subtracting the spectra at the no adsorption condition. The upshot of all these considerations is that one has to be careful when using FTIR spectroscopy to see that the information is indeed surface information. One of the most encouraging ways of doing this is to look at the frequency of a peak as a function of electrode potential. If the information is truly coming from adsorbed substances, there will usually be a slight variation of the peak with potential. However, this is only confirmatory, and not a necessary characteristic of the information.

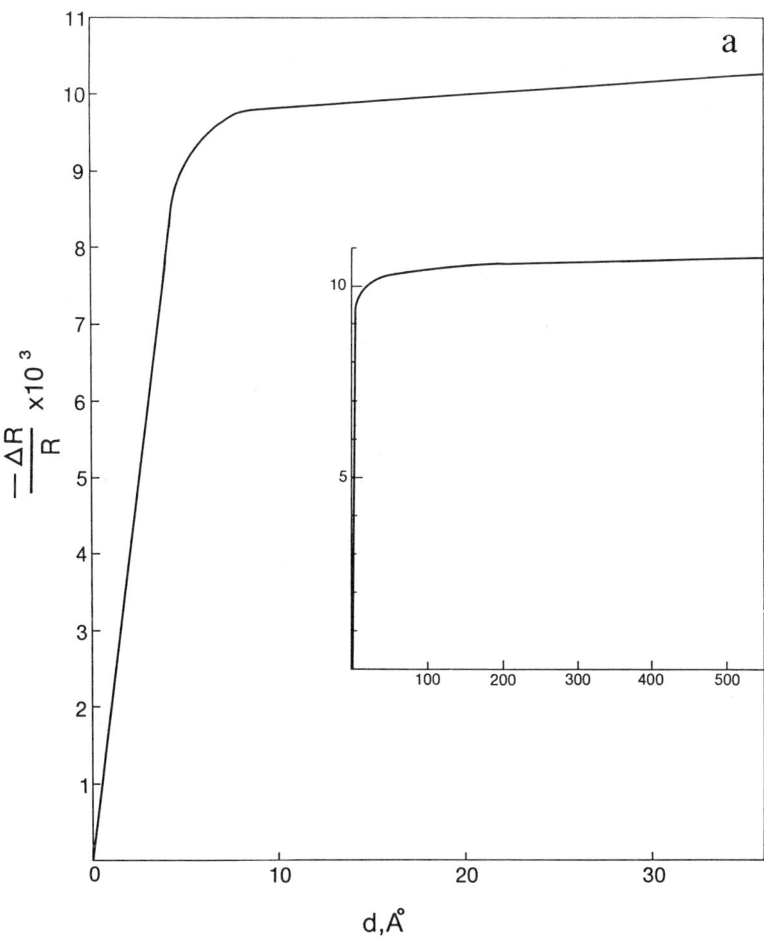

Figure 2a. Absorbance for the surface and solution as a function of distance at a surface coverage of 0.5. (Reproduced with permission from Ref. 11. Copyright 1987 North Holland.)

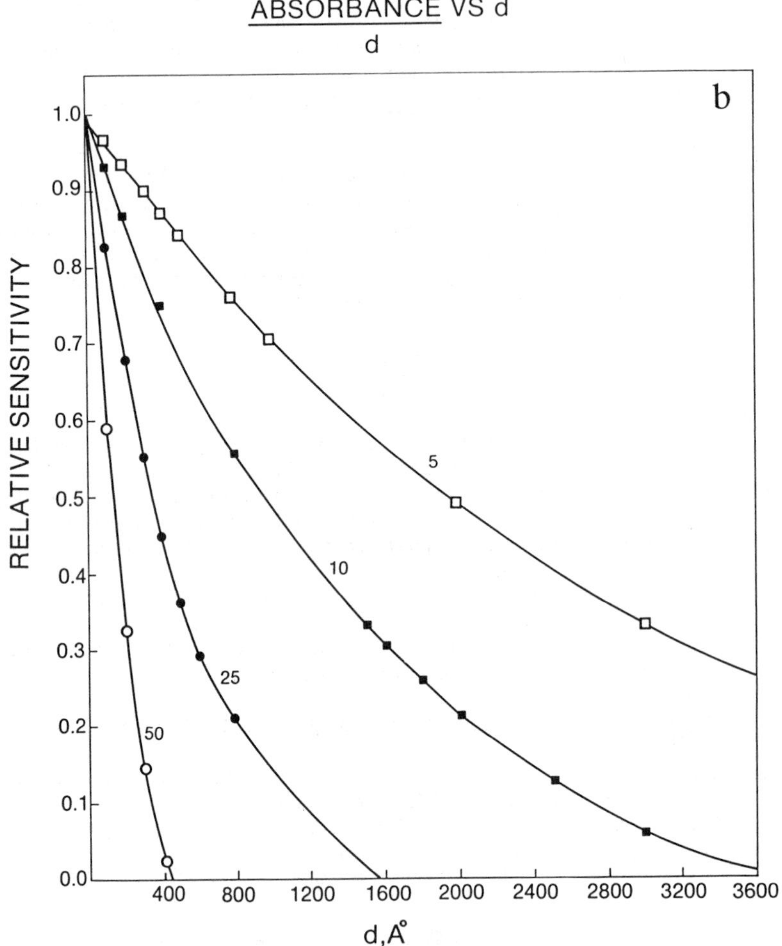

Figure 2b. Relative absorbance as a function of thickness of the film. Relative absorbance $(\Delta R/R)/d$; values normalized to unity at the surface. (Reproduced with permission from Ref. 11. Copyright 1987 North Holland.)

THE SURFACE CONCENTRATION OF HYDROGEN ON IRON DURING HYDROGEN EVOLUTION

One of the questions in the mechanism of hydrogen evolution, and ancillary topics connected with the surface hydrogen involved in corrosion, is the dependence of the surface hydrogen upon overpotential. This topic has importance not only because of the evidence it gives on the mechanism of hydrogen evolution, but also because it relates to the embrittlement aspects of hydrogen evolved during corrosion (12).

Recently, Carbajal et al. (13) in the Texas A&M Surface Electrochemistry Laboratory have been able to show two bonds for FeH, the first (at about 2060 wave numbers) is due to symmetric stretching vibration and the second (at about 980 wave numbers) to the asymmetric stretching vibration. The basic results are shown in Figure 3, where the coverage is plotted against overpotential.

From these results, it turns out that:

$$\frac{\partial \log\theta}{\partial V} \approx \frac{F}{5.6RT} \qquad (4)$$

The theoretical result for a slow discharge mechanism followed by a fast recombination reaction is F/4RT for this ratio. The discrepancy is not too serious, because the predicted value depends upon an observed Tafel slope of 2RT/F and the value obtained on iron is about twice this, which would correspond to the lower value for, $\partial\log\theta/\partial V$ observed on Fe.

This result represents the first use of FTIR measurements to obtain information about the hydrogen evolution reaction on iron. It also represents one of the first uses of FTIR to study the mechanism of the electrode kinetic reaction (14).

SURFACE RADICALS IN THE PHOTOELECTROCHEMICAL REDUCTION OF CO_2

Information on the mechanism of the reduction of CO_2 can be made with great ease when using FTIR spectroscopy.

Thus, the mechanism which has been proposed (15,16) is given by the following equations.

$$CO_{2\,aq} \rightleftharpoons CO_{2\,ads} \qquad (5)$$

$$CO_{2\,ads} + e^- \rightleftharpoons CO_{2\,ads}^- \qquad (6)$$

$$CO_{2\,ads}^- + H_2O + e^- \rightarrow HCOO^- + OH^- \qquad (7)$$

There has hitherto been no direct evidence of the presence of adsorbed CO_2^-. Some spectroscopic results are shown in Figure 4a. The frequency of 1670 (reactions carried out in acetonitrile) doubtless represents the CO_2^- radical. The IR spectrum of this radical has been recorded at -190°C, and it has a sharp maximum at 1671 cm^{-1} (17).

It is noteworthy that the adsorbed concentration of CO_2^- radical at various electrode potentials shows an increase in the CO_2^- in the more negative direction. The mechanistic significance of this observation is indicated below. Finally, in Figure 4b it

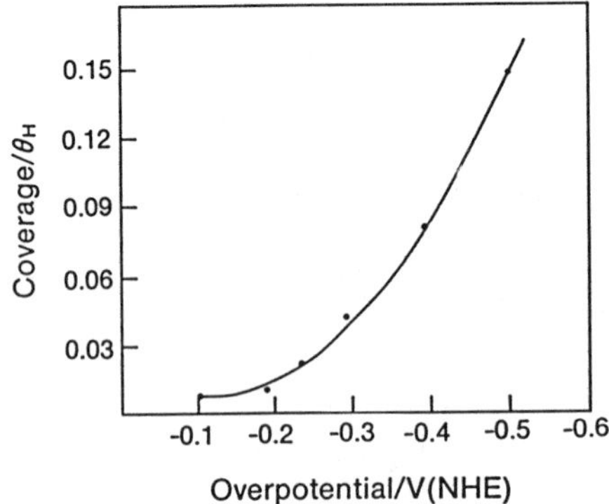

Figure 3. Coverage-overpotential relation for hydrogen adsorbed on iron corrected for reference potential. (Reproduced with permission from Ref. 13. Copyright 1987 The Electrochemical Society, Inc.)

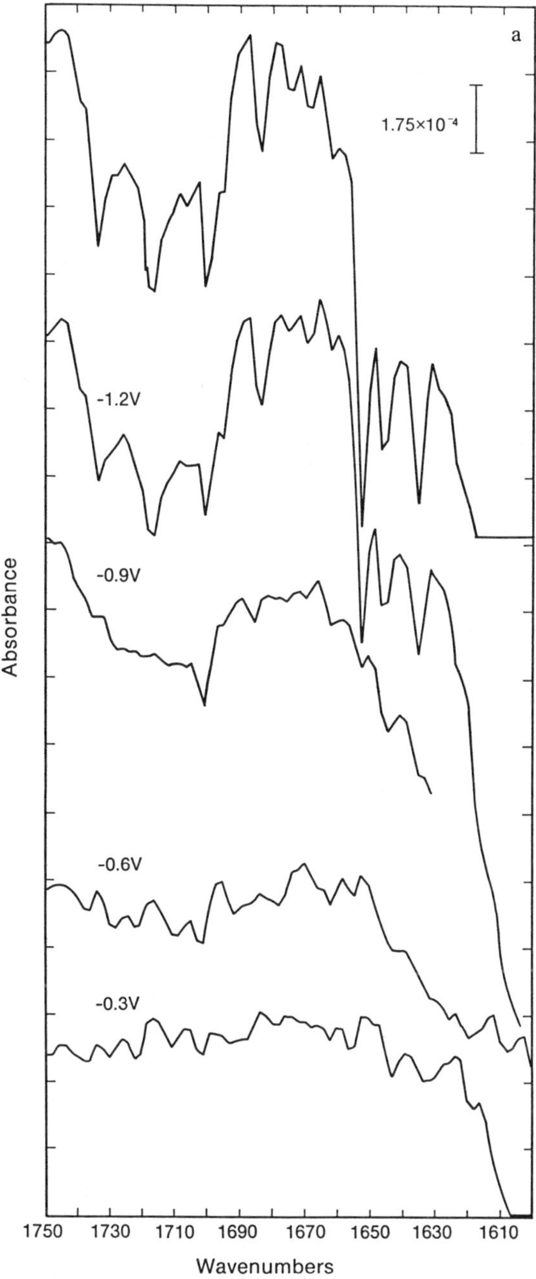

Figure 4a. Absorption spectrum of CO_2^- radical adsorbed on platinum in acetonitrile containing 0.4M $LiClO_4$. (Reproduced with permission from Ref. 11. Copyright 1987 North Holland.)

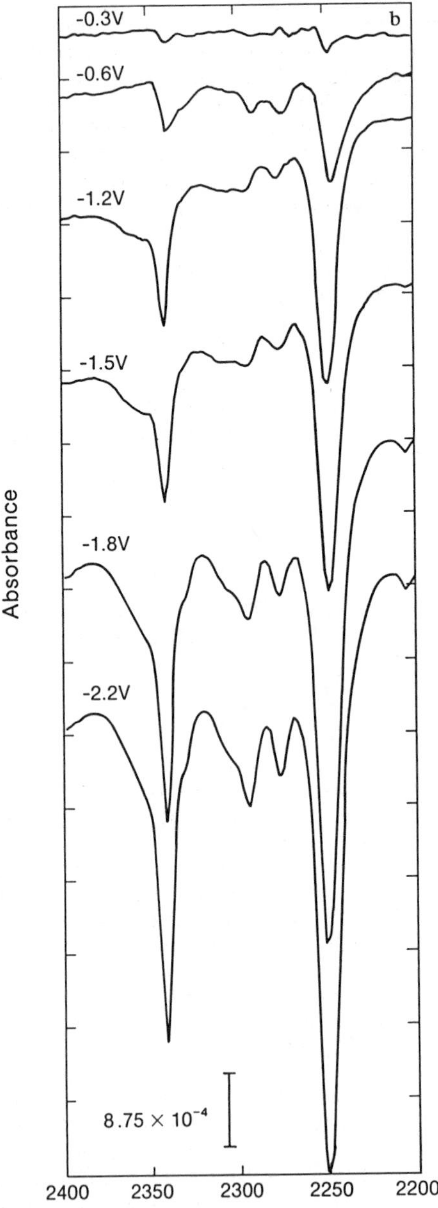

Figure 4b. Absorption spectra of adsorbed CO_2 (2340 cm^{-1}) and CH_3CN (2250 cm^{-1}) on platinum in acetonitrile containing 0.4 M $LiClO_4$. (Reproduced with permission from Ref. 11. Copyright 1987 North Holland.)

can be seen that the relative concentrations of adsorbed acetonitrile and CO_2 on platinum decrease with potential.

The information obtained can be used to give interesting information upon the CO_2 reduction mechanism. Because the radical anion increases in concentration in the negative direction, it cannot be in equilibrium with the electrode. The increase in anion concentration at cathodic potentials may, however, be explained if CO_2^- is formed as an intermediate radical. Thus from equations 5-7

$$\theta_{CO_2} = k_5 P_{CO_2} \tag{8}$$

and with equation 6 in equilibrium

$$\theta_{CO_2^-} = k_6 k_5 P_{CO_2} e^{-VF/RT} \tag{9}$$

where V is the electrode potential, k_5 and k_6 are equilibrium constants for equations 5 and 6, respectively. Thus, $\theta_{CO_2^-}$ increases with increasing negative value of V. With equation 7 rate determining, the current density, i,

$$i = 2Fk_7 \theta_{CO_2} e^{-\beta VF/RT} \tag{10}$$

$$= 2Fk_5 k_6 k_7 P_{CO_2} e^{-(1+\beta)VF/RT} \tag{11}$$

Thus, it is possible to obtain information on the rate determining step by identifying the radical and following the surface coverage with potential.

OXIDATION OF METHANOL TO CO_2

The oxidation of methanol was carried out at a potential range of 0.0 to +0.8V. The spectrum at a potential of 0.3V is shown in Figure 5a, while the spectrum at 0.8V is shown in Figure 5b.

Some interesting changes are shown with potential. Broad water peaks are observed at the potential of 0.3V, and these are considerably reduced at a potential of 0.8V. However, the most interesting change is that in the potential region at more negative than 0.2, the spectrum at about 2150 is equivalent to that of the -C≡O.

By the potential corresponding to about 0.6 V, the -C≡O spectrum has transformed to one corresponding to about 1750 wave numbers, which is typical of >C=O. Willsau et al (18) contends that COH is the intermediate in this reaction, however C-H vibrations were not detected for the adsorbed species. Further, three electrons per site are required for oxidation of this species to carbon dioxide whereas the experimentally observed value is 1.2-1.5.

The plot of the various reactants is most revealing. It is compared in Figure 6 with the radiotracer and potentiodynamic methods. It will be seen that there is a rough kind of agreement between the data here obtained and the radiotracer method data, though the potentiodynamic method is evidently less sensitive. On the other hand, the FTIR data distinguishes between the two, CO in the triple bond form and CO in the double bond form.

The most important information this gives us is on the mechanism of oxidation of CO_2.

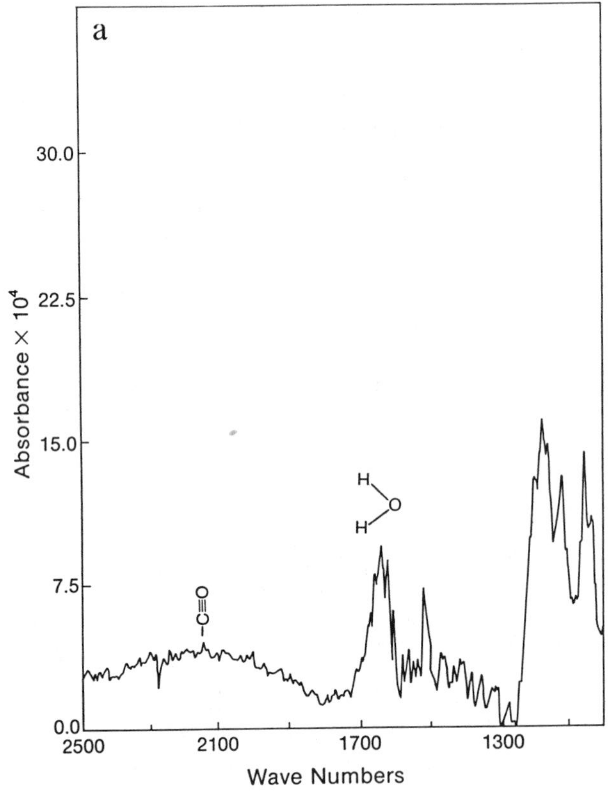

Figure 5a. Adsorption spectrum of -C≡O and H_2O on platinum electrode.

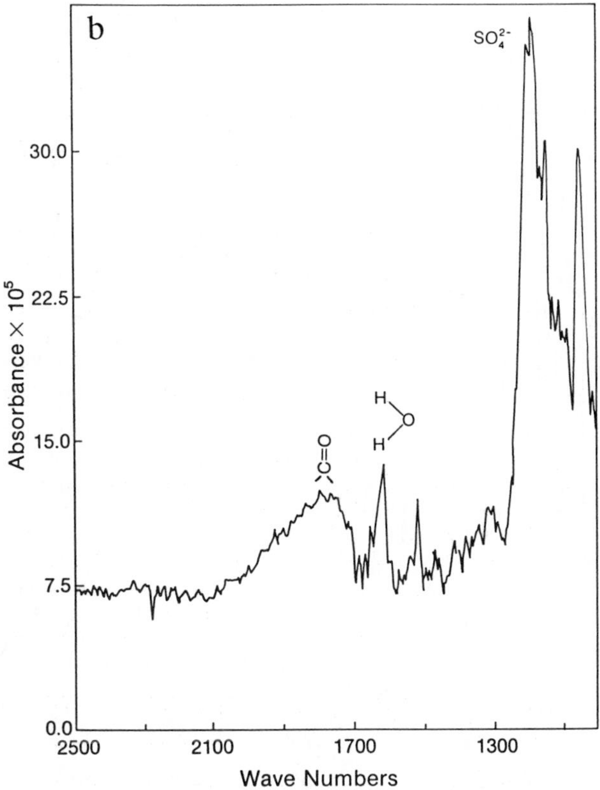

Figure 5b. Adsorption spectrum of >C=O, H_2O and SO_4^{2-} on platinum electrode.

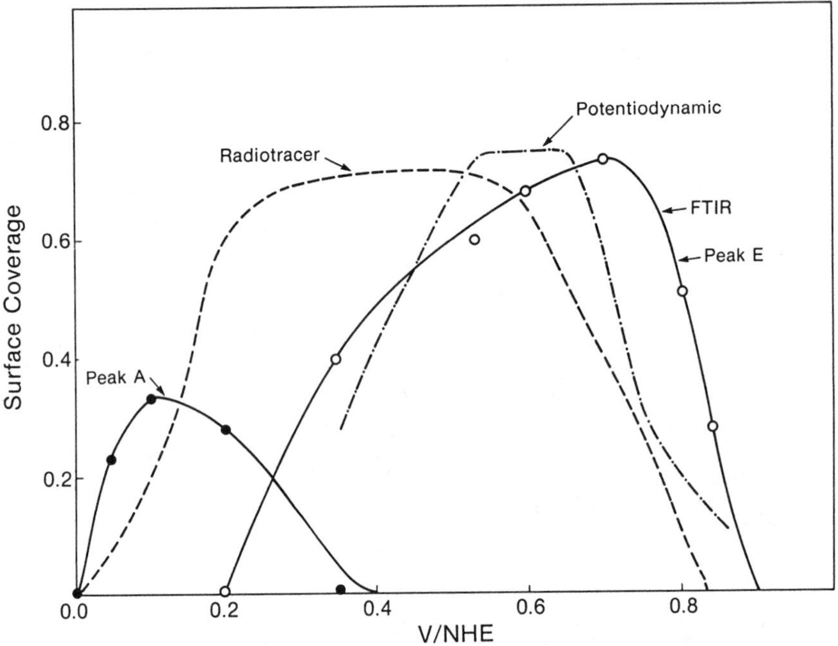

Figure 6. Comparison of various *in situ* techniques for the chemisorption of methanol on platinum.

One possible mechanism that fits the data herewith revealed by the spectroscopy. It is as follows:

$$CH_3OH \rightleftharpoons CO_{ad} + 4H^+ + 4e^- \quad (12)$$

$$H_2O \rightleftharpoons OH_{ad} + H^+ + e^- \quad (13)$$

$$CO_{ad} + OH_{ad} \rightleftharpoons COOH \qquad \textit{RDS} \quad (14)$$

$$COOH \rightleftharpoons CO_2 + H^+ + e^- \quad (15)$$

These various reactions can then be portrayed from the point of view of electrochemical parameters as follows:

$$\theta_{CO} = k_{12}\exp(4VF/RT) \quad (16)$$

$$\theta_{OH} = k_{13}\exp(VF/RT) \quad (17)$$

$$i = Fk_{14}\theta_{CO}\theta_{OH} \quad (18)$$

$$i = Fk_{14}\theta_{CO}k_{13}\exp(VF/RT) \quad (19)$$

(Langmuir) slope = 60mV

$$ln\ i = ln\ Fk_{14}\theta_{CO} - \gamma ln\ k_{13} - \gamma ln \frac{C_{H^+}}{C_{H_2O}} + \gamma VF/RT \quad (20)$$

(Tempkin) slope = 120mV

The rate-determining step is the surface combination of CO_{ads} with OH to give a carboxylic group. The remarkable slope of 120 mv, so easily confused with a discharge mechanism, can be interpreted if these adsorptions are Tempkin in type, in terms of the mechanism equations 12-15, if the θ dependence is relatively small (equation 16).

Thus, these studies again stress the multiple possibilities which arise from the use of FTIR spectroscopy and the analysis of reaction rates.

FUTURE APPLICATIONS OF FTIR SPECTROSCOPY

It has been shown in this paper particularly that the FTIR spectroscopy can identify radicals and chemical reactions, and by their potential and concentration dependence give considerable information upon the mechanism of reactions and the detailed mechanism of electrochemical reactions, including their rate-determining step. The analysis of intermediate radicals has always been a need in electrochemical research, and is clearly now here.

The absolute determination of coverage is much more difficult. The formalism is well known, i.e., it is possible to go from the intensity of a spectral ion to the concentration material on the surface, but this can only be worked out if the transition dipole moment is known. A very rough order of magnitude version of this can, however, usually be obtained by analogy (rather than by quantum mechanical calculation) so that somewhat better than an order of magnitude calculation of the coverage of the surface radicals can be given.

Other aspects of FTIR spectroscopy have not yet been put into practice. One could study the kinetics, and in particular the build-up and decay of intermediate radicals. This latter aspect would provide useful information to determine the mechanism of the reaction being studied.

At present time, these studies cannot be carried out, due to the short lifetimes involved, but with more intense sources of light it may be possible to reduce the measurement time enabling the measurement of θ in the millisecond range with FTIR ellipsometry.

ACKNOWLEDGMENTS

The authors wish to thank the Gas Research Institute for financial support and Dr. Kevin Krist for several discussions.

LITERATURE CITED

1. Hansen, W.N. In Advances in Electrochemistry and Electrochemical Engineering; Delahoy, P. and Tobias, C.W., Eds.; Wiley: New York 1973; Vol. 9, p 1.
2. Hansen, W.W., Kuwana, T. and Osteryoung, R.A. Analytical Chemistry 1966, 38, 1809.
3. Bewick, A. and Robinson, J. J. Electroanal. Chem. 1975, 60, 163.
4. Bewick, A., Mellor, J.M. and Pons, S. Electrochim. Acta 1980, 35, 931.
5. Bewick, A. and Pons, S. In Advances in IR and Raman Spectroscopy; Hexter, R.E. and Clarke, R., Eds.; Heyden: London 1984.
6. Pons, S., Davidson, T., Bewick, A. and Schmidt, P.P. J. Electroanal. Chem. 1982, 125, 237.
7. Pons, S., J. Electroanal. Chem. 1983, 150, 495.
8. Habib, M.A. and Bockris, J.O'M. J. Electrochem. Soc. 1985, 132, 108.
9. Habib, M.A. and Bockris, J.O'M. J. Electroanal. Chem. 1984, 180, 287.
10. Greenler, R.B. J. Chem. Phys. 1966, 44, 310.
11. Chandrasekaran, K. and Bockris, J.O'M. Surf. Sci. 1987, 185, 495.
12. Flitt, H.J. and Bockris, J.O'M. Int. J. Hydrogen Energy 1981, 6, 119;
13. Bockris, J.O'M., Carbajal, J.L., Scharifker, B.R. and Chandrasekaran, K. J. Electrochem. Soc. 1987, 134, 1957.
14. Golden, W.G., Kunimatsu, K. and Seki, H. J. Phys. Chem. 1984, 88, 1275.
15. Paik, W., Anderson, T.N. and Eyring, H. Electrochim. Acta 1969, 14, 1217.
16. Amatore, C. and Saveant, J.M. J. Amer. Chem. Soc. 1981, 103, 5021.
17. Hartman, K.O. and Hisatsune, I.C. J. Chem. Phys. 1966, 44, 1913.
18. Willsau, J., Walter, O. and Heitbaum, J. J. Electroanal. Chem. 1985, 185, 163.

RECEIVED May 17, 1988

Chapter 25

Vibrational Spectroscopic Studies of Adsorbate Competition During Carbon Monoxide Adsorption on Platinum Electrodes

René R. Rodriguez[1], Wade J. Tornquist[2], Francis Guillaume, and Gregory L. Griffin[3]

Department of Chemical Engineering and Material Science and Department of Chemistry, University of Minnesota, Minneapolis, MN 55455

We have compared the effect of various nitrile compounds and Sn adatoms on the vibrational frequency and oxidation kinetics of CO adsorbed on Pt electrodes. Both types of species adsorb competitively with CO, leading to a decrease in the amount of adsorbed CO and a shift of ν(CO) to lower frequency. The latter effect is consistent with the decrease in dynamic coupling expected for a uniformly reduced CO coverage. There is no evidence for significant direct interaction between CO and either co-adsorbed species. Adsorbed nitriles are able to block CO from entering its linear configuration when the adsorption step is performed at 0.05 V(SHE). Adsorbed Sn atoms behave in a manner which supports previous models for their catalytic enhancement of CO oxidation rates. Competitive adsorption of CO contributes to a loss of Sn adatoms from the Pt surface during CO oxidation cycles.

Vibrational spectroscopy is the experimentalist's most powerful tool for studying the effects of changes in local environment on individual chemical bonds. Studies of simple adsorbates like CO which have strong characteristic absorption bands have contributed greatly to our understanding of adsorption processes at surfaces (1). As shown here and in other papers in this symposium, recent experimental developments have led to a renewed effort to use the vibrational spectroscopy of adsorbates as a probe for understanding the physical chemistry of metal/electrolyte interfaces.

[1]Current address: Department of Chemistry, University of Idaho, Moscow, ID 83843
[2]Current address: Department of Chemistry, Eastern Michigan University, Ypsilanti, MI 48197
[3]Current address: Department of Chemical Engineering, Louisiana State University, Baton Rouge, LA 70803

0097–6156/88/0378–0369$06.00/0

Overview of CO Adsorption Studies

The adsorption of CO on Pt is perhaps the most throughly studied system using vibrational spectroscopy. Studies have been made using both supported catalysts (2-5) and single crystals (5-10). Sample environments have included gas phase, vacuum, and aqueous solution (11-13). The similarities between many of these results have led to a remarkably unified understanding of CO adsorption phenomena in all three environments. Features which are relevant to further studies of the metal/electrolyte interface are summarized briefly:

Configuration of Adsorbed CO. Under most conditions, CO is preferentially adsorbed in a linear configuration. This means that the CO molecule is bound to a single Pt atom on the electrode surface through a Pt-C bond that is colinear with the CO bond axis. The ν(CO) frequency generally lies in the region 2050-2110 cm^{-1}, slightly below the gas phase value of 2143 cm^{-1}. This decrease is attributed to charge transfer from the metal into the CO molecule. A simple description of the charge transfer process in molecular orbital terms is that electron density is back-donated from metal d-bands into the CO π^* orbital, which is anti-bonding in character (14). The relatively small size of the frequency shift implies that the extent of charge transfer is small. As a result, the C-O bond order is still close to 3.0 for linearly adsorbed molecules.

Molecular CO can also be adsorbed in a bridging configuration. This means that the molecule is bound to the surface through Pt-C bonds with two or more adjacent Pt atoms. The value of ν(CO) in the bridging species is 1900 cm^{-1} or less, which indicates that the C-O bond order is reduced below 2.5. In terms of the back-donation description, this means that a much larger amount of charge is transferred into the CO π^* orbital. An alternate description is that the CO σ molecular orbitals are partially rehybridized to place greater sp^2 vs. sp character on the C and O atoms. This allows more charge to be located in the lone pair orbitals on the O atom. Conditions which lead to formation of the bridging species include high CO coverage, the presence of electron donating species co-adsorbed on the surface, and the application of relatively negative potentials to a Pt electrode in a non-acidic solution. The latter two observations support the concept that formation of bridging CO can be enhanced by increasing the ability of the surface to donate charge to adsorbates.

Frequency Dependence of Adsorbed CO. The exact value of ν(CO) is affected by several factors, including CO coverage, electrode potential, and electrolyte composition. The coverage dependence has been studied primarily under vacuum conditions, using mixtures of vibrationally distinguishable ^{13}CO and ^{12}CO. These studies have shown that the coverage dependence is mainly due to dynamic coupling between neighboring CO molecules (3,15). Recently Severson et. al. showed that dynamic coupling is also responsible for the coverage dependent component of the frequency shift under electrochemical conditions, but with a coupling constant about twice as large as the value observed under vacuum conditions (16).

We have independently observed larger values for the coverage dependence in various electrolytes, but also find evidence that the composition of the electrolyte affects the magnitude of the frequency shift (17).

The potential dependence of ν(CO) has been reported for several electrolytes. Values of ∂(ν(CO))/∂E measured near pH = 1 are typically around +30 cm^{-1}/V (11-13). Recently a shift of +80 cm^{-1}/V has been reported for CO adsorbed on Au electrodes in NaOH solution (18). The positive sign of these coefficients indicates that ν(CO) increases as the potential is made more anodic. The origin of the potential dependence is still being discussed. Russell et. al. originally proposed that the shift is due to an increase in the extent of inter-molecular charge transfer between the metal and the adsorbed CO as the potential is increased (12). In contrast, Lambert has proposed that the shift is due to the Stark effect exerted by the electric field at the inner Helmholtz plane on the intra-molecular charge distribution of the adsorbed CO molecule (19).

The influence of electrolyte composition on the CO vibrational spectrum presents a complex problem which has only recently received experimental attention. Early studies using different acid electrolytes reported that the identity of the anion has a measurable effect on the potential dependence of ν(CO) (12), but later results seemed to discount this (13). In contrast, we recently studied CO spectra using several non-acidic electrolytes and observed a very pronounced dependence on pH, especially at low CO coverages (17). In 0.5 M K_2SO_4 at -0.350 V(SCE), the extrapolated zero coverage limit of ν(CO) is 1963 cm^{-1} and the saturation coverage limit is 2062 cm^{-1}. This can be compared to the corresponding limits of 2030 cm^{-1} and 2090 cm^{-1} that are measured in 0.5 M H_2SO_4 at 0.150 V(SCE), and also to the limits of 2065 cm^{-1} and 2100 cm^{-1} that are measured on the same sample under vacuum conditions. These comparisons show that the apparent coverage dependence of ν(CO) increases as a function of sample environment, in the order neutral electrolyte > acid electrolyte > vacuum. To account for the difference in the results for neutral and acid solutions, we proposed that the larger value of the coverage dependence under electrolyte conditions is due to adsorption of positive counterions on the Pt surface atoms that are not occupied by adsorbed CO.

Competitive adsorption. In the experiments which follow, we examine the influence of two other types of co-adsorbed species, namely organic nitrile compounds and Sn adatoms. We selected nitrile compounds because they might be expected to competitively adsorb on the surface with CO without significantly affecting the electronic properties of the surface. It is also possible that the ν(CN) band of the adsorbed nitrile might be observed in the IRRAS spectrum. Nitrile compounds in solution phase have a ν(CN) frequency in the range 2240-2260 cm^{-1}. Sexton et. al. reported that CH_3CN adsorbed on Pt(111) under UHV conditions has a strongly shifted band at 1615 cm^{-1}, which they attributed to CH_3CN adsorbed in a π-bonded configuration (20). This band would lie beyond the low frequency cutoff of the InSb detector on our present apparatus. Loo et. al. have recorded surface enhanced Raman spectra of $CH_2(CN)_2$ and $NC(CH_2)_2CN$ adsorbed on Cu electrodes (21-22). For both molecules, the authors observed two bands in the regions 2200-2300 cm^{-1} and 2080-2120 cm^{-1} that they assigned to non-degenerate ν(CN) modes of free and adsorbed nitrile functional groups, respectively. We can test whether similar bands might be observed on Pt electrodes.

The experiments using Sn adatoms are intended to test for a correlation between the activity of these species as promoters for CO oxidation kinetics and their influence on the CO vibrational spectrum. Watanabe et. al. have proposed an "adatom oxidation" model for the catalytic activity of these adatoms (23). They propose that the function of the Sn adatoms is to catalyze the generation of adsorbed O or OH species at a lower potential than would be required on unpromoted Pt (23). The latter species then react with neighboring adsorbed CO molecules to accomplish the overall oxidation reaction. One implication of this proposed mechanism is that the adsorbed adatom is expected to have little, if any, direct interaction with the adsorbed CO reactant partner. Vibrational spectroscopy can be used to test for such an interaction.

Experimental

Vibrational spectra were recorded using the polarization-modulated infrared reflection absorbance technique (PM-IRRAS). The spectrometer, the electrochemical cell, and the sample preparation and cleaning procedures are all described elsewhere (17). All of the measurements were performed using 0.5 M H_2SO_4 solutions, either with or without an added nitrile compound or $SnCl_4$. The solutions were saturated with CO by bubbling the gas through their storage reservoirs before admitting them into the sample cell.

Results for co-adsorbed Nitriles

In Figure 1 we show the PM-IRRAS spectra for a Pt electrode exposed to saturated CO/H_2SO_4 solutions which contain various concentrations of different organic nitriles. For comparison, we have also included a spectrum recorded in saturated CO/H_2SO_4 with no added nitrile. The adsorption step was accomplished by pulling the electrode back into the bulk solution and cycling the potential from 0.55 V(SHE) up to 1.15 V, down to 0.0 V, and back to 0.55 V. The spectra were recorded after re-positioning the electrode against the cell window while the potential was held at 0.55 V.

The top curve shows the spectrum of adsorbed CO that is observed when no nitrile compound is added to the electrolyte. The C-O stretching frequency occurs at 2085 cm^{-1}, which is characteristic of a saturated CO adlayer at this potential. The next three spectra were recorded in solutions which contain 1.0 M CH_3CN, 0.2 M C_2H_5CN, and 0.1 M $HOOCCH_2CN$, respectively. The intensity of the ν(CO) band is reduced about 50% in each case. This indicates that the amount of CO adsorbed on the electrode is reduced by the

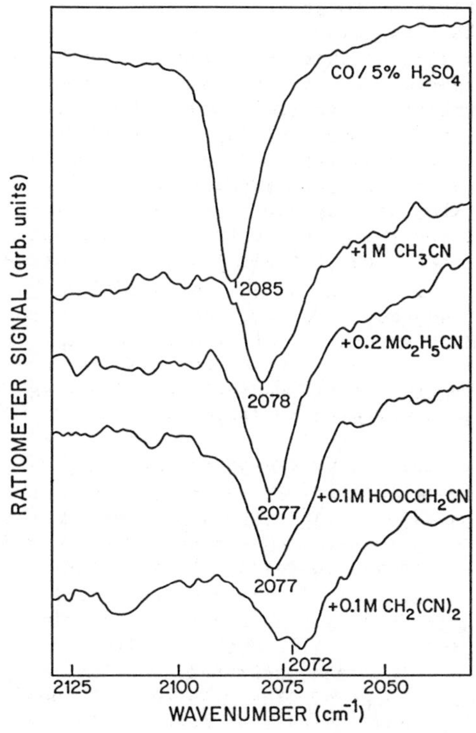

Figure 1. PM-IRRAS spectra of CO adsorbed on Pt in 0.5 M H_2SO_4 at 0.55 V(SHE) in the presence of different added nitrile compounds.

presence of the nitriles. The ν(CO) frequency is also shifted downward to around 2078 cm^{-1}. The magnitude and direction of this shift are similar to those expected for the decrease in electrodynamic coupling due to the reduced CO coverage, provided that the decrease occurs uniformly on a molecular scale. Thus we propose that a partial coverage of competitively adsorbed nitrile molecules are uniformly distributed over the surface, thereby preventing formation of a fully saturated CO adlayer.

The bottom curve shows the spectrum recorded when a di-functional nitrile, $CH_2(CN)_2$, is added to the electrolyte. The intensity of the adsorbed CO stretching band is decreased further and the frequency shifts lower, to 2072 cm^{-1}. Taken together, these observations suggest that the coverage of CO is smaller than in the previous experiments, and that the coverage of $CH_2(CN)_2$ is greater than for the other nitriles. The latter effect may be due to the presence of two -CN funcational groups in $CH_2(CN)_2$, which might be expected to increase the adsorption energy of this molecule by as much as a factor of two.

We were unable to observe any features in the region 1800-2400 cm^{-1} that could be attributed to ν (CN) of the adsorbed nitriles. A band was observed at 2260 cm^{-1} which did not shift as a function of potential, which we assign to ν(CN) of solution phase $CH_2(CN)_2$ near the electrode surface. We were unable to look for the 1615 cm^{-1} band reported by Sexton et. al. (20), since this frequency lies beyond the lower energy limit of the InSb detector in our spectrometer, and in a spectral region where there is severe background absorbance by the H_2O solvent.

Oxidation kinetics. We next measured the CO oxidation kinetic for each of these spectroscopically characterized systems. One cycle of the voltammogram was recorded for each system, working with the electrode exposed to bulk solution and starting and ending the potential sweep at 0.55 V. The results are shown in Figure 2. The CO oxidation peak occurs at 0.95 V during the anodic sweep, as seen in the voltammogram recorded in saturated CO solution with no added nitrile. The remaining curves show the results for the solutions with nitriles added. The area under the CO oxidation peak is markedly reduced in all cases. The magnitude of the decrease appears to vary in proportion to the concentration of the added nitrile. A comparison of the bottom two curves suggests that the di-functional nitrile, $CH_2(CN)_2$, may be somewhat more effective at blocking the surface than an equal concentration of a mono-nitrile.

Combining the results of the spectroscopic and kinetic experiments allows us to partially confirm one aspect of the mechanism for CO oxidation. The IRRAS results showed that the nitrile molecules are able to disrupt the continuity of the CO adlayer at the molecular scale by blocking a fraction of the surface Pt atoms. This is in contrast to the procedure of preparing a partial CO adlayer by controlling the exposure to bulk solution, which produces a uniformly distributed CO layer without blocking Pt atoms by a second adsorbate (12). The fact that there is no evidence for enhanced CO oxidation kinetics in the solutions with added nitriles indicates that the presence of an exposed Pt atom next to the

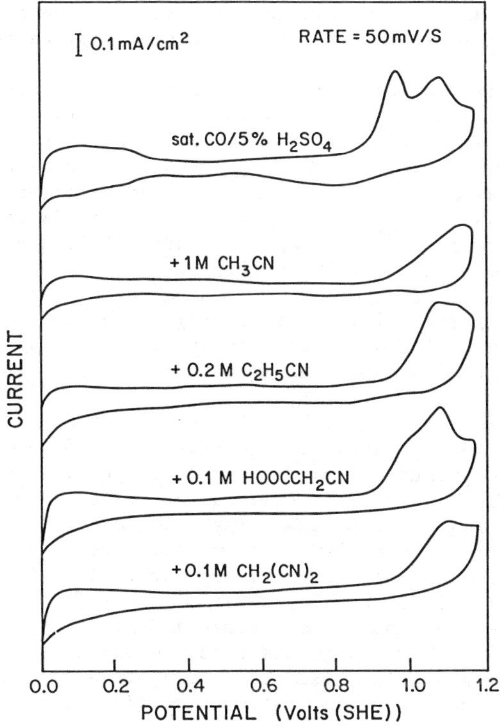

Figure 2. Cyclic voltammograms of CO oxidation in 0.5 M H_2SO_4 in the presence of different added nitrile compounds.

adsorbed CO molecule is an essential part of the "reactant-pair" mechanism for the electrochemical oxidation of CO (24).

Adsorption kinetics. We can also study the adsorption kinetics of the nitrile component. This is illustrated by the IRRAS spectra shown in Figure 3, which demonstrate the influence of electrode potential on the competitive adsorption of CO and C_2H_5CN. Curves a and b show the control experiments, in which spectra were recorded at different potentials in saturated CO electrolyte with no nitrile added. A saturated CO layer is produced in both cases, but the frequency is different at the two potentials; i.e., ν(CO) = 2085 cm^{-1} at 0.55V, vs. ν(CO) = 2070 cm^{-1} at 0.05 V. The magnitude of this shift is in agreement with the potential dependence of ν(CO) discussed above.

The lower two spectra were recorded in CO saturated solution which contained 0.1 M C_2H_5CN. The adsorbate layers were produced by cycling the potential with the electrode pulled away from the window as described above, except that a different final potential was chosen to end each cycle. Spectrum c was recorded at a final potential of 0.05 V. At this point no ν(CO) band is observed. Spectrum d was recorded at a final potential of 0.55 V (c.f. Figure 1), and shows the band at ν(CO) = 2078 cm^{-1} that we assigned to a partial coverage of adsorbed CO. We can show that this change in the spectrum is irreversible by returning the potential to 0.05 V. The ν(CO) band is still observed with the same peak area, and the frequency is shifted by only the amount predicted by the known potential dependence.

Next we demonstrate that both CO and C_2H_5CN are irreversibly adsorbed under these conditions. To prove that CO adsorption is irreversible, we prepared a saturated CO adlayer in a solution without the nitrile, and then replaced the cell contents with the solution containing 0.2 M C_2H_5CN. The ν(CO) band of the saturated adlayer remained unchanged, which shows that adsorbed CO is not displaced by solution phase C_2H_5CN. It now follows that nitrile adsorption is also irreversible, based on the fact that the partial CO layer observed in Figure 3 is stable indefinitely. If a mixed adsorbate layer is observed in the presence of both solution phase components simultaneously, then it is impossible for one species to be irreversibly adsorbed without the adsorption of the other also being irreversible.

The results in Figure 3 can now be interpreted. The nitrile molecules are able to compete with CO for Pt adsorption sites. When the potential is sufficiently negative, the adsorbed CO molecules do not enter their linear configuration. Two of our earlier studies have shown that co-adsorbed species can displace CO out of its linear configuration under appropriate conditions. For partial coverages of CO studied at low temperature under vacuum conditions, we showed that co-adsorbed adsorbed H_2O is able to displace CO away from its normally preferred linear configuration (25). In electrochemical studies using K_2SO_4, we showed that the bridging configuration becomes more favorable at negative potentials (17). In those studies, we proposed that adsorbed H_2O and alkali cations, respectively, donate electron density into the

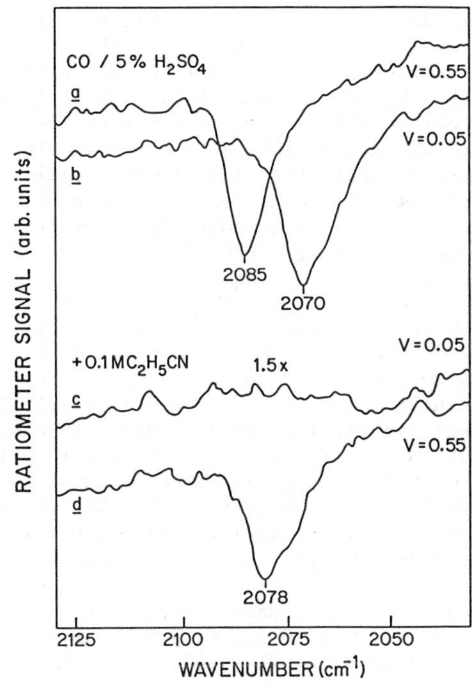

Figure 3. PM-IRRAS spectra of CO adsorbed on Pt in 0.5 M H_2SO_4 + 0.2 M C_2H_5CN solutions as a function of electrode potential during the adsorption step (see text).

metal and thereby help to convert CO into its bridging configuration.

Under the present conditions, we propose that the -CN group in the nitrile molecules acts in a similar fashion. The nitrile molecules are adsorbed on a fraction of the Pt surface atoms, and their presence forces the co-adsorbed CO molecules on the remaining sites to assume the bridging configuration. When the potential is increased, part of the driving force for this conversion is removed and the CO molecules convert back to their linear configuration. This conversion is irreversible when the potential is lowered a second time. The reason for this irreversibility is uncertain and should be explored further.

The fact that linear CO species are observed at 0.05 V in the absence of C_2H_5CN (cf. spectrum a) indicates that H_2O molecules at the inner Helmholtz plane are not able to displace CO out of its linear configuration at the same potential. This may be due to a re-orientation of the adsorbed H_2O as a function of potential, with the positive end of the molecular dipole becoming attracted to the surface as the electrode potential is made more negative. This would reduce the ability of the H_2O molecule to donate electron density from its oxygen atom, and would also increase the ability of its hydrogen atoms to compete for accepting electron density from the metal.

Results for co-adsorbed Sn atoms

Figure 4 shows a sequence of cyclic voltammograms recorded in a CO saturated solution of 0.5 M H_2SO_4 which contained 10^{-5} M $SnCl_4$. These voltammograms were recorded by increasing the upper limit of the potential during successive cycles. The major feature is the CO oxidation peak at 0.6 V(SCE) that is observed in the later scans. For cycles with an upper limit below this potential, there is a second, smaller oxidation feature at about 0.4 V. Control experiments using an $SnCl_4/H_2SO_4$ solution without CO show that this feature can be assigned to the oxidation of adsorbed Sn atoms. We note that the Sn oxidation step does not produce a well-resolved current maximum in the curves in Figure 4. We interpret this to indicate that adsorbed CO molecules are rapidly oxidized by newly generated Sn cations, thus regenerating the Sn adatoms and allowing further oxidation current to flow.

As the upper limit of the cycles approaches 0.5 V, the size of the Sn oxidation feature gradually decays. Cycles with an upper limit above 0.6 V are virtually indistinguishable from those recorded in CO saturated H_2SO_4 without added $SnCl_4$. This indicates that the Sn atoms leave the surface in their oxidized state during the anodic stage of the cycle and are unable to readsorb during the cathodic stage.

Figure 5 shows the IRRAS spectra recorded for three dosing conditions chosen to demonstrate the nature of the Sn adsorption kinetics. All three spectra were recorded at 0.0 V. The upper spectrum is the control curve recorded in CO saturated H_2SO_4. The $\nu(CO)$ band occurs at 2083 cm^{-1}, which is characteristic of a saturated CO adlayer at this potential.

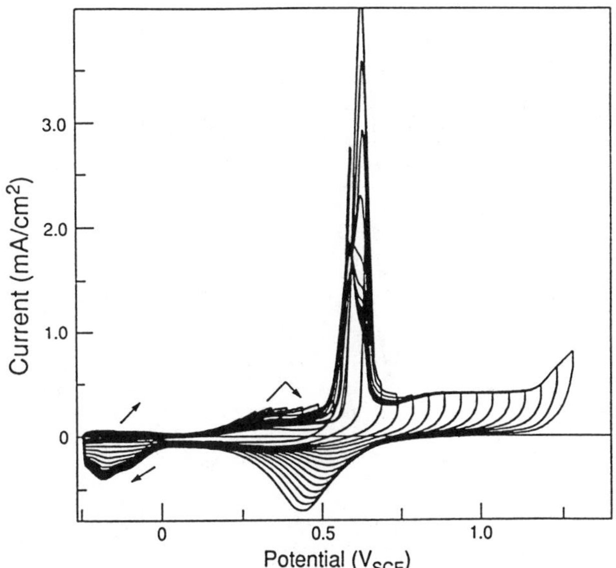

Figure 4. Cyclic voltammograms of CO oxidation in 0.5 M H_2SO_4 + 10^{-5} M $SnCl_4$ for progressively larger upper potential limits.

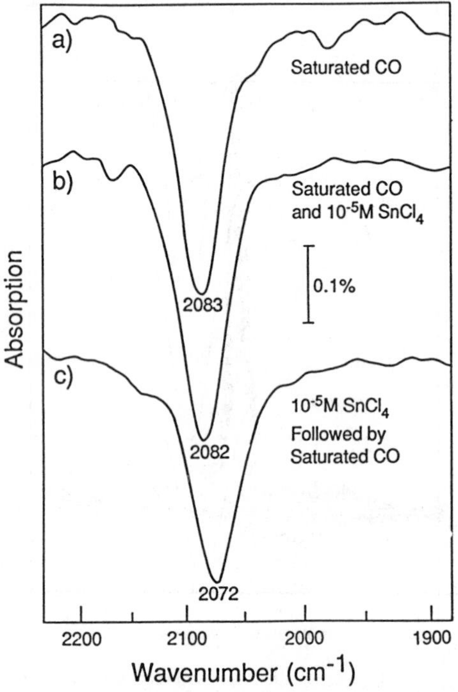

Figure 5. PM-IRRAS spectra of CO adsorbed on Pt in 0.5 M H_2SO_4 + 10^{-5} $SnCl_4$ as a function of adsorption conditions (see text).

The middle spectrum is recorded in CO-saturated $SnCl_4/H_2SO_4$ after cycling the electrode potential through 1.2 V and then returning to 0.0 V. The spectrum is virtually indistinguishable from the first curve, which shows that the CO adlayer is still saturated. This confirms that little, if any, adsorption of Sn adatoms occurs even after the initially adsorbed CO layer is removed by oxidation. Instead, competitive adsorption of new CO molecules blocks the surface.

The bottom spectrum was obtained by cycling the electrode in CO-free $SnCl_4/H_2SO_4$ solution to ensure formation of a partial Sn adlayer and then replacing the cell contents with CO-saturated solution. The ν(CO) band is still observed, which shows that the Sn adatoms do not saturate the surface even in the absence of competitive CO adsorption. The intensity and frequency of the ν(CO) band have both decreased, which confirms that the CO adlayer is only partially complete. There is no evidence for a change in ν(CO) beyond that expected for the coverage dependence expected in acid solution. This shows that there is no strong interaction between adsorbed CO molecules and neighboring Sn adatoms, in support of the assumptions used in the adatom oxidation model discussed above.

Summary

The vibrational spectroscopy of CO adsorbed on Pt electrodes has proven to be a valuable tool for probing the physical chemistry of the metal/electrolyte interface. The exact value of the vibrational frequency and the relative intensity of the linear and bridging bands both provide an indication of the ability of the electrode surface to donate charge into adsorbed species. Band intensities provide a direct, qualitative method to monitor surface coverages in experiments using co-adsorbed species. Future experiments should focus on measuring the effects of co-adsorbed species more quantitatively, using other conventional electrochemical techniques to accurately determine surface coverages.

Acknowledgments

We respectfully acknowledge both the encouragement to undertake this research and also the many contributions to the field of vibrational spectroscopy that were made by the late Professor John Overend.

This work was sponsored by the Corrosion Research Center at the University of Minnesota, which is supported by the U.S. Department of Energy, through Grants DE-FG02-84ER45173 and DE-79ER10450.

References

1. Vibrational Spectroscopy of Adsorbed Species; Bell, A. T.; Hair, M. L., Eds.; ACS Symposium Series No. 137; American Chemical Society; Washington, DC, 1980.
2. Eischens, R. P.; Francis, S. A.; Pliskin, W. A. J. Phys. Chem. 1956, 60, 194.

3. Hammaker, R. M.; Francis, S. A.; Eischens, R. P. Spectrochem. Acta 1965, 21, 1295.
4. Primet, M.; Basset, J. M.; Mathieu, M. V.; Prettie, M. J. Catal. 1973, 29, 213.
5. Sheppard, N.; Nguyen, T. T. In Advances IR and Raman Spectroscopy; Hester, R. E.; Clark, R. J. H., Eds.; Heyden: London, 1978.
6. Shigeishi, R. A.; King, D. A. Surf. Sci. 1976, 58, 379.
7. Horn, K.; Pritchard, J. J. Physique, Colloq. 1977, 38, C4-164.
8. Krebs, H. J.; Luth, H. Appl. Phys. 1977, 14, 337.
9. Hayden, B. E.; Bradshaw, A. M. Surf. Sci. 1983, 125 787.
10. Severson, M. W.; Tornquist, W. J.; Overend, J. J. Phys. Chem. 1984, 88, 469.
11. Beden, B.; Bewick, A.; Kunimatsu, K.; J. Electroanal Chem. 1982, 142 345.
12. Russel, J. W.; Severson, M.; Scanlon, K.; Overend, J.; Bewick, A. J. Phys. Chem. 1983, 87, 293.
13. Kunimatsu, K.; Seki, H.; Golden, W. G.; Gordon, J. G., II; Philpott, M. R. Langmuir 1986, 2, 464.
14. Blyholder, G.; J. Phys. Chem. 1964, 68, 2772.
15. Crossley, A.; King, D. A. Surf. Sci 1977, 68, 528.
16. Severson, M. W.; Russell, A.; Cambell, D.; Russell, J. W. Langmuir 1987, 3, 202.
17. Tornquist, W. J.; Guillaume, F.; Griffin, G. L. Langmuir 1987, 3, 477.
18. Nakajima, H.; Kita, H.; Kunimatsu, K.; Aramata, A. J. Electroanal. Chem. 1986, 201, 175.
19. Lambert, D. K. Solid State Commun. 1983, 30, 59.
20. Sexton, B. A.; Avery, N. R. Surf. Sci. 1983, 129, 21.
21. Loo, B. H.; Lee, Y. G.; Frazier, D. O. Chem. Phys. Lett. 1985, 119, 312.
22. Loo, B. H.; Lee, Y. G.; Frazier, D. O. J. Phys. Chem. 1985, 89, 4672.
23. Watanabe, M.; Shibata, M.; Motoo, S. J. Electroanal. Chem. 1985, 187, 161.
24. Gilman, S. J. Phys. Chem. 1964, 68, 70.
25. Tornquist, W. J.; Griffin, G. L. J. Vac. Sci. Technol. A 1986, 4, 1437.

RECEIVED May 17, 1988

Chapter 26

Surface-Enhanced Raman Study

Effect of pH and Electrode Potential on the Interfacial Behavior of Some Substituted Pyridines

Mark R. Anderson[1] and Dennis H. Evans[2]

[1]Department of Chemistry, University of Utah, Salt Lake City, UT 84112

[2]Department of Chemistry and Biochemistry, University of Delaware, Newark, DE 19716

The surface-enhanced Raman spectra (SERS) provide information about the extent of protonation of the species adsorbed at the silver/aqueous solution interface. The compounds investigated were 4-pyridylcarbinol (**1**), 4-acetylpyridine (**2**), 3-pyridinecarboxaldehyde (**3**), isonicotinic acid (**4**), isonicotinamide (**5**), 4-benzoylpyridine (**6**), 4-(aminomethyl)pyridine (**7**) and 4-aminopyridine (**8**). For **1**, the fraction of the adsorbed species which was protonated at -0.20 V vs. SCE varied with pH in a manner indicating stronger adsorption of the neutral than the cationic form. The fraction protonated increased at more negative potentials. Similar results were obtained with **3**. For all compounds but **4**, bands due to the unprotonated species near 1600 cm^{-1} and for the ring-protonated species near 1640 cm^{-1} were seen in the SERS spectra.

In 1974 Fleischmann *et al*. demonstrated that Raman spectra of pyridine adsorbed at a silver electrode from an aqueous solution could be obtained with excellent signal-to-noise at a roughened silver surface (1). Subsequently, Jeanmaire and Van Duyne (2) as well as Albrecht and Creighton (3) demonstrated that the Raman spectrum of adsorbed pyridine at the silver surface represented a 10^6 enhancement of the signal expected considering typical Raman parameters. Such a discovery was very exciting because it presented a relatively simple means of obtaining *in situ* the vibrational spectra of molecules adsorbed onto a surface. Since that time, considerable effort has been devoted to studying this phenomenon.

Much of this research effort has been directed to the study of the fundamental basis of surface-enhanced Raman scattering (SERS), in order to understand the underlying principles. There have also been many applications of SERS to situations in which *in situ* vibrational

0097–6156/88/0378–0383$06.00/0

spectra would provide valuable information. An electrochemical environment is one such system where SERS has been particularly valuable. SERS has been used to study a number of electrochemical phenomena, including the adsorption of simple anionic adsorbates and the mechanisms of electrode reactions (4, 5). It is also possible to study properties of surface species by systematically altering the conditions under which spectra are obtained. In this manner, spectroscopic changes may be correlated with the environmental perturbations and information about the properties of the molecule at the surface may be deduced.

In a previous study (6) we investigated the spectra of adsorbed 4-pyridinecarboxaldehyde as a function of applied electrode potential and as a function of bulk solution pH. This study demonstrated that the spectrum of 4-pyridinecarboxaldehyde was dramatically dependent upon these experimental variables. Separate spectroscopic features were identified which could be attributed to the protonated and the unprotonated 4-pyridinecarboxaldehyde species. Interestingly, in solutions near pH 7, where the concentration of protonated 4-pyridinecarboxaldehyde in solution is negligible, bands attributable to the protonated species on the surface appeared in the spectrum at negative potentials. This behavior was thought to be caused largely by the potential dependence of the adsorption coefficients, that of the cation being relatively larger at more negative potentials. A smaller contribution arises from lowering of the solution pH in the region local to the electrode surface. Such a phenomenon had previously been observed in a SERS study of phosphate species (7) and is supported by electrochemical theory. In addition, as the electrode potential was made more negative, the carbonyl feature gradually diminished in size and eventually disappeared. This observation was attributed to an increasing degree of hydration of the 4-pyridinecarboxaldehyde as the electrode potential was made more negative, in consonance with the fact that the carbonyl group of protonated 4-pyridinecarboxaldehyde is 94% hydrated in solution.

These interpretations differed somewhat from those of Bunding and Bell (8) in a similar SERS study of 4-pyridinecarboxaldehyde. These workers concluded that electron withdrawal by the electrode from the adsorbed 4-pyridinecarboxaldehyde induced the total hydration of the formyl substituent. Support for this interpretation was gained by considering the spectroscopic behavior of similar compounds within the context of the 4-pyridinecarboxaldehyde observations. Bunding and Bell, however, only investigated the spectroscopic behavior at a single, negative potential. It is the purpose of the current study to investigate systematically the spectroscopic behavior of other pyridine derivatives as a function of electrode potential and bulk solution pH and to compare the observations to those previously presented for 4-pyridinecarboxaldehyde.

Results and Discussion

Compounds **1-8** have been investigated. 4-Pyridylcarbinol, **1**, is of interest because it is one of the electrochemical reduction products of 4-pyridinecarboxaldehyde, the species for which surface protonation reactions were discovered and characterized in earlier work (6).

Characterization of Surface Species by SERS. Before presenting the results obtained with **1**, the spectral features which have proven to be useful in identifying surface species will be reviewed. Both in solution and by SERS, pyridines show a ring mode in the Raman spectrum near 1600 cm^{-1}. When the ring nitrogen is protonated, this band disappears and is replaced by a band near 1640 cm^{-1}. The

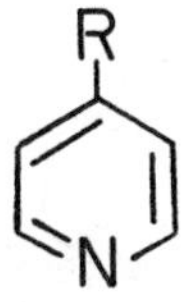

1-2, 4-8 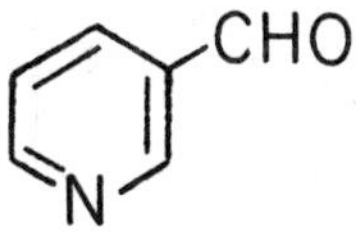   3

1 : R = $-CH_2OH$
2 : R = $-COCH_3$
4 : R = $-CO_2H$
5 : R = $-CONH_2$
6 : R = $-CH_2NH_2$
7 : R = $-NH_2$
8 : R = $-COC_6H_5$

relative intensities of these two bands provide information about the extent of protonation. If one assumes that the scattering cross sections for unprotonated species (1600 cm^{-1} band) and protonated species (1640 cm^{-1} band) are equal, the fraction of the pyridine compound which is protonated is given by $I_{1640}/(I_{1640} + I_{1600})$ where I_{1600} and I_{1640} are the heights of the two bands.

In the present research, it was confirmed for compounds **1-3** that, in the normal Raman spectra of aqueous solutions of various pH, a band near 1600 cm^{-1} appears for the neutral pyridine and a band near 1640 cm^{-1} exists for the ring-protonated species.

In cases where the substituent on the pyridine ring contains a carbonyl group, a weak band for the C-O stretch can be detected in the SERS spectra, generally near 1700 cm^{-1}.

The SERS spectra of pyridines contain many other bands but those mentioned above have proven to be particularly useful in characterizing the surface species.

4-Pyridylcarbinol, 1. This compound has been investigated previously (8) and it was noted that its SERS spectra were almost identical to those of 4-pyridinecarboxaldehyde. SERS spectra in the 1500-1700 cm^{-1} region are shown in Figure 1. Buffers with pH bracketing the pK_a of 4-pyridylcarbinol (5.76 (9)) were employed and it can be seen that the band for the unprotonated species near 1600 cm^{-1} is predominant at pH 6.88 but decreases as the band due to the protonated species (near 1640 cm^{-1}) grows in when the pH is lowered. A variety of buffers was used and the intensity ratio was measured from SERS spectra obtained at -0.20 V vs. SCE. As mentioned above, this intensity ratio is a measure of the fraction of the surface compound which is protonated. The fraction is plotted in Figure 2 along with the fraction protonated in solution (calculated from the pK_a). It is apparent that the fraction protonated at the surface lags the fraction protonated in solution as the pH is lowered. This may be caused by relatively stronger adsorption of the neutral pyridine compared to the protonated species at this potential. Considering competitive adsorption of two species according to the Langmuir isotherm at close-to-saturation surface coverage, the fraction of the surface species which is protonated, $(X_p)_s$, will be given by Equation 1, where $[H^+]$ is the hydronium ion concentration

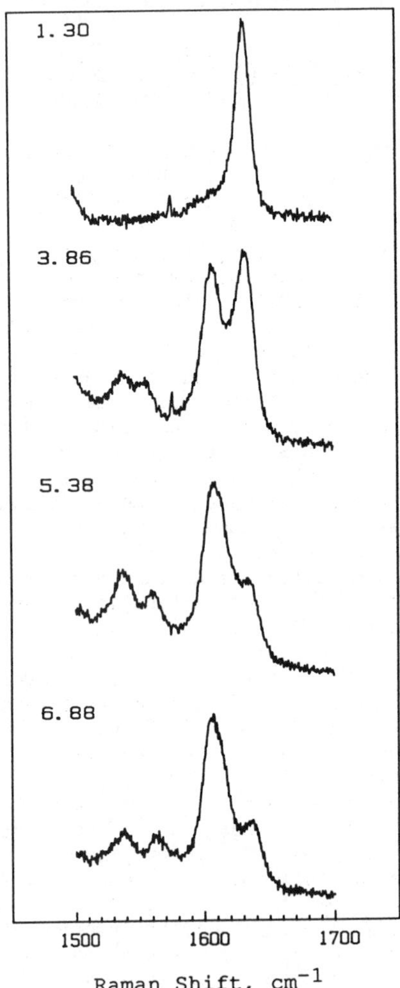

Figure 1. SERS spectra of 0.050 M 4-pyridylcarbinol in solutions of various pH. All spectra were obtained at -0.20 V vs. SCE.

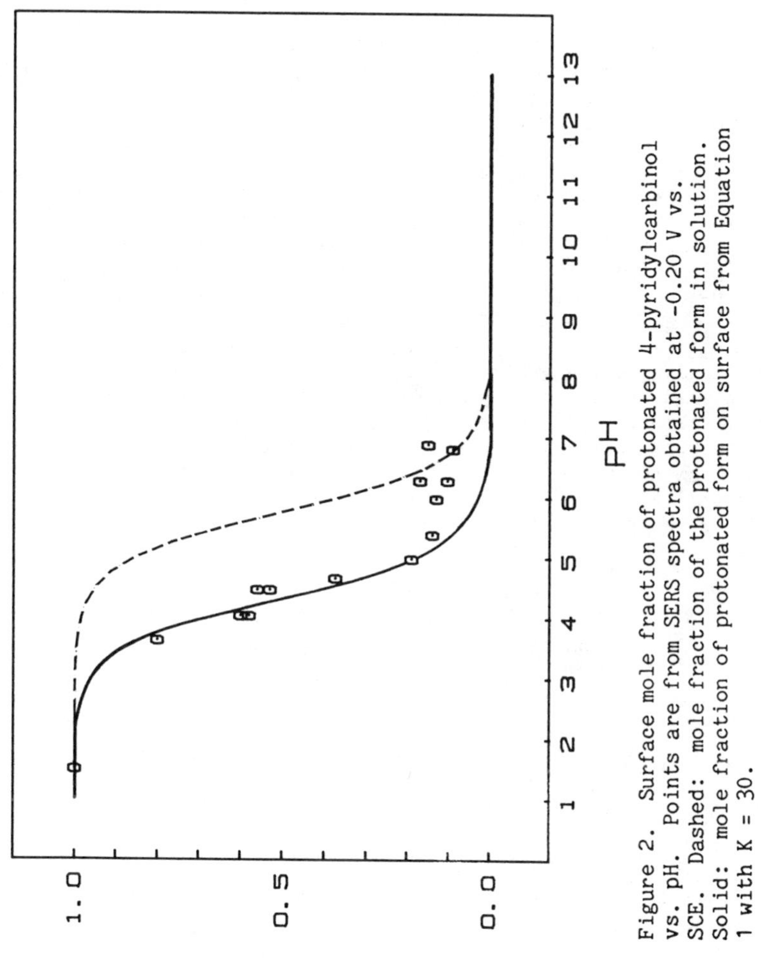

Figure 2. Surface mole fraction of protonated 4-pyridylcarbinol vs. pH. Points are from SERS spectra obtained at -0.20 V vs. SCE. Dashed: mole fraction of the protonated form in solution. Solid: mole fraction of protonated form on surface from Equation 1 with K = 30.

$$(X_p)_s = [H^+]/([H^+] + K_aK) \qquad (1)$$

in solution, K_a is the acid dissociation constant in solution and $K = \Gamma_{n,s}\beta_n/\Gamma_{p,s}\beta_p$ where Γ_s and β are saturation coverages and adsorption coefficients, respectively, for the neutral (n) and protonated (p) species. At the concentration used in this work, close-to-saturation coverage is expected based on adsorption studies of other pyridines at a mercury electrode (10, 11). Equation 1 is plotted in Figure 2 (solid curve) for K = 30 which matches the experimental data reasonably well. Introduction of interaction parameters through the Frumkin isotherm brought a perceptible improvement in the agreement but, in view of the scatter in the experimental data, this refinement is probably not meaningful.

Thus, at -0.20 V, the adsorption coefficient of the neutral form of 4-pyridylcarbinol is about 30 times that of the protonated form (assuming equal saturation coverages). This quantity is dependent upon the electrode potential, however. In Figure 3 are shown SERS spectra obtained at constant solution pH (6.88) but variable electrode potential. At -0.2 V only a small fraction is protonated (weak feature at 1640 cm^{-1}) but the fraction increases substantially as the potential is made more negative until at the most negative potential (-0.60 V) it decreases again. The same trend was observed with 4-pyridinecarboxaldehyde (6) and can be explained by relatively stronger adsorption of the cation (protonated form) which is expected at more negative potentials. A smaller contributor to the effect is the increase in the effective pH in the interfacial region as the potential is made more negative (6, 7). The intensity ratio is given in Table I along with the solution pH which would be necessary to cause the same fraction of **1** to be protonated as is present on the surface.

Table I. Fraction of 4-Pyridylcarbinol which is Protonated as Calculated from SERS Spectra[a]

Potential (V vs. SCE)	$(X_p)_s$	pH[b]
-0.20	0.15	6.6
-0.30	0.22	5.0
-0.40	0.36	4.7
-0.50	0.36	4.7

[a]Data from Figure 3, pH 6.88.
[b]pH in solution which produces fraction protonated equal to that seen on the surface.

4-Acetylpyridine, **2**. Relatively intense SERS spectra were obtained for **2** using 0.10 M KCl (8) but there was little dependence of the relative intensities of the bands on potential. A band due to the unprotonated pyridine was seen near 1600 cm^{-1} and, at pH < 6, a band due to the protonated compound appeared near 1640 cm^{-1} and increased at the expense of the 1600 cm^{-1} band until only the protonated species could be detected at pH = 1.3. The pK_a of **2** is 3.51 (9),

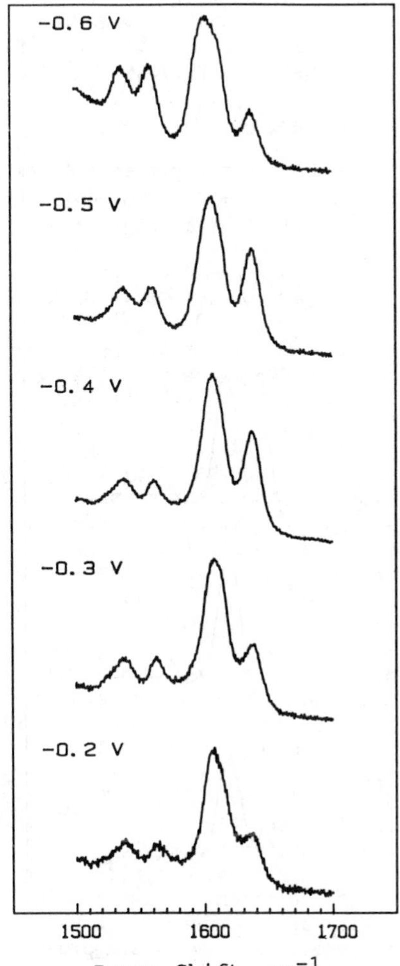

Figure 3. SERS spectra of 0.050 M 4-pyridylcarbinol in pH 6.88 buffer at various electrode potentials (vs. SCE).

about two units smaller than 1, and correspondingly lower pH values are required to generate large populations of the protonated form of 2 on the surface. Unlike 1 and 4-pyridinecarboxaldehyde, the fraction protonated at constant pH hardly changes with potential.

In 0.10 M KCl, the SERS spectra of 2 show a small band at about 1690 cm^{-1}, attributable to the carbonyl stretch. Allen and Van Duyne (12) have considered orientational effects on band intensities and have concluded that the relative intensities should be given by Equation 2, where the intensity of the carbonyl band

$$(I_{1690}/I_{1000})_{SERS} = (I_{1690}/I_{1000})_{NRS} \cdot \cos^2\Theta \qquad (2)$$

is expressed relative to the intense symmetrical ring-breathing band near 1000 cm^{-1}, in both the SERS and normal Raman solution spectra (NRS). Θ is the angle of the carbonyl group with respect to the surface normal.

For vertical orientation of 4-acetylpyridine (adsorption *via* the ring nitrogen atom), Θ will be 60°. The data for the 1690 and 1000 cm^{-1} bands in the SERS and NRS (0.10 M KCl) were analyzed according to Equation 2 and gave Θ = 61, 59 and 63° for -0.20, -0.40 and -0.60 V, respectively. Thus, the results are consistent with a vertical orientation which is independent of potential.

3-Pyridinecarboxaldehyde, 3. Possible hydration of the aldehyde group makes the aqueous solution chemistry of 3 potentially more complex and interesting than the other compounds. Hydration is less extensive with 3 than 4-pyridinecarboxaldehyde but upon protonation, about 80% will exist as the hydrate (*gem*-diol). The calculated distribution of species as a function of pH is given in Figure 4 based on the equilibrium constants determined by Laviron (9).

In comparison to 1 and 2, the SERS spectra of 3-pyridinecarboxaldehyde (3) are relatively featureless (8). The spectra are dominated by the symmetrical ring-breathing mode at 1030 cm^{-1} but the features associated with the unprotonated species (about 1600 cm^{-1}) and the protonated species (about 1640 cm^{-1}) are definitely present along with a weak carbonyl band at about 1710 cm^{-1}. The variation in the relative population of protonated species is as expected (Figure 5) though a detailed analysis reveals some surprises. As can be seen in Figure 5, about equal intensities of the 1600 and 1640 cm^{-1} bands are obtained at pH = 3.86, near the pK_a (3.73 (9)). However, the band associated with the unprotonated pyridine persists at pH = 1.3, where less than 1% of the solution species remains unprotonated. When the intensity ratios are measured for a variety of buffers, and the values are plotted vs. pH, the values approach 0.8 at low pH (Figure 6), the same as the fraction of 3 existing in solution as the protonated, hydrated form (cf. Figure 4). This could mean that the residual 1600 cm^{-1} band seen at low pH is due to protonated, unhydrated 3 on the surface.

As seen with 4-acetylpyridine, the relative intensities of the 1600 and 1640 cm^{-1} features, at constant solution pH, are insensitive to electrode potential. Again, this behavior may be associated with the relatively weaker basicity of 3 and 2 (pK_a = 3.51 and 3.73, respectively) compared to 1 (5.76) and 4-pyridinecarboxaldehyde (4.78) though the exact reason is not known. The latter two show

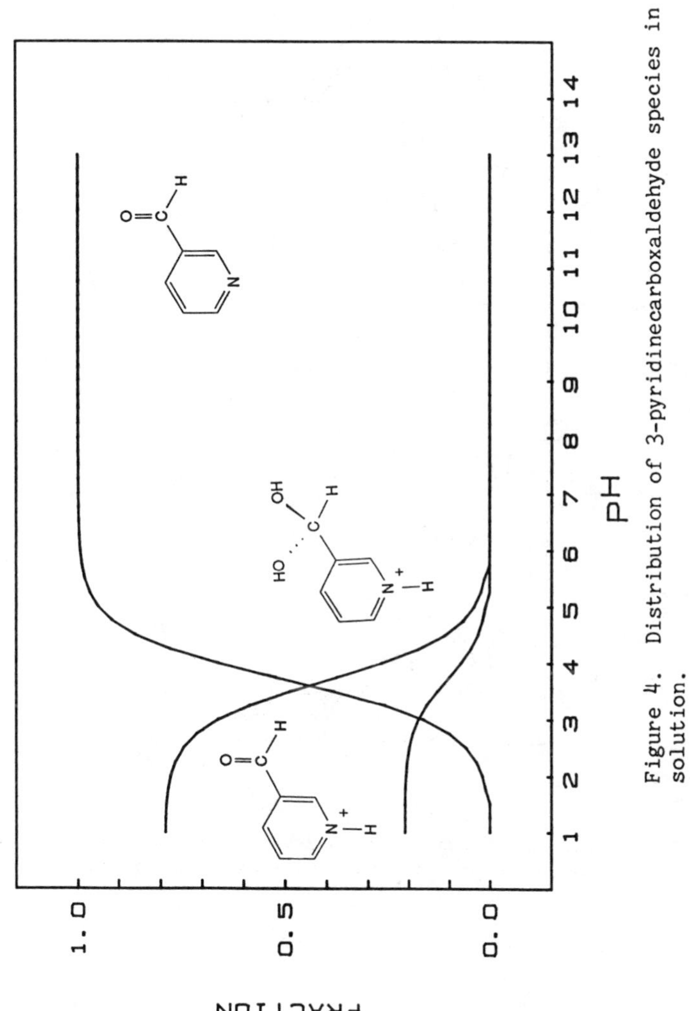

Figure 4. Distribution of 3-pyridinecarboxaldehyde species in solution.

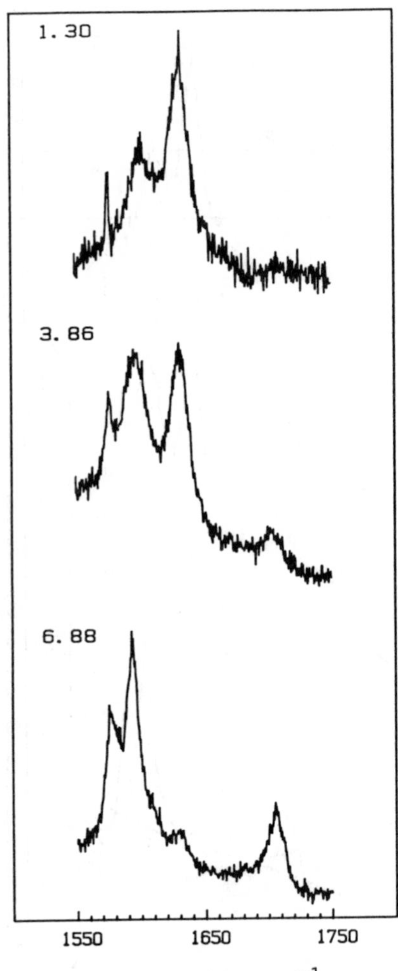

Figure 5. SERS spectra of 0.050 M 3-pyridinecarboxaldehyde in solutions of various pH. All spectra were obtained at -0.20 V vs. SCE.

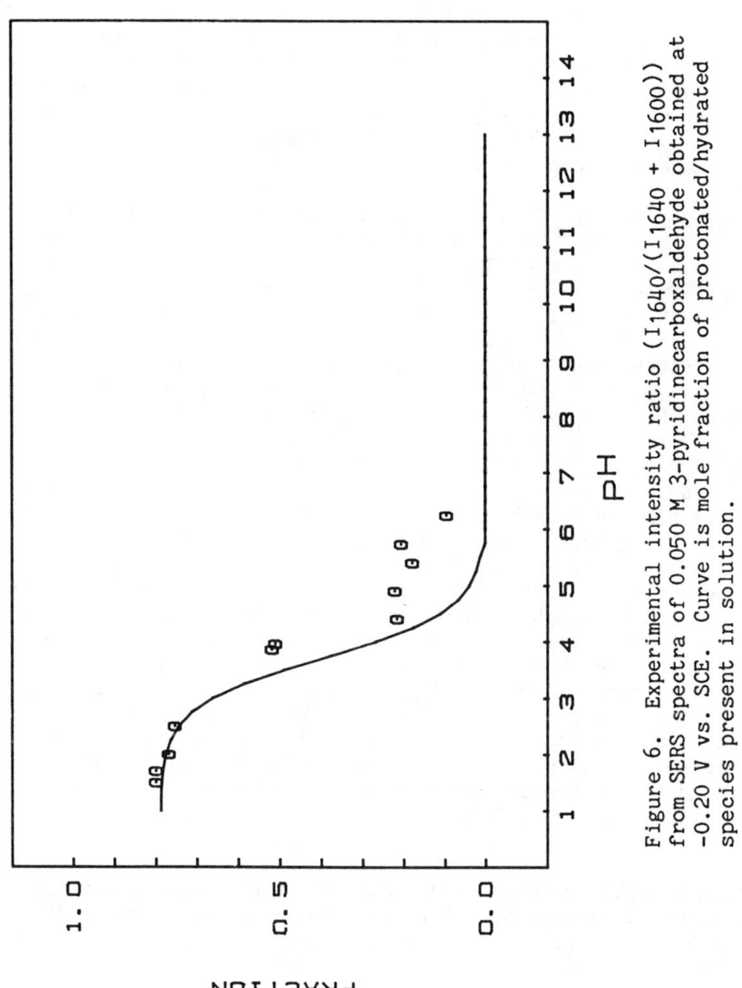

Figure 6. Experimental intensity ratio ($I_{1640}/(I_{1640} + I_{1600})$) from SERS spectra of 0.050 M 3-pyridinecarboxaldehyde obtained at -0.20 V vs. SCE. Curve is mole fraction of protonated/hydrated species present in solution.

relatively higher populations of protonated pyridine in the SERS spectra as the potential is made more negative.

At the most negative potentials, the protonated form of **3** appears to desorb as seen also with **1** and **2**. For 4-pyridinecarboxaldehyde, this desorption has been correlated with the desorption of chloride (6) suggesting that the cation and chloride ion are coadsorbed.

Isonicotinic Acid, **4**. It is difficult to obtain a spectrum of **4** because the neutral form is not very soluble. At low pH, however, the ring nitrogen is protonated (13) and the cationic isonicotinic acid is sufficiently soluble to obtain SERS spectra. A relatively intense spectrum was obtained at -0.20 V with 0.050 M isonicotinic acid, 0.10 M KCl and 0.10 M HCl. Many of the spectral features seen with other pyridines are present but the inability to vary solution pH made it impossible to investigate the relative surface populations of protonated and unprotonated forms.

Isonicotinamide, **5**. This compound was sufficiently soluble to allow SERS spectra to be obtained at the 50 mM level in 0.10 M KCl and 0.10 M KCl + 0.10 M HCl at -0.20 V. The spectra resembled those seen with other pyridines. In particular, an intense band at 1600 cm^{-1} was seen with the neutral electrolyte and it was replaced by a band at 1640 cm^{-1} in the acidic electrolyte. Of the two basic sites, only the ring nitrogen will be protonated in 0.10 M HCl (13) so, with this compound also, the 1640 cm^{-1} band appears to be due to the protonated pyridine. No carbonyl band was seen in either spectrum.

4-Benzoylpyridine, **6**. Of possible interest here is the fact that **6** contains both a pyridyl and a phenyl group. Pyridine and benzene have very similar vibrational modes which should produce bands due to both aromatic groups at approximately the same positions in the spectra of **6**. Like isonicotinic acid, **6** was relatively insoluble in water but it was sufficiently soluble in 50 % (v/v) ethanol/water to allow spectra to be obtained at the 50 mM level. Again, spectra were recorded at -0.20 V in 0.10 M KCl and in 0.10 M KCl + 0.10 M HCl (Figure 7). Under both conditions, strong bands, characteristic of the aromatic ring systems, are seen near 1000, 1200 and 1600 cm^{-1}. A carbonyl band appears at approximately 1660 cm^{-1}. The main difference between the spectrum obtained at low pH and that recorded from neutral solution, is the band near 1640 cm^{-1} which is again attributed to the protonated pyridyl ring. In this case, the strong band remaining at about 1600 cm^{-1} is probably due to the phenyl group.

4-(Aminomethyl)pyridine, **7**. The SERS spectra of most pyridines show a broad, relatively featureless background over the range of 1200-1700 cm^{-1}. This background scattering is immense for **7** in 0.10 M KCl at -0.20 V. Atop the background is the strong band near 1600 cm^{-1}, characteristic of the neutral pyridyl ring. The spectrum obtained in 0.10 M KCl + 0.10 M HCl has no band at 1600 cm^{-1} but, instead, a very strong band at about 1640 cm^{-1} is seen. In all of the other pyridines this band is associated with the protonated pyridyl ring. The aminomethyl group is the more basic site on **7** so the appearance

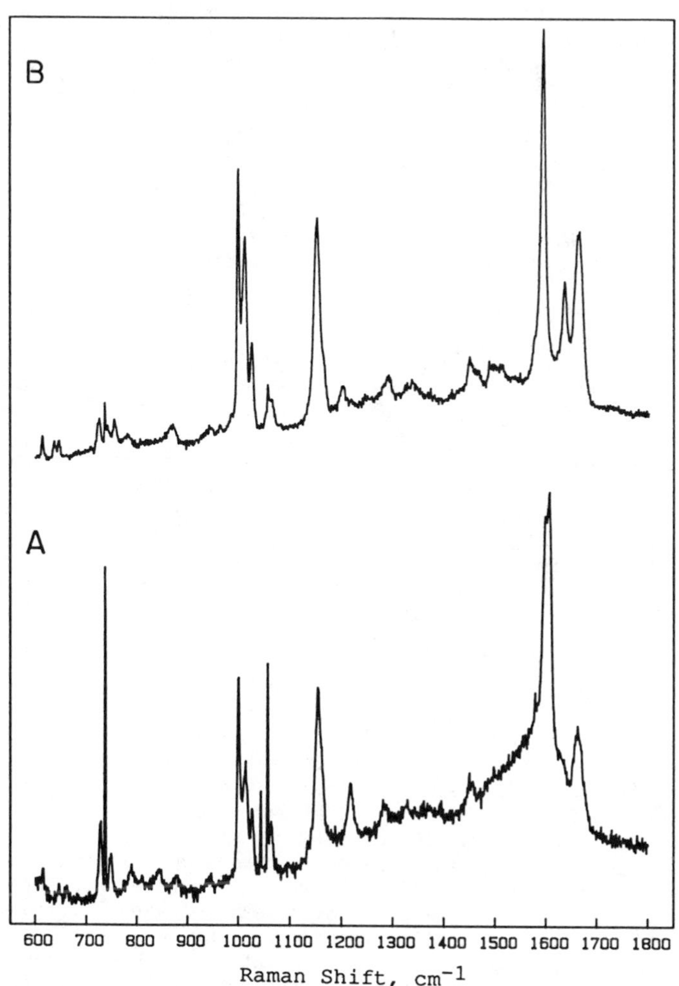

Figure 7. SERS spectra of 0.050 M 4-benzoylpyridine obtained at -0.20 V vs. SCE. 50% (v/v) ethanol/water. A: 0.10 M KCl. B: 0.10 M KCl + 0.10 M HCl. (Sharp lines at 738, 1040 and 1060 cm^{-1} are from scattered argon ion emission).

of the band at 1640 cm^{-1} probably means that both the aminomethyl group and the pyridyl nitrogen are protonated in the acidic medium.

4-Aminopyridine, **8**. Like 4-(aminomethyl)pyridine, **8** has two basic sites. The pK_a of the ring nitrogen is 9.12 while that of the amino group is estimated to be -6 (14, 15). The SERS spectrum in 0.10 M KCl has a very large background between 1200 and 1700 cm^{-1} (like that seen with **7**) with a weak band due to the unprotonated pyridine at about 1600 cm^{-1}. In this case, protonation of the pyridyl nitrogen occurs readily so that even with a buffer of pH = 6.19, the 1600 cm^{-1} band has been replaced by a band near 1640 cm^{-1}. (The broad background between 1200 and 1700 cm^{-1} is also much weaker). The spectrum obtained with 0.10 M KCl + 0.10 M HCl is almost identical to that seen with the pH = 6.19 buffer. In view of the pK_a values for **8**, protonation of the amino group is not expected in 0.10 M HCl.

Experimental

4-Pyridylcarbinol, 4-acetylpyridine, 3-pyridine-carboxaldehyde and 4-aminomethylpyridine were obtained from Aldrich Chemical Company (Milwaukee, Wisconsin) and were purified by distillation at reduced pressure. 4-Benzoylpyridine was recrystallized from ethanol. 4-Aminopyridine (G. Frederick Smith Chemical Company), isonicotinic acid (Aldrich) and isonicotinamide (Aldrich) were used as received. Triply distilled water was used. All other reagents were analytical reagent grade.

Acetate buffers were used for pH values between 3.5 and 5.5 while phosphate buffers were used for pH 5.5-7. The pH of solutions below pH 3.5 was adjusted with HCl. All solutions contained 0.10 M KCl and the ionic strength of the buffers was adjusted to 0.90 M by addition of potassium nitrate.

All SERS experiments were conducted with a polycrystalline silver working electrode prepared by press-fitting a 6 mm diameter cylinder of silver into one end of a 0.375 inch diameter Teflon rod through which a 6 mm diameter concentric hole had been drilled. Electrical contact was made *via* a copper wire soldered to the silver. The geometric area of the silver disk was 0.28 cm^2.

The SERS cell was built following the design of Brandt (16). Prior to the experiment, the silver electrode was polished with 5, 0.3 and 0.05 mμ alumina. After polishing, the electrode surface was rinsed with copious amounts of triply distilled water followed by sonication in triply distilled water. The SERS cell was then assembled and filled with an analyte solution which had been previously purged with nitrogen. The silver electrode was then subjected to an oxidation-reduction cycle (ORC), *viz.*, oxidation at 0.20 V for ten seconds followed by reduction of the generated silver salt at -0.30 V until the current decreased to about zero.

The spectra were taken using the 488.0 nm line of a Spectra Physics model 164-00 argon ion laser with an incident power of 100 mW at the electrode surface. The laser light was focused to a line image at the electrode with a cylindrical lens. The scattered light was focused onto the entrance slit of a Spex model 1401 double monochromator (2.0 cm^{-1} resolution) and detection was by photon counting (using an RCA model C31034-02 photomultiplier tube). Data collection was performed with a microcomputer system. The intensity of the SERS spectra varied considerably with changes in conditions. All spectra have been plotted with an arbitrary scale for the ordinate.

The electrode potential was controlled with an EG & G Princeton Applied Research (PAR) model 173 potentiostat/galvanostat and is referenced to a saturated calomel electrode (SCE). A PAR model 276 current-to-voltage converter allowed monitoring of current during the ORC and SERS experiments and it also provided for positive feedback iR compensation for accurate potential control.

Acknowledgment

This research was supported by the National Science Foundation, Grant CHE-8722764.

Literature Cited

1. Fleischmann, M.; Hendra, P. J.; McQuillan, A. J. Chem. Phys. Lett. 1974, 26, 163-166.
2. Jeanmaire, D. L.; Van Duyne, R. P. J. Electroanal. Chem. 1977, 84, 1-20.
3. Albrecht, M. G.; Creighton, J. A. J. Am. Chem. Soc. 1977, 99, 5215-5217.
4. Weaver, M. J.; Hupp, J. T.; Barz, F.; Gordon, J. G.; Philpott, M. R. J. Electroanal. Chem. 1984, 160, 321-333.
5. Rubim, J. C. J. Electroanal. Chem. 1987, 220, 339-350.
6. Anderson, M. R.; Evans, D. H. (submitted to J. Am. Chem. Soc.)
7. Dorain, P. B.; Von Raben, K. U.; Chang R. K. Surf. Sci. 1984, 148, 439-452.
8. Bunding, K. A.; Bell, M. I. Surf. Sci. 1983, 118, 329-344.
9. Laviron, E. Bull. Soc. Chim. Fr. 1961, 2325-2349.
10. Barradas, R. G.; Conway, B. E. Electrochim. Acta 1961, 5, 319-348.
11. Barradas, R. G.; Conway, B. E. Electrochim. Acta 1961, 5, 349-361.
12. Allen, C. S.; Van Duyne, R. P. Chem. Phys. Lett. 1979, 63, 455-459.
13. Jellinek, H. H. G.; Urwin, J. R. J. Phys. Chem. 1954, 58, 548-550.
14. Fischer, A.; Galloway, W. J.; Vaughan, J. J. Chem. Soc. 1964, 3591-3596.
15. Brignell, P. J.; Johnson, C. D.; Katritzky, A. R.; Shakir, N.; Tarhan, H. O.; Walker, G. J. Chem. Soc., B 1967, 1233-1235.
16. Brandt, E. S. Anal. Chem. 1985, 57, 1276-1280.

RECEIVED July 11, 1988

Chapter 27

Effect of Underpotentially Deposited Lead on the Surface-Enhanced Raman Scattering of Interfacial Water at Silver Electrode Surfaces

Jose C. Coria-Garcia and Jeanne E. Pemberton[1]

Department of Chemistry, University of Arizona, Tucson, AZ 85721

The influence of underpotentially deposited Pb monolayers and submonolayers on the surface enhanced Raman scattering (SERS) of interfacial H_2O at roughened Ag electrodes in aqueous chloride and bromide solutions is presented. With laser excitation at 5145 Å, the SERS intensity of the ν(OH) vibration of interfacial H_2O at 3495 cm^{-1} and 3505 cm^{-1} in aqueous chloride and bromide, respectively, decreases as the first monolayer of Pb is deposited. The rate of the decrease is essentially independent of the nature of the supporting electrolyte anion and is significantly different than the rate of decrease of the ν(Ag-X) (X=Cl^-, Br^-) vibrations in these media. The presence of interfacial H_2O at all coverages of UPD Pb precludes interpretation of these results strictly in terms of a decrease in adsorbate surface coverage. The results are interpreted as a combination of changes in electromagnetic enhancement resulting from alteration of the surface optical properties of the electrode in the presence of UPD Pb and decreased efficiency of photoassisted charge transfer resulting from alteration of the Fermi level at microscopic surface active sites in the presence of UPD Pb.

Several recent experimental investigations have been directed towards investigating the influence of underpotentially deposited foreign metal monolayers on the surface enhanced Raman scattering (SERS) ability of Ag and Au electrodes.(1-16) These studies have been undertaken with the intent of further elucidating the mechanism of the SERS phenomenon. SERS is generally thought to result from a combination of two mechanisms. The first is classical electromagnetic enhancement of the electric field at the interface between an appropriately roughened metal substrate and another medium. An additional mechanism is a resonance-like enhancement postulated to involve photoassisted charge transfer between the

[1]Address correspondence to this author.

0097–6156/88/0378–0398$06.00/0

metal and the adsorbate at unique sites on the metal surface known as active sites. These active sites have been documented to involve clusters of adatoms at the surface.(17)

In general, the deposition of monolayer and submonolayer amounts of nonenhancing foreign metals onto an enhancing metal electrode through the underpotential deposition (UPD) process results in a decrease in SERS intensity for all of the adsorbate probes studied to date. This decrease has been recently interpreted in terms of two effects, a change in substrate optical properties causing a decrease in the classical electromagnetic enhancement effect, and alteration of the energy levels of the active sites rendering the charge transfer process significantly less efficient.(14) One aspect of these studies which has yet to be adequately addressed, however, is the influence of the foreign metal layer on the surface coverage of the adsorbate during the deposition process. It is very likely that, in all of the studies performed thus far, the observed response represents a convolution of the effects noted above and a change in the adsorbate surface coverage. In fact, in several studies, a decrease in adsorbate surface coverage was proposed as a major cause of the SERS intensity decrease.(5-7,10) Although previous studies in this laboratory have been undertaken with the intent of addressing this issue using the ex-situ probe of x-ray photoelectron spectroscopy as a measure of adsorbate surface coverage (11), it is clearly more desirable to probe such changes in-situ to avoid complications resulting from removing the interface from the electrochemical environment.

The work reported here was designed to address the issue of adsorbate surface coverage in the effect on SERS of UPD Pb on Ag electrodes in aqueous chloride and bromide media using interfacial H_2O species as the probe molecule. No studies have been reported on the effect of UPD layers on the SERS of interfacial solvent molecules previously. However, the solvent is an ideal choice for such studies, because it will always remain in intimate contact with the electrode surface. Moreover, the SERS of interfacial H_2O has been characterized quite extensively in aqueous halide media (18-29) and allows the possible influence of anion on the response of the system to be assessed.

Experimental

The laser Raman system used for these studies has been described in detail in previous reports from this laboratory.(8,9) All spectra reported here were obtained with 5145 Å excitation from an Ar^+ laser. Laser power at the sample was typically 200 mW. Spectra of Cl^- and Br^- were acquired at 0.5 cm^{-1} increments over a 0.5 s integration period. Spectra of interfacial H_2O were acquired at 1.0 cm^{-1} increments over a 0.5 s integration period. All spectra were acquired as single scans with a 6 cm^{-1} bandpass. Peak areas were digitally determined assuming a straight-line background in all frequency regions.

In the low frequency region where the ν(Ag-Cl) and ν(Ag-Br) vibrations are observed, significant background intensity can obscure these vibrational features if they are weak in intensity. In order to improve the sensitivity of these measurements, the spectra acquired in this region before an oxidation-reduction cycle

(ORC) pretreatment were digitally subtracted from the spectra acquired after the ORC. The resulting spectra in this frequency region demonstrate a smaller influence from the background and allow accurate quantitation of peak areas even at low intensity.

In the frequency region where the ν(OH) vibrations of interfacial H_2O are observed, the normal Raman scattering from the bulk solution can obscure the SERS of interfacial H_2O if appropriate precautions are not taken. In the studies reported here, the SERS of interfacial H_2O was acquired with the electrode surface positioned as close to the electrochemical cell window as possible to minimize contributions from the bulk solution. When altering the electrode potential to deposit Pb onto the Ag electrode surface, the electrode was pulled away from the window several mm, the surface allowed to equilibrate at the new conditions, and the electrode repositioned near the cell window for spectral acquisition.

Electrochemical equipment and cells used for these investigations have also been described previously.(8,9) Polycrystalline Ag (Johnson Matthey, 99.9%) was mechanically polished with alumina (Buehler) to a mirror finish and sonicated in triply distilled H_2O before each run. All potentials were measured and are reported versus a saturated calomel reference electrode (SCE).

The solutions consisted of 5 x 10^{-3} M $Pb(NO_3)_2$ in either 0.1 M KCl or 0.1 M KBr made slightly acidic to pH 5.5 by the addition of HCl or HBr as appropriate, in order to maintain the solubility of the Pb^{2+}. All chemicals were reagent grade and were used as received. All solutions were prepared from triply distilled, deionized H_2O, the last distillation being from basic permanganate. All solutions were deaerated by bubbling with N_2 prior to use.

The Ag electrodes were subjected to potential sweep ORCs at 10 mV s^{-1} in either 0.1 M KCl or 0.1 M KBr from an initial potential of -0.30 V to more positive potentials. After ca. 30 mC cm^{-2} of anodic charge was passed, the direction of potential sweep was reversed to reduce the Ag halide surface to Ag metal. The roughened electrode was then removed under potential control at -0.30 V and immersed in the Pb^{2+}-containing test solution for SERS studies.

The fractional Pb monolayer coverages were calculated by comparing the charge under the stripping wave obtained at different potentials with that obtained for a full monolayer on a given electrode surface. A value of 300 μC cm^{-2} was used for a complete Pb monolayer as reported by Dickertmann, Koppitz, and Schultze.(30)

Results and Discussion

SERS of H_2O in the Absence of Pb^{2+}. The SERS of H_2O at Ag electrodes has been investigated thoroughly.(18-29) The ν(OH) vibration can be easily seen in electrochemical SERS studies if the electrode surface is separated from the cell window by only a thin film of solution. Using this approach, the SERS spectra of H_2O at Ag electrodes in 0.1 M KCl and 0.1 M KBr were obtained. The ν(OH) vibration is observed at 3495 cm^{-1} and 3505 cm^{-1} in Cl^- and Br^-, respectively. The relatively high position of this symmetric O-H vibration has been observed previously and has been interpreted as evidence for disruption of hydrogen bonding between water molecules by these anions, and weak hydrogen bonding between the H_2O and the anion.(18-20,26) Such interactions between the anion and the H_2O species give rise to a considerably narrower linewidth of this vibration than observed in bulk solution. Moreover, the acquisition of these spectra in K^+ solutions, a low hydration energy cation, implies that the H_2O molecules are preferentially aligned with their O ends facing the positively charged Ag electrode, as noted in earlier reports.(27)

It has been previously observed that the nature of the anion has a marked effect on the frequency of this band. In 1.0 M and 4.0 M KCl, bands are observed at 3498 cm^{-1} and 3433 cm^{-1}, respectively.(29) In 1.0 M KBr and 1.0 M KI, bands are observed at 3523 cm^{-1} and 3553 cm^{-1}, respectively.(18) In 0.5 M KCN, this band is observed at 3521 cm^{-1}.(18) These results show that at a similar halide concentration of 1.0 M, a decrease in frequency of the band is observed from 3553 cm^{-1} to 3523 cm^{-1} to 3498 cm^{-1} for I^-, Br^-, and Cl^-, respectively. The results obtained here show a similar decrease in position from 3505 cm^{-1} in Br^- to 3495 cm^{-1} in Cl^-. These bands are also shifted with respect to those observed in 1.0 M solutions of these anions, consistent with the anion concentration results noted above.

Evidence that H_2O species also interact with the Ag electrode independent of adsorbed anions comes from the potential dependence of the ν(OH) intensity as compared with the ν(Ag-X) (X=Cl^-, Br^-) intensities. The normalized intensities of the ν(Ag-X) (X=Cl^-, Br^-) vibrations in 0.1 M KCl and 0.1 M KBr are shown in Figure 1a, and the corresponding intensities of the ν(OH) vibration shown in Figure 1b. The observation that the intensity of the ν(OH) vibration reaches a maximum at more negative potentials than the ν(Ag-X) (X=Cl^-, Br^-) vibrations has been interpreted as indication that the H_2O molecules can become maximally adsorbed on the surface when the positive charge has decreased to allow partial desorption of the anions.(23) Obviously, the potential at which this occurs depends on the strength of interaction of the anion with the electrode.

The frequency of the ν(OH) band is also observed to decrease with potential in both KCl and KBr environments, as shown in Figure 2. This effect can be interpreted in terms of a weakening of the interaction strength of the H_2O with the Ag electrode surface as the potential and the charge on the electrode are made more negative. As the H_2O interacts less with the electrode, it can hydrogen bond more with other H_2O molecules, giving rise to the observed decrease in frequency of the ν(OH) vibration. This interpretation is consistent with the orientation of the H_2O molecule in K^+ solutions discussed above.

SERS of H_2O in the Presence of Pb^{2+}. The presence of 5 x 10^{-3} M Pb^{2+} in 0.1 M KCl and 0.1 M KBr in the potential region positive of UPD has no effect on the frequency of the ν(OH) band in these media. The only effects observed are that the 3495 cm^{-1} band in Cl^- has measurable intensity to more negative potentials than in the absence of Pb^{2+}, and the absolute intensities of this band are lower in both halide solutions than in the absence of Pb^{2+}. The fact that the 3495 cm^{-1} band in Cl^- can be observed at potentials where it is not observed in the absence of Pb^{2+} can be explained in terms of the Pb^{2+}/X^- (X=Cl^-, Br^-) species present in the interface. Previous work in this laboratory has demonstrated the existence of Pb^{2+}

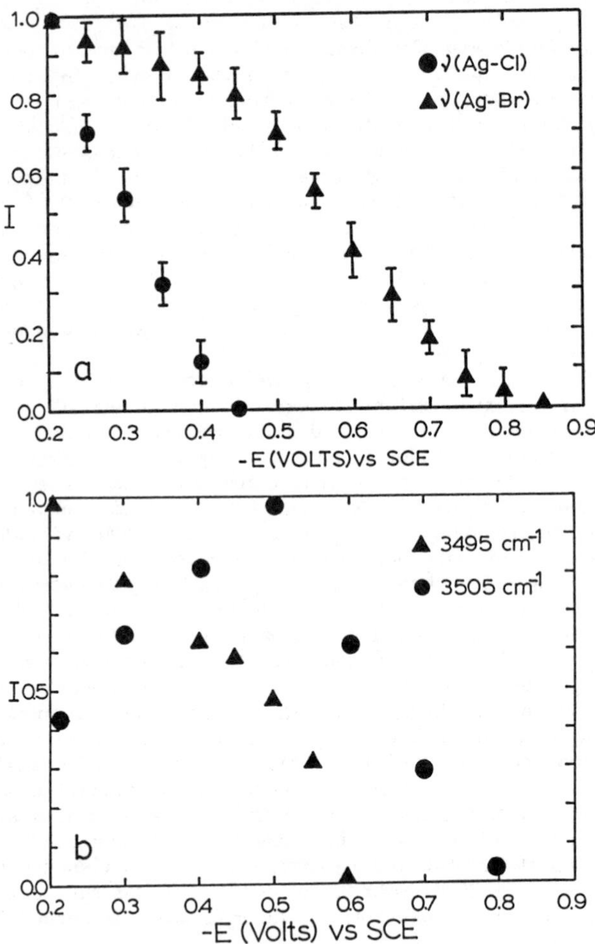

Figure 1. Intensity as a function of potential for a) ν(Ag-Cl) band (238 cm^{-1}) and ν(Ag-Br) band (163 cm^{-1}), and b) ν(OH) bands in Cl^- (3495 cm^{-1}) and Br^- (3505 cm^{-1}).

halide complexes in the interface at potentials more positive than Pb UPD.(12) These complexes stabilize the halides in the interface such that they remain at more negative potentials. The decrease in absolute intensity of the H_2O band in the presence of Pb^{2+} most likely results from the fact that there are fewer free halide ions in the interface with which the H_2O can interact.

The potential regions within which the UPD of Pb occur are shown in the cyclic voltammograms in Figure 3. In Cl^- media, Pb UPD occurs between -0.40 and -0.48 V. In Br^- media, Pb UPD is observed at slightly more negative potentials between -0.43 and -0.52 V. Considering all of the data discussed above, it is unlikely that drastic decreases in surface coverage by H_2O species occur in these relatively narrow potential regions. In fact, it is known that during the UPD of Pb on Ag, the surface coverage of Cl^- and Br^- decreases. Therefore, the surface coverage of H_2O should actually increase during Pb UPD as halide ions are partially desorbed. These considerations are critical in understanding the SERS response of the ν(OH) band during UPD of Pb.

Figure 4 shows the normalized SERS intensities of the 3495 cm^{-1} ν(OH) band in Cl^- media and the 3505 cm^{-1} band in Br^- media as a function of Pb coverage from zero to one monolayer. Also shown are the corresponding normalized intensities of the ν(Ag-Cl) and ν(Ag-Br) vibrations as a function of Pb coverage for comparison. As these data demonstrate, the intensity of the ν(OH) vibration decreases as the Pb coverage increases in both halide environments. The intensities of the ν(Ag-X) (X=Cl^-, Br^-) vibrations also decrease with increasing Pb coverage. These decreases are consistent with the results of other studies involving the effect of UPD thin films on the SERS of other adsorbate probes in this (8-14) and other (1-7,15,16) laboratories.

The data in Figure 4 show different behavior in three different coverage regions. The first region occurs between zero coverage and ca. 30% of a Pb monolayer. In this region, the intensities of all vibrations decrease. The rate of increase of the ν(Ag-Br) intensity is greater than that for the ν(Ag-Cl) band. This result has been observed and explained previously.(13) The rate of intensity decrease in the ν(OH) band in this region in both halide environments is significantly less than that of either ν(Ag-X) (X=Cl^-, Br^-) vibration. Moreover, the rate of decrease is independent of the anion. This latter observation is further evidence that the H_2O species being monitored are interacting with the Ag electrode independent of the specifically adsorbed halide ions.

The second region of distinct behavior occurs between ca. 40% and 80% of a Pb monolayer. In this region, the intensities of the ν(Ag-X) (X=Cl^-, Br^-) bands decrease to unmeasurable values. In contrast, the intensities of the ν(OH) vibrations remain essentially constant in both media. The third region of behavior occurs for Pb coverages greater than 80% of a monolayer. The intensities of the ν(Ag-X) (X=Cl^-, Br^-) bands remain at unmeasurable levels. The intensities of the ν(OH) bands decrease rapidly to unmeasurable levels between 80% of a monolayer and one complete monolayer. The behavior of the ν(OH) bands in the second and third regions are observed to depend only slightly on the nature of the anion.

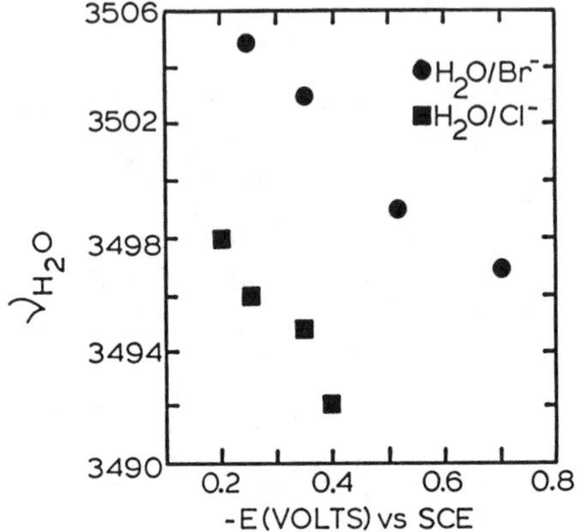

Figure 2. Frequency of the ν(OH) vibration as a function of potential in Cl^- and Br^- media.

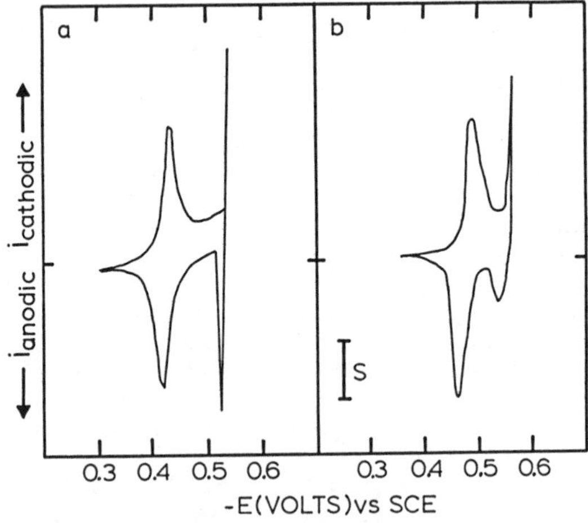

Figure 3. Cyclic voltammetry of Pb UPD on Ag electrode in a) 0.1 M KCl, and b) 0.1 M KBr. Sweep rate 10 mV s^{-1}, S = 47 μA cm^{-2}.

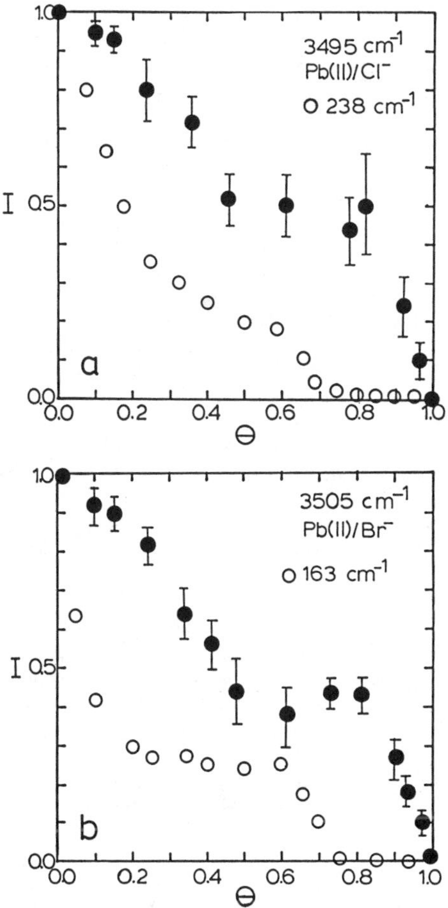

Figure 4. Normalized SERS intensity as a function of Pb coverage for a) 3495 cm^{-1} ν(OH) band and 238 cm^{-1} ν(Ag-Cl) band in 0.1 M KCl, and b) 3505 cm^{-1} ν(OH) band and 163 cm^{-1} ν(Ag-Br) band in 0.1 M KBr.

It is clear from these H_2O results that the decreases in SERS intensities as UPD layers grow on electrodes cannot be solely attributed to decreases in adsorbate coverage. This report represents the first concrete evidence for this fact in the electrochemical SERS environment.

It is likely that the decreases observed can be rationalized in terms of two contributions. Changes in surface optical properties resulting from modification by the foreign metal have been shown to decrease the electromagnetic enhancement contribution to SERS. However, for the case of Pb UPD on Ag, this effect has been shown to account for only ca. 40% of the decrease in going from zero coverage to one monolayer.(14) Moreover, this model does not account for the relatively rapid decrease in intensity observed with the deposition of small (i.e., less than 20% of a monolayer) amounts of Pb on the Ag surface.

The second contribution to these observed decreases that is proposed to be particularly important at small Pb coverages is alteration of the microscopic Fermi energy at Ag active sites, presumably adatom clusters, rendering photoassisted charge transfer less efficient.(14) In this model, the Fermi energy is proposed to increase as the Pb coverage increases so that, at a constant excitation energy, the charge transfer from the metal at sites of atomic scale roughness to acceptor levels of the adsorbate gradually goes out of resonance. It is proposed that this type of mechanism is responsible for the quenching of SERS of H_2O in the first coverage region reported here. Due to the fact that the H_2O species remain at the surface, the quenching of the charge transfer process is not complete, and some atomic scale roughness remains, the SERS intensity of the $\nu(OH)$ vibration is retained at ca. 50% throughout the second coverage region.

In the third region of coverage, most of the atomic scale roughness has been proposed to be irreversibly destroyed as the Pb layer rearranges to assume the final hexagonal close packed configuration of the monolayer.(9) This loss of atomic scale roughness results in the irreversible decrease in $\nu(OH)$ intensity to essentially unmeasurable levels. This observation further emphasizes the importance of the chemical enhancement mechanism contribution to SERS in electrochemical systems.

Conclusions

The effect of UPD Pb on the SERS of H_2O at Ag electrodes is presented in this report. The work reported here demonstrates that a significant decrease in adsorbate coverage is not responsible for the quenching of SERS at electrodes in the presence of UPD layers. These results lend credence to a previously postulated model in which the quenching effect is interpreted in terms of both a decrease in the electromagnetic enhancement at these surfaces resulting from a change in the macroscopic optical properties of the surface as the UPD layer is formed, and a decrease in the efficiency of photoassisted charge transfer resulting from alteration of the Fermi energy at microscopic active sites on the surface at which charge transfer occurs.

Acknowledgments

The authors gratefully acknowledge the financial support of this work by the National Science Foundation (CHE-8614955).

Literature Cited

1. Loo, B. H.; Furtak, T. E. Chem. Phys. Lett. 1980, 71, 68.
2. Pettinger, B; Moerl, L. Solid State Commun. 1982, 43, 315.
3. Pettinger, B.; Moerl, L. J. Electron Spectrosc. Relat. Phenom. 1983, 29, 383.
4. Watanabe, T.; Yanigahara, H.; Honda, K.; Pettinger, B.; Moerl, L. Chem. Phys. Lett. 1983, 96, 649.
5. Kester, J. J. J. Chem. Phys. 1983, 78, 7466.
6. Furtak, T. E.; Roy, D. Phys. Rev. Lett. 1983, 50, 1301.
7. Roy, D.; Furtak, T. E. J. Chem. Phys. 1984, 81, 4168.
8. Guy, A. L.; Bergami, B.; Pemberton, J. E. Surf. Sci. 1985, 150, 226.
9. Guy, A. L.; Pemberton, J. E. Langmuir 1985, 1, 518.
10. Kellogg, D. S.; Pemberton, J. E. J. Phys. Chem. 1987, 91, 1127.
11. Guy, A. L.; Pemberton, J. E. Langmuir 1987, 3, 125.
12. Coria-Garcia, J. C.; Pemberton, J. E.; Sobocinski, R. L. J. Electroanal. Chem. 1987, 219, 291.
13. Pemberton, J. E.; Coria-Garcia, J. C.; Hoff, R. L. Langmuir 1987, 3, 150.
14. Guy, A. L.; Pemberton, J. E. Langmuir 1987, 3, 777.
15. Leung, L. H.; Weaver, M. J. J. Electroanal. Chem. 1987, 217, 367.
16. Fleischmann, M.; Tian, Z. Q. J. Electroanal. Chem. 1987, 217, 385.
17. Roy, D.; Furtak, T. E. Phys. Rev. B 1986, 34, 5111.
18. Fleischmann, M.; Hendra, P. J.; Hill, I. R.; Pemble, M. E. J. Electroanal. Chem. 1981, 117, 243.
19. Pettinger, B.; Philpott, M. R.; Gordon, J. G. J. Chem. Phys. 1981, 74, 934.
20. Chen, T. T.; Owen, J. F.; Chang, R. K.; Laube, B. L. Chem. Phys. Lett. 1982, 89, 356.
21. Macomber, S. H.; Furtak, T. E.; Devine, T. M. Surf. Sci. 1982, 122, 556.
22. Owen, J. F.; Chen, T. T.; Chang, R. K.; Laube, B. L. Surf. Sci. 1983, 125, 679.
23. Owen, J. F.; Chen, T. T.; Chang, R. K.; Laube, B. L. J. Electroanal. Chem. 1983, 150, 389.
24. Fleischmann, M.; Hill, I. R. J. Electroanal. Chem. 1983, 146, 367.
25. Pettinger, B.; Moerl, L. J. Electroanal. Chem. 1983, 150, 415.
26. Owen, J. F.; Chang, R. K. Chem. Phys. Lett. 1984, 104, 510.
27. Chen, T. T.; Smith, K. E.; Owen, J. F.; Chang, R. K. Chem. Phys. Lett. 1984, 108, 32.
28. Chen, T. T.; Chang, R. K. Surf. Sci. 1985, 158, 325.
29. Furtak, T. E. J. Electroanal. Chem. 1983, 150, 375.
30. Dickertmann, D.; Koppitz, F. D.; Schultze, J. W. Electrochim. Acta 1976, 21, 967.

RECEIVED May 19, 1988

Chapter 28

Synthesis and Electrochemical Properties of a Polythiophene–Viologen Polymer

Ching-Fong Shu and Mark S. Wrighton

Department of Chemistry, Massachusetts Institute of Technology, Cambridge, MA 02139

This article reports the synthesis and electrochemical properties of poly(I) (I ≡ 1-methyl-1'-(3-thiophene-3-yl)-propyl-4,4'-bipyridinium bis-hexafluorophosphate), a polymer with a conducting polymer, polythiophene, as the backbone and a redox couple, viologen (V^{2+} ≡ N,N'-dialkyl-4,4'-bipyridinium), as the pendant group. Poly(I) can be grown by the oxidation of I and thereby deposited onto the electrode surface. The cyclic voltammetry of poly(I) in CH_3CN/0.1 M [n-Bu_4N]PF_6 shows an oxidation wave at 0.6 V vs Ag^+/Ag and the resistivity of the polymer declines by a factor of 10^5 upon changing the potential from 0.4 V to 0.8 V, consistent with the behavior of polythiophene and its derivatives. Poly(I) also exhibits two successive one-electron reduction waves at -0.6 V and -1.1 V vs. Ag^+/Ag, as expected for the V^{2+} group. An array of eight closely-spaced microelectrodes can be coated and connected by poly(I). Analogous to generation and collection experiments at a rotating ring/disk electrode, charge transport occurs from one microelectrode (generator) to another (collector)via charge transport in the polymer. The mechanism of charge transport via the $V^{2+/+}$ system is consistent with the expectations from "redox conduction" The value of the diffusion coefficient for charge transport, D_{CT}, for the $V^{2+/+}$ system is measured to be ~7 x 10^9 cm^2/s. The temperature dependence of the steady state generator-collector current shows an Arrhenius activation energy of 9.8 kcal/mole. A poly(I)-based microelectrochemical transistor gives an I_D-V_G characteristic with a sharp maximum in I_D at $V_G = E^{o'}$ of $V^{2+/+}$, and for V_D = 25 mV the full width at half-height of the I_D-V_G curve is only 56 mV. The transistor also shows turn on at V_G corresponding to oxidation of the polythiophene backbone, as reported earlier for poly(3-methylthiophene)

0097–6156/88/0378–0408$06.75/0

In this article we report the synthesis and electrochemical properties of the polymer derived from oxidation of I, poly(I), and the characteristics of a microelectrochemical transistor based on the polymer. Poly(I), which is formed by electrochemical oxidation of I, Equation 1, consists of a conducting polymer backbone, polythiophene,

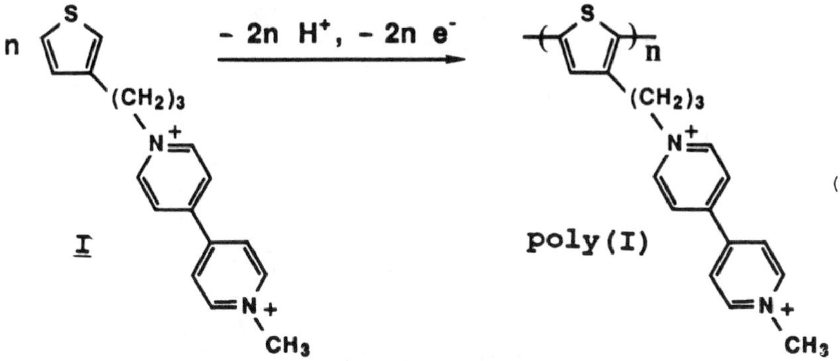

$$n\ \text{I} \xrightarrow{-2n\,H^+,\ -2n\,e^-} \text{poly(I)} \qquad (1)$$

and a pendant viologen redox group, V^{2+}. An array of eight closely-spaced microelectrodes coated with poly(I) can function as a microelectrochemical transistor.(1-2)

Many electrochemical studies have been devoted to the development of new electrode coating materials.(3-4) It has been shown that electropolymerization can be used to synthesize conducting polymer films such as polypyrrole containing conventional redox couples like viologen, ruthenium complexes, ferrocene, and quinone.(5-12) Recent studies (13-14) show that poly(3-methylthiophene) is very durable in aqueous solution and has a superior transconductance compared to that of polypyrrole(1) or polyaniline.(15) The chemical ruggedness and high conductivity make poly(3-alkylthiophene) a good backbone upon which to attach conventional redox centers. The V^{2+} group has been demonstrated to be very useful in electrocatalysis and photoelectrocatalysis of redox reactions and in electrochromism.(3,4,16-20) The electrochemical oxidation of I produces a surface-confined film of electroactive polythiophene containing one V^{2+} redox center per repeat unit of the polymer. Studies of the conductivity of poly(I) vs. potential reveal fundamental information regarding the amount of charge withdrawn per repeat unit to achieve a given degree of conductivity. The $V^{2+/+}$ centers represent a kind of internal calibration since the amount of V^{2+} present can be assessed electrochemically.

Recently, several molecule-based microelectrochemical devices have been developed by the Wrighton group.(14,15,21-22) A microelectrode array coated with poly(I) results in a microelectrochemical transistor with the unique characteristic that shows "turn on" in two gate potential, V_G, regimes, one associated with the polythiophene switching from an insulator to a conductor upon oxidation and one associated with the $V^{2+/+}$ conventional redox centers.

Experimental Section

Chemicals. CH_3CN was distilled from CaH_2 and stored under N_2. [n-Bu_4N]PF_6 was prepared by adding an aqueous solution of [n-Bu_4N]Br to an aqueous solution of NH_4PF_6. The precipitate was collected and purified by recrystallization from 95% EtOH. The [n-Bu_4N]PF_6 was dried under vacuum at 100 °C overnight. 3-(Bromomethyl)thiophene (3-thenylbromide) was prepared via the method reported by Gronowitz and Frejd.(23) 2,2'-Azobis(2-methylpropionitrile), CH_3I, 4,4'-bipyridine, N-bromosuccinimide, diethyl malonate, 3-methylthiophene, PBr_3, $[Ru(NH_3)_6]Cl_3$, and K_2PtCl_4 were used as received from commercial sources.

Preparation of Diethyl 3-thenylmalonate. 2.3 g (0.1 mole) of Na was added to 100 ml of absolute EtOH in a 250 ml three-necked flask. When all the Na had reacted, 16.8 g (0.105 mole) of diethyl malonate was added through an addition funnel. This was followed by the dropwise addition of 17.7 g (0.1 mole) of 3-thenyl bromide. The mixture was refluxed under N_2 overnight. The amount of solvent was reduced to about 50 ml by rotary evaporation. The residue was then treated with 40 ml of H_2O. The ester was extracted with Et_2O and dried over Na_2SO_4. The Et_2O was removed by rotary evaporation and the product was purified by vacuum distillation: yield, 13 g; bp 74 °C (0.05 torr); ^1H NMR (60 MHz, CCl_4) δ 6.93 (m, 1 H), 6.70 (m, 2 H), 4.00 (q, 4H, J = 7 Hz), 3.47 (t, 1 H, J = 7 Hz), 3.07 (d, 2H, J = 7 Hz), 1.16 (t, 6 H, J = 7 Hz).

Preparation of 3-(Thiophene-3-yl)propionic Acid. 13 g (0.05 mole) of diethyl 3-thenylmalonate was added dropwise through an addition funnel to a solution of 10 g of NaOH in 100 ml of H_2O. The mixture was then refluxed overnight. The resulting solution was cooled with an ice bath and concentrated HCl(aq) was added until the pH of the solution became 1. The product was extracted with Et_2O and dried over Na_2SO_4. The Et_2O was removed by rotary evaporation and the white solid 3-thenylmalonic acid resulted: yield, 10 g; mp 137 °C; ^1H NMR (60 MHz, CD_3CN) δ 8.00 (br, 2 H), 7.10 (m, 1 H), 6.82 (m, 2 H), 3.58 (t, 1 H, J = 7 Hz), 3.17 (d, 2 H, J = 7 Hz).

3-Thenylmalonic acid (9.8 g) was decarboxylated by heating at a temperature of 140-150 °C in an oil bath. The reaction was run overnight under N_2. The solid 3-(thiophene-3-yl)propionic acid is formed cleanly: yield, 7.5 g; mp 56 °C; ^1H NMR (60 MHz, $CDCl_3$) δ 10.11 (br, 1 H), 7.01 (m, 1H), 6.77 (m, 2 H), 2.78 (m, 4 H).

Preparation of 3-(Thiophene-3-yl)propanol. A solution of 7.5 g (0.05 mole) of 3-(thiophene-3-yl)propionic acid in 50 ml of dry THF was added dropwise to a solution of 2.8 g (0.075 mole) of $LiAlH_4$ in 100 ml of dry THF. The resulting solution was refluxed for 4 h under N_2. After careful addition of H_2O, 7.5 ml of concentrated H_2SO_4 in 75 ml of H_2O was added to the mixture. The product was extracted with Et_2O. The Et_2O solution was washed successively with aqueous $NaHCO_3$ solution and H_2O ,then dried over Na_2SO_4. The Et_2O was removed by rotary evaporation and the product was obtained by vacuum distillation: yield, 6.19 g; bp 68 °C (0.05 torr); ^1H NMR (300 MHz, $CDCl_3$) δ 7.24 (m, 1 H), 6.94 (m, 2 H), 3.65 (t, 2 H, J = 7.5 Hz), 2.72 (t, 2 H, J = 7.5 Hz), 2.03 (br, 1 H), 1.88 (quintet, 2 H, J = 7.5 Hz).

Preparation of 3-(Thiophene-3-yl)propyl Bromide. 2.5 g of PBr_3 was added slowly to 3.55 g of 3-(thiophene-3-yl)propanol in a 50 ml three necked flask with stirring at 0°C. The resulting solution was stirred at room temperature overnight. The product was isolated from the reaction mixture by vacuum distillation: Yield, 2.4 g; bp 60 °C (0.05 torr); ^{1}H NMR (300 MHz, CD_3Cl) δ 7.25 (m, 1 H), 6.98 (d, 1 H, J = 3 Hz), 6.94 (d, 1 H, J = 5Hz), 3.39 (t, 2 H, J = 7 Hz), 2.80 (t, 2 H, J = 7 Hz, 2.15 (quintet, 2 H, J = 7 Hz).

Preparation of 1-(3-Thiophene-3-yl)propyl-4,4'bipyridinium Hexafluorophosphate. A solution of 1.45 g (7 mmol) of 3-(thiophene-3-yl)propyl bromide and 3.12 g (20 mmol) of 4,4'-bipyridine in 50 ml of CH_3CN was refluxed overnight. The amount of CH_3CN was reduced to 5 ml by rotary evaporation and the Br^- salt was precipitated by adding 100 ml of Et_2O. The Br^- salt was isolated from the unreacted 4,4'-bipyridine by filtration and converted to the PF_6^- salt by dissolving in H_2O and addition to a solution of 2.28 g of NH_4PF_6 in 20 ml of H_2O. The PF_6^- salt precipitated and was isolated by filtration. The product was recrystallized from H_2O: yield, 3.0 g; mp 179 °C dec; ^{1}H NMR (250 MHz, $(CD_3)_2SO$) δ 9.20 (d, 2 H, J = 7 Hz), 8.87 (dd, 2 H, J = 4.6 Hz, 1.6 Hz), 8.59 (d, 2 H, J = 7 Hz), 8.02 (dd, 2 H, J = 4.6 Hz, 1.6 Hz), 7.45 (dd, 1 H, J = 5 Hz, 3 Hz), 7.19 (dd, 1 H, J = 3 Hz, 1 Hz), 6.98 (dd, 1 H, J = 5 Hz, 1 Hz), 4.66 (t, 2 H, J = 7.4 Hz), 2.65 (t, 2 H, J = 7.5 Hz), 2.32 (quintet, 2 H, J = 7.3 Hz).

Preparation of Compound I. A solution of 3.55 g of CH_3I and 2.13 g of 1-(3-thiophene-3-yl)-4,4'-bipyridinium hexafluorophosphate in 50 ml of CH_3CN was refluxed overnight. The volume of the solution was reduced to 5 ml by rotary evaporation and the I^- salt was precipitated by adding 100 ml of Et_2O. The I^- salt was isolated from the solution by filtration and converted to the PF_6^- salt by dissolving in H_2O and adding to NH_4PF_6(aq) solution. The product precipitated and was isolated by filtration. The product was recrystallized from H_2O: yield, 2.3 g; mp 234 °C dec; ^{1}H NMR (250 MHz, $(CD_3)_2SO$) δ 9.35 (d, 2 H, J = 6.8 Hz), 9.27 (d, 2 H, J = 6.8 Hz), 8.73 (d, 2 H, J = 6.8 Hz), 8.72 (d, 2 H, 6.8 Hz), 7.45 (dd, 1 H, J = 5 Hz, 3 Hz), 7.19 (dd, 1 H, J = 3 Hz, 1 Hz), 6.99 (dd, 1 H, J = 5 Hz, 1 Hz), 4.72 (t, 2 H, J = 7.2 Hz), 4.42 (s, 3 H), 2.70 (t, 2 H, J = 7.2 Hz) 2.33 (quintet, 2 H, J = 7.2 Hz); MS, $(M-PF_6)^+$ 441.1012 (obsd), 441.0989 (calcd), composition $C_{18}H_{20}N_2SPF_6$; $(M-2PF_6)^+$ 296.1354 (obsd), 296.1347 (calcd), composition $C_{18}H_{20}N_2S$.

Electrochemical Equipment. Electrochemical experiments were performed using either a PAR Model 175 universal programmer and a PAR Model 363 potentiostat/galvanostat, or a Pine Instruments RDE-4 bipotentiostat, coupled with a Kipp and Zonen BD 91 X-Y-Y' recorder. The current-time response for the chronoamperometry experiments was recorded with a Nicolet 4094 digital oscilloscope. All potentials were measured vs. a Ag/10^{-2} M Ag^+ reference electrode.

Preparation of Microelectrode Arrays. The microelectrode arrays used in the work were arrays of microelectrodes each ~80 µm long, 2.3 µm wide and 0.1 µm thick and spaced 1.7 µm apart. Fabrication and encapsulation of the microelectrode arrays has been described previously.(14,15,21-22) Prior to use, arrays of microelectrodes were cleaned by an rf O_2 plasma etch to remove residual photoresist, followed by cycling the potential of each electrode between -1.5 V and -2.0 V vs. SCE in 0.1 M aqueous K_2HPO_4 to evolve H_2. The microelectrodes were then tested by examining their electrochemical behavior in aqueous 0.1 M LiCl containing 5 mM $Ru(NH_3)_6^{3+}$. Such a test reveals, for a good microelectrode, a sigmoidal current-voltage curve with a plateau current of ~25nA.

The Pt microelectrodes were modified by electrochemical deposition of Pt onto their surface to "activate" the electrodes and reduce the inter-electrode gap.(21-22,24) The Pt was deposited from aqueous 0.1 M K_2HPO_4 containing 2mM K_2PtCl_4. The electrodes were platinized one at a time by holding all electrodes not to be plated at 0.2 V vs. SCE while the one to be plated was held at a negative potential to deposit Pt. Two different arrangements of platinized electrodes were used. As shown in Scheme I(a), one arrangement is that all eight electrodes in an array were lightly platinized and the deposition of the Pt typically involved ~150 nC. The arrays thus fabricated had a spacing of ~1.3 µm between microelectrodes. The second arrangement, as shown in Scheme I(b), is that electrodes #2,4, 6, and 8 were lightly platinized and electrodes #1,3,5, and 7 were heavily platinized. The deposition of the Pt involved ~5 µC of charge per heavily platinized electrode. The spacing of microelectrodes by this procedure is ~0.2 µm.

Electrochemical Growth of Poly(I). Poly(I) can be grown onto electrode surfaces by oxidizing I in CH_3CN/0.1 M [n-Bu_4N]PF_6 solution. The typical procedure involves cycling (200 mV/s) the potential of an electrode to be derivatized between 0.0 V and 1.5 V vs. Ag^+/Ag in the presence of 0.2 M I. The deposited polymer exhibits cyclic voltammetry waves at 0.6 V, -0.6 V and -1.1 V vs Ag^+/Ag.

Thickness Measurement. The thickness of poly(I) at different coverages was obtained using a Tencor alpha-step 100 surface profile measuring system. Electrodes used were glass slides coated with Pt by electron beam evaporation. In order to produce a "step" across which the stylus of the surface profiler was drawn, Apiezon N grease was applied to part of the electrode surface and was removed with CH_2Cl_2 after derivatization with poly(I).

Results and Discussion

Synthesis of Compound I. As shown in Scheme II, 3-(thiophene-3-yl)propyl bromide can be prepared by a two-carbon homologation(25) of 3-thenyl bromide via reaction with diethyl malonate to form diethyl 3-thenylmalonate. This is followed by saponification, decarboxylation, reduction of acid to alcohol,(26) and replacement of the hydroxyl group with bromide by reacting with PBr_3.(27) Compound I is synthesized by mono-quaternization of an excess of 4,4'-bipyridine with 3-(thiophene-3-yl)propyl bromide followed by N-methylation with CH_3I. All the intermediates in Scheme II have been identified by NMR spectroscopy. I has been characterized by NMR and high resolution mass spectroscopy and by electrochemistry.

Preparation and Electrochemical Behavior of Macroelectrodes Modified with Poly(I). The electrochemical behavior of I was investigated in CH_3CN/0.1 M [n-Bu_4N]PF_6 via cyclic voltammetry at a Pt electrode. There are two reversible waves due to the successive, reversible one-electron reductions of V^{2+}, $E^{\circ\prime}$ ($V^{2+/+}$) = -0.72 V, $E^{\circ\prime}$ ($V^{+/o}$ = -1.15 V vs. Ag^+/Ag. In addition there is one anodic wave at ~1.5 V vs.

(a) Array of eight lightly platinized microelectrodes.

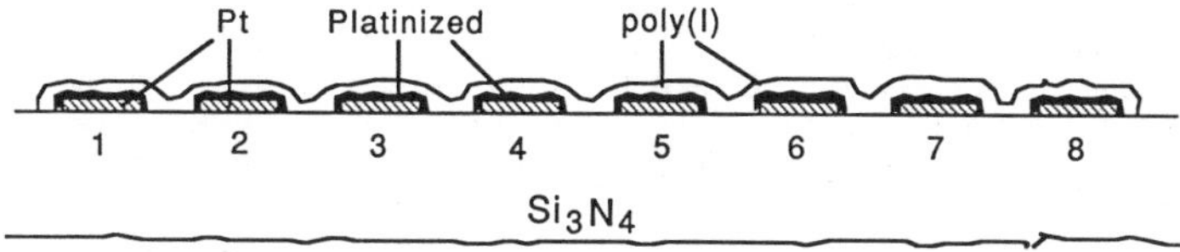

(b) Array of eight alternating heavily and lightly platinized microelectrodes

Pt Platinized poly(I)

1 2 3 4 5 6 7 8

Si_3N_4

Scheme I. Arrangements of platinized microelectrode arrays used for experimentation in this work.

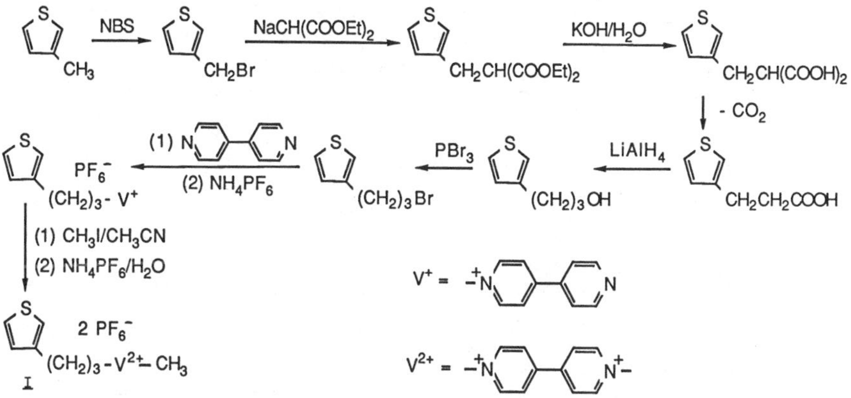

S, CH_3 → NBS → S, CH_2Br → $NaCH(COOEt)_2$ → S, $CH_2CH(COOEt)_2$ → KOH/H_2O → S, $CH_2CH(COOH)_2$

↓ $- CO_2$

S, CH_2CH_2COOH → $LiAlH_4$ → S, $(CH_2)_3OH$ → PBr_3 → S, $(CH_2)_3Br$ → (1) N⟨⟩–⟨⟩N (2) NH_4PF_6 → S, PF_6^- $(CH_2)_3$- V^+

↓ (1) CH_3I/CH_3CN (2) NH_4PF_6/H_2O

S, 2 PF_6^- $(CH_2)_3$-V^{2+}- CH_3

I

V^+ = $-\overset{+}{N}$⟨⟩–⟨⟩N

V^{2+} = $-\overset{+}{N}$⟨⟩–⟨⟩$\overset{+}{N}-$

Scheme II. Synthetic procedure used to prepare compound I.

Ag+/Ag, in the range of the oxidation potential of 3-methylthiophene,(14) due to the irreversible oxidation of the monomeric thiophene unit. Figure 1 shows the typical cyclic voltammetry upon repeatedly scanning the potential of a Pt electrode between 0.0 V and 1.5 V vs. Ag^+/Ag in a solution of 0.2 M I. Repeated scanning results in a continuous increase in the size of the wave at 0.6 V vs. Ag^+/Ag, which is believed to be the oxidation wave of the polythiophene backbone of poly(I). The amount of polymer produced per unit of charge in the deposition has not been measured, but is qualitatively similar to polythiophene itself. A film of poly(I) on the Pt electrode can be seen with the naked eye. After deposition of poly(I), the polymer-modified electrode is transferred, after thorough rinsing, to clean electrolyte for further characterization.

The cyclic voltammetry of a poly(I)-coated Pt electrode with a coverage of ~8 x 10^{-8} mol cm^2 in CH_3CN/0.1 M [n-Bu_4N]PF_6 as a function of sweep rate is shown in Figure 2. As expected, there is a cyclic voltammetry wave in the range of the oxidation potential of polythiophene, at about +0.6 V vs. Ag^+/Ag. The peak height of the anodic peak varies linearly with sweep rate at least up to 200 mV/s. In the region expected for the reduction of V^{2+}, there are two successive one-electron waves associated with the V^{2+} centers, $E^{\circ\prime}$ ($V^{2+/+}$) = -0.62 V, $E^{\circ\prime}$ ($V^{+/o}$) = -1.11 V vs. Ag^+/Ag. The peak heights of the cyclic voltammetry waves do not vary linearly with sweep rate or with (sweep rate)$^{1/2}$. Rather, for the sweep rates shown, the sweep rate dependence is intermediate between square-root and linear. These data suggest that the redox cycling of the $V^{2+/+/o}$ system is sluggish compared to oxidation/reduction of the polythiophene backbone. For the potential regime where $V^{2+/+/o}$ redox cycling is found, the polythiophene is non-conducting and cannot assist the rate of redox cycling of the $V^{2+/+/o}$. Therefore, charge transport via $V^{2+/+/o}$ is expected to be as in other viologen-based polymers; i.e. via redox conduction.(28-30)

The ratio of the integrated currents for the first reduction wave of V^{2+} and the oxidation wave of the polythiophene from 0.4 V to 1.0 V vs. Ag^+/Ag is about 4. This value means that upon oxidation of poly(I) one electron is withdrawn from four repeat units in the backbone of the polymer upon scanning to +1.0 V vs. Ag^+/Ag. At this potential, the polythiophene achieves its maximum conductivity (*vide infra*). The level of oxidation to achieve maximum conductivity is consistent with the result reported by Garnier and co-workers (31-33) that the doping level of oxidized polythiophene is about 25%, but the Garnier work did not establish that the 25% doping level corresponds to maximum conductivity. Scheme III illustrates the electrochemical processes of poly(I) showing reversible oxidation of the polythiophene backbone and reversible reduction of the pendant V^{2+} centers.

Preparation and Electrochemical Behavior of Microelectrodes Modified with Poly(I). A microelectrode array consisting of eight Pt microelectrodes is cleaned and platinized as described in the Experimental Section. Poly(I) is deposited on the platinized microelectrode array by scanning the potential of the electrode(s) from 0 V to 1.5 V vs. Ag^+/Ag in a solution of 0.2 M I in CH_3CN/0.1 M [n-Bu_4N]PF_6. The resulting array is illustrated in Scheme I. The

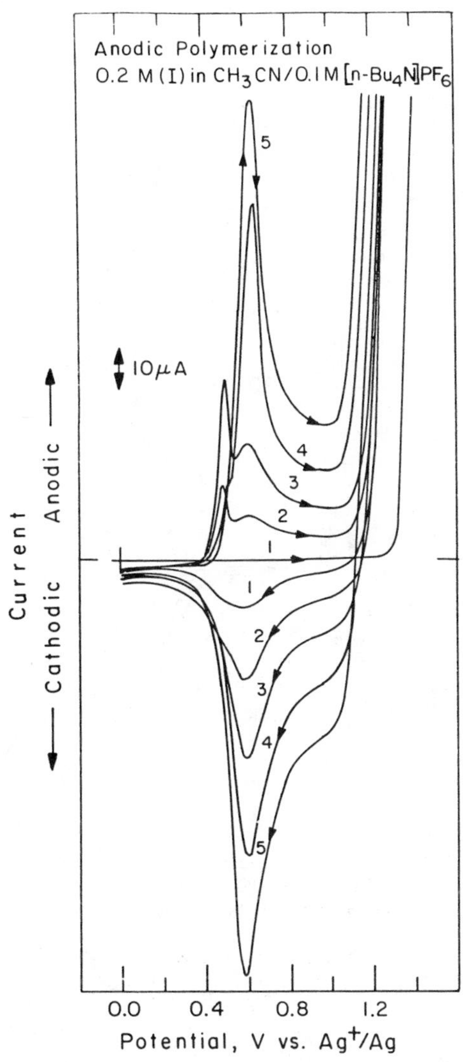

Figure 1. Cyclic voltammetry (200 mV/s) accompanying repeated scanning of the potential of a Pt electrode between 0.0 V and 1.5 V vs. Ag^+/Ag in a solution of 0.2 M I in CH_3CN/0.1 M [n-Bu_4N]PF_6.

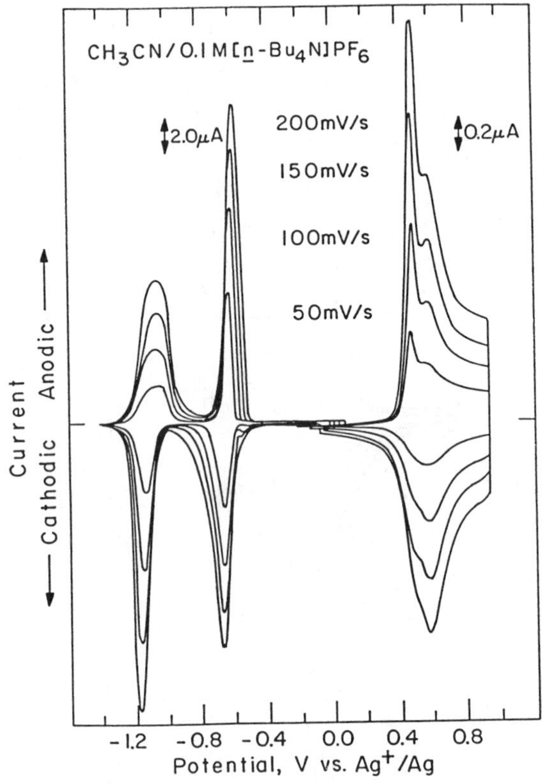

Figure 2. Cyclic voltammetry as a function of scan rate of a poly(I)-coated Pt electrode with a coverage of 8×10^{-8} mol/cm^2 of V^{2+} centers in CH_3CN/0.1 M [n-Bu_4N]PF_6.

cyclic voltammogram of the poly(I)-coated microelectrode shows well-developed cyclic voltammetry waves. The cyclic voltammograms for a pair of adjacent, poly(I)-coated microelectrodes in CH_3CN/0.1 M [n-Bu_4N]PF_6 in the range of the oxidation potential of polythiophene are shown in Figure 3. The essential features of the cyclic voltammetry waves for the derivatized microelectrodes are the same as for macroscopic electrodes coated with poly(I). The importance of the data in Figure 3 is that it shows the two microelectrodes are electrically connected with poly(I) when the potential of the electrodes are in the range of the oxidation potential of polythiophene. The key fact is that the cyclic voltammograms have nearly the same area when wire A alone is driven, when wire B alone is driven, and when wires A and B together are driven. These facts are consistent with the conclusion that wires A and B are connected by poly(I) and that all of the polymer is accessible by addressing A or B or by addressing A and B together. If the wires were not connected by poly(I), or if charge transport were slow on the timescale of the potential sweep, the cyclic voltammogram from driving wires A and B together would have an area equal to the area from A alone *plus* the area from B alone.

In contrast, to the results from a positive potential excursion, when the potential of the electrodes is moved to the range of the first one-electron reduction of V^{2+}, the electrodes are apparently not electrically connected. The cyclic voltammograms for a pair of adjacent, poly(I)-coated microelectrodes in CH_3CN/0.1 M [n-Bu_4N]PF_6 in the vicinity of the first one-electron reduction potential of V^{2+} are shown in Figure 4. The integral of the cyclic voltammetry wave of wires A and B driven together is equal to the sum of the integrals for wire A and wire B driven individually. This result indicates that the microelectrodes are not electrically connected on the timescale of the potential sweep and that only the V^{2+} near a given microelectrode is detected. The inability to rapidly access all of the polymer-bound V^{2+} using a single microelectrode is a manifestation of the low value of the diffusion coefficient for charge transport, D_{CT}, for the redox polymer, as shown recently for a microelectrode array coated with ferrocyanide-loaded, protonated poly(4-vinylpyridine).(34)

Charge Transport from One Microelectrode to Another via Pendant $V^{2+/+}$ Subunits. Just as charge transport occurs in generator and collector experiments at a rotating ring/disk,(35) it can occur from one microelectrode to another in a microelectrode array coated with redox polymer. This phenomenon has been recently illustrated for microelectrodes coated with polyvinylferrocene (21) or with ferrocyanide-loaded, protonated poly(4-vinyl-pyridine).(34) Figure 5 shows representative data for an array like that illustrated in Scheme Ia. Electrode #4 of the array is regarded as the "generator" (analogous to the disk electrode), and the pairs #3 and 5, #2 and 6, and #1 and 7 are regarded as "collectors" (analogous to the ring electrode). The potential of the collector electrodes is held constant at 0.0 V vs. Ag^+/Ag while the potential of the generator electrode is swept linearly in time, from 0.0 V to -0.9 V vs. Ag^+/Ag. The poly(I) is thus initially all at the V^{2+} state and has no electron transport. When the generator is capable of reducing V^{2+} to V^+, charge passes to the collector via the polymer-bound $V^{2+/+}$ redox system. The current at the collector increases to a plateau for

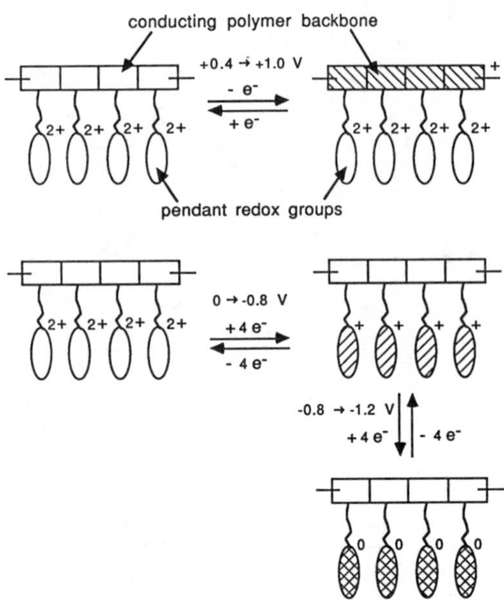

Scheme III. Electrochemical processes of poly(I) showing reversible oxidation of the polythiophene backbone and reversible reduction of the pendant viologen centers.

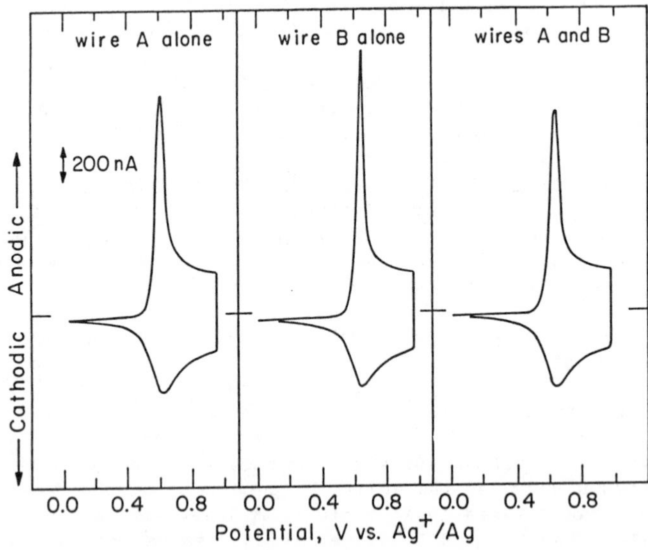

Figure 3. Cyclic voltammetry of adjacent electrodes of a poly(I)-coated microelectrode array driven individually and together at 200 mV/s in the region of the oxidative potential of polythiophene in CH_3CN/0.1 M [n-Bu_4N]PF_6.

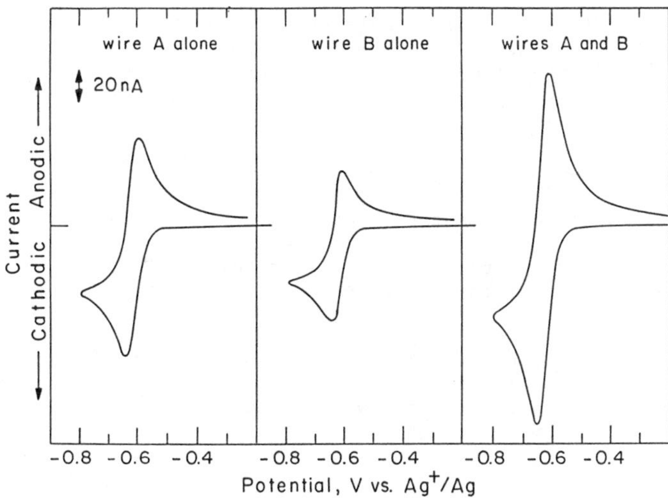

Figure 4. Cyclic voltammetry of adjacent electrodes of a poly(I)-coated microelectrode array driven individually and together at 200 mV/s in the region of the first reduction potential of V^{2+} in CH_3CN/0.1 M [n-Bu_4N]PF_6.

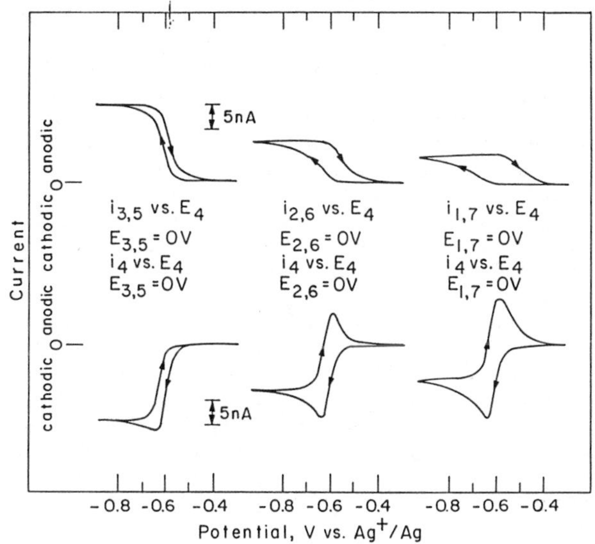

Figure 5. Generation/collection experiment with poly(I)-coated microelectrodes in CH_3CN/0.1 M [n-Bu_4N]PF_6 at 10 mV/s. The lower cyclic voltammograms are for the generator electrode as its potential is swept between -0.2 V and -0.9 V vs. Ag^+/Ag while the potential of the collector electrodes is held at 0.0 V vs. Ag^+/Ag.

generator potentials significantly negative of E°' ($V^{2+/+}$). Thus, the maximum rate of charge transport occurs at the maximum concentration gradient when [V^{2+}] = 0 at the generator and [V^{+}] = 0 at the collector. Also, the potential of the generator for half-maximum current is at E°' ($V^{2+/+}$), when [V^{+}] and [V^{2+}] at the generator are equal.

The second feature of the data in Figure 5 is that the maximum current at the collector falls as the collector is moved geometrically farther from the generator and the magnitude of the current is inversely proportional to the distance across which the charge must be transported. The final feature in Figure 5 is the increase in the hysteresis of the curves and the increase in the time required to achieve a steady state generator-collector current as the distance between generator and collector increases. These results are consistent with the fact that the steady-state charge transport from generator to collector is by a succession of self-exchange electron transfer reactions between neighboring oxidized and reduced sites in the redox polymer, (36) and the charge transport can be treated as a diffusion of electrons which obeys Fick's diffusion laws. (28-30)

The current as a function of generator potential at different fixed collection potentials is shown in Figure 6 which further illustrates the fact that charge transport is driven not by the electrical potential gradient but by the linear concentration gradient between the collector and generator electrodes. When the potential of the collector electrodes is held at 0.0 V or -0.5 V vs. Ag^{+}/Ag, the electron flow across the film begins as the generation potential is scanned near E°'($V^{2+/+}$) and reaches a plateau at a generator potential significantly negative of E°' ($V^{2+/+}$). If the collector potential is moved close to or equal to E°'($V^{2+/+}$), electrons flows from generator to collector or from collector to generator depending on whether the generator potential is more or less positive than the collector potential. When the collector potential is significantly negative of E°' ($V^{2+/+}$), nearly all of the polymer in contact with collector is in the V^{+} (reduced) state and electron flow occurs only from collector to generator. In all cases, [V^{+}] near the collector does not change, and the net change of [V^{+}] near the generator while it is being scanned from 0 V to -0.9 V vs. Ag^{+}/Ag is the same: from 0 to 100% V^{+}. So the net change of concentration gradient of V^{+} between generator and collector is the same for all collection potentials within the limits of the generator potential excursion, and the net current change is constant. To summarize, the data illustrated in Figures 5 and 6 are qualitatively in accord with expectation for "redox conduction" (28-30) associated with the pendant $V^{2+/+}$ centers.

A microelectrode array coated with redox polymers has recently been used to measure D_{CT} for the polymers. (13,37) Using this method, D_{CT} can be calculated from the data of the total amount of the redox couple on the electrodes and the limiting current in generation/collection experiments with the polymer-coated microelectrode array. The cyclic voltammetry of an array of eight poly(I)-coated microelectrodes in CH_3CN/0.1 M [n-Bu_4N]PF_6 is presented in Figure 7a. The total charge calculated from the integration of the current of the voltammogram is about 3.3 x 10^{-7} C. The cyclic voltammetry of generator and collector in the

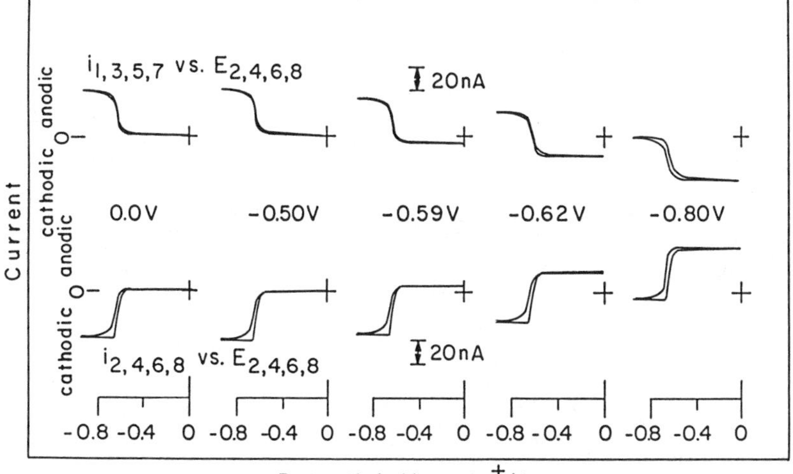

Figure 6. Generation/collection experiments for different fixed collector potentials in an interdigitated array of microelectrodes coated with poly(I) in CH_3CN/0.1 M [n-Bu_4N]PF_6. The potential of the collector electrodes is held at 0.0 V, -0.50 V, -0.59 V, -0.62 V, or -0.80 V vs. Ag^+/Ag while the potential of the generator electrodes is swept between 0.0 V and -0.9 V vs. Ag^+/Ag at 20 mV/s.

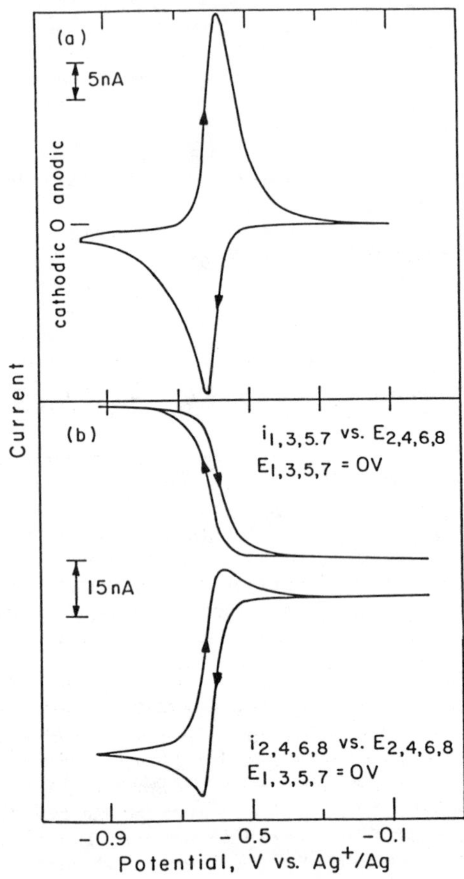

Figure 7. Cyclic voltammetry of an interdigitated array of microelectrodes coated with poly(I) in CH_3CN/0.1 M [n-Bu_4N]PF_6. (a) The potential of all eight electrodes is scanned together at 10 mV/s. (b) The potential of electrodes #2,4,6, and 8 is scanned at 10 mV/s while the potential of electrodes #1,3,5, and 7 is held at 0 V vs. Ag^+/Ag.

interdigitated array microelectrodes is shown in Figure 7b. The Steady-state current is 40 nA. From the ratio of the steady-state current, the total charge, and the distances between two electrodes, we find a value of D_{CT} ~7 x 10^{-9} cm^2/s for charge transport via the pendant $V^{2+/+}$ in poly(I) at 298 °K in CH_3CN/0.1 M [n-Bu_4N]PF_6. For comparison, the value of D_{CT} of the $V^{2+/+}$ has also been measured by potential step chronoamperometry. The value of $D_{CT}^{1/2}C$ from chronoamperometry is 1.6 x 10^{-7} $mol/cm^2s^{1/2}$. The value of C, the concentration of redox centers in the polymer, can be determined by measuring the polymer coverage from electrochemical determination and the film thickness as measured by a surface profile measuring device. The thickness of 3.0 x 10^{-7} mol/cm^2 of poly(I) is ~1.3 µm at 0.0 V vs. Ag^+/Ag in CH_3CN/0.1 M [n-Bu_4N]PF_6. The dry thickness of the same film of poly(I) is ~1.2 µm. There is a linear relationship between coverage and thickness for coverages in the range 4-30 x 10^{-8} mol/cm^2,and the intercept is zero for a plot of thickness against coverage. From the slope of the straight line, the value of C can be calculated and is ~2.4 M. The value of D_{CT} obtained from the potential step data is thus 4.6 x 10^{-9} cm^2/s, which is smaller than that obtained with steady-state current in an interdigitated microelectrode array. A similar difference between potential step and steady state data has been observed for poly-[Os(bpy)$_2$(vpy)$_2$]-(ClO_4)$_2$ films.(37)

Charge transport via a redox polymer is treated as a thermally activated diffusion.(38) Figure 8 illustrates the temperature dependence for the generator-collector experiment showing that the limiting current increases significantly at higher temperatures. The logarithm of the limiting current is linear with 1/T, as shown in Figure 9. The slope of the straight line gives an Arrhenius activation energy of 9.8 kcal/mol, which is slightly larger than the value of 8.9 kcal/mol reported for a viologen based polysiloxane.(39-40)

Poly(I)-based Microelectrochemical Transistors. Research in the Wrighton group has already demonstrated some of the principles of microelectrochemical devices based on redox active materials. These materials include conducting polymers,(1,14,15) electroactive oxides,(41-42) and conventional redox polymers.(34,43-44) Scheme IV illustrates the operation of a microelectrochemical transistor based on poly(I). This device combines characteristics that have previously been demonstrated for devices based on conducting polymers and devices based on redox polymers. A change in the gate voltage, V_G, results in a change in the state of charge of the electroactive material connecting the two microelectrodes called source and drain. Thus, the drain current, I_D, at a fixed source-drain potential, V_D, can be turned on or off by variation of V_G, because for some redox state the electroactive material is insulating and for some other redox state the electroactive material is conducting. The data in Figures 1, 2, and 3 are consistent with the fact that the backbone of poly(I) is similar to that of poly(3-methylthiophene). The resistance of poly(I) connecting microelectrodes as a function of V_G in the regime of the oxidation potential of polythiophene is shown in Figure 10. The microelectrode array used is like that shown in Scheme Ib where microelectrodes #1, 3, 5, and 7 are regarded as the "source" and microelectrodes #2, 4, 6, and 8 are regarded as the

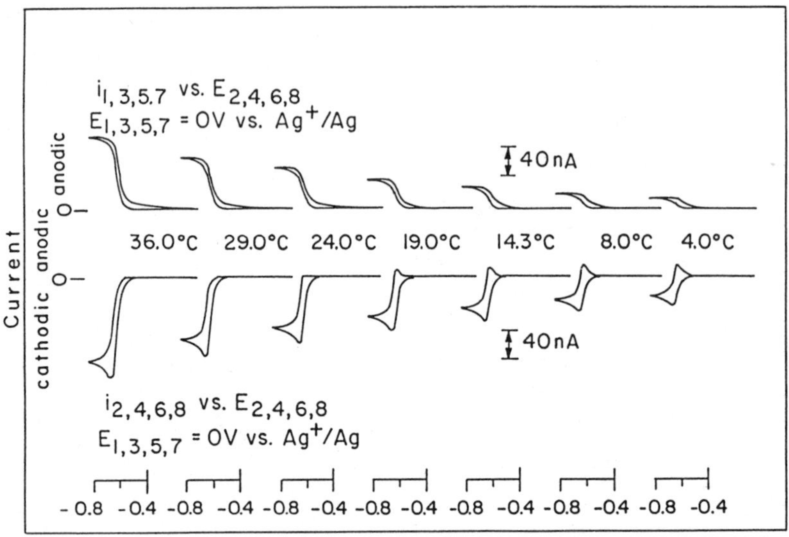

Figure 8. Generation/collection experiments as a function of temperature at an interdigitated array of microelectrodes coated with poly(I) at 50 mV/s in CH_3CN/0.1 M [n-Bu_4N]PF_6.

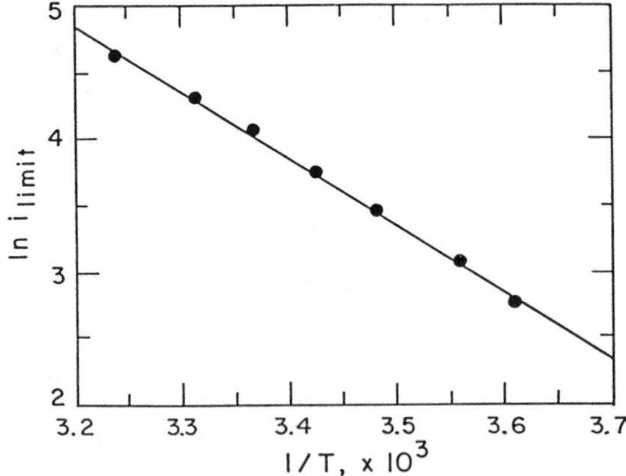

Figure 9. Log of the limiting steady-state generator/collector current of an interdigitated array of microelectrodes coated with poly(I) vs. 1/T.

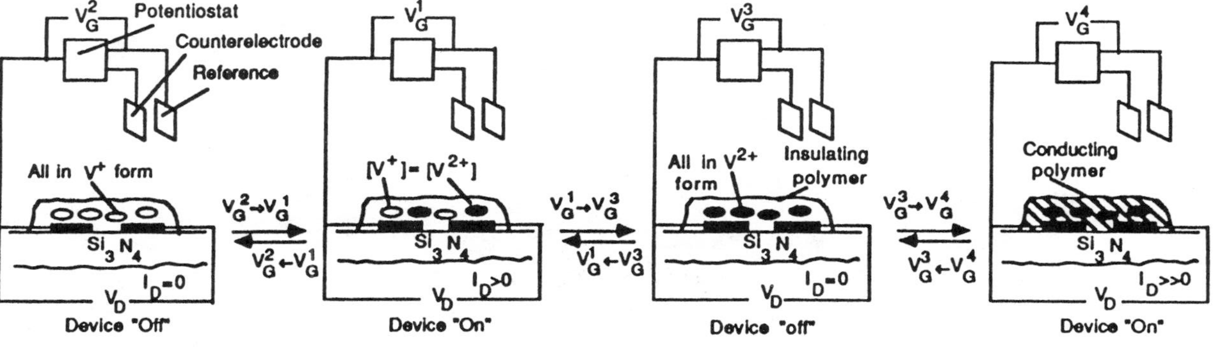

Scheme IV. A poly(I)-based microelectrochemical transistor that turns on when V_G is moved from $V_G{}^3$ (ie +0.4 V vs. Ag^+/Ag) where polythiophene is reduced and insulating to $V_G{}^4$ (ie +0.7 V vs. Ag^+/Ag) where polythiophene is oxidized and conducting. This transistor also turns on to a smaller extent at $E°'(V^{2+/+})$, $V_G{}^1 = -0.63$ V vs. Ag^+/Ag. At V_G significantly (>0.2 V) more negative ($V_G{}^2 < -0.8$ V vs. Ag^+/Ag) or positive ($+0.4 \text{ V} > V_G{}^3 > -0.4$ V vs. Ag^+/Ag) of $E°'(V^{2+/+})$ only the reduced or oxidized form of viologen redox centers is present, respectively, and this device is off.

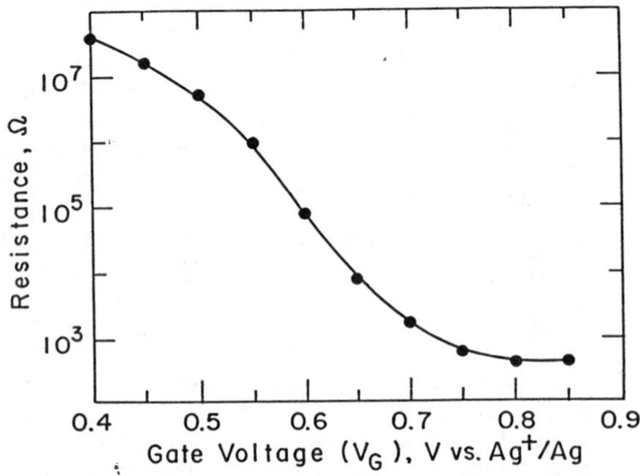

Figure 10. Resistance of a poly(I) as a function of V_G in the region of the oxidative potential of polythiophene.

"drain". In the reduced state, poly(I) is insulating and the device is "off"; in the oxidized state, poly(I) is conducting and the device is "on". The resistance of the poly(I) declines by a factor of 10^5 upon changing the value of V_G from 0.4 V to 0.8 V vs. Ag^+/Ag.

Microelectrochemical transistors based on conventional redox polymers, in contrast to those based on conducting polymers, will give an I_D-V_G characteristic with a sharp maximum in I_D at $V_G = E°'$ of the redox polymer. This characteristic arises because the conductivity of the conventional redox polymer is due to electron self-exchange between oxidized and reduced sites(36) and the rate of self-exchange is proportional to [ox] and [red]. Since [ox] + [red] is fixed in the polymer and [ox]/[red] changes steeply with variation in potential of the polymer, a conventional redox polymer will show a narrow region of conductivity with its maximum conductivity when [ox] equals [red], which occurs when the potential of the polymer is held at E°'. The conductivity will fall sharply when the potential of the polymer is moved away from E°'.(28-30) Figure 11 shows the I_D-V_G characteristic of a poly(I)-based transistor for V_G in the range of the first one-electron reduction potential of V^{2+}. The microelectrochemical transistor shows only a narrow region of V_G where the device is "on". For V_D = 25 mv, the width at half-height is only 56 mV with the peak in I_D occurring very close to the E°' of the polymer-bound $V^{2+/+}$. The height and width become greater for larger V_D. Whenever V_D exceeds -0.3 V the maximum I_D does not increase, because the maximum difference in concentration of V^{2+} and V^+ is no longer influenced by V_D.

Conclusions

Electrochemical oxidation of I produces a polymer film with polythiophene as the backbone and viologen centers as pendant redox groups. The electrochemical properties of the polymer are the combination of polythiophene and viologen. Using viologen subunits as the internal standard (one per repeat unit of the polymer), the "doping level" of the oxidized polythiophene backbone at its maximum conductivity can be measured and is about 25%. The charge transport via the pendant $V^{2+/+}$ of poly(I) has been studied by generation/collection experiments using poly(I)-coated microelectrode arrays. The redox conduction associated with the $V^{2+/+}$ redox system can be treated as a diffusion of electrons which follows Fick's diffusion laws. The D_{CT} of $V^{2+/+}$ of poly(I) has been measured to be 4-7 x 10^{-9} cm^2/s, somewhat larger than for a viologen-based polysiloxane.(39-40)

The poly(I)-based transistor is the first illustration of a microelectrochemical transistor based on a combination of a conducting and a conventional redox polymer as the active material. The transistor "turns on" at V_G corresponding to oxidation of the polythiophene backbone. The resistivity of poly(I) declines by a factor of 10^5 upon changing V_G from 0.4 V to 0.8 V vs. Ag^+/Ag. When V_G is moved close to the one-electron reduction potential of $V^{2+/+}$, the conventional redox conductivity gives a small degree of "turn on". A sharp I_D-V_G characteristic results, with an I_D(peak) at V_G = E°' ($V^{2+/+}$). Though the microelectrochemical devices based on conventional redox conduction have both slow switching speed and a

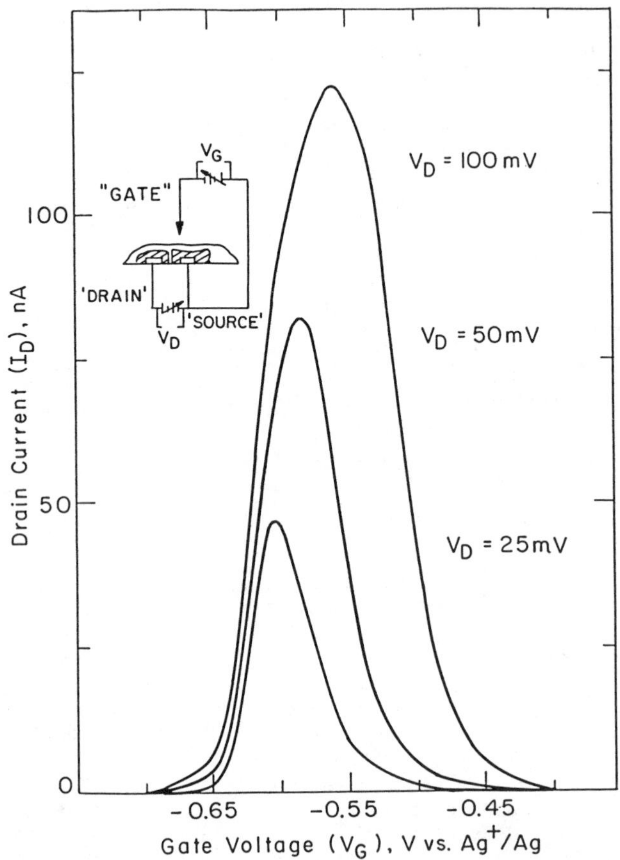

Figure 11. Drain current, I_D, vs. gate voltage, V_G, for various drain voltages, V_D (25, 50, 100 mV) for a poly(I)-based microelectrochemical transistor. The gate voltage, V_G was scanned at 1 mV/s.

small value of I_D, in comparison to devices based on so-called conducting polymers, the narrow region of V_G where the device is turned on is an advantage that may prove useful in sensor applications.(41,46) Also smaller microelectrodes, more closely spaced will yield microelectrochemical devices having better electrical characteristics.(47)

Acknowledgments

We wish to thank the Office of Naval Research and the Defense Advanced Research projects Agency for partial support of this research.

Literature Cited

1. White, H.S.; Kittlesen, G.P.; Wrighton, M.S. J. Am. Chem. Soc., 1984, 106, 5375.
2. Kittlesen, G.P.; White, H.S.; Wrighton, M.S. J. Am. Chem. Soc., 1984, 106, 7389.
3. Murray, R.W. in Electroanalytical Chemistry, Bard, A.J. Ed.; Marcel Dekker: New York, 1984; Vol. 13, p 191.
4. Wrighton, M.S. Science, 1986, 231, 32.
5. Bidan, G.; Deronzier, A.; Moutet, J.C. J. Chem. Soc. Chem. Commun., 1984, 1185.
6. Coche, L.; Deronzier, A.; Moutet, J.C. J. Electroanal. Chem., 1986, 198, 187.
7. Coche, L.; Moutet, J. C. J. Electroanal. Chem., 1987, 224, 111.
8. Coche, L.; Moutet, J. C. J. Am. Chem. Soc., 1987, 109, 6887.
9. Bidan, G.; Deronzier, A.; Moutet, J. C. Nouv. J. Chim., 1984, 8, 501.
10. Eaves, G.; Munro, H.S.; Parker, D. Inorg. Chem., 1987, 26, 644.
11. Haimerl, A.; Merz, A. Angew. Chem. Int. Ed. Engl., 1986, 25, 180.
12. Audebert, P.; Bidan, G.; Lapkowski, M. J. Chem. Soc. Chem Commun., 1986, 887.
13. Waltman, R.J.; Bargon, J.; Diaz, A.F. J. Phys.Chem., 1983, 87, 1459.
14. Thackeray, J.W.; White, H.S.; Wrighton, M.S. J. Phys. Chem., 1985, 89, 5133.
15. Paul, E.W.; Ricco, A.T.; Wrighton, M.S. J. Phys. Chem., 1985, 81, 1441.
16. Lewis, N.S.; Wrighton, M.S. Science, 1981, 211, 944.
17. Elliott, C.M.; Martin, W.S. J. Electroanal.Chem., 1982, 137, 377.
18. Bruce, J.A.; Murahashi, T.; Wrighton, M.S. J. Phys. Chem., 1982, 86, 1552.
19. Dominey, R.N.; Lewis, N.S.; Bruce, J.A.; Bookbinder, D.C.; Wrighton, M.S. J. Am. Chem. Soc., 1982, 104, 467.
20. Bookbinder, D.C.; Wrighton, M.S. J. Electrochem. Soc., 1983, 130, 1080.
21. Kittlesen, G.P.; White, H.S.; Wrighton, M.S. J. Am. Chem. Soc., 1985, 107, 7373.
22. Kittlesen, G.P.; Wrighton, M.S. J. Mol. Electron., 1986, 2, 23.
23. Gronowitz, S.; Frejd, T. Synth. Commun., 1976, 6, 475.
24. Kittlesen, G.P.; Ph.D. Thesis, Massachusetts Institute of Technology, Cambridge, MA, 1985.

25. Marvel, C.S. Org. Synth., Coll. Vol. 3, 705.
26. Blicke, F.F.; Sheet, D.J. J. Am. Chem. Soc., 1949, 71, 2857.
27. Noller, C.R.; Dinsmore, R. Org. Synth., Coll. Vol 2, 358.
28. Pickup, P.G.; Murray, W.R. J. Am. Chem. Soc., 1983, 105, 4510.
29. Pickup, P. G.; Kutner, W.; Leidner, C. R.; Murray, R.W. J. Am. Chem. Soc., 1984, 106, 1991.
30. Pickup, P.G.; Murray, R.W. J. Electrochem. Soc., 1984, 131, 833.
31 Tourillon, G.; Garnier, F. J. Electroanal. Chem., 1982, 135, 173.
32. Touillon, G.; Garnier, F. J. Phys. Chem., 1983, 87, 2289.
33. Tourillon, G.; Garnier, F. J. Electoanal. Chem., 1984, 161, 407.
34. Belanger, D.; Wrighton, M.S. Anal. Chem., 1987, 59, 1426.
35. Bard, A.J.; Faulkner, L.R. Electrochemical Methods; Wiley: New York, 1980.
36. Kaufman, F.B.; Engler, E.M. J. Am. Chem. Soc., 1979, 101, 547.
37. Chidsey, C.E.; Feldman, B.J.; Lundgren, C.; Murray, R. W. Anal. Chem., 1986, 58, 601.
38. Daum, P.; Lenhard, J.R.; Rolison, D.; Murray, R.W. J. Am. Chem. Soc., 1980, 102, 4649.
39. Lewis, T.J.; White, H.S.; Wrighton, M.S. J. Am. Chem. Soc., 1984, 106, 6947.
40. Lewis, T.J.; Ph.D. Thesis, Massachusetts Institute of Technology, Cambridge, MA, 1984.
41. Natan, M.J.; Mallouk, T.E.; Wrighton, M.S. J. Phys. Chem., 1987, 91, 648.
42. Natan, M.J.; Belanger, D.; Carpenter, M.S.; Wrighton M.S. J. Phys. Chem., 1987, 91, 1834.
43. Smith, D.K.; Lane, G.A.; Wrighton, M.S. J. Am. Chem. Soc., 1986, 108, 3522.
44. Smith, D.K.; Lane, G.A.; Wrighton, M.S. J. Phys. Chem.,1988, 92, 2616.
45. Thackeray, J.W.; Wrighton, M.S. J. Phys. Chem., 1986, 90, 6674.
46. Wrighton, M.S.; Thackeray, J.W.; Natan, M.J.; Smith, D.K.; Lane, G.A.; Belanger, D.; Phil. Trans. Royal Soc. Lond., 1987, B316, 13.
47. Jones, E.T.T.; Chyan, O.M.; Wrighton, M.S. J. Am. Chem Soc., 1987, 109, 5526.

RECEIVED May 19, 1988

Chapter 29

Electroactive Bipyridiniums in Self-Assembled Octadecylmercaptan Monolayers on Gold Electrodes

H. O. Finklea, J. Fedyk, and J. Schwab

Department of Chemistry, West Virginia University, Morgantown, WV 26506

N-Octadecyl-n'-methyl-4,4'-bipyridinium dichloride and the n'-ethyl analog are co-deposited with octadecylmercaptan onto a gold substrate during self-assembly of an organized monolayer from a chloroform/methanol solution. Surface redox waves are observed in aqueous electrolyte corresponding to the reduction of the bipyridiniums. The bipyridiniums also assemble from aqueous electrolyte to form an electroactive monolayer on both bare gold and gold coated with a mercaptan monolayer. The electrochemical behavior suggests that in both cases the bipyridiniums reside on but not in the mercaptan monolayer. Evidence is given for penetration of the mercaptan monolayer by bipyridiniums.

Long chain alkylmercaptans and disulfides readily self-assemble on gold surfaces to form compact organized monolayers in which the sulfur is chemisorbed to the gold and the hydrocarbon tail is extended away from the surface (1-5). The mercaptan monolayers strongly inhibit gold oxidation in dilute sulfuric acid and also block diffusion of aqueous ions (e.g. $Fe^{2+/3+}$, $Fe(CN)_6^{3-/4-}$, $Ru(NH_3)_6^{2+/3+}$) to the gold surface (4-5). The unique anisotropy of the organized monolayer provides an opportunity to explore the effect of both orientation and distance on electron transfer between a molecule and a metal electrode (6-9).

Our strategy is to incorporate a prolate redox molecule into the hydrocarbon phase of the organized monolayer. Steric restraints imposed by a close-packed monolayer would presumably force the redox molecule to adopt an orientation parallel to the hydrocarbon tails. Spacing can then be controlled by a short hydrocarbon chain between the redox center and the metal. A class of molecules fitting these requirements are the assymetric 4,4'-bipyridiniums:

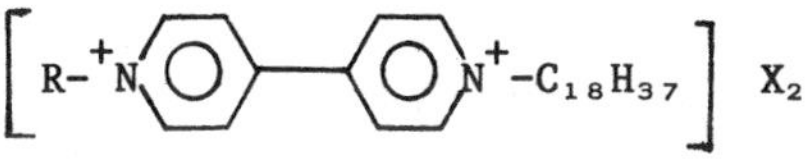

$$\left[R{-}^{+}N\langle\bigcirc\rangle{-}\langle\bigcirc\rangle N^{+}{-}C_{18}H_{37} \right] X_2$$

0097-6156/88/0378-0431$06.00/0

R is the spacer group which should also terminate in an anchor e.g. a mercaptan. We report here intermediate results in which R is the methyl and ethyl moiety; the case of R=H is the subject of a separate paper (K. A. B. Lee, R. Mowry, G. McLennan and H. O. Finklea, in press). Our primary concern is whether the bipyridiniums are incorporated into the mercaptan monolayer and whether they are located at a fixed distance relative to the gold surface.

Experimental

Synthesis. N-Octadecyl-4-pyridinium-4'-pyridyl iodide (mp 125°C) was synthesized by stirring stoichiometric amounts of 4,4'-bipyridine and 1-iodooctadecane in acetone for several days. The precipitate was collected and recrystallized from methanol/ether. N-Methyl-n'-octadecyl-4,4'-bipyridinium diiodide [$(C_{18}bpyMe)I_2$] (mp 270°C with decomposition) and N-ethyl-n'-octadecyl-4,4'-bipyridinium diiodide [$(C_{18}bpyEt)I_2$] (mp 255°C with decomposition) were synthesized by refluxing the mono-alkylated bipyridine with methyl iodide in methanol. The iodide was replaced with chloride by dissolving the biipyridiniums in a minimum amount of boiling water and adding excess saturated KCl solution. After chilling the yellow solid was collected and subjected to a second metathesis step to remove the last traces of iodide.

Deposition of the mixed monolayer. Deposition solutions were prepared by dissolving octadecylmercaptan [$C_{18}SH$] and the respective bipyridinium in a mixture of chloroform and methanol. The electrode was cleaned by heating it in a gas-air flame. After cooling, the electrode was immersed in the deposition solution for 15 - 30 minutes, withdrawn, and rinsed in clean methanol or chloroform. Qualitatively the most reproducible surface redox waves and lowest charging currents during cyclic voltammetry were obtained with a freshly-prepared deposition solution containing 50 mM $C_{18}SH$ and 10 mM of the bipyridinium in a 1:1 volume ratio of chloroform and methanol.

Electrochemistry. Cyclic voltammetry was performed using gold flag electrodes (area 0.7 cm^2) and a SCE reference electrode. The electrolyte was 0.1 M KCl buffered to pH8 with 0.01 M Na_2HPO_4. Data acquisition, manipulation and plotting were performed using a Zenith computer interfaced to a potentiostat via a Metrabyte DASH-16 board and running MacMillan ASYSTANT+ software.

Results and Discussion

Mixed monolayers. An electrode coated with a mixed monolayer of $C_{18}SH$ and $C_{18}bpyMe^{2+}$ exhibits a surface redox wave assignable to the reduction of the bipyridinium to the radical cation (Figure 1) (in all 3 figures the cyclic voltammograms at 0.1 V/s (solid line), 1 V/s (dashed line) and 10 V/s (dotted line) are plotted with the current axis scaled in proportion to the scan rate). A second reduction wave at -0.9 V (not shown) is obscured by increasing background currents. The electrochemical parameters for the surface redox wave given in Table I along with typical values for the surface redox wave of a mixed monolayer containing $C_{18}bpyEt^{2+}$. At scan rates less than 0.1 V/s the redox wave obeys the normal relationships for a surface bound species i.e. peak splitting (ΔE_p) is close to zero, peak currents are proportional to scan rate, and the peak full width at half maximum (ΔE_{FWHM}) is close to the theoretical 90 mV. The coverage represents ca. 10% of a close packed monolayer ($3x10^{-10}$ mol/cm assuming 50 Å^2 per bipyridinium). At higher scan rates the peaks broaden (particularly the anodic peak), the peak currents are not proportional to scan rate, and the apparent coverage decreases.

Table I. Electrochemical Parameters for the First Redox Wave of Deposited or Adsorbed Bipyridiniums

Coating	Scan Rate (V/s)	E°' (	ΔE_p mV	ΔE_{FWHM} cat	ΔE_{FWHM} an)	j_p cat (μA/cm²)	j_p an	Γ cat ($x10^{-10}$	Γ an mol/cm²)
mixed monolayer	.10	-471	24	116	126	2.3	2.2	.30	.30
$C_{18}bpyMe^{2+}$	1.0	-470	61	147	181	15	16	.24	.31
+ $C_{18}SH$	10.	-475	63	-	-	56	63	.08	.09
mixed monolayer $C_{18}bpyEt^{2+}$ + $C_{18}SH$	.10	-480	30	110	120	3.1	2.9	.36	.31
adsorbed $C_{18}bpyMe^{2+}$	.10	-459	4	124	116	15	14	1.8	1.2
clean gold	1.0	-461	16	119	112	164	163	1.9	1.6
	10.	-467	86	123	134	1300	1270	1.5	1.4
adsorbed $C_{18}bpyMe^{2+}$	.10	-445	9	153	153	17	17	2.0	2.0
$C_{18}SH$-	1.0	-458	54	128	170	180	114	1.9	1.5
coated gold	10.	-	-	185	-	890	-	1.2	-

The first two entries refer to mixed monolayers deposited prior to the electrochemical measurements; the last two entries refer to bipyridinium monolayers adsorbed from the electrolyte. J_p is the peak current in the cathodic and anodic directions for the first redox wave of the bipyridiniums; Γ is the coverage found by integration of the respective cathodic and anodic peaks. The other headings have been defined in the preceding text. Data is omitted where the surface wave is not well-defined relative to the background current.

If the bipyridiniums are located at a fixed distance from the electrode surface, the surface coverage would be independent of the scan rate. The loss of apparent coverage at fast scan rates indicates that the bipyridiniums are not rigidly fixed near the electrode surface. Daifuku et. al. (10) have also observed an apparent loss of coverage at fast scan rates for a Langmuir-Blodgett monolayer of surfactant osmium bipyridine complexes. They postulated that parts of the monolayer were not in contact with the electrode and that the remote $Os(bpy)_3$ complexes were oxidized or reduced via lateral electron exchange in the monolayer.

Further experiments with the coated electrode suggests that the surfactant bipyridiniums are mobile within the mercaptan monolayer. If the electrode is transferred to a fresh electrolyte, the coverage (measured at 0.1 V/s) greatly diminishes. Likewise rinsing the coated electrode with a stream of pure water between cyclic voltammograms removes most of the electroactive bipyridiniums. These treatments do not remove the mercaptan monolayer (5).

We therefore hypothesize that once the mixed monolayer is in contact with water, the surfactant bipyridiniums reside not within the mercaptan monolayer but adsorbed to its exterior. The adsorbed bipyridiniums communicate with the electrode by a small fraction of bipyridiniums which penetrate the monolayer. A short-chain analog, methyl viologen (N,N'-dimethyl-4,4'-bipyridinium dichloride), can penetrate the mercaptan monolayer at sites not accessible to the ferric ion or other inorganic ions (3,5). For example, in a 1 mM methyl viologen solution the cathodic peak current on a $C_{18}SH$-coated electrode is attenuated only by 30% relative to the peak current on a clean electrode (0.1 V/s). In a more dilute solution (0.05 mM) the cathodic peak currents are identical for clean and coated electrodes. Yet the same coated electrode is strongly blocking to Fe^{3+} reduction in dilute sulfuric acid; the cathodic peak current is reduced by a factor of 500 in the presence of the monolayer.

Additional support for the preceding model is found in experiments in which the surfactant bipyridinium is dissolved in the electrolyte rather than co-deposited with the mercaptan monolayer.

Adsorbed bipyridiniums. Surfactant bipyridiniums adsorb onto bare gold from aqueous solution; saturation is reached at concentrations of 10^{-4} M or greater for $C_{14}bpyMe^{2+}$ (11,12). Figure 2 illustrates the cyclic voltammogram of a $1x10^{-4}$ M solution of $C_{18}bpyMe^{2+}$ using a clean gold electrode; Table I contains the electrochemical parameters of the first redox wave near −0.46 V. The data in Table I clearly indicates that the currents are due to an adsorbed monolayer of $C_{18}bpyMe^{2+}$. The peak currents due to diffusion of the bipyridinium to the electrode are predicted (based on a diffusion coefficient of $7x10^{-6}$ cm^2/s (13) to be 23, 71 and 230 $\mu A/cm^2$ at the three scan rates. As expected, the coverage is constant with increasing scan rate (0.1 and 1.0 V/s). At 10 V/s the coverage decreases, possibly because the monolayer is not fully re-assembled after desorption at the positive potential limit (see below). Analysis of the cathodic peak at 10 V/s assuming irreversible reduction of an adsorbed reactant (14) yields values of 70/s for the standard rate constant and 0.6 for the transfer coefficient.

Reduction to the neutral radical appears as an irreversible wave at -0.9 V. Neither anodic peak exhibits the shape characteristic of stripping a solid coating from the electrode; hence precipitation of the radical cation or neutral radical on the electrode is not evident (11-13). The sharp peaks at +0.46 V are tentatively assigned to desorption and adsorption of the $C_{18}bpyMe^{2+}$; there are no anticipated redox reactions at that potential.

The voltammetry of $C_{18}bpyMe^{2+}$ ($1x10^{-4}$ M) in the presence of a $C_{18}SH$ monolayer is striking (Figure 3 and Table I). The mercaptan monolayer displaces electrolyte from the electrode/electrolyte interface; consequently the interfacial capacitance and charging current decrease dramatically. For comparison, the interfacial capacitance at 0.0 V is 120, 100, and 10 $\mu f/cm^2$ for clean gold in the KCl electrolyte, gold with the adsorbed layer of $C_{18}bpyMe^{2+}$ and gold with the mercaptan monolayer. The adsorption/desorption wave at +0.46 V is no longer visible. Most notably, a monolayer $C_{18}bpyMe^{2+}$ is absorbed on the electrode (compare the first reduction peak of Figure 3 with the corresponding peak in Figure 2; the current scales are identical). At 0.1 V/s the coverage on the mercaptan-coated electrode is identical to that obtained on the clean electrode (Table I). It is likely that the bipyridinium monolayer assembles on the exterior of the mercaptan monolayer to form a bilayer structure (15). At 0.1 V/s the bipyridiniums penetrate the $C_{18}SH$ monolayer at a sufficient rate that the entire adsorbed $C_{18}bpyMe^{2+}$ monolayer is reduced. At faster scan rates transportation of charge to the remote $C_{18}bpyMe^{2+}$ monolayer limits the current and the surface coverage apparently decreases.

The fate of the bipyridinium radical cation in the mercaptan monolayer is not clear. The second reduction wave (Figure 3) possesses structure suggesting a precipitated phase; similar sharp peaks are seen during the reductive precipitation of $C_{14}bpyMe^{2+}$ (11). Yet at 0.1 V/s the anodic peak due to the oxidation of the radical cation does not exhibit the shape characteristic of stripping of a solid phase. At faster scan rates the anodic peak broadens considerably and splits into two peaks; the same behavior is noticeable in Figure 1. We do not have an explanation for this phenomenon. A recent theoretical treatment of redox molecules attached to electrode surfaces predicts that under certain conditions an anodic surface wave can broaden and split with increasing scan rate in a manner shown in Figure 3 (16). However the same theory predicts that the corresponding cathodic peak normalized to constant scan rate will increase with increasing scan rate. The latter prediction is not observed in our system.

Conclusion. Our major conclusion is that surfactant bipyridinium dications do not remain at a fixed location within a mercaptan monolayer. In order to control the spacing and orientation of the bipyridiniums in the monolayer, it will be necessary to anchor them within the monolayer or to the electrode. We are currently pursuing that goal by synthesizing a bipyridinium with a terminal mercaptan.

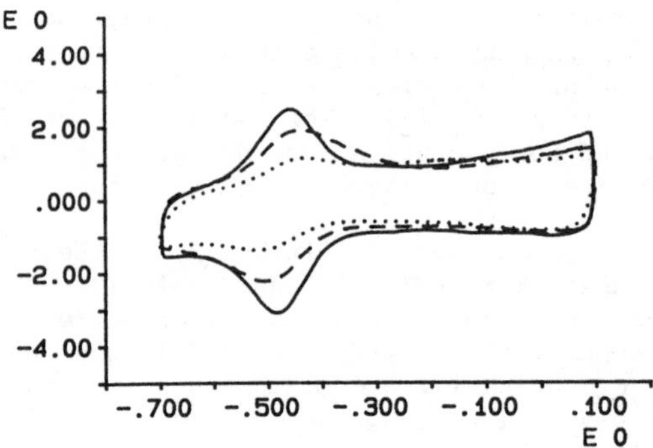

Figure 1. Cyclic voltammograms of a mixed monolayer of $C_{18}SH$ + $(C_{18}bpyMe)Cl_2$.

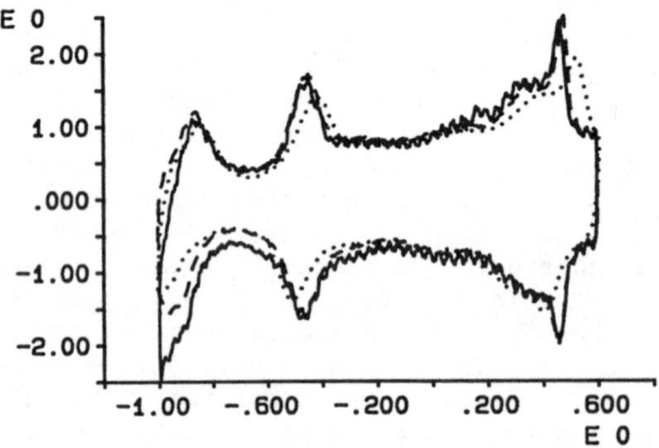

Figure 2. Cyclic voltammograms of $C_{18}bpyMe^{2+}$ adsorbed from solution on a clean gold electrode.

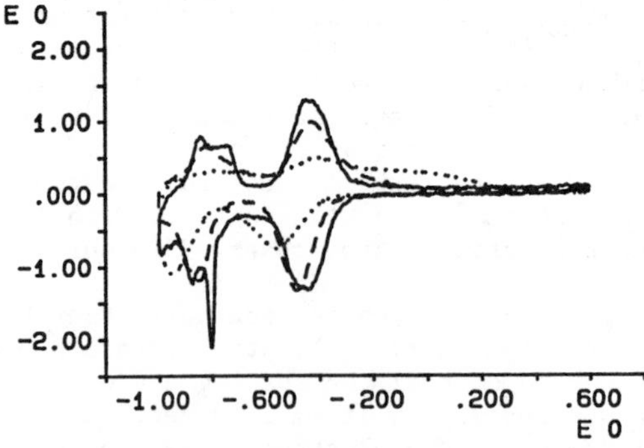

Figure 3. Cyclic voltammograms of $C_{18}bpyMe^{2+}$ adsorbed from solution on a gold electrode coated with a $C_{18}SH$ monolayer.

Acknowledgments

Acknowledgement is made to the donors of the Petroleum Research Fund, administered by the American Chemical Society, for support of this research.

Literature Cited

1. Nuzzo, R. G.; Allara, D. L. J. Am. Chem. Soc. 1983, 105, 4481.
2. Nuzzo, R. G.; Fusco, F. A.; Allara, D. L. J. Am. Chem. Soc. 1987, 109, 2358.
3. Porter, M. D.; Bright, T. B.; Allara, D.; Chidsey, C. E. D. J. Am. Chem. Soc. 1987, 109, 3559.
4. Sabitini, E.; Rubinstein, I.; Maoz, R.; Sagiv, J. J. Electroanal. Chem. 1987, 219, 365.
5. Finklea, H. O.; Avery, S.; Lynch, M.; Furtsch, T. Langmuir 1987, 3, 409.
6. Lane, R. F.; Hubbard, A. T. J. Phys. Chem. 1973, 77, 1401.
7. Bravo, B. G.; Mebrahtu, T.; Soriaga, M.; Zapien, D. C.; Hubbard, A. T.; Stickney, J. L. Langmuir 1987, 3, 595.
8. Li, T. T.-T.; Liu, H. Y.; Weaver, M. J. Am. Chem. Soc. 1984, 106, 1233.
9. Li, T. T.-T.; Weaver, M. J. Am. Chem. Soc. 1984, 106, 6107.
10. Daifuku, H.; Aoki, K.; Tokuda, K.; Matsuda, H. J. Electroanal. Chem. 1985, 183, 1.
11. Enea, O. Electrochim. Acta 1986, 31, 789.
12. Crouigneau, P.; Enea, O.; Beden, B. J. Electroanal. Chem. 1987, 218, 307.
13. Bruinink, J.; van Zanten, P. J. Electrochem. Soc. 1977, 124, 1232.
14. Bard, A. J.; Faulkner, L. R. Electrochemical Methods: Fundamentals and Applications, J. Wiley & Sons: New York, 1980: p. 525.
15. Miller, C. J.; Majda, M. J. Am. Chem. Soc. 1986, 108, 3118.
16. Matsuda, H.; Aoki, K.; Tokuda, K. J. Electroanal. Chem. 1987, 217, 1 and 15.

RECEIVED May 17, 1988

Chapter 30

Ideal Polarizable Semiconductor–Solution Interfaces

Carl A. Koval[1], John B. Olson[1], and Bruce A. Parkinson[2]

[1]Department of Chemistry and Biochemistry, University of Colorado, Boulder, CO 80309–0215
[2]Central Research and Development Department, Experimental Station E328/216B, E. I. du Pont de Nemours and Company, Wilmington, DE 19898

Ideal polarizable interfaces are critical for the interpretation of electrochemical kinetic data. Ideality has been approached for certain metal electrode-solution interfaces, such as mercury-water, allowing for the collection of data that can be subjected to rigorous theoretical analysis. Herein, criteria are developed for ideal polarizable semiconductor electrode-solution interfaces. A variety of experimental studies involving metal dichalcogenide-solution interfaces are discussed within the context of these criteria. These interfaces approach ideality in most respects and are well suited for fundamental studies involving electron transfer to solution species or adsorbed dyes.

Interfacial electron transfer is the critical process occurring in all electrochemical cells in which molecular species are oxidized or reduced. While transfer of an electron between an electrode and a solvated molecule or ion is conceptually a simple reaction, rates of heterogeneous electron transfer processes depend on a multitude of factors and can vary over many orders of magnitude. Since control of interfacial electron transfer rates is usually essential for successful operation of electrochemical devices, understanding the kinetics of these reactions has been and remains a challenging and technologically important goal.

Prior to the 1970's, electrochemical kinetic studies were largely directed towards faradaic reactions occurring at metal electrodes. While certain questions remain unanswered, a combination of theoretical and experimental studies has produced a relatively mature picture of electron transfer at the metal-solution interface (1-4). Recent interest in photoelectrochemical processes has extended the interest in electrochemical kinetics to semiconductor electrodes (5-15). Despite the pioneering work of Gerischer (11-14) and Memming (15), many aspects of electron transfer kinetics at the semiconductor-solution interface remain controversial or unexplained.

The authors propose that a major difficulty in interpreting kinetic current flow at the semiconductor-solution interface lies in the inability of experimentalists to prepare interfaces with ideal and measurable properties. In support of this hypothesis, the importance of ideal interfacial properties to metal electrode kinetic studies is briefly reviewed and a set of criteria for ideality of semiconductor-solution interfaces is developed. Finally, the use of semiconducting metal dichalcogenide electrodes as ideal interfaces for subsequent kinetic studies is explored.

0097–6156/88/0378–0438$06.00/0

Kinetics of Molecular Processes at Ideal Polarizable Metal-Solution Interfaces

Describing the current in electrochemical cells due to the faradaic processes is the greatest concern of electrochemical kinetics. There are three general rate processes controlling this current: mass transfer, electron transfer, and homogeneous chemical reactions accompanying electron transfer (16). Since mass transfer is often controlled by hydrodynamics and since diffusion coefficients for different species in solution do not vary greatly, mass transfer is not usually considered to be a molecular process. In kinetic studies, the effects of mass transfer must either be controlled or accounted for mathematically. The processes of electron transfer and associated chemical reactions are molecular in nature and the major goal of the kineticist is to understand why different molecules react at widely different rates.

In general, understanding the rates of electrode reactions that involve more than one electron per molecule is extremely difficult. These processes are usually the combination of several single electron transfers that proceed through unstable intermediates(16). Furthermore, multi-electron electrode processes can be accompanied by chemical reactions which can also be rate limiting. Single electron, outer sphere reactions for many molecules, especially complexed metal ions, are less complicated because the bonding in the oxidized and reduced species is very similar. Following the theoretical treatment of Marcus, the molecular factors that control the rates of single electron transfers are thought to be well understood (17). However, widespread acceptance of the Marcus model required the acquisition of a considerable body of experimental data, involving both homogeneous and heterogeneous (electrochemical) systems (1-4,18,19).

Even in the absence of mass transfer and chemical reactions, relating the current flowing across an interface to a molecular factor is extremely complicated. In order for net current to flow, the electrode potential must be different from the solution potential. The dependence of the current on this overpotential is reasonable well understood (16). The difficulty lies in describing the spatial distribution of this potential drop across the interface. At the metal-solution interface, the spatial distribution of charges that constitutes the potential drop can extend from a few to several hundred Å into the solution. Thus, ionic reactants near the surface develop a concentration vs. distance profile which can be dramatically different from the bulk concentration. A description of the spatial distribution of the potential drop combined with the energetic distribution of electrode and molecular orbitals as a function of applied electrode potential can be called the **interfacial energetics (IE)**. Since rates of electron transfer reactions depend exponentially on distance, driving force and activation energy, knowledge of the interfacial energetics is essential in order to apply theoretical treatments to experimental kinetic data.

Fortunately, a model describing the metal-solution interface has been developed together with a methodology for experimentally measuring the necessary parameters (1-4,16). The basic requirement for application of this model is interfaces that resist faradaic reactions over a wide range of applied potentials in the absence of the electroactive species. Such an electrode is referred to as an ideal polarizable electrode (IPE) (16). At an IPE, the interfacial capacitance as a function of applied potential (which is controlled by the interaction of the metal with the solvent and electrolyte) can be used to calculate the interfacial energetics. In subsequent experiments, small amounts of an electroactive species can be added to the solution and the resulting faradaic current can be analyzed. It should be noted that this model is most applicable in situations that are dominated by electrostatics. Chemical interactions between solution species and the electrode which result in specific adsorption greatly complicate the determination of the interfacial energetics and the kinetic analysis.

Metal-Solution Interfaces that Approach Ideality

By far, most of electrochemical kinetic data that can be subjected to detailed analysis has been obtained at the mercury-aqueous electrolyte interface or at mercury electrodes in other solvents (1-4,16). The main advantages of mercury over other metals are that mercury is a liquid, it is relatively chemically inert, and that a new interface can be prepared every few seconds. The latter property helps to minimize complications due to contamination of the surface by organic molecules and other impurities present in the solvent. Mercury electrodes are very useful for studying reduction reactions because reduction of protons in protic solvents is very slow (it has a high overpotential for hydrogen reduction). However, mercury electrodes are not useful for studying oxidative molecular reactions because mercury itself is easily oxidized.

Solid metal electrodes such as platinum, gold and carbon are less easily oxidized than mercury and many faradaic processes can be observed. Unfortunately, accurate determination of the interfacial energetics at solid metal electrodes interfaces is rarely possible. Unlike mercury which has a smooth surface, the surfaces of most polycrystalline solid metal electrodes are microscopically rough with crystal faces of various orientations exposed. Due to the microscopic roughness, the surface area is not well defined and molecules reacting at these surfaces can experience a variety of sites with differing energics. With the exception of gold, metal surfaces react spontaneously with oxygen to form an oxide layer which can have dramatic and unpredictable effects on kinetic processes. Preparation of ideal solid metal electrode surfaces by typical physical and/or chemical treatments cannot be accomplished rapidly or reproducibly. Within the past few years, Hubbard and co-workers have developed ultra high vacuum and ultrapurification techniques that allow them to investigate electrochemical reactions at clean metal electrodes with specific crystallographic orientations (20-22). While these procedures are elaborate and not generally available, they do afford the possibility of extending detailed kinetic studies to solid metals.

Criteria for the Ideal Polarizable Semiconductor-Solution Interface

Based on the discussion above, it seems evident that a detailed understanding of kinetic processes occurring at semiconductor electrodes requires the determination of the interfacial energetics. Electrostatic models are available that allow calculation of the spatial distributions of potential and charged species from interfacial capacitance vs. applied potential data (23,24). Like metal electrodes, these models can only be applied at ideal polarizable semiconductor-solution interfaces (25). In accordance with the behavior of the mercury-solution interface, a set of criteria for ideal interfaces is:

1. The electrode surface is clean or can be readily renewed within the timescale of kinetic measurements.
2. The electrode surface is uniform with respect to a known crystal lattice orientation and the surface area can be accurately measured.
3. The electrode is chemically unreactive with solvents containing electrolytes and redox couples over a wide range of applied potential.
4. The interfacial capacitance can be determined experimentally as a function of applied potential.

The potential distribution at the semiconductor-solution interface is quite different than at the metal-solution interface. Metals do not support internal electric fields. Excess charge on a metal electrode is localized at the surface and the interfacial potential drop occurs in the solution. These considerations indicate that energies of electronic states associated with the metal are continuously variable with respect to states associated with redox couples in the solution. For semiconductors, excess charge is contained in the space charge region and determines the concentration of electrons and holes at the surface. For ideal semiconductors, the energies of electronic states associated with redox processes are determined by the position of the conduction and valence band edges. Several processes such as charging of surface states and adsorption of ions cause the band edges to shift which produces non-ideal behavior. Thus, for an ideal polarizable semiconductor-solution interface:

5. The band edge postions should be stable over a wide range of applied potential.
6. It should be possible to prepare the material doped both p- and n- type over a wide range of doping densities.
7. Materials with different band gaps and/or band positions should be available.

Criteria 6 and 7 are important because in order to vary the kinetically relevant parameters such as the spatial distribution of charge carriers and the thermodynamic driving force, semiconductor electrodes with different majority carrier types, doping densities and band gaps must be used.

Metal Dichalcogenides (MDC)

A class of semiconducting solids that has received considerable attention with respect to photoelectrochemistry is the layered metal dichalcogenides (MDC) (26-33) which have the formula AB_2 where A is a transition metal, formally in the +4 oxidation state, and B is a chalcogenide ion, S^{2-}, Se^{2-} or Te^{2-}. Many of these materials display a layer type structure in which the metal is sandwiched between two layers of the chalcogenide and occupies either an octahedral or trigonal prismatic hole (Figure 1). The sandwiches themselves are assembled into stacks which are held together by weak van der Waals forces. Single crystals of many MDC's which have surface areas as large as one cm^2 can be prepared by chemical vapor transport or by Bridgeman techniques.

Semiconductor electrodes often suffer from the same properties that result in the non-ideality discussed previously for solid metals. MDC crystals have properties that are especially attractive for the formation of well-defined semiconductor-solution interfaces. Not surprisingly, the physical properties of these crystals can be highly anisotropic with respect to the directions parallel and perpendicular to the van der Waals surfaces. The van der Waals surfaces can be nearly molecularly smooth and fresh surfaces can be prepared by cleaving the crystal with an adhesive tape. When the edges of the crystal are masked with epoxy, only the van der Waals surface is exposed to the solution. Thus, redox couples involved in electron transfer reactions experience a single set of interfacial energetics. Furthermore, the area of this surface can be measured to a high degree of accuracy using photographic procedures. An alternative cell design where the electrochemistry is done on a small drop of electrolyte positioned on a selected portion of the van der Waals surface allows for both cleavage of the sample and elimination of dark currents and chemical reactions associated with the unsaturated chemical bonds at edges and steps (34).

While freshly cleaved surfaces clearly contain some steps and other defects, the surfaces are remarkably stable to corrosion. Dramatic evidence of the stability of $MoSe_2$ surfaces was provided by Stickney et al. who obtained LEED and Auger spectra of surfaces that had been exposed to the atmosphere and a variety of oxidizing solutions (35). Except for the presence of a ubiquitous carbon which was attributed to the epoxy resin or cleaving tape, no evidence for surface reactions was found. Long term stability tests of a photoelectrochemical cell made with a WSe_2 electrode, where over 400,000 coulombs/cm^2 were passed through the cell with no detectable photocorrosion, also attest to the durability of these surfaces (36).

Based on results discussed above and others that have appeared in the literature, it appears that interfaces derived from MDC's are likely to satisfy the first three criteria for ideality (at least for certain solvents and redox couples). The remaining criteria must be examined utilizing data obtained from specific systems.

Tungsten Diselenide-Acetonitrile Interfaces

In order to obtain fundamental kinetic data on outer-sphere electron transfer processes at the semiconductor-nonaqueous solution interface, Olson investigated the WSe_2-CH_3CN interface (37). Chemical vapor transport was used to prepare WSe_2 single crystals. By varying the transporting agent (Cl_2 or I_2), amount of dopant (Re or Nb) and the transport conditions, n-type and p-type crystals of good to excellent quality were prepared (38). Crystals from each growth batch were mounted as electrodes and examined using impedance and photoelectrochemical techniques in acetonitrile solutions containing 0.5 M supporting electrolyte. The impedance data was analyzed in the form of Mott-Schottky plots from which the doping densities and flatband potentials were derived. Doping densities were proportional to the amount of

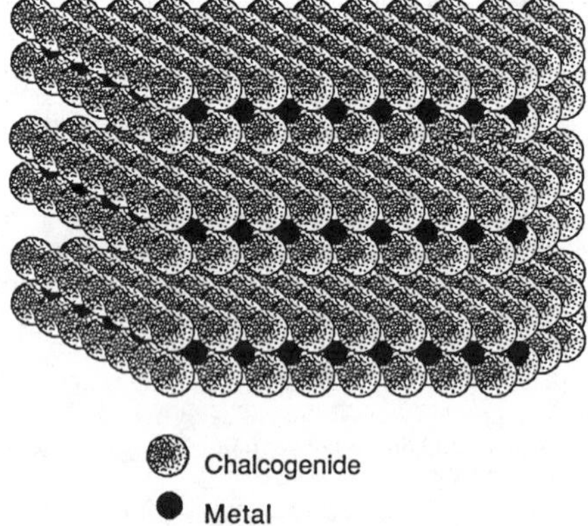

Figure 1. Schematic drawing showing the structure of the trigonal prismatic variety of the metal dichalcogenide structure. Note the structure is not drawn to scale but to emphasize the layered structure of the materials.

dopant used in the synthesis and ranged from 4 x 10^{15} to 3 x 10^{19} cm^{-3} for the n-type crystals and from 3 x 10^{16} to 2 x 10^{19} cm^{-3} for the p-type crystals (38). Within a given growth batch, the deviations in the doping densities for different crystals was quite small compared to the range for the different batches, especially for the lower doped crystals. These two observations taken together with the fact that crystals were cleaved several times prior to mounting support the conclusion that the dopant was homogeneously distributed in the crystals.

Flat band potentials and doping densities obtained from Mott-Schottky plots can be used to calculate the position of the conduction band edge (CBE) for n-type electrodes and the valence band edge (VBE) for p-type crystals (39). The average position of the CBE for nine batches of n-type electrodes was -730 ± 90 mV vs $E^{0'}$ for ferricenium/ferrocene (40). The average position of the VBE for seven batches of p-type electrodes was +670 ± 120 mV. Standard deviations of the band edges for electrodes from a given growth batch were less than 40 mV. The difference in the average band edges for the p- and n-type electrodes is 1.4 V which is close to the direct bandgap of WSe_2, 1.35 eV, although the indirect band gap is slightly lower (1.20 eV). These data support the contention that the interfacial energetics for the WSe_2-CH_3CN interface is not a function of majority carrier type or concentration, i.e. controlled by the interaction of the van der Waals surface with the solvent.

Clearly, the WSe_2-CH_3CN interface satisfies the 6th criterium for ideality. It would seem that similar synthetic efforts would produce ranges in doping densities for other MDC's as well. For WSe_2-CH_3CN, the range of doping densities implies that for 0.5 V of band bending the width of the space charge layer would range from 7.5 to 240 nm. The impedance data obtained in solutions containing only electrolyte indicate that criteria 3,4 and 5 are met for this interface in the absence of redox couples.

Electrochemistry of Ferrocenes at the WSe_2-CH_3CN interface

Although different electrodes within a growth batch yielded similar interfacial energetics, the quality of these electrodes varies greatly with respect to photoconversion efficiency. Reduced photoconversion efficiencies have been attributed to high densities of edge sites and lattice defects which act as recombination centers (26,28). Parkinson and co-workers used a scanning laser spot system to show that photocurrents are greatly diminished on regions of MDC surfaces which are macroscopically flawed and presumably contain a high density of defect sites (41). Reasoning that surfaces containing many defects would be least likely to display ideal behavior, Olson measured photocurrent vs. potential curves in order to calculate fill factors for the WSe_2 electrodes (37). From approximately 100 candidates, the electrode from each batch with the highest fill factor and the most ideal impedence behavior was selected as the "best" electrode. The fill factors for the best electrodes ranged from 0.37 to 0.65 (38).

The selected electrodes (five n-type and four p-type) were used to obtain kinetic current vs. potential data in solutions containing poised ferrocene redox couples (50% oxidized, 50% reduced) (37,39). The electrode potential was varied over a range of at least 0.5 V to over 1.0 V. Three couples were examined: ferrocene (FER) itself, decamethylferrocene (DFER) and acetylferrocene (AFER). The reduction potentials of DFER and AFER with respect to FER (which is assigned a value of 0.0) are -0.50 and +0.25 V, respectively. The reduction potentials for all three couples are located between the CBE and VBE of the WSe_2-CH_3CN interface.

Three concentrations of each redox couple that ranged over two orders of magnitude were examined as well as a solution containing only electrolyte. The details of these comprehensive experiments will be published elsewhere (39); however, several pertinent features are described here. The kinetic currents were measured at constant potential. In order to eliminate mass transfer limitations to the current, a jet electrode configuration was utilized (42). The capacitance of the space charge layer (C_{SC}) was measured at the same potentials simultaneously with the kinetic currents.

Repeated visual examination and impedence measurements indicated no degradation of the electrode over the course of the experiments (3 weeks). During this time, the selected electrodes were stored in the inert-atmosphere glove box in which the electrochemical experiments were performed. No electrode pretreatment procedures were used and the crystals were not recleaved.

The C_{SC} vs. potential data was used to calculate band edge positions for each of the nine selected electrodes in each of the ten solutions. These values are presented in Figure 2. Except for the electrodes with doping densities greater than 10^{19} cm^{-3} (which are degenerately doped), the band edge positions were remarkably stable. These results demonstrate that for outer-sphere redox couples the WSe_2-CH_3CN interface satisfies criteria 3 through 5 and therefore approaches ideality in every respect.

Redox Processes at Non-illuminated MDC-Aqueous Interfaces

An example of the type of kinetic data which can be obtained with a MDC electrode with a nearly perfect surface can be seen in Figure 3. The currents were all obtained in 1 mM solutions of the oxidized form of the respective redox couple in a 1M $HClO_4$ solution which had been exhaustively deoxygenated with argon bubbling. The electrode was a n-WSe_2 electrode with an area of 0.12 cm^2, a doping level of about 3 x 10^{16} cm^{-3}, a flatband potential of about -0.3 V vs. SCE and a nearly defect free surface as determined from microscopic examination and from the very small (<1 nA) dark currents in the blocking region of the diode curve (more positive than -0.2 V). Like mercury, the material has a very high hydrogen overpotential allowing for the study of reduction reactions in strongly acidic electrolytes. The Tafel lines shown in Figure 3 were remarkably reproducible upon repeated scanning in either anodic or cathodic directions.

The current-voltage behavior of the various redox couples can be grouped into four classes. Class one are the very positive redox potential species (Ce^{4+} and $IrCl_6^{2-}$) which can inject holes directly into the valence band over a wide potential range but show small (> 500 mV) non-linear Tafel slopes at very negative potentials even though the driving force for the Ce^{4+} reduction exceeds 2.0 V at these potentials. (These redox couples were actually investigated last and some degradation of the electrode could be observed in the Ce^{4+} solutions undoubtably due to some oxidation of the surface by this highly oxidizing couple). It is interesting that the hole injection current for the two couples is nearly equal despite the large difference in their redox potentials (0.55 V, Figure 4) suggesting that the hole injection current is not dependent on the driving force of the reduction reaction.

Class two redox couples have potentials located in the bandgap region (Fe^{3+}, $Fe(CN)_6^{3-}$, $Co(phen)_3^{3+}$, phen = o-phenanthroline) and show parallel tafel lines with very low slopes (450 mV/decade) with the more positive redox species showing earlier onsets of reduction. The currents appear to be due to a tunneling process where holes can be injected into the valence band of the semiconductor as the bulk valence band potential passes the redox level of the couple during formation of an accumulation layer. The tunneling process results in very low currents due to the low number of surface defects and the large space charge layer width through which the electron must tunnel in the moderately low doped material.

The third class of redox species are couples located near the conduction band of WSe_2. The only outer-sphere example found, which is suitable for use in aqueous electrolytes, is $Ru(NH_3)_6^{3+}$. Its reduction is characterized by an immediate onset upon accumulation in the semiconductor and a tafel slope of 130 mV/decade. The reduction mechanism appears to be direct reduction of the $Ru(NH_3)_6^{3+}$ by electrons from the accumulation layer. The only member of the forth class of redox species is triiodide ion. It is characterized by adsorption onto the semiconductor surface as was demonstrated by the first application of chronocoulometry to a semiconductor electrode (another demonstration of the reproducibility and low background currents on

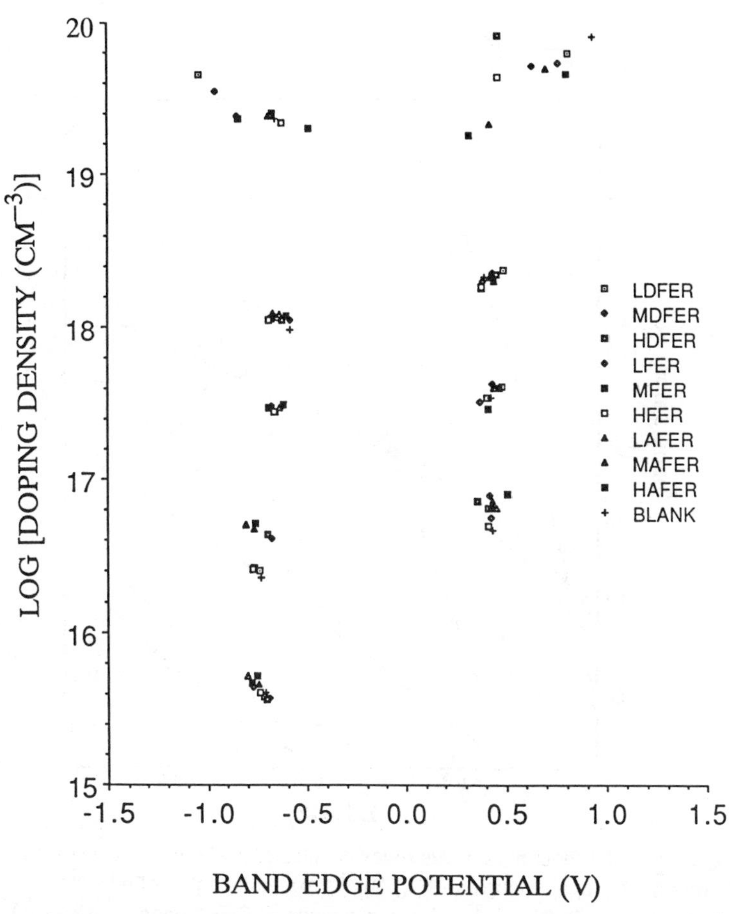

Figure 2. Band edge positions obtained over a period of three weeks for p-and n-type WSe_2 -CH_3CN interfaces containing metallocene redox couples (ferrocene, FER; decamethylferrocene, DFER; and acetylferrocene, AFER) each at three concentrations (preceding letter refers to high,H; medium,M; and low, L). Two different electrodes were used to obtain the data for n-WSe_2 with doping densities between 10^{16} - 10^{17} cm^{-3}.

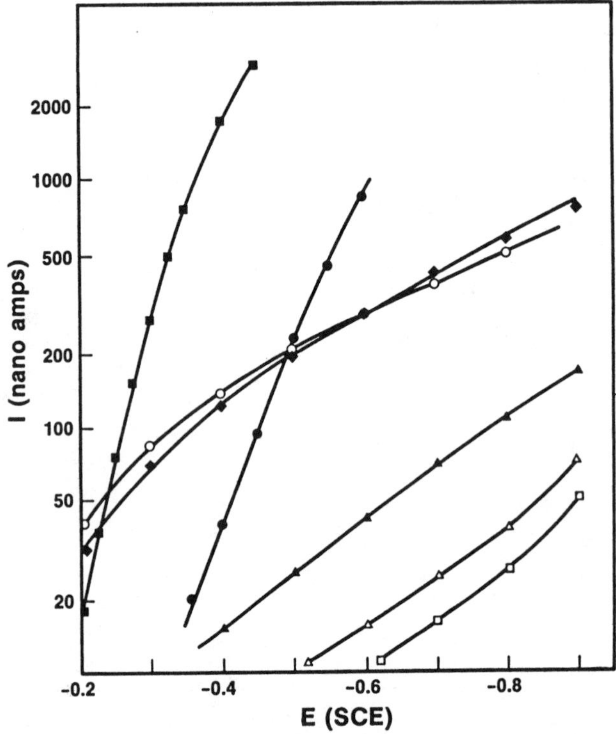

Figure 3. Tafel Plots of aqueous redox couples at n-WSe_2: open squares - $Co(phen)_3^{3+}$, open triangles - $Fe(CN)_6^{3-}$, closed triangles - Fe^{3+}, closed circles - $Ru(NH_3)_6^{3+}$, closed diamonds - $IrCl_6^{2-}$, open circles - Ce^{4+}, closed squares - I_3^-.

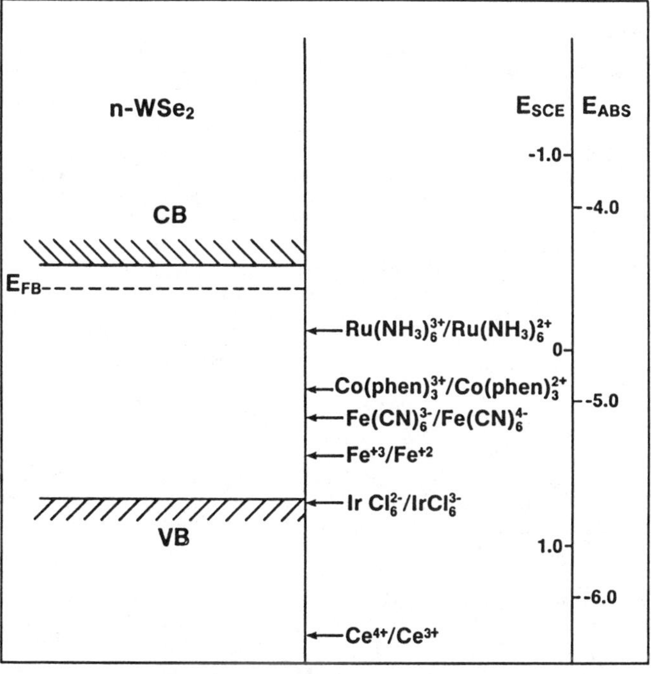

Figure 4. Band diagram showing the relative positions of the conduction and valence bands in WSe_2 with respect to the reduction potentials in aqueous solutions for the redox couples shown in Figure 3.

these surfaces) (27). Triiodide reduction is characterized by an early onset of current due to the positive shift in the band edge as a result of the negative charge on the surface due to adsorbed triiodide. The Tafel slope is higher (85 mv) than for the other systems perhaps due to the inner sphere pathway for reduction of the adsorbed species by electrons in the accumulation layer. Although a quantitative interpretation of the kinetic currents for reduction of various aqueous redox couples on a WSe_2 electrode has not yet been formulated, this section has demonstrated that due to the low background currents on the nearly perfect surface-state free surfaces, qualitative information about the different interfacial charge transfer mechanisms can be obtained even when the kinetic currents are below $1\mu A/cm^2$.

Adsorption of Dyes on Metal Dichalcogenides

An interesting approach to measuring rates of electron transfer reactions at electrodes is through the study of surface bound molecules (43-45). Molecules can be attached to electrode surfaces by irreversible adsorption or the formation of chemical bonds (46). Electron transfer kinetics to and from surface bound species is simplified because there is no mass transport and because the electron transfer distance is controlled to some degree.

Photoinduced electron transfer reactions between surface bound dye molecules and semiconductor electrodes are important for practical as well as fundamental reasons. Absorption of light by the dye can improve the spectral response of the semiconductor and these systems are models for the photographic process (47-51). MDC surfaces are excellent substrates for studying electron injection into the conduction band of the semiconductor.

Understanding of dye and dye aggregate sensitization efficiency, the excited states involved in injection and the energetic thresholds for electron injection has mostly resulted from photoelectrochemical and spectroscopic studies of oxide semiconductors because of their large bandgaps and photocorrosion resistant surfaces. The quantum yield, defined as electrons per photon absorbed by the dye on oxide semiconductor single crystal surfaces was consistently limited to less than 0.5%. The low quantum yield has been attributed to recombination of the photogenerated carriers via back reaction of the injected electron with the dye radical, perhaps through surface states located in the bandgap. Surface states are virtually unavoidable on materials which have a three dimensional structure due to the need to terminate the bonding at every low index crystal face.

It has recently been shown that van der Waals (001) surfaces of two dimensional semiconductors such as WS_2, $MoSe_2$ and WSe_2 are superior substrates for studying sensitization (52). High quantum yields (>4% per incident photon or >80% per absorbed photon were measured for the sensitization of the WSe_2 surface by an infrared absorbing pentathiacyanine dye. The high quantum yields were attributed to the absence of interface states due to the lack of bond termination, characteristic of van der Waals surfaces.

Metal Dichalcogenides with Band Gaps Greater than 2 eV

Although semiconductors with band gaps ranging from 1 to 1.6 eV are most suited to solar energy conversion, materials with wider band gaps are useful for fundamental studies because situations involving larger differences between band edge positions and reduction potentials for redox species can be examined.

Recently, MDC's and other semiconductors with a layered structure having large bandgaps have been used in electrochemical studies. The photoanode properties of n-SnS_2 (indirect transition at 2.2 eV) in acidic and basic solutions has been reported (53) and this material was also shown to have high quantum yields for electron injection from adsorbed dyes (54) GaS has a band gap of approximately 2.4 eV and has a double sandwich structure with two metal layers sandwiched between chalcogenide layers (S-Ga-Ga-S repeat unit along the c axis). Preliminary dye-sensitization studies indicate that GaS (or Ga_2S_2) has an extremely negative conduction band edge (-1.3

volts vs SCE). $CdPS_3$ is an example of another class of layered semiconductors. It is colorless with a bandgap of about 3.1 eV. Synthetic procedures for controlling the type and level of doping for these wide bandgap materials have not been developed and further electrochemical characterization is necessary to determine band edge positions and the surface stability of these materials. Nevertheless, they represent a large class (55) of potentially ideal interfaces for use in electrochemical kinetic studies.

Acknowledgment

Research performed by CAK and JBO was supported by the Department of Energy (Division of Chemical Sciences) Contract No. DE-FG02-84ER13247. The authors thank Arthur Nozik, Mark Peterson and John Turner (Solar Energy Research Institute, Golden, CO) for the generous use of their equipment and for valuable discussions.

Literature Cited

1. Albery, W.J. Electrode Kinetics, Clarendon Press, Oxford, 1975.
2. Bockris, J. O'M.; Reddy, A.K.N. Modern Electrochemistry, Plenum Press, New York, 1970.
3. Vetter, K.J. Electrochemical Kinetics, Academic Press, New York, 1967.
4. Delahay, P. Double-Layer and Electrode Kinetics, Interscience Publishers, New York, 1965.
5. Bard, A.J.; J. Phys. Chem. 1982, 86, 172.
6. Wrighton, M.S.; Acc. Chem. Res. 1979, 9, 303.
7. Rajeshwar, K.; Singh, P.; Dubow, J.; Electrochim. Act. 1978, 23, 1117.
8. Nozik, A.J.; Ann. Rev. Phys. Chem. 1978, 29, 189.
9. Wilson, R.H.; CRC Crit. Rev. Sol. State Mat. Sci. 1980, 10, 1.
10. S. R. Morrison, Electrochemistry at Semiconductor and Oxidized Metal Electrodes, Plenum Press, N.Y., 1980.
11. Gerischer, H. Z. Physik. Chem. 1960, 26, 223.
12. Gerischer, H. Z. Physik. Chem. 1961, 27, 48 1961.
13. Gerischer, H. Adv. Electrochem. Electrochem. Engr. 1961, 1, 139;
14. Gerischer, H. Top. Appl. Phys. 1979, 31, 115.
15. Memming, R. Electroanal. Chem. 1979, 11, 1.
16. Bard, A.J.; Faulkner, L.R. Electrochemical Methods, Wiley, New York, 1980, Chapters. 1, 3,.11, 12.
17. Marcus, R.A.; J. Chem. Phys. 1956, 24, 966; Disc. Far. Soc. 1960, 29, 21; J. Chem. Phys. 1965, 43, 2654.
18. Marcus, R.A.; Sutin, N. Biochim. Biophys. Act. 1985, 811, 265.
19. J. Phys. Chem. (R.A. Marcus Commemorative Issue), 1986, 90(16), 3453-3862.
20. Baltruschat, H.; Martinez, M.; Lewis, S.K.; Frank, L.; Dian, S.; Stern, D.A.; Datta, A.; Hubbard, A.T. J. Electroanal. Chem. (1987) 217, 111.
21. Stern, D.A.; Baltruschat, H.; Martinez, M.; Stickney, J.L.; Song, D.; Lewis, S.K.; Frank, D.G.; Hubbard, A.T. ibid. (1987) 217, 101.
22. Schardt, B.C.; Stickney, J.L.; Stern, D.A.; Wieckowski, A.; Zapien, D.C.; Hubbard, A.T. Langmuir (1987) 3, 239.
23. Mott, N.F. Proc. R. Soc. London 1939, A171, 27.
24. Schottky, W. Z. Phys.1942, 118, 539.
25. Dutoit, E.C.; Meirhaeghe, R.L.; Cardon, F.; Gomes, W.P.; Ber. Bunsenges. Physik. Chem.1975, 79, 1206.
26. Tributsch, H. Struct. & Bonding 1982, 49, 127.
27. Turner, J.A. and Parkinson, B.A. J. Electroanal. Chem. 1983, 150, 611.
28. Kline, G.; Kam, K.; Canfield, D.; Parkinson, B.A. Sol. Ener. Mat. 1981, 4, 301.
29. Parkinson, B.A.; Furtak, T.E.; Canfield, D.; Kam, K.K.; Kline, G., Faraday Disc. 1980, 70, 233.

30. White, H.S.; Fan, F.F.; and Bard, A.J.; J. Electrochem. Soc. 1981, 128, 1045; ibid. 1980, 127, 518; J. Am. Chem. Soc. 1980, 102, 5142.
31. Nagasubramanian, G.; Bard, A.J. J. Electrochem. Soc., 128, 1055 (1981).
32. Wrighton, M.S.; Schneemeyer, L.F. J. Am. Chem. Soc. 1979, 101, 6496; ibid. 1980, 102, 6946.
33. Lewerenz, H.J.; Heller, A.; DiSalvo, S.J. J. Am. Chem Soc. 1980, 102, 1877.
34. Parkinson, B.A., unpublished results.
35. Stickney, J.L.; Rosasco, S.D.; Schardt, B.C.; Solomun, T.; Hubbard, A.T.; Parkinson, B.A. Surface Science 1984, 136, 15.
36. Kline, G.; Kam, K.; Ziegler, R.; Parkinson, B.A. Solar Energy Materials 1982, 6, 337.
37. Olson, J.B. Ph.D. Thesis, University of Colorado, 1987.
38. Koval, C.A.; Olson, J.B. J. Electroanal. Chem. 1987, 234, 133 .
39. Koval, C.A.; Olson, J.B. , submitted for publication.
40. The difference in the band edge positions for n-WSe_2-CH_3CN discussed in this section and the value stated later for n-WSe_2-H_2O is due to problems associated with relating reference electrode potentials in different solvents. See Gagne', R.R.; Koval, C.A.; Lisensky, G.C. Inorg. Chem. 1980, 19, 2855.
41. Furtak, T.; Canfield, D.; Parkinson, B.A. J. Appl. Phys. 1980, 51, 6018.
42. Olson, J.B.; Koval, C.A. Anal. Chem., 1988, 60, 88.
43. Li, T.-T.; Weaver, M.J. J. Am. Chem. Soc. 1984, 106, 6107.
44. Li, T.-T.; Liu, Y.Y.; Weaver, M.J. J. Am. Chem. Soc. 1984, 106, 1233.
45. Guyer, K.L.; Weaver, M.J. Inorg. Chem. 1984, 23, 1664.
46. Murray, R.W. Acc. Chem. Res. 1980, 13, 135.
47. The Theory of the Photographic Process, 4th Edition, T. H. James, Ed.; Macmillan, New York, 1977.
48. Berriman, R.W.; Gilman, P.B., Jr. Photographic Science and Engineering, 1973, 17, 235.
49. Gerischer, H.; Tributsch, H. Ber. Bunsenges. Phys. Chem., 1968, 72, 437.
50. Memming, R.; Tributsch, H. J. Phys. Chem.,1971, 75, 562.
51. Memming, R. Photochemistry and Photobiology, 1972, 16, 325.
52. Spitler, M.; Parkinson, B.A. Langmuir, 1986, 2, 549.
53. Katty, A.; Fotouhi, B.; Gorochov, O. J. Electrochem. Soc. 1984, 131, 2806.
54. Parkinson B. A. Proceedings of the Electrochemical Society, Honolulu, in press
55. Physics and Chemistry of Layered Materials, Vol. 1-6 , D. Reidel, Dordrecht, Holland.

RECEIVED May 17, 1988

Chapter 31

Electrosynthesis and Electrochemistry of Metalloporphyrins Containing a Metal–Carbon σ-Bond

Reactions of Rhodium, Cobalt, Germanium, and Silicon Complexes

K. M. Kadish, Q. Y. Xu, and J. E. Anderson[1]

Department of Chemistry, University of Houston, Houston, TX 77004

The electrosynthesis of metalloporphyrins which contain a metal-carbon σ-bond is reviewed in this paper. The electron transfer mechanisms of σ-bonded rhodium, cobalt, germanium, and silicon porphyrin complexes were also determined on the basis of voltammetric measurements and controlled-potential electrooxidation/reduction. The four described electrochemical systems demonstrate the versatility and selectivity of electrochemical methods for the synthesis and characterization of metal-carbon σ-bonded metalloporphyrins. The reactions between rhodium and cobalt metalloporphyrins and the commonly used CH_2Cl_2 is also discussed.

Metalloporphyrins containing a metal-carbon σ-bond are currently limited to complexes with eight different transition metals (Ti, Ni, Fe, Ru, Co, Rh, Ir and Zn) and seven different non-transition metals (Al, Ga, In, Tl, Si, Ge, and Sn). These compounds have been the subject of several recent reviews(1-3) which have discussed their synthesis and physicochemical properties.

The synthesis of metalloporphyrins which contain a metal-carbon σ-bond can be accomplished by a number of different methods(1,2). One common synthetic method involves reaction of a Grignard reagent or alkyl(aryl) lithium with (P)MX or $(P)M(X)_2$ where P is the dianion of a porphyrin macrocycle and X is a halide or pseudohalide. Another common synthetic technique involves reaction of a chemically or electrochemically generated low valent metalloporphyrin with an alkyl or aryl halide. This latter technique is similar to methods described in this paper for electrosynthesis of cobalt and rhodium σ-bonded complexes. However, the prevailing mechanisms and the chemical reactions

[1]Current address: Department of Chemistry, Boston College, Chestnut Hill, MA 02167

0097–6156/88/0378–0451$06.00/0

following electrogeneration may differ due to the difference in reaction conditions for generation of the reactive species.

Two aspects of porphyrin electrosynthesis will be discussed in this paper. The first is the use of controlled potential electroreduction to produce metal-carbon σ-bonded porphyrins of rhodium and cobalt. This electrosynthetic method is more selective than conventional chemical synthetic methods for rhodium and cobalt metal-carbon complexes and, when coupled with cyclic voltammetry, can be used to determine the various reaction pathways involved in the synthesis. The electrosynthetic method can also lead to a simultaneous or stepwise formation of different products and several examples of this will be presented.

The second type of porphyrin electrosynthesis discussed in this paper is controlled potential electrooxidation of σ-bonded bis-alkyl or bis-aryl porphyrins of Ge(IV) and Si(IV). This electrooxidation results in formation of σ-bonded mono-alkyl or mono-aryl complexes which can be isolated and characterized in situ. Again, cyclic voltammetry can be coupled with this method and will lead to an understanding of the various reaction pathways involved in the electrosynthesis.

Dozens of electrochemical and spectroelectrochemical papers on transition metal and main group metal-carbon σ-bonded metalloporphyrins were published between 1984 and 1987 and a summary of these results are well covered in three recent reviews(1-3). Therefore, a characterization of chemically synthesized metal-carbon porphyrins will not be discussed in this paper.

Rhodium Porphyrins. Chemical syntheses of $[(P)Rh]_2$ and (P)Rh(R) complexes are well known(4-11). Electrochemical techniques have also been used to synthesize dimeric metal-metal bonded $[(TPP)Rh]_2$ as well as monomeric metal-carbon σ-bonded (TPP)Rh(R) and (OEP)Rh(R)(12-16). The electrosynthetic and chemical synthetic methods are both based on formation of a highly reactive monomeric rhodium(II) species, (P)Rh. This chemically or electrochemically generated monomer rapidly dimerizes in the absence of another reagent as shown in Equation 1.

$$(P)Rh \rightleftharpoons 1/2\ [(P)Rh]_2 \qquad (1)$$

However in the presence of hydrogen gas, (P)Rh(H) can be formed as shown in Equation 2.

$$(P)Rh + 1/2\ H_2 \rightleftharpoons (P)Rh(H) \qquad (2)$$

Reactions 1 and 2 involve an equilibrium which is shifted to the right. The methods generally utilized for chemical generation of (P)Rh initially involve a dimeric Rh(II) species or the Rh(III) hydride, and relatively high temperatures and/or long reaction times are required to synthetically generate (P)Rh(4-9). For this reason, much of the mechanistic information involving (P)Rh reactions is lost and a detection of reaction intermediates in the synthesis of the σ-bonded (P)Rh(R) complexes is generally not possible.

The method for electrosynthetic generation of (P)Rh(R) involves an initial formation of (P)Rh and is accomplished by the one electron reduction of an ionic Rh(III) porphyrin species(14). The generated Rh(II) porphyrin can dimerize as shown in Equation 1 but, in the presence of an alkyl halide RX, will react to form (P)Rh(R). The overall reaction between RX and $(P)Rh^{II}$ is given by Equation 3, where S is an electron source other than the electrode.

$$(P)Rh^{II} + RX + S \rightarrow (P)Rh(R) + X^- + S^+ \quad (3)$$

The reactions associated with Equation 3 can be monitored by cyclic voltammetry which give current voltage curves of the type shown in Figures 1a and 1b for the case of RX = CH_3I. The first reduction in Figure 1b is the Rh(III) → Rh(II) transition and the second is the reversible reduction of $(TPP)Rh(CH_3)$. An analysis of these cyclic voltammograms and other similar voltammetric/spectroelectrochemical data gave the following reaction sequence where L is dimethylamine(14).

Scheme I

$$[(TPP)Rh(L)_2]^+ + e \rightarrow (TPP)Rh + 2L \quad (4)$$

$$(TPP)Rh + RX \rightarrow (TPP)Rh(R) + X^\bullet \quad (5a)$$

$$(TPP)Rh + RX \rightarrow [(TPP)Rh(R)]^+ + X^- \quad (5b)$$

$$X^\bullet + S \rightarrow X^- + S^+ \quad (6a)$$

$$[(TPP)Rh(R)]^+ + S \rightarrow (TPP)Rh(R) + S^+ \quad (6b)$$

After electrogeneration of (TPP)Rh (Equation 4) and the attack at the R-X bond, either a loss of $X^\bullet$ and formation of (TPP)Rh(R) (Equation 5a) or a loss of X^- with formation of $[(TPP)Rh(R)]^+$ (Equation 5b) will occur. Reaction pathways 5a and 6a occur for RX complexes where X = I or Br while reactions 5b and 6b occur for RX complexes where X = Cl or F.

The mechanism shown in Scheme I was supported by coulometric determinations, the formation of X^- as one of the reaction products, and the absence of any (TPP)RhX as a final product(14). Equations 6a and 6b in Scheme I indicate that a second electron source other than the electrode is involved in the overall reaction and was postulated on the basis of coulometric data which indicated that only one electron was electrochemically transferred. This was true despite the fact that the overall reaction requires a total of two electrons. A full discussion of these results are presented in the original report(14).

An electrosynthetic method was used to generate the 25 different (P)Rh(R) complexes listed in Table I(14,16). Many of the complexes in this Table had not been previously reported, especially (P)Rh(RX). In all cases, bulk electrolysis of

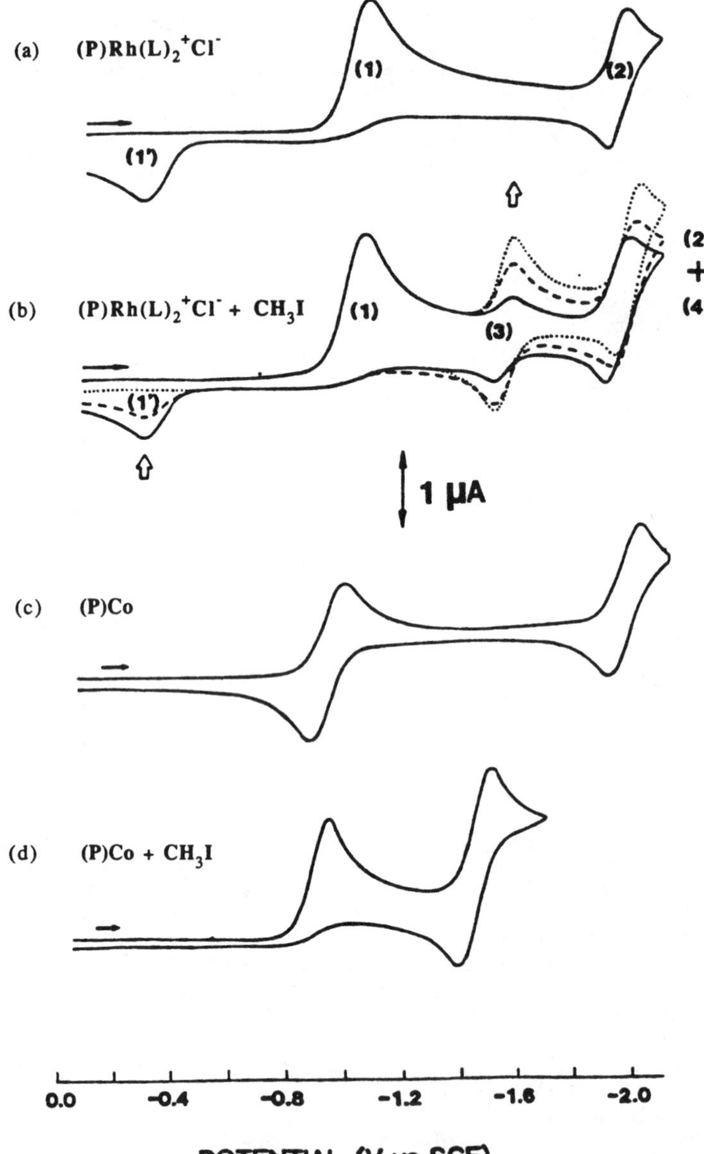

Figure 1. Cyclic voltammograms obtained at Pt electrode: a, 9.9 × 10^{-4} M $[(TPP)Rh(L)_2]^+Cl^-$; b, 9.9 × 10^{-4} M $[(TPP)Rh(L)_2]^+Cl^-$ and 1.0 equivalent of CH_3I; c, (TPP)Co; and d, (TPP)Co and 1.5 equivalent of CH_3I in THF containing 0.1 M TBAP. Adapted from refs. 14 and 26.

Table I. The Redox Potentials and UV-vis Spectral Data for the Electrogenerated (TPP)Rh(R) Complexes in THF Containing 0.2 M TBAP [a]

Complex	R Group	$E_{1/2}$ (Ox)[b]		$E_{1/2}$ (Red)			UV-Vis[c]	
(R)Rh(R)	CH_3[d]	--	--	-1.45		-1.90	418(22.9)	525(2.3)
	C_2H_5	1.12	--	-1.41		-1.90	411(21.6)	524(1.9)
	C_3H_7	0.97	1.33	-1.42		-1.91	411(17.7)	524(1.5)
	C_4H_9	0.97	1.34	-1.42		-1.92	411(21.5)	524(1.7)
	C_5H_{11}	0.98	1.34	-1.42		-1.91	411(25.2)	524(2.3)
	C_6H_{13}	0.97	1.34	-1.42		-1.91	411(27.9)	524(2.8)
	$CH(CH_3)CH_2CH_3$[e]	--	--	--		--	417	530
	$C(CH_3)_3$[e]	--	--	--		--	415	
(P)Rh(RX)	C_3H_6Cl	--	--	-1.41		-1.86	411(27.2)	524(2.5)
	C_4H_8Cl	--	--	-1.41		-1.88	411(27.9)	524(2.7)
	$C_5H_{10}Cl$	--	--	-1.42		-1.90	411(24.9)	524(2.0)
	C_3H_6Br	--	--	-1.39	-1.81[f]	-1.83[f]	411(24.7)	524(2.4)
	C_4H_8Br	1.14[g]	--	-1.40	-1.85[f]	-2.01[f]	411(29.2)	524(3.2)
	$C_5H_{10}Br$	1.14[g]	--	-1.40		-1.89[e]	411(23.7)	524(1.9)
	C_3H_6I	0.98	1.35	-1.38[f]		--	411(26.9)	524(2.2)
	C_4H_8I	0.97	1.34	-1.39	-1.78[f]	-1.99[f]	411(30.9)	524(2.6)
	$C_5H_{10}I$	0.98	1.35	-1.40	-1.81[f]	-1.88	411(27.7)	524(2.2)
	$C_6H_{12}I$	0.99	1.35	-1.41	-1.82[f]	-1.88	411(27.2)	524(2.2)
	CH_2Cl	--	--	-1.42			415(20.0)	527(2.0)
	CH_2Br	--	--	-1.43			413(21.9)	526(2.2)
	CH_2I	--	--	-1.43			415(18.5)	527(1.8)
(P)Rh(CHX_2)	$CHCl_2$	--	--	-1.41			418(20.8)	529(2.2)
	CHI_2	--	--	-1.43			418(15.5)	528(2.0)
(P)Rh(CX_3)	CCl_3	--	--	-1.41			421(21.2)	533(2.1)
	CI_3	--	--	-1.43			416(11.8)	526(1.7)

[a] From reference 12, 13, 14.

[b] Measured in PhCN.

[c] nm ($\varepsilon \times 10^{-4}$, M^{-1} cm^{-1}).

[d] Measured in PhCN.

[e] Complete formation of (P)Rh(R) is not observed.

[f] E_p measured at 0.1 V/s.

[g] Measured in THF.

$[(TPP)Rh(L)_2]^+Cl^-$ in the presence of an alkyl halide leads to a given (P)Rh(R) or (P)Rh(RX) complex. The yield was nearly quantitative (>80%) in most cases based on the rhodium porphyrin starting species. However, it should be noted that excess alkyl halide was used in Equation 3 in order to suppress the competing dimerization reaction shown in Equation 1. The ultimate (P)Rh(R) products generated by electrosynthesis were also characterized by ^1H NMR, which demonstrated the formation of only one porphyrin product(14). No reaction is observed between (P)Rh and aryl halides but this is expected from chemical reactivity studies(10,15). Table I also presents electronic absorption spectra and the reduction and oxidation potentials of the electrogenerated (P)Rh(R) complexes.

Cyclic voltammetry was used to monitor the mechanism for (TPP)Rh(R) formation according to Equation 3 where RX is a terminal alkyl halide. The current for reduction of electrogenerated (P)Rh(R) or (P)Rh(RX) species is a measure of the concentration of this product at the electrode surface, and hence a measure of its rate of formation. The current must be standardized for experimental factors such as scan rate, initial concentration of $[(TPP)Rh(L)_2]^+Cl^-$ (or other initial Rh(III) species), electrode area, and concentration of the alkyl halide reactant.

The results of kinetic experiments with terminal alkyl halides demonstrate that the overall reaction rate between (P)Rh and RX is dependent upon the size of the alkyl halide, the number of halide groups (RX or RX_2) and on the type of halide. The reaction rate of RX follows the trend I > Br > F = Cl(14). Also, the larger the alkyl group, the slower is the chemical reaction(14).

Mechanistic studies of homogenous chemical reactions involving formation of (P)Rh(R) from (P)Rh and RX demonstrate a radical pathway(9). These studies were carried out under different experimental conditions from those in the electrosynthesis. Thus, the difference between the proposed mechanism using chemical and electrochemical synthetic methods may be due to differences related to the particular investigated alkyl halides in the two different studies or alternatively to the different reaction conditions between the two sets of experiments. However, it should be noted that the electrochemical method for generating the reactive species is under conditions which allow for a greater selectivity and control of the reaction products.

The highly reactive nature of (P)Rh is perhaps best demonstrated by the reaction of electrochemically generated (TPP)Rh with terminal alkenes and alkynes. The overall reaction with alkynes is given by Equation 7 and the suggested mechanism

$$(P)Rh + HC{\equiv}CR \rightarrow (P)Rh(R) + products \qquad (7)$$

is presented in Scheme II(15). A similar reaction mechanism was also found to occur with alkenes.

Scheme II

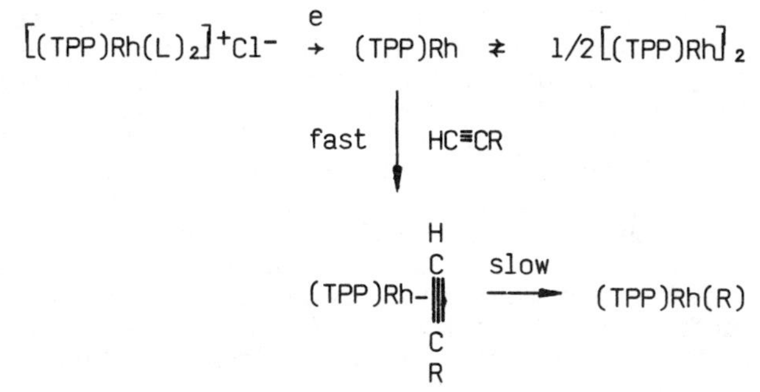

$$[(TPP)Rh(L)_2]^+Cl^- \xrightarrow{e} (TPP)Rh \rightleftarrows 1/2[(TPP)Rh]_2$$

fast ↓ HC≡CR

$$(TPP)Rh-\overset{H}{\underset{R}{C \equiv C}} \xrightarrow{slow} (TPP)Rh(R)$$

The reactive (TPP)Rh in Scheme II is electrogenerated from $[(TPP)Rh(L)_2]^+Cl^-$ in the presence of an alkene or an alkyne. The formation of an intermediate is observed. This intermediate is not detected by chemical reaction methods and was tentatively assigned as a π complex(15). Similar π complexes have been reported for ruthenium porphyrin species(17,18).

The reaction of (TPP)Rh with terminal alkenes or alkynes is of special interest due to the cleavage of the carbon-carbon bond adjacent to either the alkene or the alkyne functionality and results in the ultimate formation of (TPP)Rh(R). This overall reaction implies activation of a relatively inert carbon-carbon bond, especially for the case of the terminal alkene. However, the ultimate formation of (P)Rh(R) is not surprising if one considers the relative stability of the rhodium carbon bond in this species(17).

Cobalt Porphyrins. The primary synthetic method for generating cobalt porphyrins with a metal carbon σ-bond is to react a chemically or electrochemically generated cobalt(I) anion, $[(P)Co]^-$, with an alkyl or an aryl halide(19-26). $[(P)Co]^-$ is stable and the spectroscopic properties of isolated $Na[(TPP)Co]\cdot 5THF$ has been reported(27). The solution spectra of $[(TPP)Co]^-$ in toluene and benzene are also known (Kadish, K. M.; Mu, X. H., submitted for publication).

The generated $[(P)Co]^-$ is stable in numerous solvents but reacts with alkyl halides as shown in Equation 8(1).

$$[(P)Co]^- + RX \rightleftarrows (P)Co(R) + X^- \quad (8)$$

The above reactivity is different from that of rhodium porphyrins in two respects. The first difference is that (P)Rh is not stable, as is the case for electrogenerated $[(P)Co^I]^-$. In addition, the Rh(I) state is not accessible for (P)Rh and at low temperature only a porphyrin ring reduction to generate the Rh(II) π anion radical, $[(P)Rh]^-$ will occur(12). Hence, the electrogenerated cobalt species can be viewed as a two electron reagent while the rhodium species is an one electron reagent.

The electrogeneration of $[(TPP)Co]^-$ from (TPP)Co, and the reaction of this species with CH_3I can be followed by cyclic voltammetry as shown in Figures 1c and 1d. In the absence of any added reagent, there are two reversible reduction waves which occur at $E_{1/2}$ = 0.85 and -1.86 V (see Figure 1c). These are due to the formation of $[(TPP)Co]^-$ and $[(TPP)Co]^{2-}$, where the second reduction has occurred at the porphyrin π ring system. The first reduction of (TPP)Co is not reversible in the presence of CH_3I, and occurs at E_p = -0.86 V (see Figure 1d). A new reversible reduction also appears at $E_{1/2}$ = -1.39 V. This process is due to $(TPP)Co(CH_3)$ which is formed as shown by Equation 8. The formation of $(TPP)Co(CH_3)$ as the final product of the electrosynthesis was confirmed by spectroelectrochemical experiments which were carried out under the same experimental conditions(26).

Reactions of Low Valent Rh and Co with Methylene Chloride. A number of porphyrin electrochemical studies have been carried out in CH_2Cl_2(3). However, this solvent is not as unreactive as once believed and will react with reduced rhodium and cobalt complexes to generate $(P)Rh(CH_2Cl)$(13) and $(P)Co(CH_2Cl)$(26), respectively. These reactions can be monitored by cyclic voltammetry and give current-voltage curves very similar to those obtained during the reaction of (P)Rh or $[(P)Co]^-$ with RX (See Figure 1).

Figure 2 shows cyclic voltammograms of (TPP)Co and $[(TPP)Rh(L)_2]^+$ in CH_2Cl_2. The formation of both $[(TPP)Co]^-$ and (TPP)Rh is irreversible and for the case of the cobalt complex, the overall reaction is shown in Scheme III(26).

Scheme III

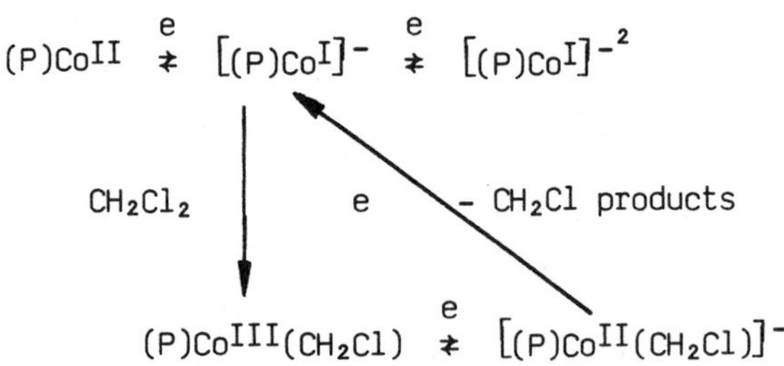

As seen in Figure 2, similar reactions occur between CH_2Cl_2 and (P)Rh or $[(P)Co]^-$ but the current-voltage curves differ in that the electrosynthesized $(P)Rh(CH_2Cl)$ is reversibly reduced at $E_{1/2}$ = -1.43 V (see Table I).

Germanium Porphyrins. The electrosynthesis of σ-bonded mono-alkyl or mono-aryl germanium porphyrins involves the conversion of $(P)Ge(R)_2$ to (P)Ge(R)X where X is an anionic ligand. A standard Grignard reaction between $(P)Ge(Cl)_2$ and RMgX leads to the σ-bonded bis-alkyl or bis-aryl complexes, $(P)Ge(R)_2$. These complexes were initially synthesized as ^1H NMR shift reagents(28,29). However, almost no reactivity of these species was reported until several years later when it was shown that the

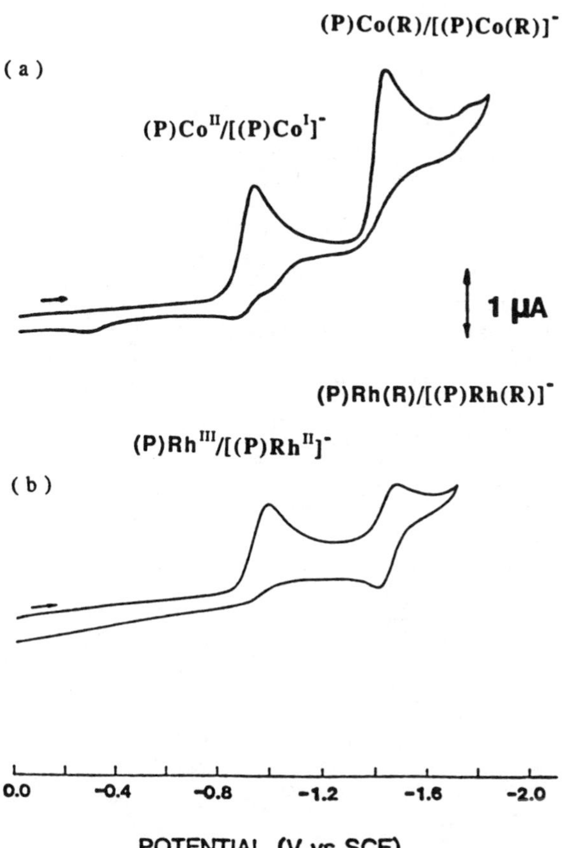

Figure 2. Cyclic voltammograms obtained at Pt electrode: a, (TPP)Co; and b, 4.8×10^{-4} M $[(TPP)Rh(L)_2]^+Cl^-$ in CH_2Cl_2 containing 0.1 M TBAP. Adapted from refs. 13 and 26.

Ge(IV) porphyrins underwent a photochemical insertion of O_2 into the germanium-carbon bond to give peroxides(30-34). The σ-bonded compounds also undergo photohomolytic cleavage of the germanium-carbon bond in the absence of oxygen to give alkyl or aryl radicals and porphyrin radical anions(Xu, Q. Y.; Guilard, R.; Kadish, K. M., submitted for publication and 35).

The overall mechanism for electrochemical conversion of $(P)Ge(R)_2$ to (P)Ge(R)X is presented in Scheme IV where P = OEP(35).

In Scheme IV, a $[(OEP)Ge(C_6H_5)_2]^+$ cation radical is initially generated upon oxidation of $(OEP)Ge(C_6H_5)_2$ but one germanium carbon bond rapidly cleaves to give $(OEP)Ge(C_6H_5)ClO_4$. The ClO_4^- counterion is from the supporting electrolyte. This cleavage of the germanium-carbon bond results in an overall irreversible oxidation process which occurs at E_p values between 0.64 and 0.95 V depending upon the nature of R and the specific porphyrin macrocycle. An example of the resulting cyclic voltammogram is shown in Figure 3.

The electrosynthesized $(OEP)Ge(C_6H_5)ClO_4$ was characterized in situ by thin-layer spectroelectrochemistry. The final product of electrosynthesis was spectrally compared with the same compounds which were synthesized using chemical and photochemical methods(35). $(OEP)Ge(C_6H_5)Cl$ and $(OEP)Ge(C_6H_5)OH$ were also electrochemically generated by the use of specific solvent/supporting electrolyte systems(35).

The electroreduction/oxidation reactions of $(OEP)Ge(C_6H_5)ClO_4$, $(OEP)Ge(C_6H_5)Cl$ and $(OEP)Ge(C_6H_5)OH$ were investigated in PhCN containing 0.1 M TBAP(35). Under these experimental conditions, the overall reaction mechanism shown in Scheme V is demonstrated to occur.

Half-wave potentials for each of the electron transfer steps shown in Scheme V are listed in Table II. The first oxidations of $(OEP)Ge(C_6H_5)Cl$ and $(OEP)Ge(C_6H_5)OH$ are irreversible and occur at E_p = 1.00 V for X = OH^-. The second oxidations of these complexes are reversible and both occur at $E_{1/2}$ = 1.40 V. These second oxidation processes occur at identical to the $E_{1/2}$ values for oxidation of $(OEP)Ge(C_6H_5)ClO_4$ and this was presented as strong evidence for an oxidation of Cl- on $(OEP)Ge(C_6H_5)Cl$ and OH^- on $(OEP)Ge(C_6H_5)OH$(35).

Silicon Porphyrins. The electrosynthesis of σ-bonded mono-alkyl or mono-aryl silicon porphyrins parallels that of the germanium complexes in that $(P)Si(R)_2$ can be converted to (P)Si(R)X after electrooxidation. However, $(OEP)Si(C_6H_5)_2$ and $(OEP)Si(CH_3)_2$ show slightly different oxidative behavior than the germanium analogues with the same R groups. For example, (OEP)Si(R)OH is generated immediately following oxidation of $(OEP)Si(R)_2$ which differs from $[(OEP)Ge(R)_2]^+$ which produces $(OEP)Ge(R)ClO_4$ after the cleavage of one metal-carbon bond. All of the electrosynthesized (P)Si(R)X complexes undergo reversible reductions and half-wave potentials for these reactions are summarized in Table II. The overall oxidation-reduction mechanisms of $(P)Si(R)_2$ and (P)Si(R)X are summarized in Scheme VI(36).

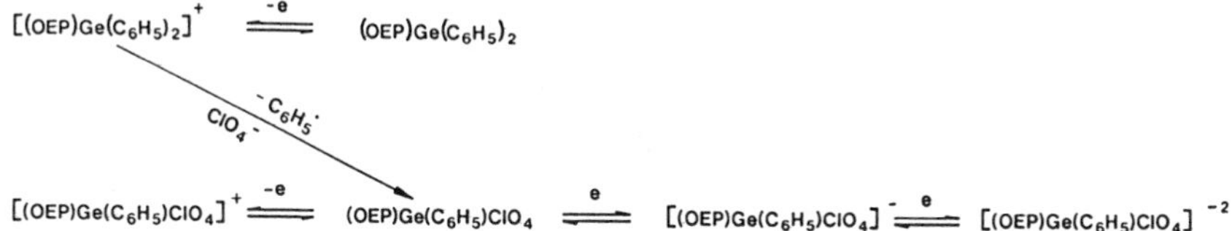

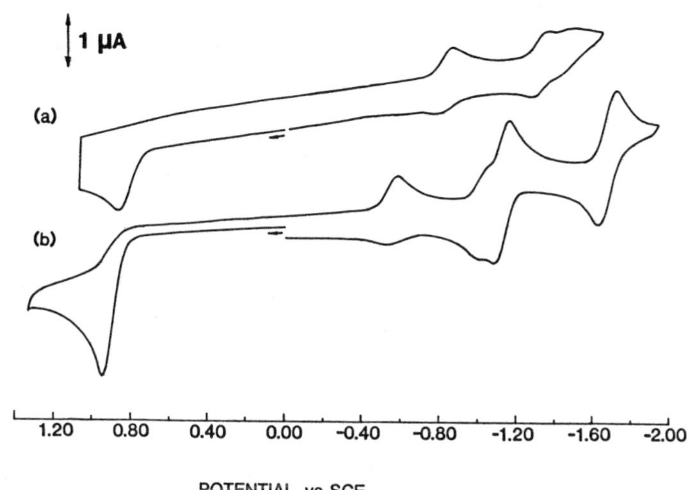

Figure 3. Cyclic voltammograms obtained at Pt electrode: a, $(OEP)Ge(C_6H_5)_2$; and b, 10^{-3} M $(TPP)Ge(C_6H_5)_2$ in PhCN containing 0.1 M TBAP. Adapted from ref. 35.

Scheme V

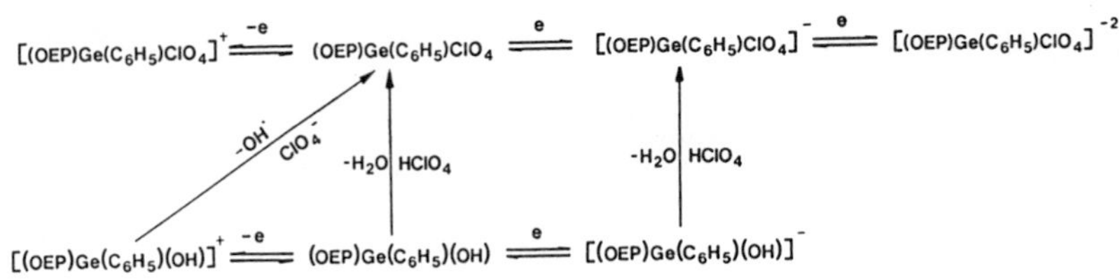

Table II. Half-wave and Peak Potentials for Metal-Carbon σ-bonded Germanium[a] and Silicon[b] Porphyrins in PhCN or CH_2Cl_2 (0.1 M TBAP)

Complex	Oxidation: Axial Ligand	Oxidation: Porphyrin Ring		Reduction: Porphyrin	Reduction: Ring
$(OEP)Ge(C_6H_5)ClO_4$	---	---	1.40	-0.82	-1.30
$(OEP)Ge(C_6H_5)OH$	1.00[c,d]	---	1.40	---	-1.38
$(OEP)Ge(C_6H_5)Cl$	1.10[c,e]	---	1.40	-0.89	-1.30
$(OEP)Ge(CH_2C_6H_5)_2$	---	0.64[c]	1.34[c]	-1.54	--
$(OEP)Ge(CH_3)_2$	---	0.75[c]	1.37	-1.48	--
$(OEP)Ge(C_6H_5)_2$	---	0.88[c]	1.39	-1.40	--
$(TPP)Ge(CH_2C_6H_5)_2$	---	0.73[c]	1.34[c]	-1.17	-1.72
$(TPP)Ge(C_6H_5)_2$	---	0.95[c]	1.45[c]	-1.10	-1.65
$(OEP)Si(C_6H_5)OH$	1.14[c,d]	0.86[c]	---	-1.43	--
$(OEP)Si(C_6H_5)_2$	---	0.80[c]	---	-1.42	--
$(OEP)Si(CH_3)_2$	1.12[c,d]	0.72[c]	---	-1.46	--

[a] From reference 35.
[b] From reference 36.
[c] E_{pa} measured at 0.1 V/s.
[d] Corresponds to oxidation of OH^- ligand.
[e] Corresponds to oxidation of Cl^- ligand.

Scheme VI

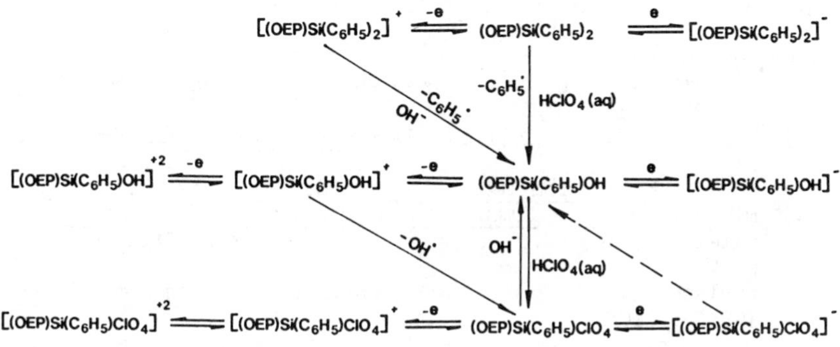
[(OEP)Si(C6H5)2]+
-e
(OEP)Si(C6H5)2
e
[(OEP)Si(C6H5)2]-
-C6H5·
HClO4 (aq)
-C6H5·
OH-
[(OEP)Si(C6H5)OH]+2
-e
[(OEP)Si(C6H5)OH]+
-e
(OEP)Si(C6H5)OH
e
[(OEP)Si(C6H5)OH]-
-OH·
OH-
HClO4(aq)
[(OEP)Si(C6H5)ClO4]+2
[(OEP)Si(C6H5)ClO4]+
-e
(OEP)Si(C6H5)ClO4
e
[(OEP)Si(C6H5)ClO4]-

A comparison of Schemes V and VI show that the electrochemistry of (OEP)Ge(C_6H_5)OH and (OEP)Si(C_6H_5)OH are similar. The first oxidation of (OEP)Si(C_6H_5)OH occurs at the hydroxyl ligand and generates (OEP)Si(C_6H_5)ClO_4 as a final product. This compound can be reversibly oxidized by up to two electrons at more positive potentials and gives a porphyrin π cation radical and dication(36).

In summary, the four chemical systems described in this paper demonstrate the versatility and selectivity of electrochemical methods for synthesis and characterization of metal-carbon σ-bonded metalloporphyrins. The described rhodium and cobalt systems demonstrate significant differences with respect to their formation, stability and to some extend, reactivity of the low valent species. On the other hand, properties of the electrochemically generated mono-alkyl or mono-aryl germanium and silicon systems are similar to each other.

Acknowledgment. The support of the National Science Foundation (Grant CHE-8515411) is gratefully acknowledged. We also acknowledge numerous discussions with our collaborator, Dr. Roger Guilard, from the University of Dijon in France.

Literature Cited

(1) Guilard, R.; Lecomte, C.; Kadish, K. M. Struct. Bond. **1987**, 64, 205-268.
(2) Brothers, P. J.; Collman, J. P. Acc. Chem. Res. **1986**, 19, 209.
(3) Kadish, K. M. Prog Inorg. Chem. **1986**, 34, 435-605.
(4) Highes, R. P. Comprehensive Organometallic Chemistry; Wilkinson, G. Ed.; Pergamon Press: N. Y., **1982**, Vol. 5, p. 277.
(5) Wayland, B. B.; Woods, B. A.; Coffin, V. L. Organometallics **1986**, 5, 1059.
(6) Del Rossi, K.; Wayland, B. B. J. Am. Chem. Soc. **1985**, 107, 7941.
(7) Wayland, B. B.; Woods, B. A.; Pierce, R J. Am. Chem. Soc. **1982**, 104, 302.
(8) Setsune, J.; Yoshida, Z.; Ogoshi, H. J. Chem. Soc. Perkin Trans. **1982**, 983.
(9) Paonessa, R. S.; Thomas, N. C.; Halpern, J. J. Am. Chem. Soc. **1985**, 107, 4333.
(10) Aoyama, Y.; Yoshida, T.; Sakurui, K.-I.; Ogoshi, H. J. Chem. Soc. Chem. Comm. **1983**, 478.
(11) Aoyama, Y.; Yoshida, T.; Sakurui, K.-I.; Ogoshi, H. Organometallics **1986**, 5, 168.
(12) Kadish, K. M.; Yao, C.-L.; Anderson, J. E.; Cocolios, P. Inorg. Chem. **1985**, 24, 4515.
(13) Anderson, J. E.; Yao, C.-L.; Kadish, K. M. Inorg. Chem. **1986**, 25, 718.
(14) Anderson, J. E.; Yao, C.-L.; Kadish, K. M. J. Am. Chem. Soc. **1987**, 109, 1106.

(15) Anderson, J. E.; Yao, C.-L.; Kadish, K. M. Organometallics **1987**, 6, 706.
(16) Anderson, J. E.; Liu, Y. H.; Kadish, K. M. Inorg. Chem. **1987**, 26, 4174.
(17) Collman, J. P.; Barnes, C. E.; Swepston, P. N.; Ibers, J. J. Am. Chem. Soc. **1984**, 106, 3500.
(18) Farnor, M. D.; Woods, B. A.; Wayland, B. B. J. Am. Chem. Soc. **1986**, 108, 3659.
(19) Dolphin, D.; Halko, D. J.; Johnson, E. Inorg. Chem. **1981**, 30, 4348.
(20) Lexa, D.; Saveant, J.-M.; Soufflet, J.P. J. Electroanal. Chem. Interfacial Electrochem. **1979**, 100, 159.
(21) Perree-Fauvet, M.; Gaudemer, A.; Boucly, P.; Devynck, J. J. Organomet. Chem. **1976**, 120, 439.
(22) Clarke, D. A.; Grigg, R.; Johnson, A. W.; Pinnock, H. A. J. Chem. Soc. Chem. Comm. **1967**, 305.
(23) Clarke, D. A.; Dolphin, D.; Grigg, R.; Johnson, A. W.; Pinnock, H. A. J. Chem. Soc. **1968**, 881.
(24) Ogoshi, H.; Wantanabe, E.; Koketsu, N.; Yoshida, Z. Bull. Chem. Soc. Jpn. **1976**, 49, 2529.
(25) Momenteau, M.; Fournier, M.; Rougee, M. J. Chim Phys. **1970**, 67, 926.
(26) Kadish, K. M.; Lin, X. Q.; Han, B. C. Inorg. Chem. **1987**, 26, 4161.
(27) Kobayashi, H.; Hara, T.; Kaizu, Y. Bull. Chem. Soc. Jpn. **1972**, 45, 2148.
(28) Maskasky J. E.; Kenney, M. E. J. Am. Chem. Soc. **1971**, 93, 2060.
(29) Maskasky, J. E.; Kenney, M. E. J. Am. Chem. Soc. **1973**, 95, 1443.
(30) Cloutour, C.; Lafargue, D.; Richards, J. A.; Pommier, J. C. J. Organomet. Chem. **1977**, 137, 157.
(31) Cloutour, C.; Debaig-Valade, C.; Pommier, J. c.; Dabosi, G.; Marthineau, J. J. Organomet. Chem. **1981**, 220, 21.
(32) Cloutour, C.; Lafargue, D.; Pommier, J. C. J. Organomet. Chem. **1983**, 190, 35.
(33) Cloutour, C.; Debaig-Valade, C.; Gacherieu, C.; Pommier, J. C. J. Organomet. Chem. **1984**, 269, 239.
(34) Cloutour, C.; Lafargue, D.; Pommier, J. C. J. Organomet. Chem. **1978**, 161, 327.
(35) Kadish, K. M.; Xu, Q. Y.; Barbe, J.-M.; Anderson, J. E.; Wang, E.; Guilard, R. J. Am. Chem. Soc. **1987**, 109, 7705.
(36) Kadish, K. M.; Xu, Q. Y.; Barbe, J.-M.; Guilard, R. Inorg. Chem. **1988**, 27, 1191.

RECEIVED July 5, 1988

Chapter 32

Formation of Metal–Oxygen Surface Compounds on Copper, Silver, and Gold Electrodes

Reaction with Dioxygen and Electrochemical Oxidation of OH^- in Acetonitrile

Donald T. Sawyer and Pipat Chooto

Department of Chemistry, Texas A&M University, College Station, TX 77843

The oxidation of hydroxide ion in acetonitrile at copper, silver, gold, and glassy-carbon electrodes has been characterized by cyclic voltammetry. In the absence of bases the metal electrodes are oxidized to their respective cations (Cu^+, Ag^+, and Au^+) at potentials that range from -0.2V vs. SCE for Cu to +1.3 V for Au. At glassy carbon ^-OH is oxidized to $O^-\cdot$ (+0.35 V vs SCE) and then to $\cdot O \cdot$ by an ECE mechanism. In contrast, with clean metal electrodes the oxidation of ^-OH to $O^-\cdot$ and $\cdot O \cdot$ is facilitated via formation of metal-s/oxygen-p covalent bonds to give a series of surface compounds (MOH, MOM, MOOM, and MO^-). The shift to less positive oxidation potentials has been used to obtain an estimate of the metal-oxygen covalent bond energies (about 50-70 kcal). Concurrent reduction of Cu^+ plus O_2, and of Ag^+ plus O_2 yields a series of surface compounds (MOM, MOOM, and MO^-) that parallel those that result from ^-OH oxidation at the metal surfaces.

A recent study (1) has demonstrated that the electrochemical oxidation of hydroxide ion yields hydroxyl radical ($\cdot OH$) and its anion ($O^-\cdot$). These species in turn are stabilized at glassy carbon electrodes by transition-metal ions via the formation of metal-oxygen covalent bonds (unpaired d electron with unpaired p electron of $\cdot OH$ and $O^-\cdot$). The coinage metals (Cu, Ag, and Au), which are used as oxygen activation catalysts for several industrial processes (e.g., Ag/O_2 for production of ethylene oxide) (2-10), have an unpaired electron ($d^{10}s^1$ or d^9s^2 valence-

0097–6156/88/0378–0466$06.00/0

shell) and their mode of activation may involve covalent bond formation with dioxygen as a first step towards its atomization. Likewise, the oxidation of hydroxide ion at coinage-metal electrodes is facilitated by the presence of an unpaired electron to couple with ·OH and/or O^{-}· to form metal-oxygen adducts, which appear to be similar to those that result from metal-dioxygen surface reactions. The present paper summarizes the results of an electrochemical investigation in acetonitrile of (a) the oxidation of ^{-}OH at Cu, Ag, and Au electrodes, (b) the potentiometric titration of Cu^{+}, Ag^{+}, and Au^{+} with ^{-}OH, and (c) the concerted reduction of O_2 and Cu^{+} or Ag^{+} at a glassy carbon electrode.

Experimental Section

Equipment. Cyclic voltammetry and controlled-potential electrolysis were accomplished with a Bioanalytical Systems Model CV-27 and a Houston Instruments Model 2000 XY recorder. The electrochemical measurements were made with a 10-mL microcell assembly that was adapted to use a working electrode, a platinum-wire auxilliary electrode (contained in a glass tube with a medium-porosity glass frit and filled with a concentrated solution of supporting electrolyte), and a Ag/AgCl reference electrode (filled with aqueous tetramethylammonium chloride solution and adjusted to 0.000 V vs SCE) (11) with a solution junction via a glass tube closed with a cracked-glass bead that was contained in a luggin capillary. Glassy carbon, gold, and silver working electrodes were obtained from Bioanalytical Systems, the copper electrode was prepared by electrodeposition of copper on a glassy-carbon electrode.

Chemicals and Reagents. Acetonitrile (MeCN), "distilled-in-glass" grade (0.002 % H_2O) from Burdick and Johnson, was used without further purification. Tetraethylammonium perchlorate (TEAP) was vacuum-dried for 24 h prior to use. Tetrabutylammonium hydroxide (TBAOH) was obtained from Aldrich as a 25% solution in methanol and its concentration was determined by acid-base titration. $Cu(H_2O)_4(ClO_4)$ was prepared from reduction of copper(II) perchlorate (G.F. Smith Chemicals) or from the dissolution of Cu_2O with $HClO_4$ (12). The resulting salt was recrystalled from dry MeCN four times to give $Cu^{I}(MeCN)_4^{-}(ClO_4)$. Gold(I) solutions were prepared by anodic electrolysis of a gold foil in acetonitrile (0.1M TEAP) (13). $AgClO_4$ was obtained from G.F. Smith Chemicals.

Results

Oxidation of ^{-}OH. The voltammetric oxidation of hydroxide ion at glassy carbon (GC), copper, silver, and gold electrodes is illustrated by Figures 1, 2, and 3. For each metal the presence of excess ^{-}OH results in an initial oxidation at a potential that is less positive than that (a) for ^{-}OH at a glassy carbon electrode and (b) for the oxidation of the metal electrode. After a negative scan to -1.5V vs. SCE each of the metal electrodes exhibits (after the scan direction is reversed) an oxidation peak ($E_{p,a}$: Cu, -0.80 V vs SCE; Ag, -0.27 V; Au -0.35V) that is due to the presence of 3 mM ^{-}OH. In each case the product of the anodic peak is reduced upon reversal of the scan direction. Each electrode exhibits a second oxidation peak ($E_{p,a}$: Cu, -0.47 V vs SCE; Ag, 0.00 V; Au, +0.20 V), which produces a product species that is reduced as an adsorbed film upon scan reversal ($E_{p,c}$: Cu, -0.90 V vs SCE; Ag, -0.75 V; Au, -0.60 V).

The potentiometric titrations of $[Cu^{I}(MeCN)_4](ClO_4)$, $Ag^{I}ClO_4$, and $Au^{I}ClO_4$ with $(Bu_4N)OH$(in MeOH) are illustrated in Figure 4, and demonstrate that each process has one-to-one stoichiometry. The three systems form precipitates such that all of the metal is removed from solution at the equivalence point. Addition of excess ^{-}OH causes some dissolution of the CuOH and AgOH precipitates, and appears as a second step for the titration curve of Ag(I) (Figure 4b).

<u>Concerted Reduction of O_2 and Cu^+ or Ag^+</u>. Figure 5 illustrates the cyclic voltammograms for O_2 in MeCN(0.1M TEAP) at glassy carbon, Cu, Ag, and Au electrodes (each polished immediately prior to exposure to O_2). The drawn out reduction waves and the absence of significant anodic peaks upon scan reversal for the three metal electrodes indicate that O_2 reacts with the surface prior to electron transfer.

The cyclic voltammograms at a glassy carbon electrode for $[Cu^{I}(MeCN)_4]ClO_4$ and $AgClO_4$, respectively, in the absence and presence of O_2 are consistent with the initial deposition of metal atoms (Figures 6 and 7). In the absence of O_2 , both metal ions are reduced to metal films on the glassy carbon electrode ($E_{p,c}$: Cu^+, -0.45 V vs SCE; Ag^+, -0.10 V) and exhibit typical stripping-dissolution peaks upon reversal of the voltage scan. However, the presence of O_2 causes the reduction peaks for an initial negative voltage scan to be somewhat less intense and at -0.6 V for Cu^+; scan reversal does not yield a stripping peak for copper and a smaller stripping peak for silver. Continuation of the initial negative scan yields two additional one-electron reduction peaks ($E_{p,c}$: Cu, -1.02 V and -1.80 V; Ag, -0.95 V and -1.6 V), and with cycling of the scan direction, results in the appearance of a chemically-reversible couple (Cu: $E_{p,a}$, +0.20 V and $E_{p,c}$, -0.40 V; Ag: $E_{p,a}$, -0.18 V and $E_{p,c}$,

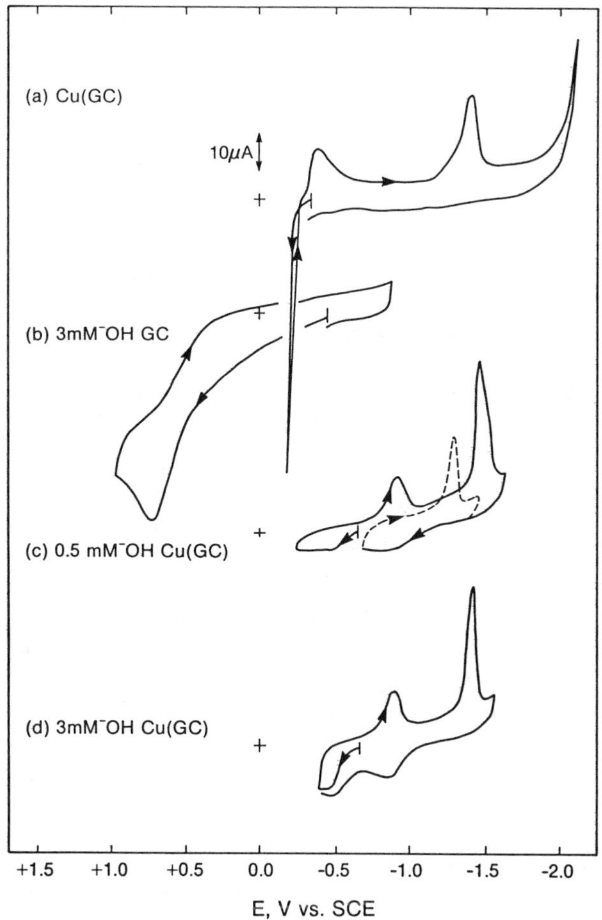

Figure 1. Cyclic voltammograms in MeCN(0.1M tetraethylammonium perchlorate) for the oxidation of (a) a copper electrode, (b) 3 mM ^{-}OH at a glassy carbon electrode, (c) 0.5 mM ^{-}OH at a copper electrode, and (d) 3 mM ^{-}OH at a copper electrode. Scan rate, 0.1V s^{-1}; electrode area, 0.08 cm^2; copper electrode prepared by electroplating $Cu(ClO_4)$ onto a glassy carbon electrode (GCE).

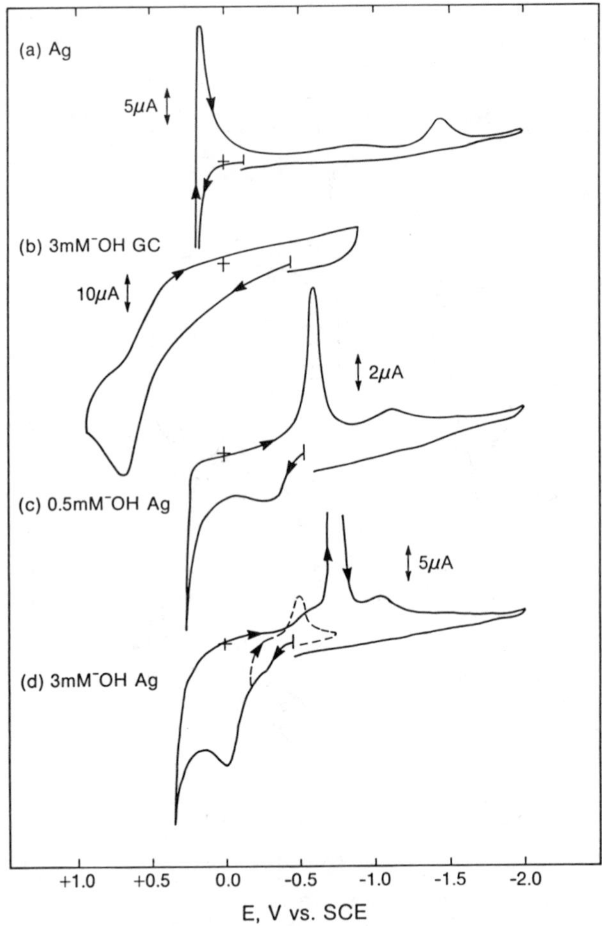

Figure 2. Cyclic voltammograms in MeCN(0.1M TEAP) for the oxidation of (a) a silver electrode, (b) 3 mM ^{-}OH at a GCE, (c) 0.5 mM ^{-}OH at a silver electrode, and (d) 3 mM ^{-}OH at a silver electrode. Scan rate, 0.1 V s^{-1}; GCE area, 0.08 cm^2; Ag electrode area, 0.03 cm^2.

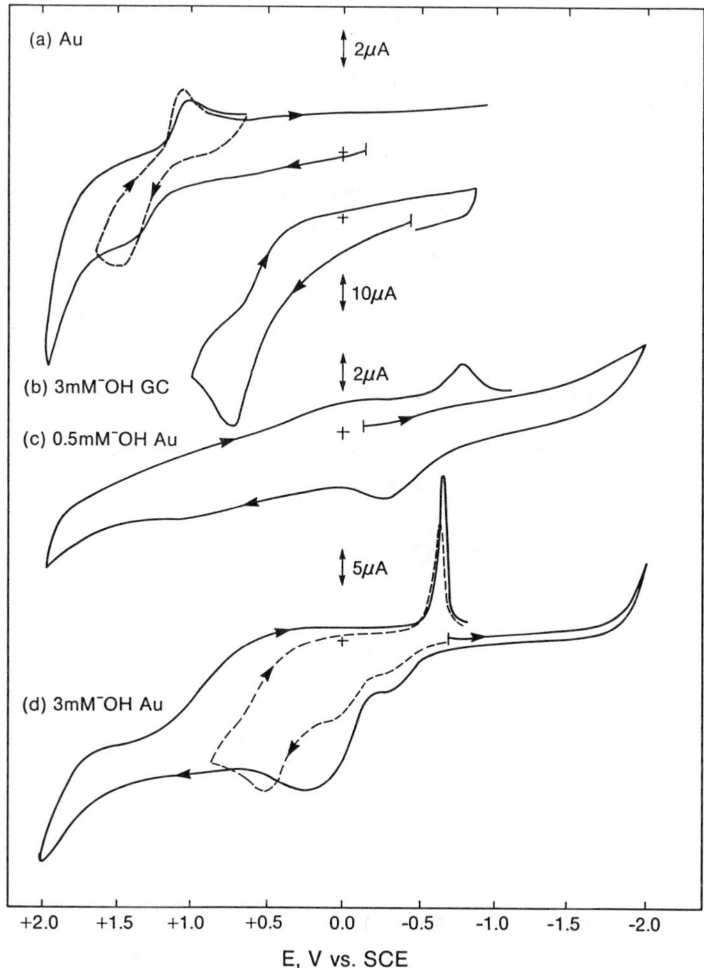

Figure 3. Cyclic voltammograms in MeCN(0.1M TEAP) for the oxidation of (a) a gold electrode, (b) 3 mM ^{-}OH at a GCE, (c) 0.5 mM ^{-}OH at a gold electrode, and (d) 3 mM ^{-}OH at a gold electrode. Scan rate, 0.1V s^{-1}; GCE area, 0.08 cm^2; Au electrode area, 0.03 cm^2.

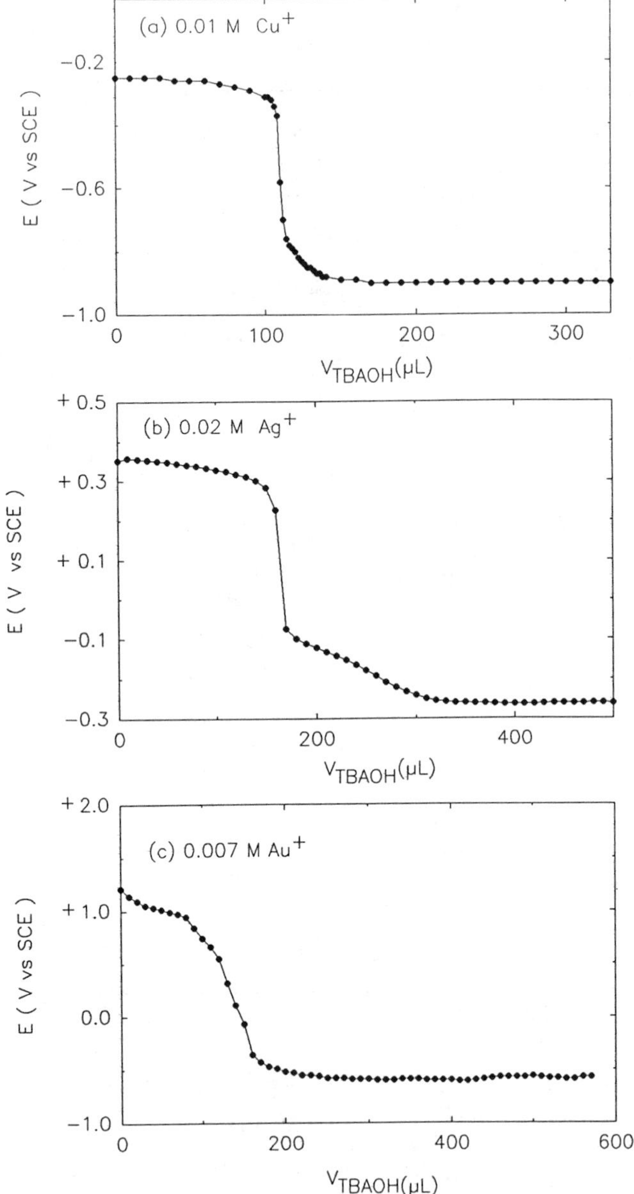

Figure 4. Potentiometric titrations in MeCN (0.1M TEAP) of (a) 0.012 M $Cu^{I}(MeCN)_4ClO_4$ with 1.1M ^{-}OH (Cu indicator electrode), (b) 0.020M $AgClO_4$ with 1.2 M ^{-}OH (Ag indicator electrode), and (c) 0.0072 M Au^{+} with 0.11 M ^{-}OH (Au indicator electrode).

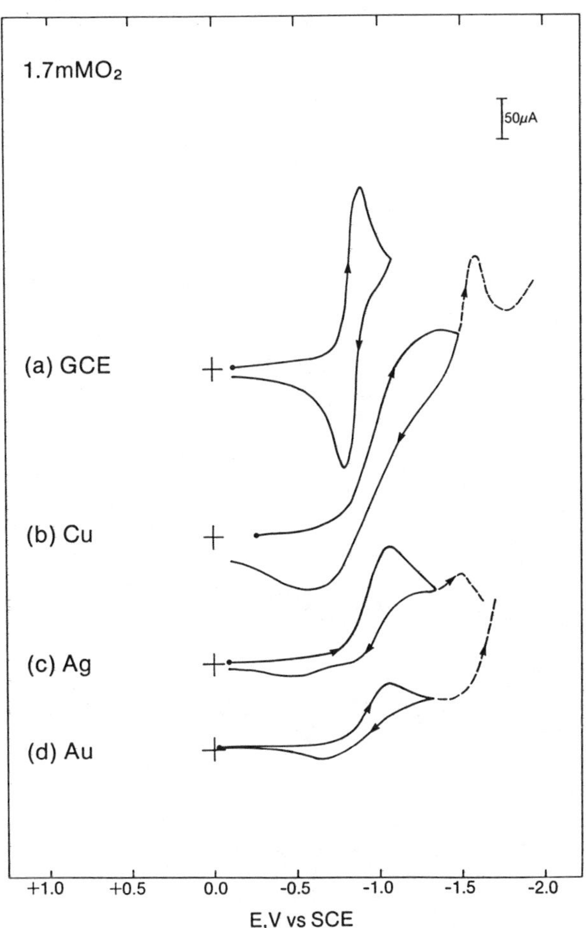

Figure 5. Cyclic voltammograms in MeCN(0.1M TEAP) of 1.7 mM O_2 (0.21 atm) at (a) glassy carbon, (b) copper, (c) silver, and (d) gold electrodes. Scan rate, 0.1 V s^{-1}; electrode areas: GCE, 0.08 cm^2; Cu, 0.08 cm^2; Ag, 0.03 cm^2; Au, 0.03 cm^2.

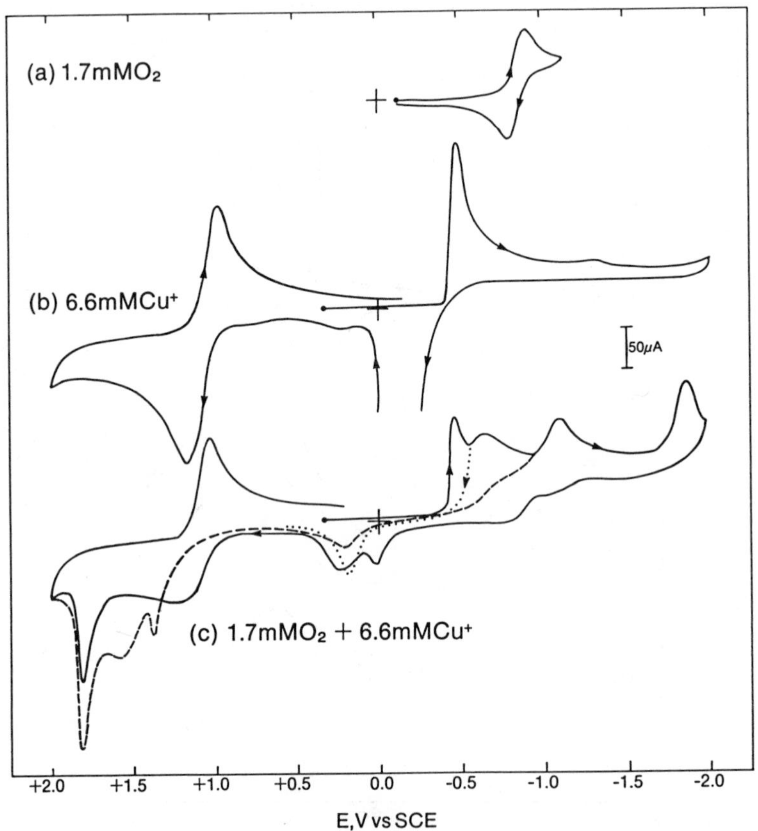

Figure 6. Cyclic voltammograms in MeCN(0.1M TEAP) at a glassy carbon electrode (area, 0.08 cm^2) of (a) 1.7 mM O_2, (b) 6.6 mM $Cu^I(MeCN)_4ClO_4$, and (c) 1.7 mM O_2 + 6.6 mM $Cu^I(MeCN)_4ClO_4$. Scan rate, 0.1 V s^{-1}.

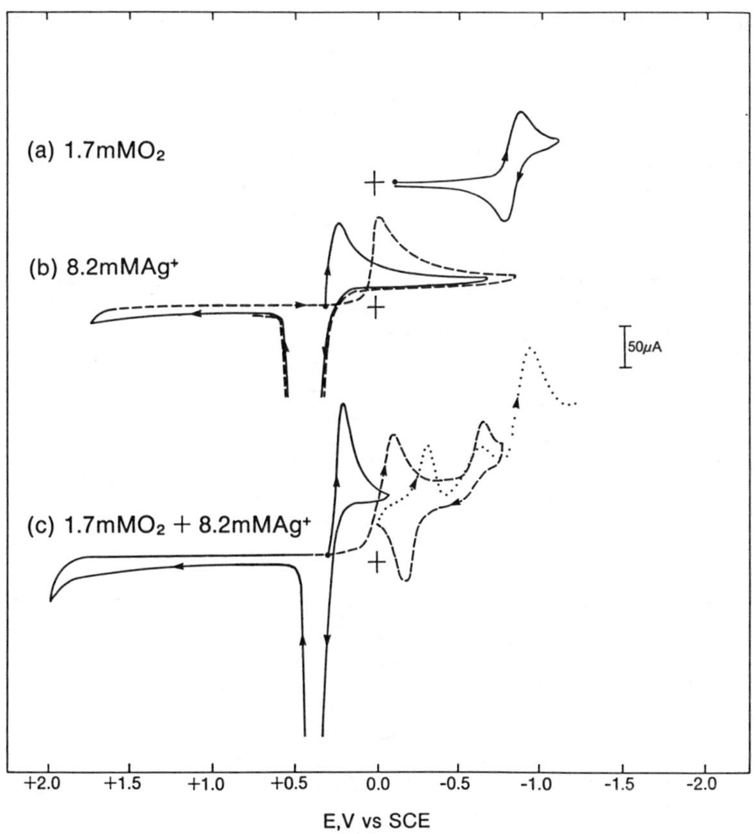

Figure 7. Cyclic voltammograms in MeCN(0.1M TEAP) at a glassy carbon electrode (area, 0.08 cm^2) of (a) 1.7 mM O_2, (b) 8.2 mM $AgClO_4$, and (c) O_2 + 1.7 mM 8.2 $AgClO_4$. Scan rate, 0.1V s^{-1}.

-0.31 V). Attempts to observe similar effects from the presence of O_2 for the reduction of Au^+ have been unsuccessful.

Discussion and Conclusions

The oxidation of ^-OH in acetonitrile at glassy carbon elec-trodes yields a hydroxyl radical ($\cdot OH$) in the primary step (1)

$$^-OH \rightarrow \cdot OH + e^- \qquad E^{\circ\prime},\ + 0.68V \text{ vs SCE} \qquad (1)$$

With excess ^-OH the process is facilitated via formation of the anion radical.

$$2\ ^-OH \rightarrow O^-\cdot + H_2O + e^- \qquad E^{\circ\prime},\ + 0.56V \qquad (2)$$

The oxidation of excess ^-OH in the presence of Zn(TPP) (TPP = tetraphenylporphyrin) is somewhat impeded

$$2\ ^-OH + Zn(TPP) \rightarrow Zn(TPP)(O^-\cdot)^- + H_2O + e^- \qquad E^{\circ\prime},\ + 0.49V \qquad (3)$$

but it is facilitated in the presence of Co(TPP)

$$2\ ^-OH + Co(TPP) \rightarrow (O^-\cdot)Co(TPP)^- + H_2O + e^- \qquad E^{\circ\prime},\ -0.22V \qquad (4)$$

The latter is attributed to the formation of a covalent bond by the unpaired electron of $O^-\cdot$ with one of the unpaired d electrons of cobalt (1).

The oxidation of ^-OH at copper, silver, and gold electrodes (Figures 1-3) also occurs at substantially less positive potentials than that at a glassy carbon electrode. This appears to be the result of coupling the unpaired electron of the $\cdot OH$ product with the s electron of metallic (atomic) copper, silver, or gold ($d^{10}s$ valence shell).

$$M + ^-OH \rightarrow M\text{-}OH + e^- \qquad E_{p,a} \qquad (5)$$

The titration curves of Figure 4 also are consistent with the proposition that Cu^+, Ag^+, and Au^+ ions react with ^-OH via electron transfer to give a covalent M-OH product via the coupling of the unpaired electron of $\cdot OH$ with the s electron of the metal atom.

$$M^+ + ^-OH \rightarrow [M\cdot + \cdot OH] \rightarrow M\text{-}OH \qquad (6)$$

The metal indicator electrodes for the potentiometric titrations respond to the M^+/M couple prior to the equivalence point, and to the MOH/M, ^-OH couple after the equivalence point

$$M^+ + e^- \rightarrow M \qquad E^{\circ\prime}{}_{M^+/M} \tag{7}$$

$$M\text{-}OH + e^- \rightarrow M + {}^-OH \qquad E^{\circ\prime}{}_{MOH/M,{}^-OH} \tag{8}$$

The latter is the reverse of the half reaction for the oxidation of ^-OH at the metal electrodes (eq. 5).

The data of Figures 1-4 have been used to calculate formal potentials for Reactions 7 and 8 for copper, silver, and gold; these are summarized in Table I. Analogous redox parameters for aqueous solutions also are included (14).

Whereas the reduction potentials for the three metal ions range from +0.19V vs NHE(Cu) to +1.58V(Au), the potentials for oxidation of ^-OH in their presence are -0.79V vs NHE(Cu), -0.30V(Ag), and -0.19V(Au). This is compatible with the proposition that oxidation occurs via the facilitated removal of an electron from ^-OH and formation of an M-OH covalent bond. The only exception to the close agreement between gas-phase and redox-derived M-OH bond energies is the Cu-OH bond energy from aqueous redox data. This may be due to an inaccurate formal potential for the CuOH/Cu, ^-OH couple (a value of 0.0V vs NHE rather than -0.36V would result in a more consistent bond-energy estimate).

Consideration of these primary processes together with the voltammetric results for the M/^-OH systems (Figures 1-3), the potentiometric titration data (Figure 4), and the voltammetric data for O_2 reduction at metal electrodes (Figure 5) and in the presence of metal ions at a glassy carbon electrode (Figures 6 and 7), prompts the formulation of self-consistent reaction Schemes for the three metals in combination with ^-OH and O_2 (Schemes I, II, and III).

The shift in the formal oxidation potential for the $\cdot OH/{}^-OH$ couple from its value at glassy carbon to that at copper, silver, and gold is a measure of the stabilization of $\cdot OH$ via covalent bond formation ($\Delta E^{\circ\prime}$ x 23.1 kcal/eV). However, the atoms of the Cu, Ag, and Au electrodes exist as dimer molecules that are held together via a $d^{10}s$-sd^{10} covalent bond (the bond energies for these dimers are listed in Table I) (15). Evidence for this dimeric bond is provided from the voltammetric data for Ag^+ at a glassy carbon electrode (Figure 7). Reversal of an initial positive scan gives a reduction peak at -0.10 V vs SCE ($Ag^+/Ag\cdot$), but an initial negative scan gives a reduction peak at +0.20 V that is analogous to reduction

Table I. Redox Potentials for the M^+/M and $MOH/M, {}^-OH$ Couples of Copper, Silver, and Gold in Acetonitrile and in Water, and Bond-Energies for Their M-M and M-OH Bonds

M	E°', V vs. NHE[a] M^+/M[b]	E°', V vs. NHE[a] $MOH/M, {}^-OH$[b]	B.E., kcal mol^{-1} M-M[c]	B.E., kcal mol^{-1} M-OH[d]	B.E., kcal mol^{-1} M-O[c]
A. Acetonitrile					
Cu	+0.19 (CV)	-0.79 (CV)	46.6	70.6	64.7
	+0.05 (Pot.)	-0.78 (Pot.)		70.4	
Ag	+0.54 (CV)	-0.30 (CV)	39.0	55.5	52.6
	+0.64 (Pot.)	-0.12 (Pot.)		51.3	
Au	+1.58 (CV)	-0.19 (CV)	53.8	60.3 (>62.5)[e]	53.0
	+1.57 (Pot.)	-0.43 (Pot.)		54.8	
$\cdot OH/{}^-OH$		+0.92 (CV)			
B. Water[f]					
Cu	+0.52	-0.36		59.8	64.3
Ag	+0.80	+0.34		43.6	52.6
Au	+1.7				53.0
$\cdot OH/{}^-OH$		+1.89			

a SCE = +0.24V vs NHE

b For voltammetric data from Figure 1: $E°'_{M^+/M} = E_{1/2} + 0.059 \log ([M^+]_{bulk}/2)$ and $E°'_{MOH/M} = E_{1/2} - 0.059 \log ([{}^-OH]/2)$. For potentiometric data from Figure 2: $E°'_{M^+/M} = (E_{obs})_{initial} + 0.059 \log [M^+]_{initial}$ and $E'_{MOH/M} = (E_{obs})(2\ {}^-OH/M^+) - 0.059 \log [{}^-OH]$

c Ref. 15

d $B.E. = (-\Delta G_{BF}) + T\Delta S_{DBE} + B.E._{M-M}/2 = [E°'_{\cdot OH/{}^-OH} - E°'_{MOH/M}] \times 23.1 + 7.8 + B.E._{M-M}/2$

e Gas phase value, Ref. 16; f Ref. 14.

Scheme I.

A. $Cu/^{-}OH$ ELECTROCHEMISTRY IN MeCN

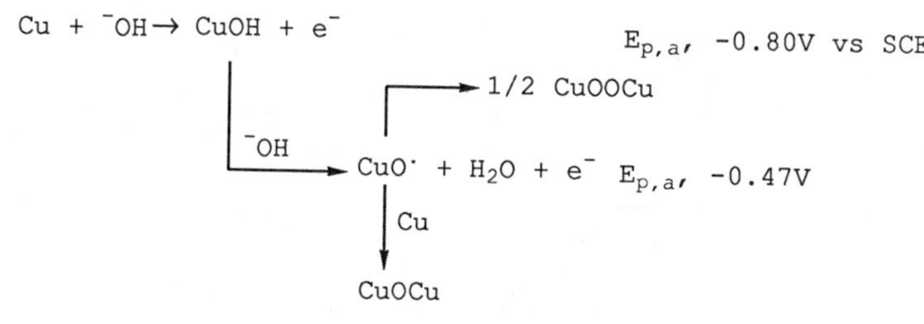

$CuOCu + 2e^- + H_2O \rightarrow 2\ Cu + 2\ ^{-}OH$ $E_{p,c}$, -0.90V

$CuOOCu + 2e^- \rightarrow 2\ CuO^-$ $E_{p,c}$, -1.45V

$CuOOCu + 2H_2O + 2e^- \rightarrow 2\ CuOH + 2\ ^{-}OH$ $E_{p,c}$, -1.30

$CuO^- \rightarrow CuO^\cdot + e^-$ $E_{p,a}$, -0.87V

→ 1/2 CuOOCu

B. $Cu(I)/O_2$ ELECTROCHEMISTREY IN MeCN AT GC ELECTRODE

$2\ Cu^+ + 2e^- \rightarrow Cu_2$ $E_{p,c}$, -0.45V vs SCE

$2\ Cu^+ + O_2 + 2e^- \rightarrow CuOOCu$ $E_{p,c}$, -0.60V vs SCE

↓ $2e^-$ $E_{p,c}$, -1.02V

$2\ CuO^-$

$CuO^- + H_2O + e^- \rightarrow Cu + 2\ ^{-}OH$ $E_{p,c}$, - 1.80V

↓ $-e^-$ $E_{p,a}$, -0.80V

$CuO^- + H_2O$

↓ $-e^-$ $E_{p,a}$, -0.85V

$CuO^\cdot$

→ 1/2 CuOOCu

$CuOOCu \rightarrow CuOO^\cdot + Cu^+ + e^-$ $E_{p,a}$, + 0.20V

↓ $E_{p,c}$, - 0.40V

CuOOCu

$CuOO^\cdot \rightarrow Cu^{2+} + O_2 + 2\ e^-$ $E_{p,a}$, + 1.68V

Scheme II

A. $Ag/^{-}OH$ ELECTROCHEMISTRY IN MeCN

$Ag + {}^{-}OH \rightarrow AgOH + e^{-}$ $E_{p,a}$, -0.27V vs SCE

$\xrightarrow{e^{-}} Ag + {}^{-}OH$ $E_{p,c}$, -0.50V

$\xrightarrow{^{-}OH} AgO^{\cdot} + H_2O + e^{-}$ $E_{p,a}$, 0.00V

$AgO^{\cdot} \rightarrow 1/2\ AgOOAg$

$AgO^{\cdot} \xrightarrow{Ag} AgOAg$

$AgOAg + e^{-} \rightarrow Ag + AgO^{-}$ $E_{p,c}$, -0.75V

$AgOOAg + 2e^{-} \rightarrow 2AgO^{-}$ $E_{p,c}$, -1.04V

$AgO^{-} + H_2O + e^{-} \rightarrow Ag + 2\ {}^{-}OH$ $E_{p,c}$, -1.43V

B. $Ag(I)/O_2$ ELECTROCHEMISTRY IN MeCN AT GC ELECTRODE

(1) $4\ Ag^{+} + O_2 + 4e^{-} \rightarrow 2\ AgOAg$ $E_{p,c}$, -0.10V vs SCE

$AgOAg \xrightarrow{2e^{-}} 2\ AgO^{-} + Ag_2$ $E_{p,c}$, -0.65V

$AgO^{-} + H_2O + e^{-} \rightarrow Ag + 2\ {}^{-}OH$ $E_{p,c}$, -1.55V

(2) $Ag^{+} + e^{-} \xrightarrow{Ag} Ag_2$ $E_{p,c}$, +0.20V

$Ag^{+} + e^{-} \rightarrow Ag^{\cdot} \rightarrow 1/2\ Ag_2$ $E_{p,c}$, -0.10V

$Ag^{\cdot} \xrightarrow{O_2,\ e^{-}} AgOO^{-}$ $E_{p,c}$, -0.95V

$AgOO^{-} \rightarrow AgOO^{\cdot} + e^{-}$ $E_{p,a}$, -0.18V

$AgOO^{\cdot} \xrightarrow{e^{-}} AgOO^{-}$ $E_{p,c}$, -0.31V

$AgOO^{\cdot} \xrightarrow{Ag} AgOOAg$

(3) $2\ Ag^{+} + O_2 + 2e^{-} \rightarrow AgOOAg$ $E_{p,c}$, -0.10V

$AgOOAg + 2e^{-} \rightarrow 2\ AgO^{-}$ $E_{p,c}$, - 1.04V

$AgOOAg \rightarrow O_2 + 2\ Ag^{+} + 2e^{-}$ $E_{p,a}$, +0.90V

$Ag_2 \rightarrow 2Ag^{+} + 2e^{-}$ $E_{p,a}$, + 0.40V

Scheme III

Au/$^-$OH ELECTROCHEMISTRY IN MeCN

$Au + {}^-OH \rightarrow AuOH + e^-$ $\quad E_{p,a}$, −0.35V vs SCE

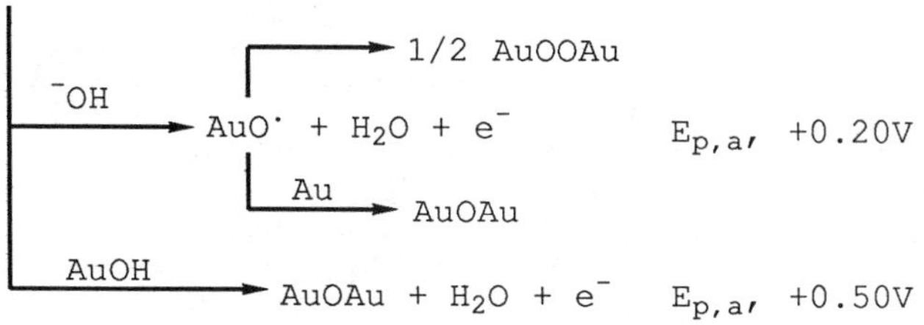

AuOH + $^-$OH → $AuO^{\cdot} + H_2O + e^-$ $\quad E_{p,a}$, +0.20V

- $AuO^{\cdot}$ → 1/2 AuOOAu
- $AuO^{\cdot}$ + Au → AuOAu

AuOH + AuOH → $AuOAu + H_2O + e^-$ $\quad E_{p,a}$, +0.50V

$AuOAu + e^- \rightarrow Au + AuO^-$ $\quad E_{p,c}$, −0.60V

$AuOOAu + 2e^- \rightarrow 2\ AuO^-$ $\quad E_{p,c}$, −0.76V

at a Ag electrode ($Ag^+ + Ag + e^- \rightarrow Ag_2$). This bond energy must be overcome to provide an unpaired s-electron for bonding with the ·OH from the oxidation of ^-OH.

$$M\text{-}M + 2\ ^-OH \rightarrow 2\ M\text{-}OH + 2e^- \qquad E^{\circ\prime}_{MOH/M,^-OH} \qquad (9)$$

$$(B.E.)_{M\text{-}OH} = [E^{\circ\prime}{}_{\cdot OH/^-OH} - E^{\circ\prime}_{MOH/M,^-OH}]23.1 + 1/2\ (B.E.)_{M\text{-}M} \qquad (10)$$

Table I summarizes the bond energies for the M-OH bond that result from the electrochemical data and this calculation. The values compare closely with those for gas-phase M-O bond energies (15). A recent gas-phase study estimates that the Au-OH bond energy is greater than 62.5 kcal (16).

Although the initial steps of Schemes IA, IIA, and IIIA are strongly supported by the experimental data, the subsequent reactions and electron-transfer steps are based solely on the electrochemical measurements of Figures 1-3, 6 and 7. Intermediates have not been detected or isolated, but there is self consistency in the redox thermodynamics between the $M/^-OH$ systems and the M^+/O_2 systems. The cyclic voltammograms also indicate the presence of common intermediates between the two systems. One of these, MOOM, is especially intriguing and is the focus of current efforts to detect and characterize it via surface-analysis techniques.

Although each of the metal electrodes has been polished immediately prior to the measurements that are illustrated in Figures 1-3, the initial negative scans often exhibit significant reduction peaks due to autoxidation by O_2. The polishing process undoubtedly atomizes some of the M-M surface molecules, which probably initiates the process via a mechanism similar to that for the autoxidation of reduced iron porphyrins (17)

$$M\cdot + \cdot O_2\cdot \rightarrow M\text{-}OO\cdot \xrightarrow{M\cdot} MOOM \xrightarrow{2M\cdot} 2\ MOM \qquad (11)$$

The final step may be slow and represent the path for passivation of the surfaces of the coinage metals.

Acknowledgment

This work was supported by the National Science Foundation under Grant No. CHE-8516247.

Literature Cited

1. Tsang, P.K.S.; Cofré, P.; Sawyer, D.T. Inorg. Chem. 1987, 26, 3604.
2. Karlin, K.S., Zubieta, J., Eds. Biological and Inorganic Copper Chemistry, Adenine: Guilderland, New York, 1986; Vol. 1 and 2.
3. Verykios, K.E.; Stein, F.P.; Coughlin, R.W. Catal. Rev. Sci. Eng. 1980, 22, 197.
4. Schwank, J. Gold Bull. 1983, 16, 4.
5. Cu: Spitzer, A.; Luth, H. Surf. Sci. 1982, 118, 121.
6. Campbell, C.T. Surf. Sci., 1986, 173, L641.
7. Gutka, D.A.; Madix, R.J. J. Am. Chem. Soc. 1987, 109, 1708.
8. C. Pyun, C.-H.; Park, S.M. J. Electrochem. Soc. 1986, 133, 2024.
9. Stonehart, P. Electrochim. Acta. 1968, 13, 1789.
10. Herman, G.E.; Alonso, C.; Gonzalez-Velasco, J. J. Electroanal. Chem., 1987, 223, 277.
11. Sawyer, D.T., Robert, J.L., Jr. Experimental Electro-chemistry For Chemists, Wiley-Interscience, New York, 1974, pp. 44-46, 144-145, 336-339.
12. Simmons, M.G.; Merrill, C.L.; Wilson, L.J.; Bottomley, L.A.; Kadish, K.M. J. Chem. Soc. Dalton Trans. 1980, 10, 1827.
13. Goolsby, A.D.; Sawyer, D.T. Anal. Chem. 1968, 40, 1979.
14. Parsons, R. Handbook of Electrochemical Data, Butters-worth, London, 1952.
15. CRC Handbook of Chemistry and Physics, 68th Ed.; CRC: Boca Raton, FL; pp. F-168 - F-177.
16. Weil, D.A.; Wilkins, C.L. J. Am. Chem. Soc. 1985, 107, 7316.
17. Chin, D.-H.; LaMar, G.N.; Balch, A.L. J. Am. Chem. Soc. 1980, 102, 4344.

RECEIVED May 17, 1988

Chapter 33

Effect of Surface Crystallography on Electrocatalytic Oxidation of Carbon Monoxide on Platinum Electrodes

B. Love and J. Lipkowski

Guelph–Waterloo Centre for Graduate Work in Chemistry, University of Guelph, Guelph, Ontario N1G 2W1, Canada

The kinetics of CO oxidation from $HClO_4$ solutions on the (100), (111) and (311) single crystal planes of platinum has been investigated. Electrochemical oxidation of CO involves a surface reaction between adsorbed CO molecules and a surface oxide of Pt. To determine the rate of this reaction the electrode was first covered by a monolayer of CO and subsequently exposed to anodic potentials at which Pt oxide is formed. Under these conditions the rate of CO oxidation is controlled by the rate of nucleation and growth of the oxide islands in the CO monolayer. By combination of the single and double potential step techniques the rates of the nucleation and the island growth have been determined independently. The results show that the rate of the two processes significantly depend on the crystallography of the Pt surfaces.

Carbon monoxide adsorbs strongly on platinum surfaces thereby acting as a poison of platinum catalysts. The adsorption of CO has a particularly adverse affect on the performance of low temperature hydrogen/oxygen and methanol/oxygen fuel cells. In the first case CO is frequently present as an impurity in the hydrogen fuel, in the second case it is produced as an intermediate in the methanol oxidation reaction. In both cases the presence of CO is responsible for an increase of the overvoltage for the anodic reactions. Therefore, the electrosorption and electrooxidation of CO on Pt have been investigated extensively. The bulk of the previous research was carried out on polycrystalline platinum, see [1-3] for recent reviews of the literature, and only a few papers have been published on CO adsorption/oxidation on well defined single crystal surfaces [3-6].

It is known that CO oxidation involves a surface reaction between adsorbed CO molecules and an adsorbed oxygen species, which we shall refer to as a surface oxide, but which may be an adsorbed oxygen

0097–6156/88/0378–0484$06.00/0

atom, OH radical or even a water molecule. At Pt surfaces partially covered by CO the oxygen species forms islands, the reaction between the adsorbed CO molecules and the surface oxide proceeds at the perimeter of the islands [7]. If the electrode surface is initially covered by a monolayer of CO and subsequently exposed to the conditions at which the oxidation of CO occurs, the rate of reaction is controlled by the rate of nucleation and growth of the oxide islands. It has been shown by Pletcher and McCallum that potential step experiments provide the best way to study this reaction [8].

The objective of the present work was to assess the influence of surface cyrstallography on the mechanism and kinetics of CO oxidation at Pt surfaces. Three crystallographic orientations; (111), (100) and (311), were selected for this investigation. The (111) and (100) planes have smooth surfaces build up from three and four fold coordination centers, respectively. The (311) plane is a stepped surface possessing an equal number of the three fold and four fold coordination centers. In previous investigations of CO oxidation at Pt single crystal surfaces the rate of the reaction was followed by cyclic voltammetry [4-6]. Under such conditions the currents are a complex function of the rates of nucleation and growth. Thus, the results provide only qualitative information. In the present work potential step techniques were applied such that the nucleation and growth rates were measured independently. In this way the effect of surface morphology on the rate of CO oxidation has been quantitatively assessed.

Experimental

The single crystal electrodes were prepared according to the procedure described elsewhere [9]. The electrode surfaces were cleaned using Clavilier's flaming and quenching technique [10]. Contact between the investigated surface and the electrolyte was made using the hanging electrolyte method [11]. A conventional, three electrode cell was used in the electrochemical investigations. An external reversible hydrogen electrode (RHE) was used as the reference electrode. The 0.1 M $HClO_4$ solutions used in both the cell and the external RHE compartment were prepared from suprapure 65% $HClO_4$ (Seastar Chemicals) and Mill-Q water (17 MΩ cm). Argon was used for degassing the solutions. All experiments were carried out at room temperature (21 ± 1 °C).

Cyclic voltammetry (CV) curves were recorded on a Kipp and Zonen BD91 X-Y recorder and the current-time transients resulting from the potential step experiments were recorded digitally. The apparatus used included a PAR Model 173 potentiostat and an IBM XT computer equipped with a IS-16 A-D interface (RC Electronics, California). The interface had its own memory buffer and allowed for the acquisition of 2048 points at a variable A/D conversion frequency which can be as high as 1 MHz with a resolution 12 bits over an input range of ±2.5 volts.

Results

Cyclic Voltammetry. Figure 1 shows the cyclic voltammograms (CV's) determined for the three crystal planes in a clean (CO free)

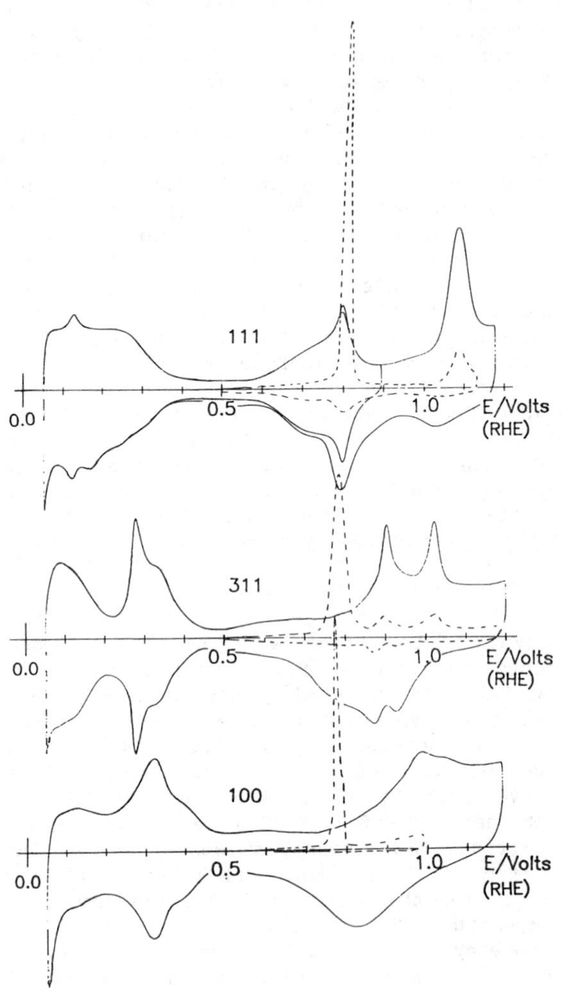

Figure 1. Solid lines - cyclic voltammetric curves for hydrogen and oxygen adsorption at the three Pt single crystal planes investigated determined from the pure (CO free) 0.1 M $HClO_4$ solution. Dotted lines - cyclic voltammetric curves of adsorbed CO oxidation (note: the sensitivity for the CO oxidation curves has been ten fold attenuated. Sweep rate 50 mv/s.

electrolyte solution. The shape of the CV's is virtually identical to the curves which have been obtained for similarly oriented faces of platinum single crystals, cleaned and characterized by low-energy electron diffraction (LEED) and Auger-electron spectroscopy (AES) in a combined UHV-electrochemistry experiment [12-14].

The dotted curves are CV's determined for Pt electrodes covered by a monolayer of CO in an electrolyte free from bulk CO. This condition was achieved by imposing a potential of 0.40 V on the electrode immersed in a solution saturated with CO for 2 min and subsequently removing the CO from the bulk of the electrolyte by purging the solution with argon for a further 2 min. The main feature of these curves is the sharp peak corresponding to CO oxidation. Following oxidation of the adsorbed CO at potentials less than 1.10 V the shape of CV's returned to that described by a solid line and characteristic for the CO free electrolyte. This indicates that the adsorption and the oxidation of CO do not perturb the crystallographic structure of the surface. The figure also shows a small affect of the surface crystallography on the rate of CO oxidation revealed by a systematic shift of the potential of the CO oxidation peak in the positive direction from the (100) surface to the (111) surface.

Potential Step Experiments

A. Single Step Experiments. Potential step experiments were performed in order to determine the reaction mechanism and the reaction rate. As described above, the platinum surface was initially covered by a monolayer of CO at a controlled potential, E_i = 0.40 V (referred to as the initial potential) and then CO was removed from the bulk of the solution. Next, the electrode potential was suddenly changed to a more positive value, E_f, (referred to as the final potential) where the adsorbed CO was oxidized and the rate of oxidation was followed by recording the resulting current transients.

Representative current-time transients obtained for a Pt(100) surface are presented as a three-dimensional i-t-E plot in Figure 2. The graph is constructed in such a way that the time and current axis are in the plane of the figure while the E_f axis is normal to this plane. All transients display the characteristic maximum, the coordinate of which will be denoted as (i_{max}, t_{max}). As the final potential becomes progressively more positive i_{max} increases and t_{max} decreases. The area under the transients gives the charge of adsorbed CO (σ_{CO}). After correction for charging of the double layer and the charge of the oxide formation, the value of σ_{CO} was found to be equal to 350 $\mu C cm^{-2}$ and to be within an error of ±10%, independent of the electrode protential. Thus, the ratio of σ_{CO} to the doubled charge of adsorbed hydrogen is approximately equal to 0.7. Assuming that the ratio of adsorbed hydrogen atoms to platinum atoms at the surface is equal to one, the above result is equal to the degree of coverage of the Pt surface by CO molecules. Incidentally, the same value of the saturation coverage was found for CO adsorption on Pt from the gas phase [15]. The shape of the transients changes with potential. Figure 3a shows three current transients recorded at low potentials and Figure 3b shows three

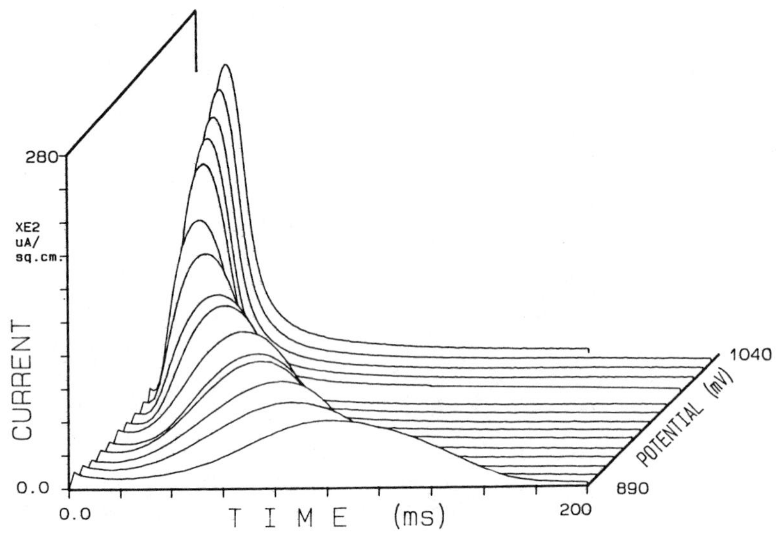

Figure 2. Three-dimensional plots of the CO oxidation transients determined in the single potential step experiments on the Pt(100) electrode in 0.10 M $HClO_4$ solution. The potential step was applied from E_i = +0.40 V to E_f displayed on the third axis of the figure.

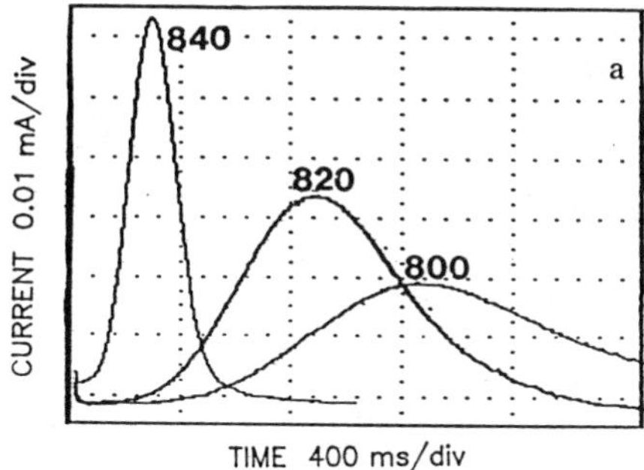

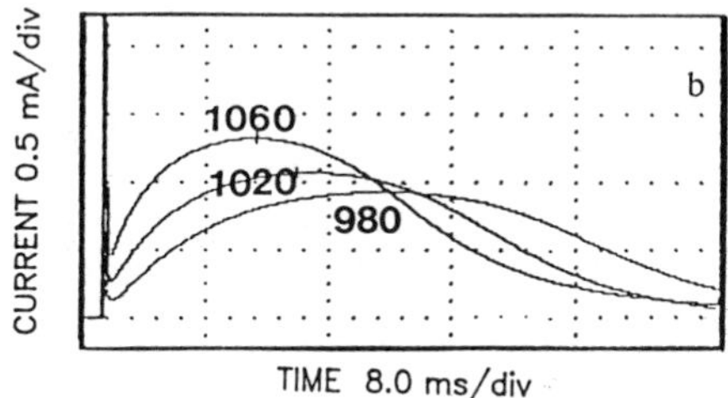

Figure 3 a. Three current transient selected from Figure 2 for the low values of the final potential. The values of E_f in mV are indicated above the corresponding curves. b. Three current transients selected from Figure 2 for the high values of the potential. The values of E_f are indicated above the corresponding curves.

transients observed at high potentials. The curves presented in Figure 3a have a symmetric bell shape. Integration of these curves gives a charge which is approximately equal to the product of i_{max} and t_{max}. The curves have a form consistent with that produced by a reaction controlled by progressive nucleation and growth of oxide islands in the CO monolayer, and can be described by the expression [16, 17]:

$$i = \sigma_T \pi k_N k_G^2 t^2 \exp(-\pi k_N k_G^2 t^3/3) \quad 1$$

where σ_T is the charge corresponding to the area contained under the transient, k_N is the rate of nucleation, and k_G is the rate of nuclei growth.

The transients given in Figure 3b are asymmetric. The area under the curves is almost two times higher than the product of i_{max} and t_{max}. In the initial segment of the curves, the current varies linearly with time. These features are consistent with the process controlled by the instantaneous nucleation and growth of a fixed number of oxide islands N_N in the CO monolayer. The transients are well described by the expression [16, 17]:

$$i = i(0) + \sigma_T 2\pi N_N k_G^2 t \exp(-\pi N_N k_G^2 t^2) \quad 2$$

where current i(0) corresponds to the reaction proceeding at the perimeter of the oxide islands formed during the time of the charging of the double layer. If the time constant of the cell can be substantially reduced i(0) could reach values close to zero.

The changes in the transient shapes reflect changes in the reaction mechanism. At low potentials, the reaction is controlled simultaneously by the rate of nucleation and the growth of the oxide islands, at high potentials the reaction is controlled by the rate of the island growth.

The inverse of the time corresponding to the current maximum can be taken as a measure of the reaction rate [16-18]. For a reaction controlled by progressive nucleation and growth t_{max} is given by the expression:

$$t_{max} = (2/\pi k_N k_G^2)^{\frac{1}{3}} \quad 3$$

and if the rate is controlled by island growth, t_{max} is equal to:

$$t_{max} = \left(\frac{1}{2\pi N_N k_G^2}\right)^{\frac{1}{2}}$$

Figure 4 shows a plot of the logarithm of t_{max} against the final potential E_f determined for the three crystallographic orientations investigated. The plots are nonlinear which is consistent with a change in the reaction mechanism discussed above as the value of E_f is increased. They also display an apparent affect due to surface crystallography on the rate of the oxidation reaction. This affect is particularly significant in the range of low potentials where the reaction rate is controlled simultaneously by the rate of nucleation and the rate of growth of the oxide islands. To get further insight into the mechanism of the investigated reaction an attempt has been

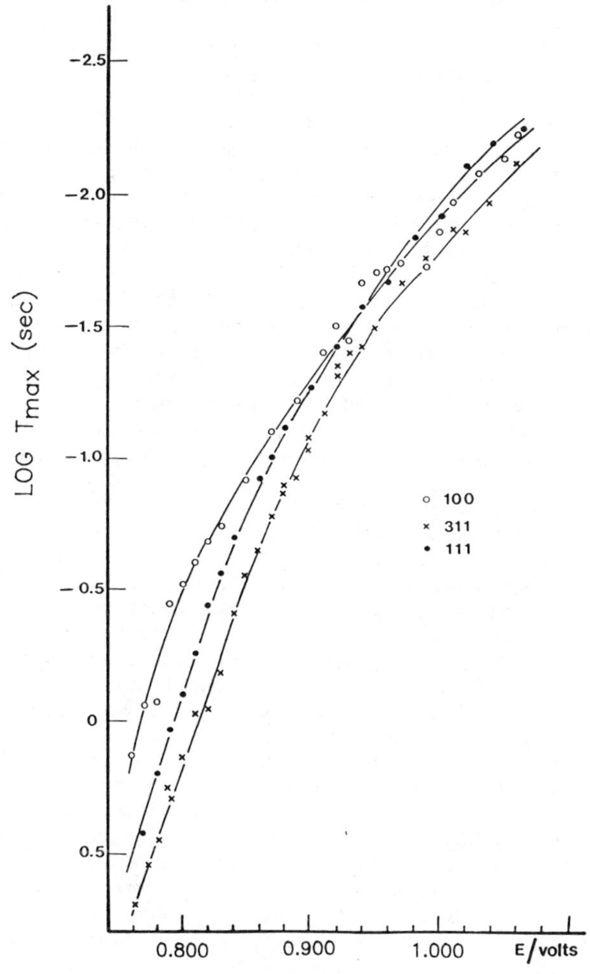

Figure 4. A plot of the logarithm of the time corresponding to the maximum on the current transients plotted against the final potential for the three single crystal surfaces investigated.

made to determine the rate of nucleation and the rate of the growth independently. This required double potential step experiments [7, 16].

B. Double Step Experiments To determine the rates of oxide island growth over the broad range of electrode potentials, the following double step experiments were performed. After the surface was covered by a monolayer of CO using the procedure described in the preceding paragraphs a short duration pulse (2 ms) to E = 1.10 V was applied to the electrode. Next the potential was stepped back to a lower value E_f and the chronoamperometric transient was recorded. During the short pulse to 1.10 V a large number of nuclei were generated so that the number of nuclei formed during the subsequent period at potential E_f contributed negligibly to the total number of nuclei present on the surface. Therefore the oxidation of CO at potential E_f was assumed to be proceeding from a fixed number of nuclei and was expected to be controlled only by the rate of the island growth. It should be stressed that in the present case, the formation of a nuclei is equivalent to the formation of a hole in the CO monolayer of sufficient size to initialize the island growth. The number of the holes made at E = 1.10 V definitely depends on the type of the surface oxide formed at this potential. However, the subsequent expansion of the holes at potential E_f is controlled by the structure of the oxide overlayer which exists at the particular value of E_f.

The shape of the transients determined in the double step experiments is compared with the shape of the transients obtained from the single step experiments, for three selected values of E_f, in Figure 5. As expected, the shape of the transient determined in the double step experiments changes from that characteristic for the progressive nucleation and growth to that characteristic for the instantaneous nucleation and growth. The initial segments of the transients determined in the double step experiments depend linearly on time. The slope of this section of the curves, according to equation 2, is proportional to the square of the growth constant k_G and the number of nuclei formed at the surface during the short pulse to E = 1.10 V. The logarithm of the square root of these slopes is plotted as a function of the final potential E_f, in Figure 6, for the three crystallographic orientations investigated. The plots are nonlinear and their slope changes from about 80 mV/decade at $E_f < 850$ mV to 240 mV/decade at $E_f > 850$ mV. Actually the potential range 750 mV $< E_f <$ 1000 mV, for which the growth constants were determined, corresponds to the initial stage of the oxidation of the platinum surface, see Figure 1. In this range, the coverage of the electrode surface by the water oxidation product, probably absorbed OH species is smaller than 30% and changes strongly with potential. Therefore, the islands which are formed in the monolayer of CO on the surface must be composed of adsorbed water and the OH species. Obviously, the concentration of OH within the island, and at the same time at the perimeter of the island, vary with the electrode potential. Assuming that the island growth is controlled by the reaction between CO and the OH species present at the perimeter line, the variation of the growth rate with potential can be explained, at least qualitatively, by the variation of the OH coverage.

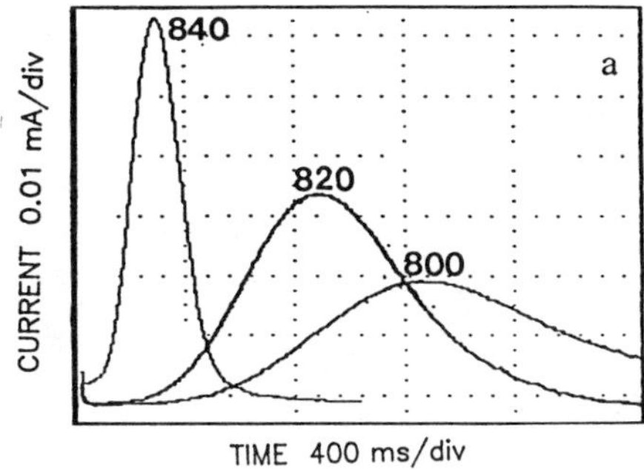

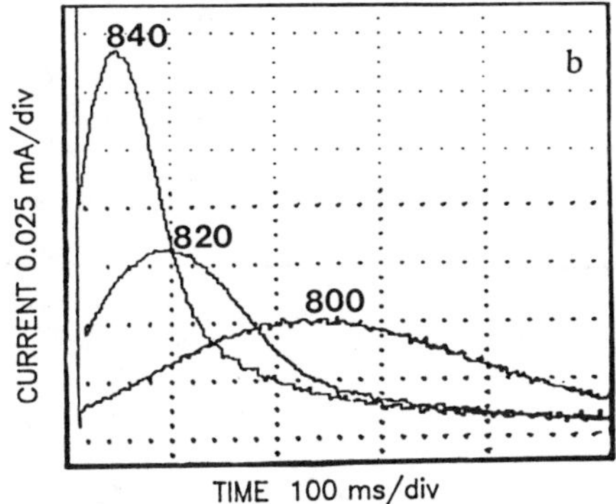

Figure 5. Comparison of the current transients obtained in the single step and the double step experiments. a. the current transient determined in a single step experiments, the same plot as in Figure 3a. b. the current transients recorded in the double step experiments when the potential was first stepped to E = 1.10 V for a period of 2 ms and then stepped back to E_f which values are indicated at corresponding curves.

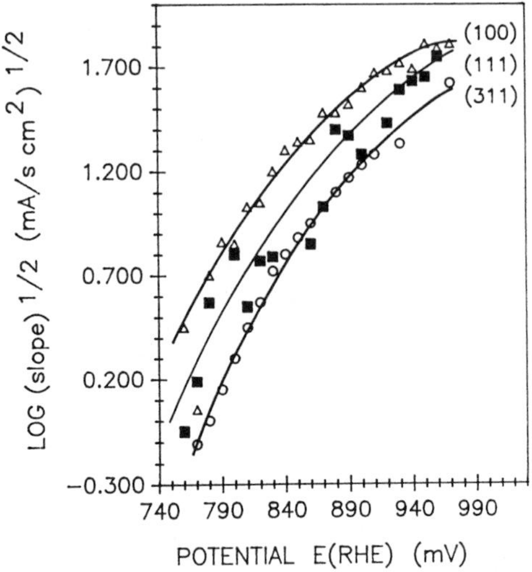

Figure 6. A plot of the logarithm of the square root of the initial slope of the current time transients determined in the double step experiments against the final potential for the three single crystal surfaces investigated.

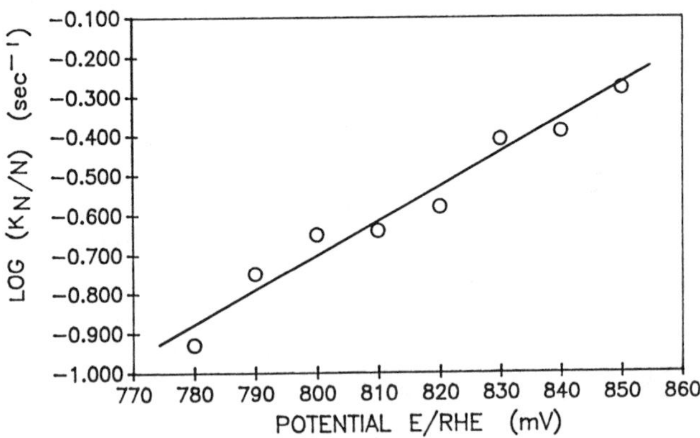

Figure 7. A plot of the logarithm of the ratio k_N/N_N against the electrode potential determined for the (311) surface.

The vertical distance between the plots in Figure 6 is a measure of the influence of surface crystallography on the product k_G/N_N. Provided the same number of nuclei were generated at each surface during the prepulse to E = 1.10 V, the magnitude of the vertical separation between the plots illustrates the variation of the rate of growth. The maximum distance between the plots in Figure 6 amounts to half an order of magnitude and this suggests that the influence of the surface crystallography on the rate of growth is significant.

The single step and the double step experiments were finally combined to determine the rate constants for nucleation k_N. By fitting equation 1 to the single step transients or by using the BFT analysis [19] on the transients the product $k_N k_G^2$ was obtained. The double step experiments allowed for the determination of the product $N_N k_G^2$. From the ratio of the two results, the quotient k_N/N_N was calculated. Figure 7 shows a plot of the logarithm of k_N/N_N against the electrode potential determined for a Pt (311) electrode. The relationship is linear with the slope equal to 120 mV/decade. This suggests that the formation of the island of the oxygen species in the CO monolayer involves a charge transfer as the rate determining step. The calculations of the nucleation rate constant for the two other surfaces are in progress and will be reported shortly, the results obtained so far indicate that surface cyrstallography has a significant influence on rate of nucleation as well.

Summary

Oxidation of CO at the interface between (111), (100) and (311) Pt single crystal surfaces and a solution of an aqueous electrolyte has been investigated. A monolayer of CO has been adsorbed onto the Pt electrode surface and then oxidized by applying an anodic potential pulse. The onset of CO oxidation coincides with the beginning of oxygen adsorption (OH radicals or O atoms) at the free platinum surfaces. The plots of the oxidation rate versus time display maximum and their shape is characteristic for a reaction controlled by nucleation and growth phenomena. Therefore, the overall rate of the reaction can be expressed in terms of the rate of nucleation (formation of oxide islands) and the rate of nuclei growth (expansion of the oxide island). By a combination of the single potential step and the double potential step experiments the nucleation and the nuclei growth rates have been determined independently. The results have shown that the rate of the nuclei growth (island expansion) significantly depends on the crystallography of the Pt surface. Preliminary calculations have shown that the rate of the nucleation depends on the surface structure as well.

Acknowledgments

This work was supported by a grant from Natural Sciences and Engineering Research Council of Canada. We thank J. Richer for his help in writing software for the data acquisition and processing.

Literature Cited

1a) Kunimatsu, K.; Golden, W.G.; Seki, H.; Philpott, M.R. Langmuir, 1985, 1, 245.

b) Kunimatsu, K.; Seki, H.; Golden, W.G.; Gordon, J.D.; Philpott, M.R. Langmuir, 1986, 2, 464.

2. Sobkowski, J.; Czerwinski, A. J. Phys. Chem. 1985, 89, 365.

3. Wagner, F.T.; Moylan, T.E.; Schnieg, J.S. Surface Science, in press.

4a) Beden, B.; Bilmes, S.; Lamy, C.; Leger, J.M.; J. Electroanal. Chem. 1983, 149, 295.

b) Lamy, C.; Leger, J.M.; Clavilier, J.; Parsons, R. J. Electrochem. Chem. 1983, 150, 71.

c) Leger, J.M.; Beden, B.; Lamy, C.; Bilmes, S. J. Electroanal. Chem. 1984, 170, 305.

5. Wagner, F.T.; Ross, Jr., Ph.N.; J. Electroanal. Chem., 1983, 50, 141.

6a) Palaikis, L.; Zurawski, D.; Hourani, M.; Wieckowski, A. Surface Science, in press.

b) Santo, E.; Leiva, E.P.M.; Vielstich, W.; Linke, V. J. Electroanal. Chem. 1987, 227, 199.

7a) Gilman, S. J. Phys. Chem. 1963, 67, 78.

b) Gilman, S. J. Phys. Chem. 1964, 68, 70.

8. McCallum, C.; Pletcher, D. J. Electroanal. Chem. 1976, 70, 277.

9. Seto, K.; Iannelli, A.; Love, B.; Lipkowski, J. J. Electroanal. Chem. 1987, 226, 351.

10. Clavilier, J.; Faure, R.; Guinet, G.; Durand, R. J. Electroanal. Chem. 1980, 107, 205.

11. Dickertmann, D.; Koppitz, F.D.; Schultze, J.W. Electrochim. Acta. 1976, 21 967.

12. Ross, P.N.; Wagner, F.T. In Advances in Electrochemistry and Electrochemical Engineering; Gerischer, H.; Tobias, C.W., Eds.; Wiley; New York, 1984; Vol. 13, p. 69.

13. Aberdam, D.; Durand, R.; Faure, R.; El-Omar, F. Surf. Sci. 1986, 171, 303.

14. Markovic, N.; Hanson, M.; McDougall, G.; Yeager, E.; J. Electroanal. Chem. 1986, 214, 555.

15. Norton, P.R.; Davis, J.A.; Jackman, T.E. Surf. Sci., 1982, 122, L593.

16. Fleischmann, M.; Thirsk, H.R. In Advances in Electrochemistry and Electrochemical Engineering; Delahay, P., Ed.; Interscience; New York, 1963; Vol. 3.

17. Deutscher, R.E.; Fletcher, S. J. Electroanal. Chem. 1983, 153, 67.

18. Quarin, G.; Buess-Herman, U.; Gierst, L. J. Electroanal. Chem. 1981, 123, 35.

19. Bewick, A.; Fleischmann, M.; Thirsh, H.R. Trans. Faraday. Soc. 1962, 58, 2200.

RECEIVED May 17, 1988

Chapter 34

Hydrogen Electrosorption and Oxidation of Formic Acid on Platinum Single-Crystal Stepped Surfaces

N. M. Marković, A. V. Tripković, N. S. Marinković, and R. R. Adžić

Institute of Electrochemistry, Institute of Chemistry, Technology, and Metallurgy, and Center for Multidisciplinary Studies, University of Belgrade, P.O. Box 815, Belgrade, Yugoslavia

Hydrogen adsorption and oxidation of formic acid show a pronounced dependence on the structure of single crystal surfaces. The influence of the terrace and step orientation and step density is reflected in both reactions on step surfaces. The multiple states of hydrogen adsorption can be correlated with the nature of adsorption sites. There is a negligible effect of adsorbate-adsorbate interaction on step surfaces. Some lateral repulsion of hydrogen adsorbed on Pt(111) could be inferred. A strong adsorption of bisulphate and sulphate anions on the (111) oriented terraces and step sites considerably affects both reactions. These data show that each crystallographic orientation of the electrode surfaces gives a different electrochemical entity.

A remarkable progress has been made in the last several years in electrocatalysis on single crystal surfaces. This parallels the progress in surface science and it has been partly stimulated by developments in that field, mostly regarding the preparation and characterization of surfaces. New advances in preparation of surfaces outside of high vacuum, achieved in electrocatalytic studies, also helped this trend.

Electrocatalysis is a complex area which has lacked so far a molecular-level understanding of elementary steps in electrocatalytic reactions. This lack has been due to unsuitable techniques for in situ identification of reaction intermediates and products and to poor characterization of surfaces, despite the high sensitivity of electrochemical techniques. All existing

0097–6156/88/0378–0497$06.25/0

concepts, however, are based on the data obtained with polycrystalline electrodes. It already appears that the concepts resulting from studies with single crystal surfaces will cast doubts on earlier established ideas. Many interesting features, not recognizable with polycrystalline surface, have been observed. These studies, with in situ spectroscopic investigations, could provide a rapid development of this exciting area. Since the first observation of the structural dependence of hydrogen adsorption on Pt single crystal electrodes (1), many experiments have been reported on this topic including the transfer of the samples from the UHV chambers into the electrochemical cell (2-5). Another important development in surface preparation is the flame annealing technique (6) with its modifications (7). Only recently have data from various laboratories started to agree.

Oxidation of formic acid shows a pronounced structural dependence which is well illustrated by data on the low index planes (8-10) and preliminary data on stepped single crystal surfaces (11). In this work further investigations of hydrogen adsorption and oxidation of formic acid on single crystal Pt surfaces with 15 orientations are reported.

Experimental

Single crystal electrodes were obtained from Metal Crystal Ltd. (Cambridge, England) oriented and cut to better than 1o. The surfaces have been prepared by H_2/O_2 flame annealing at 1200K, followed by cooling in H_2 down to 400K, and were protected by a drop of water while being transferred into the cell. The surface structures of several electrodes have been checked by LEED. However, no direct transfer from the UHV chamber into the cell has been achieved. The solutions have been prepared from the organic-free water obtained from 18 MΩ Millipore water irradiated by a UV-radiation for 24 h or distilled from alkaline permanganate to remove traces of organics. The electrolytes were Merck reagent grade acids. Platinum foil served as a counter electrode while $Hg/HgCl_2$ and $Hg/HgSO_4$ were used as a reference. All potentials are given against the saturated calomel electrode.

Results and Discussion

Hydrogen Adsorption on the Low-Index Surfaces. The data on the hydrogen electrosorption on the low-index planes of Pt differ in some detail, as will be discussed below with the help of results for stepped surfaces.

Voltammetry curves for all three low-index surfaces are given in Fig. 1. Hydrogen adsorption at Pt(111), the process at $-0.25 < E < -0.05$ V in Fig. 1, is not affected by the nature of the anion (such as SO_4^{2-}, ClO_4^- or F^-) (12). The lack of a well defined peak, in the drawn-out curve of Fig. 1 clearly indicates a strong lateral repulsion between adsorbed hydrogen adatoms. This is probably a consequence of a partially charge on the adsorbed hydrogen adatoms which, in turn, does not allow the

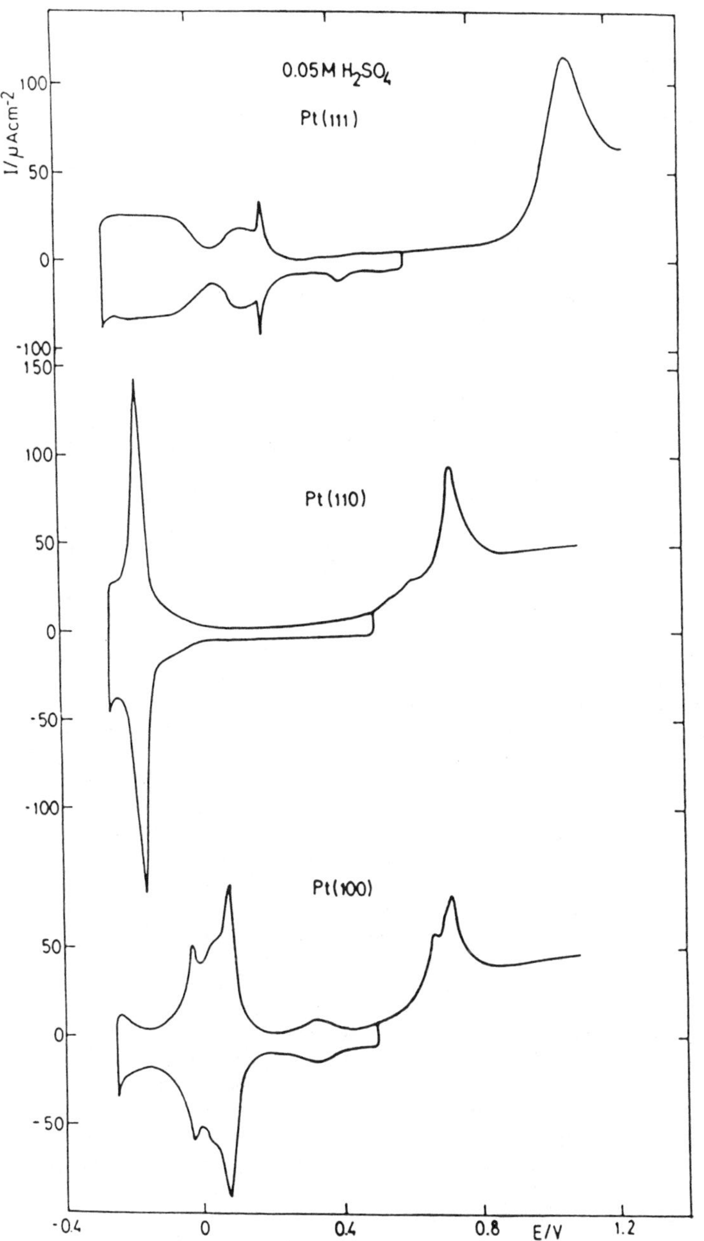

Figure 1. Cyclic voltammetry for Pt single crystals in 0.05 M H_2SO_4. Sweep rate 50 mV/s.

coverage to reach a full monolayer (12). Hydrogen adsorption on the Pt(111) plane could be described by the following equation:

$$H^{+} + \lambda e- \longrightarrow H_{ad}^{(1-\lambda)+} \qquad (1)$$

where λ is a partial charge transfer coefficient.

In contrast to curve in Fig. 1, two small peaks on a flat portion of the curve at E=-0.025 V and E=-0.15 V have been observed by several authors (4-6). Our data to be shown in the next section strongly suggest that such peaks are due to a stepped surface with a small step density, while a well-ordered Pt(111) surface should give a curve as in Fig. 1.

The interpretation of voltammetry curve for the Pt(100) surface poses some problems, e.g. the origin of the peak at E=-0.15 V (Fig. 1). Marković et al. (12) ascribed this peak to hydrogen adsorption on particular surface imperfections, the (111)-oriented step sites. The height of this peak varies from one set of data to another, indicating a lack of control of the surface structure. Further support of this view will be shown below with the data for stepped surfaces.

The sharp and narrow peak at -0.15 V on the Pt(110) plane indicates on a coupling of hydrogen adsorption/desorption and HSO_4- and SO_42- desorption/adsorption. This will be further discussed in connection with the data for stepped surfaces.

Positive to the hydrogen adsorption region, the curve for the Pt(111) surface shows "anomalous peaks" which, according to the predominant view, are due to sulphate adsorption (12, 13). Clavilier ascribed these peaks to adsorption of 1/3 of a monolayer of strongly-bound hydrogen. The charge associated with these peaks is 75 μC cm-2, which corresponds to 1/3 of a monolayer of a completely discharged univalent species. Such a discharge is possible for oxy-anions such as HSO_4- and SO_42-, since their adsorption is very strong at (111) sites. This is due to the compatibility of their tetrahedral structure with (111) surface symmetry (14). It was assumed that one oxy-anion can cover at least 3 Pt atoms. Such adsorption can result in almost complete discharge of the overlayer of anions (14). A sizable discharge of SO_42-, HSO_4- on Au(111) can be also deduced from the shape of C-E curves (15), which suggests that the (111) symmetry is more important in this particular case than the nature of the electrode material. A strong adsorption of SO_42- and HSO_4- on Pt(111) is also reflected in the beginning of oxide formation which, on this plane, is shifted to high positive potentials (Fig. 1). A sharp reversible pair of peaks at E=0.2 V (which merge with the process at E=0.15 V) require a two-dimensional order, since they do not occur at stepped surfaces with (111)-oriented terraces (12). This unusual pair of peaks require further study which is outside of scope of this paper.

A direct evidence of the way of tetrahedral anion adsorption at three-fold sites and the degree of hydratation is not available at present. However, a strong indication of such adsorption of sulphates is found in voltammetry on gold (14) and in our data for platinum surfaces (12). A pronounced difference between the sulphate and perchlorate adsorption effects is partially due to their different hydratation on the surface. There are indications that sulphates are hydrated by only two, while perchlorates by 7-8 water molecules (14). A high hydratation of perchlorates may be cause of a small difference between adsorption of these anions and fluorides.

Two other small peaks appear at E=0.45 and 0.7 V on the Pt(111) surface. They are shifted to more negative potentials with increasing H_2SO_4 concentration, suggesting their link with anion adsorption. A peak at 0.35 V for the Pt(100) surface, where one would expect the so-called double-layer region for Pt, is also intriguing. It remains to be seen whether or not it is due to sulphate adsorption on the (111)-oriented imperfections.

The curve for Pt(110) surface shows a wide double layer region, indicating that the anion adsorption is completed concurrently with hydrogen desorption. The anions are strongly held in the "troughs" of the Pt(110) surface, i.e. in the 2(111)-(111) terrace-step sites. This point will be touched upon connection with the stepped surfaces. The C-E curves for Au(110) in H_2SO_4 clearly show a completion of anion adsorption on that surface at very negative potentials (15).

A pronounced structural sensitivity of the oxidation of Pt surfaces is also seen in Fig. 1. The reaction takes place at the most positive potential on Pt(111). This is probably due to effective blocking of the surface by oxy-anions with the trigonal symmetry, compatible with the (111) orientation. A detailed analysis of this reaction on Au(111) has been recently performed by Angerstein-Kozlowska et al. (14). No such blocking is possible for the Pt(100) and Pt(110) surfaces with four-fold and two-fold symmetries. Consequently, the oxidation commences at more negative potentials, probably predominantly determined by the surface energy as found with Au (16).

Hydrogen Adsorption on Platinum Single Crystal Stepped Surfaces. Few data are available for hydrogen adsorption on stepped surface (12, 17, 20-22). The reaction on several surfaces from the [110] zone has recently been analyzed in some detail (12). Fig. 2 shows voltammetry curves for hydrogen adsorption in all three zones of the stereographic triangle. For the [110] zone (Fig. 2a) the reaction in the region $-0.25 < E < -0.075$ V has been ascribed to hydrogen adsorption on the (111)-oriented terraces. The peak at -0.025 V has been found due to adsorption on the step sites of the (111)-(100) or (100)-(111) intersections of terraces and steps for surfaces vicinal to the (111) and (100) planes, respectively. The trigonal sites formed by intersections on the (111) and (100)-oriented steps and terraces are responsible for a strong adsorption of oxy-anions at steps. It appears that upon anion desorption, hydrogen adsorbs first at the (100)-oriented sites (12). This size of this peak at -0.025 V is proportional to the step density. Hydrogen adsorption is concurrent with the bisulphate/sulphate desorption. The similarity of these sites, which can bee seen from models in Fig. 3, causes the appearance of the peak at almost the same potential in the whole zone.

For surfaces vicinal to the (111) plane (Fig. 2a) in the potential region $0.15 < E < 0.5$ V a broad peak has been ascribed

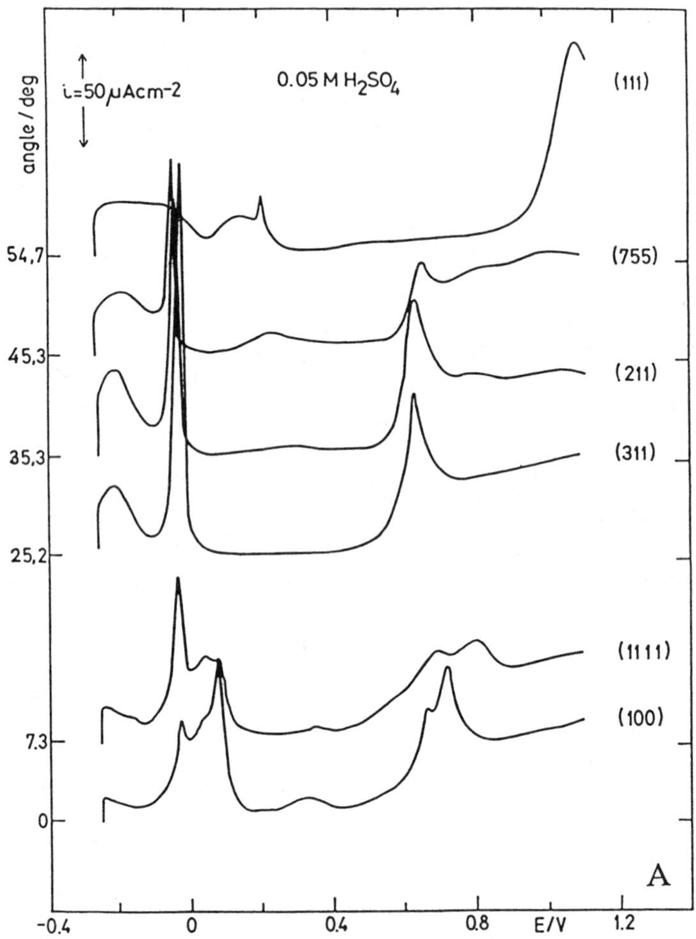

Figure 2A. Cyclic voltammetry for the single crystal surfaces laying on the 111 zone line. Sweep rate 50 mV/s.

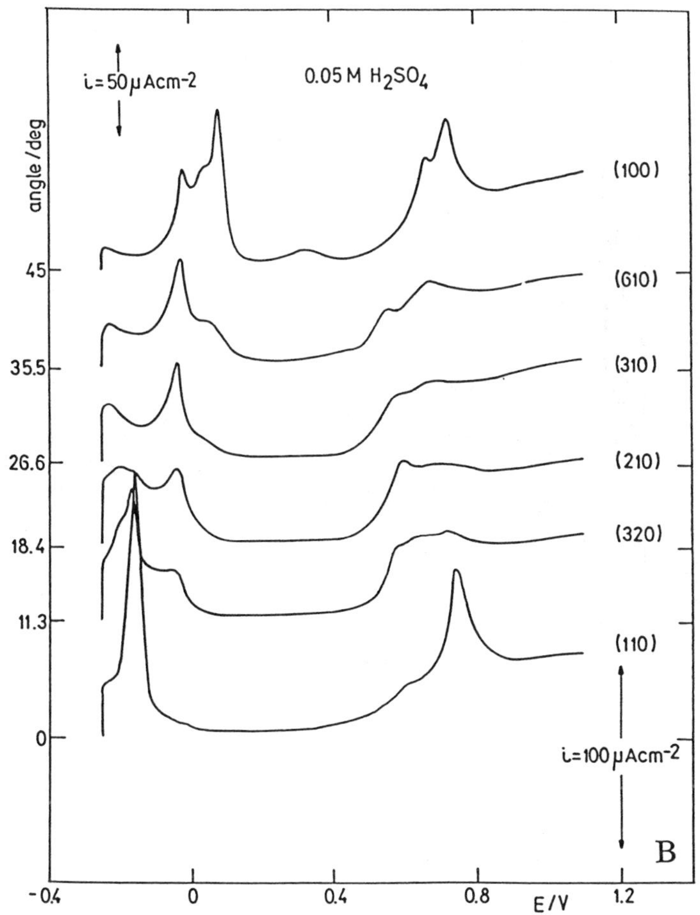

Figure 2B. Cyclic voltammetry for the single crystal surfaces laying on the 100 zone line. Sweep rate 50 mV/s.

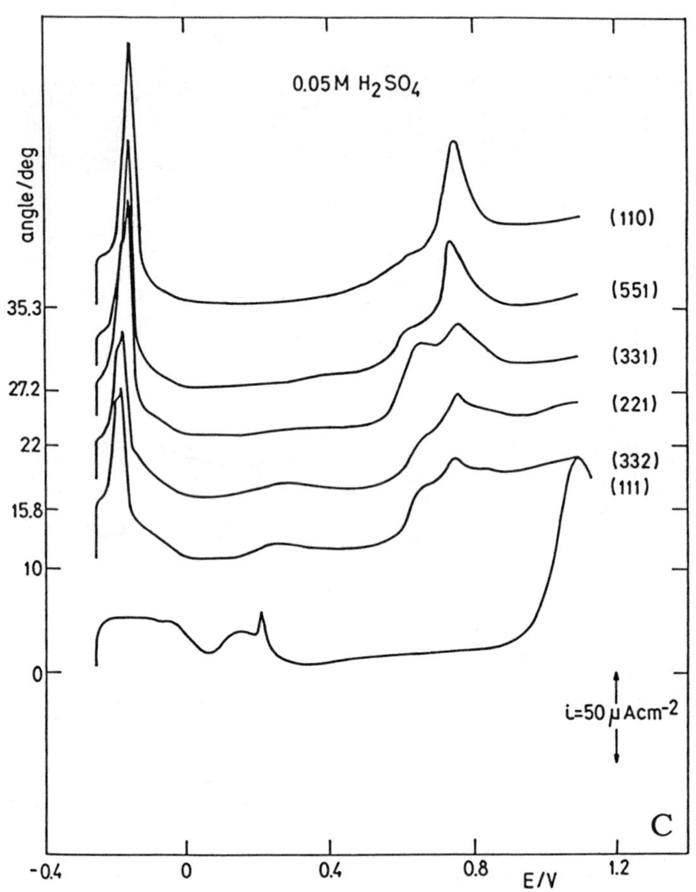

Figure 2C. Cyclic voltammetry for the single crystal surfaces laying on the 110 zone line. Sweep rate 50 mV/s.

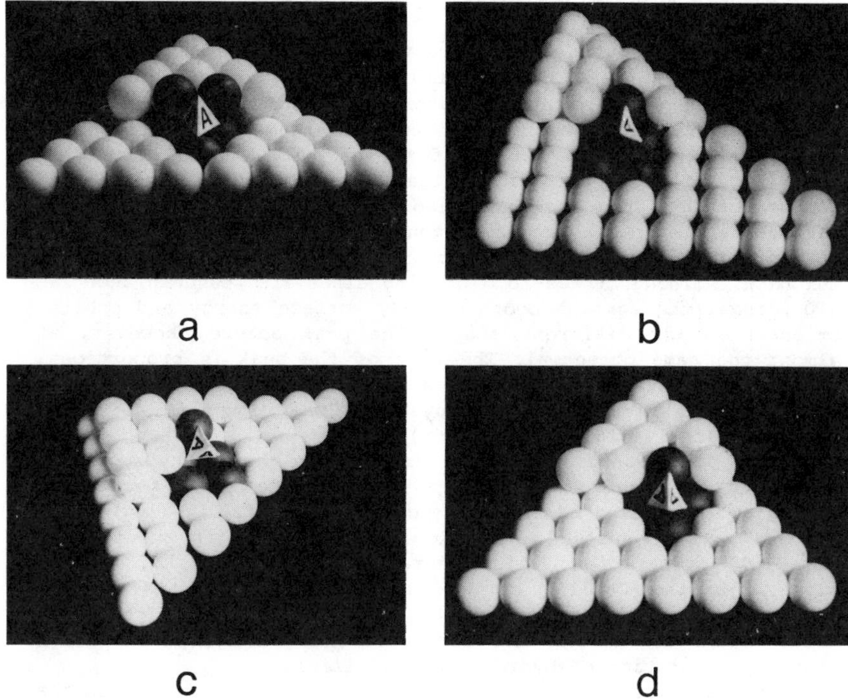

Figure 3. Models for the stepped surfaces, with tetrahedral anions adsorbed on the (111) trigonal sites for a) n(111)-(100), b) n(100)-(111), c) n(100)-(100) and d) n(111)-(111) terrace-step orientations.

to bisulphate/sulphate adsorption on the (111)-oriented terraces (12). Its size is inversely proportional to the step density. No such peak is seen for the Pt(311) surface whose structure, being 2(111)-(100), cannot accommodate the tetrahedral anions in the terrace. Such anions, adsorbed in the step, block the whole 2-atom long terrace (12).

Voltammetry curves for surfaces from the [100] zone are displayed in Fig. 2b. These are the (210), (310) and (610) faces, which in Lang et al. (23) notation are 2(100)-(100), 3(100)-(100) and 6(100)-(100), respectively. The curve for the (320) surface, on the other side of the (210) orientation which is the "turning point" of the zone, is also shown.

A common feature of these curves is a peak at -0.025 V, the same potential as for the [110] zone. This peak has been explained above by particular atomic arrangements in the step. The question arises whether the same arrangements are formed in this zone? A close inspection of the stepped surfaces from this zone (Fig. 3) shows that geometrically similar atomic arrangement can be obtained by a combination of the (100) and (110) oriented terraces and steps. Again, a trigonal symmetry of Pt atoms can be formed, providing sites for a strong adsorption of bisulphate or sulphate. Although three-fold binding sites are found in the [100] zone, the surface coordination, surface energy and orbital orientation are different shape. The peak occurs, however, at almost the same potential. The size of the peak is proportional to the step density, as has been checked by combining our data with the data of Motoo and Furuya (22) who reported additional three surfaces vicinal to the (100) plane.

The peak associated with a strongly bound hydrogen at $E \sim 0.15$ V in Fig. 2b decreases as the terrace width decreases, giving no indication of its presence for the Pt(210)=2(100)-(100) or 2(110)-(100) surface. The highly active site is provided by the (100) orientation, which decreases with decreasing terrace width. For the Pt(210) surface adsorption on the terrace is apparently strongly affected by the anion adsorption in the step, which blocks the 2-atom wide terrace. Consequently, no sites for a strong adsorption of hydrogen are available.

On the basis of the models in Fig. 3 and comparison with the peak for the [110] zone, the reaction in the potential range $-0.25 < E < -0.15$ V could be ascribed to the hydrogen adsorption on the (111)-oriented step sites.

The behaviour of the Pt(110) surface, as discussed above, is largely determined by a strong adsorption of anions in the steps of its 2(111)-(111) structure. The same is valid for vicinal surfaces such as Pt(320), which gives a sizable peak at the same potential as Pt(110).

For most of the surfaces from the [$1\bar{1}0$] zone, Fig. 2c, hydrogen adsorption is predominantly associated with the single peak occurring at almost the same potential for all the surfaces investigated. That peak, in the potential region $-0.25 < E < -0.1$ V, is determined by hydrogen adsorption, coupled with anion desorption, from the (111)-(111) or (110)-(111) terrace-step sites (Fig. 2c). They coincide with the peak for the

Pt(110) surface. As explained above, this surface behaves as the "fully stepped" surface with the orientation 2(111)-(111).

The coupling of H adsorption with HSO_4^- and SO_4^{2-} desorption can be written in the following way:

$$yH^+ + A_{ad}^{(z-\lambda_A)-} + (y\lambda_H+\lambda_A)e^- \text{ ----> } yH_{ad}^{(1-\lambda_H)+} + A^{z-} \qquad (2)$$

Evidence for this coupling was obtained by a comparison of the data for sulphuric and perchloric acids (12).

The reaction at $-0.1 < E < 0.15$ V seems to be associated with hydrogen adsorption on the (111) oriented terrace sites for the surfaces vicinal to the (111) plane. This can be deduced from a comparison with the Pt(111) surface. A small peak at E~0.35 V, seen for the Pt(332) and Pt(221) surfaces, is due to bisulphate/sulphate adsorption on the (111)-oriented terraces. This point has been discussed in detail for the Pt(332) and Pt(755) surfaces (12).

<u>Identification of Peaks for Hydrogen Adsorption on the Disordered Low Index Planes.</u> Besides the major objective for studying electrocatalysis on single crystal stepped surfaces mentioned above, these studies offer a wealth of information on the behaviour of polycrystalline surfaces, of preferentially oriented surfaces and, as we suggested recently (12), of disordered low-index surface.

A single sweep into the oxide formation with the Pt(111) surface causes a considerable change of voltammetry curve for a well-ordered surface, as can be seen in Fig. 4a. Two new small peaks are seen on the curve of the original Pt(111). These peaks can be seen on the curves for surfaces by believed to be "well-ordered" by several authors (10, 17, 22). To unravel the origin of the two peaks at E=-0.15 and -0.075 V, it is necessary to be able to introduce, in a controlled manner, deviations from the perfect well-ordered lattice structure. This is offered by the use of the stepped surfaces. Fig. 4b shows that the sharp peaks of the Pt(755)=6(111)-(100) and Pt(332)=6(111)-(111) coincide with the above peaks. This strongly suggests that these peaks are due to the introduction of the (100)- and (111)-oriented steps upon cycling into the oxide formation region. Introduction of steps, i.e., more active sites on the well-ordered (111) surface, is also seen in the oxide formation region which now commences at much less positive potential. It appears that the surface energy plays important role in surface oxidation (15), although the lack of sulphate adsorption at the top step sites certainly has an effect. Also, the sharp peak at E=0.2 V is lost and the peak of sulphate adsorption is diminished. All these processes are now typical for a stepped surface.

The curve for Pt(100) shows some features which could be due to the presence of the surface imperfections. Defect formation on the Pt(100) surface by an H_2/O_2 flame or by electrochemical oxidation may not be surprising, considering results from the gas-phase (24) which showed that the reaction of O_2 with surfaces

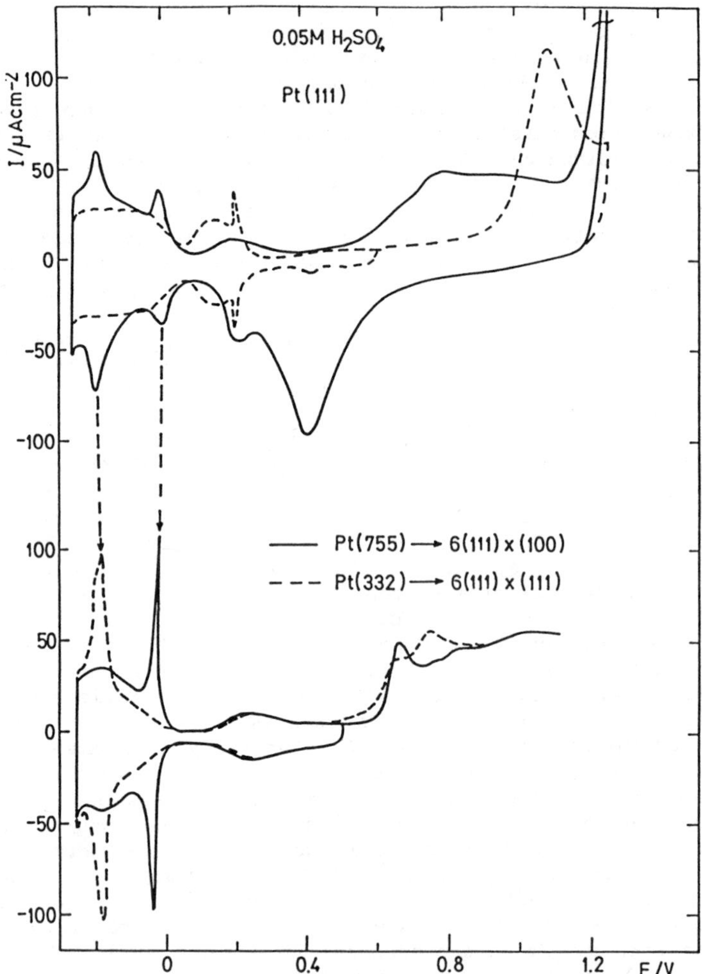

Figure 4. Cyclic voltammetry for well ordered (- - - -) and disordered (——) Pt(111), for well ordered Pt(755) and Pt(332) in 0.05 M H_2SO_4. Sweep rate 50 mV/s.

from the [100] zone produced reconstructions. It remains to be seen how much cooling in H_2 prevents reconstructing.

The effect of a single potential excursion into the oxide region causes a considerable decrease of the peak of strongly bound hydrogen and an increase of the peak at E=-0.15 V (Fig. 5a). At the same potential a small peak is seen for the Pt(100), prepared as described above. A similar, or even larger, peak is seen with all the curves published so far (21, 22, 23). The curves for vicinal (Pt(11,1,1)=6(100)-(111) and Pt(610)=6(100)-(100)) surfaces (Fig. 5b) show much larger peaks at the same potentials. These data strongly suggest that a small peak at E=-0.16 V for Pt(100) is due to the contribution of the (111) step sites.

Oxidation of Formic Acid. The oxidation of formic acid on Pt has been the subject of numerous studies on polycrystalline (26) and single crystal (8-11) electrodes. However, no consensus on the mechanism has been reached so far.

It is accepted that in the first step the C-H bond is broken:

HCOOH ----> -COOH + H+ + e-

It is also agreed that further reaction goes through two parallel pathways,

$$-COOH \nearrow CO_2 + H^+ + e^-$$
$$-COOH \searrow \text{poisoning species}$$

one leading to a formation of CO_2, the other to a formation of strongly bound species, which may be eventually be oxidized to CO_2 at high positive potentials. The H-C bond is broken first, unlike in the gas phase where O-H bond reacts first (27). This electrochemical behaviour has been confirmed by comparison of the kinetics of DCOOD, DCOOH and HCOOH on polycrystalline Pt modified by Pb (28) and on Pt(100) and Pt(110) electrodes (29).

A long disputed issue of the nature of strongly bound species in this reaction has been recently revived with the vibrational spectroscopy studies of Bewick et al. (30) using EMIRS technique and of Kunimatsu and Kita (31) using polarization modulation IR-reflection-absorption technique. These data indicated the only CO is a strongly bound intermediate. Heitbaum et al. (32) on the other hand advocate COH, and most recently HCO (33), as the poisoning species on the basis of differential electrochemical mass spectroscopy (DEMS).

To allow for CO formation in the H adsorption region one can modify the reaction proposed earlier for formation of COH (31):

$$COOH_a + H_a \text{ ----> } CO_a + H_2O \qquad (3)$$

Outside the hydrogen adsorption region CO can be formed by the reaction (3):

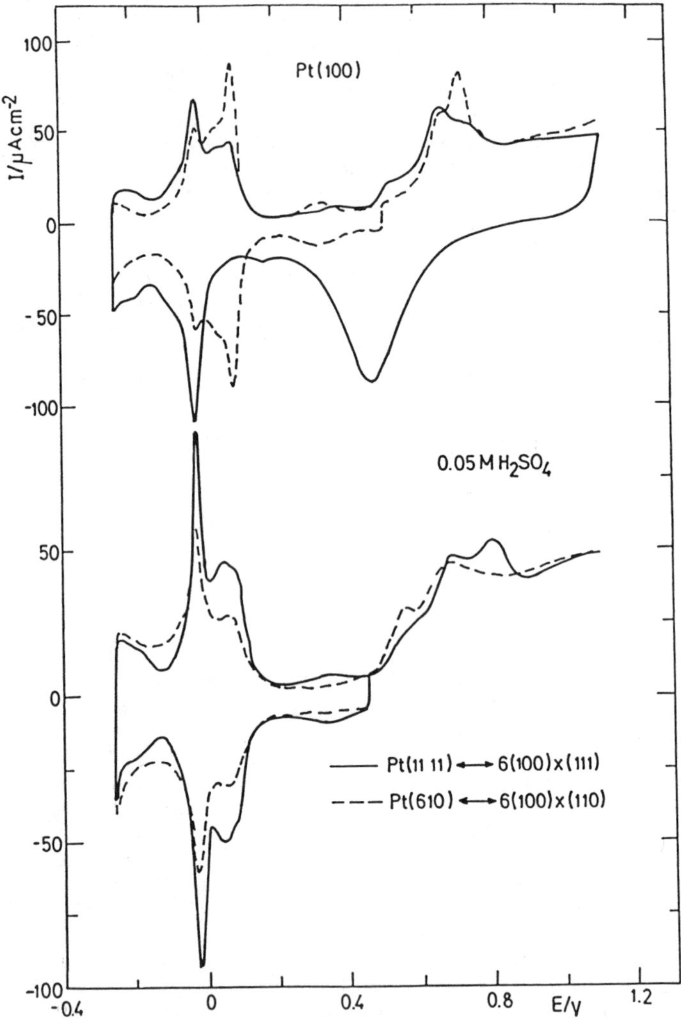

Figure 5. Cyclic voltammetry for Pt(100) before (- - - -) and after (——) disordering of the surface and for ordered Pt(11,1,1) and Pt(610) surfaces in 0.05 M H_2SO_4. Sweep rate 50 mV/s.

$$2COOH_a \text{ ----> } -CO_a + H_2O + CO_2 \quad (4)$$

The oxidation of formic acid was the first electrocatalytic reaction which clearly showed a pronounced influence of the structure on its kinetics and mechanism (8-10). Despite these studies, a thorough understanding of the reaction mechanism on various planes has not been obtained. This applies especially to the negative-going sweeps for the Pt(100) and Pt(110) planes after reversal at 1.1 - 1.2 V. Upon "activation" of the electrode at these potentials, the Pt(110) surface on the negative going sweep immediately (once the oxide layer is reduced) gives high current density (Fig. 6). For Pt(100), on the other hand, one obtains a gradual increase of the current as the potential becomes more negative, i.e. the current increases with decreasing the overpotential for the reaction. The Pt(100) surface (Fig.6a) is blocked, most probably by CO, up to the potential of the oxide formation. Upon sweep reversal, the reduction of a partially oxidized surface takes place gradually. Instead of the expected fast rise of the current for the oxidation of HCOOH on clean surface, one sees a gradual increase of the current as the potential becomes more negative.

The Pt(110) surface exhibits a quite different behaviour. First, the positive-going sweep gives a large peak commencing at ~0.65 V. This indicates that upon oxidation of the blocking species the reaction takes place at a high rate at that high overpotential. Upon sweep reversal similar high currents are observed. This behaviour, especially the cathodic sweep, contrasts with the behaviour of the Pt(100) surface. The Pt(111) plane shows no activation effect, i.e. no poisoning species seem to be formed on that plane. The sweeps in anodic and cathodic directions almost retrace, with the current peak at 0.5 V. This peak was found a consequence of anion adsorption (29). It is noteworthy that the oxidation of CH_3OH is less structure sensitive (9).

Anion adsorption is apparently determining the negative going sweeps for Pt(111) and Pt(100). This will be illustrated by following discussion. For Pt(110), the indirect arguments given in the H_2SO_4 voltammetry section indicate that sulphate desorption occurs at very negative potentials, while the oxidation of HCOOH in the cathodic sweep takes place at much more negative potentials on this surface than on the others. The adsorption of sulphates at those potentials (0.7 V) is clearly very strong. There is no direct data available on the adsorption of sulphate on Pt single crystal surfaces. For Au, on the other hand, there exists a considerable body of experimental work (15, 34). These data show a pronounced role of crystallographic orientation on adsorption of sulphate. The data of Štrbac et al. (15) corroborate the conclusions reached for Pt based on hydrogen adsorption and its relation to adsorption of anions. These data show that at Pt(110) sulphate adsorption is largely completed at $E < 0.05$ V.

The oxidation of HCOOH at Pt(110) at E~0.65 V in sweeps in both directions now appears as a contradiction to the above analysis. Why doesn't sulphate block the reaction, since they are

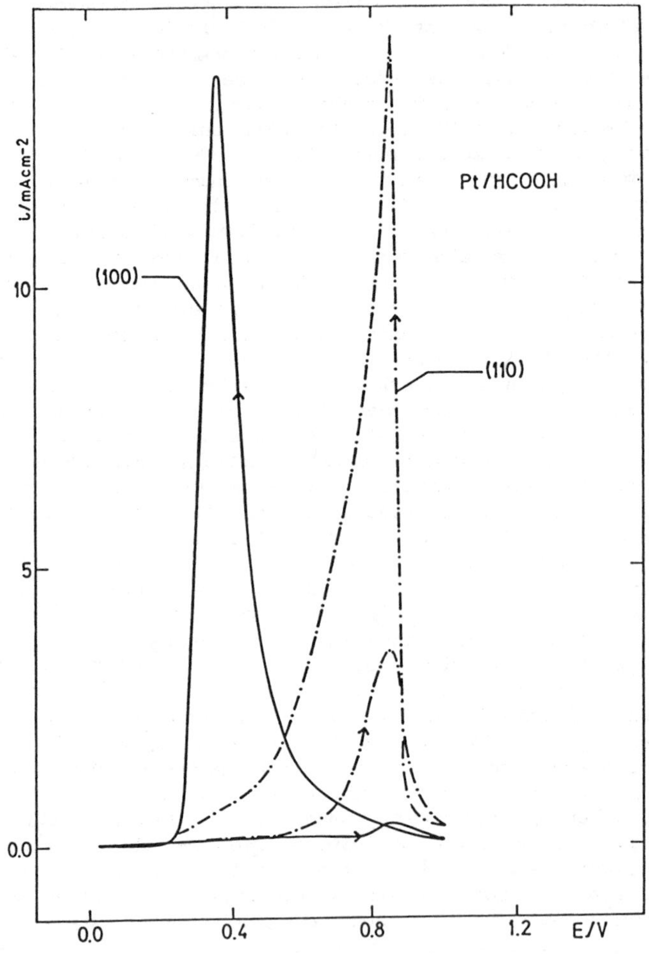

Figure 6. Oxidation of HCOOH on the Pt(100) and Pt(110) single crystal surfaces in 0.5 M H_2SO_4. Sweep rate 50 mV/s.

strongly bound at these potentials? Oxide formation requires more positive potentials and can be ruled out as a positive "trigger" of the reaction. From the above data it follows that the Pt(110) surface behaves as "fully" stepped surface with the orientation 2(111)-(111), the anions are probably strongly adsorbed in "troughs" of that surface. Their repulsion prevents the additional adsorption of tetrahedral anions on "ridges" the top sites. HCOOH can, however, react with these sites, resulting in current peaks at E 0.65 V. The question arises as to why the reaction "waits" for such positive potentials? This is due to immediate poisoning with CO at lower potential. CO is oxidized at the same potential as HCOOH on that plane (35).

From gas phase measurements CO is known to prefer top sites on all three low index faces, with the CO molecule perpendicular to the surface and bonded through the carbon end of the molecule except at high coverages (27). It is likely that HCOOH and COOH are adsorbed in a similar way. It is not likely that they could "enter" the "troughs", which seems to be possible for anions. For Pt(100) on the other hand, upon sweep reversal and gradual oxide reduction, anions are immediately adsorbed on that "flat" surface. They block adsorption of HCOOH. Adsorption of anions decreases as potential becomes more negative. The oxidation of HCOOH commences and the rate increases as at more negative potentials, i.e. at lower overpotential. A competition between anions and HCOOH adsorption explains this apparently anomalous behaviour. The explanation of the "anomalous" behaviour of the Pt(110) surface can be also found in the data for stepped surface vicinal to the (100) and (110) orientations.

Fig. 7 gives the curve for the Pt(11,1,1)=6(100)-(111). Introduction of steps in the (100) flat surface causes the appearance of a new peak in cathodic sweep at E 0.75 V. The peak grows as the step density increases (29).

Sulphate is strongly adsorbed in the steps of these surfaces just as in the "troughs" of the (110) plane. The top step sites should be again available for the oxidation of HCOOH, as strongly indicated by Fig. 7. The data for the Pt(332)=6(111)-(111) surface provide further support of this analysis. As with Pt(111) this surface shows no poisoning effect (as do other low index planes) (Fig. 7), giving a peak in anodic sweep. However, the (111)-(111) terrace-step combination gives the sites of the (110) geometry. This causes the appearance of the peak at 0.7 V, as on Pt(110). Fig. 8 shows a strong structural dependence on surfaces from the 3 zones.

General discussion and conclusions.

Both the data on hydrogen adsorption and formic acid oxidation show pronounced structural sensitivity, thus confirming a paramount role of surface structure in electrocatalytic reactions. It can be concluded that each crystallographic orientation represents a distinct electrochemical (chemical) entity. The investigation of stepped surfaces seems to be necessary to reach an understanding of these systems on a molecular level. Hydrogen adsorption shows dependences on the terrace orientation, step orientation, and step density. All the

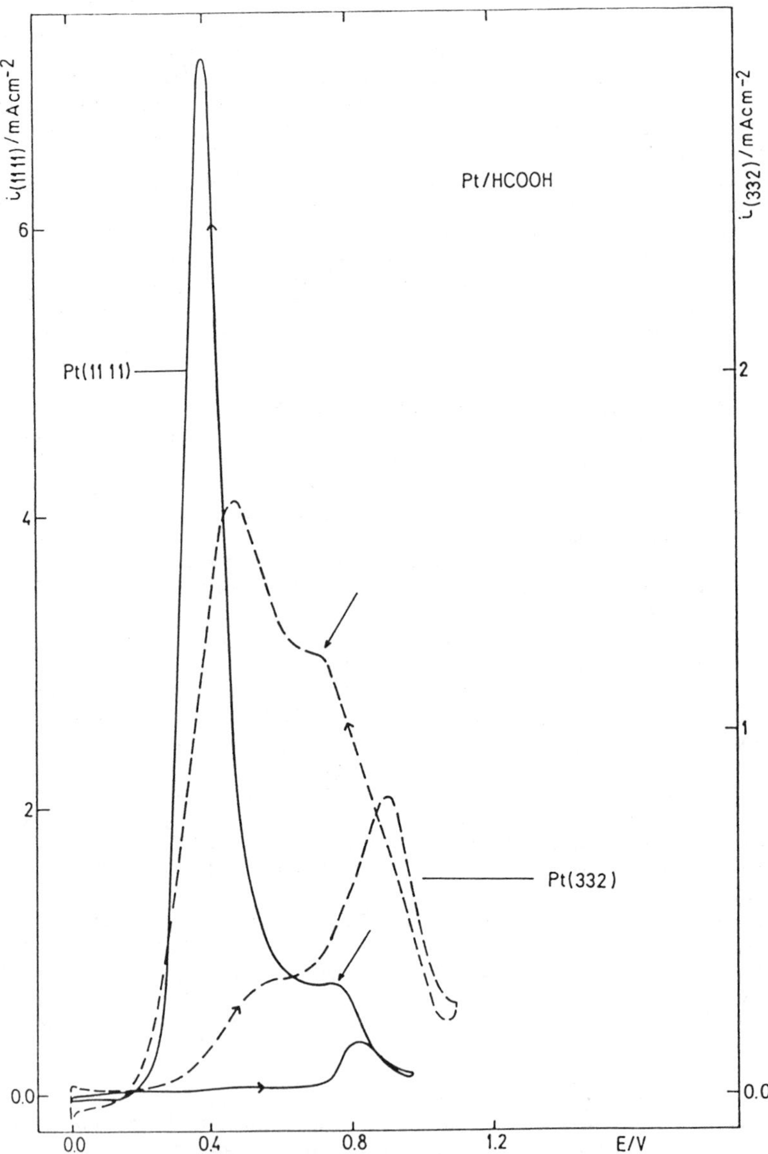

Figure 7. Oxidation of HCOOH on the Pt(11,1,1) and Pt(332) single crystal stepped surfaces in 0.5 M H_2SO_4. Sweep rate 50 mV/s.

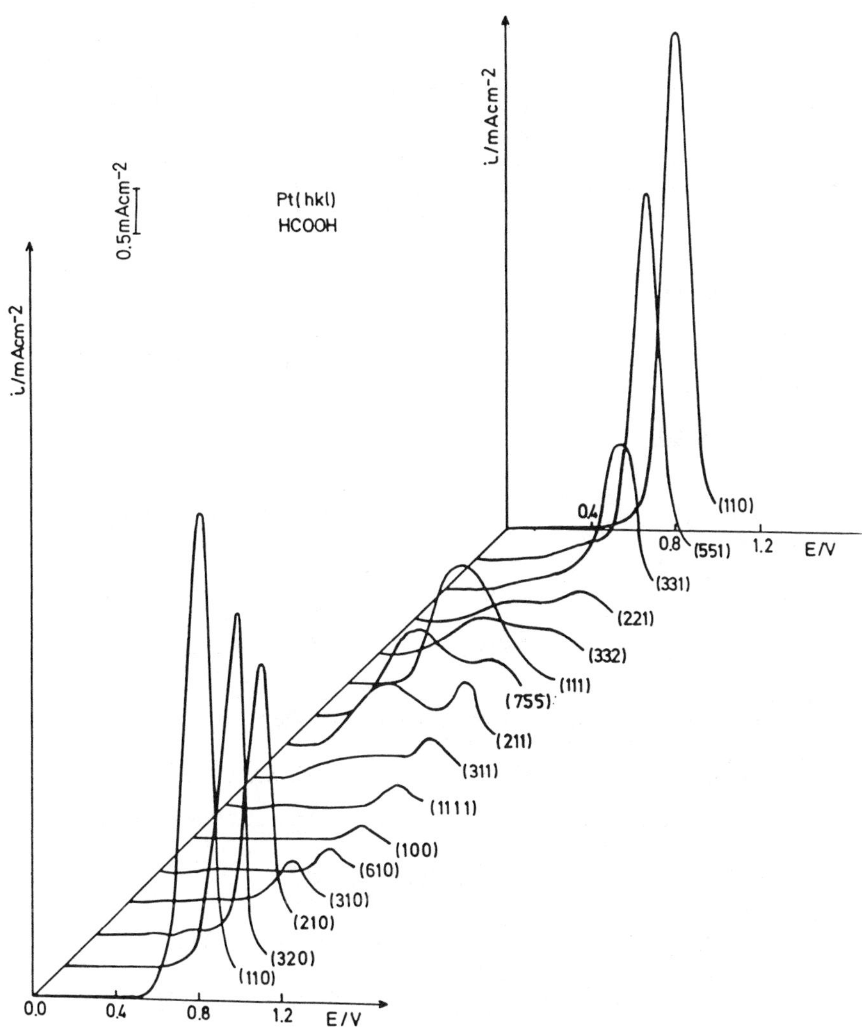

Figure 8. Comparison of the positive-going sweeps for the oxidation of HCOOH on Pt single crystal surfaces in 0.1 M $HClO_4$ + 0.3 M HCOOH. Sweep rate 50 mV/s.

peaks observed with voltammetry are directly due to surface structure, rather than to induced heterogeneity. The same conclusions hold for the oxidation on formic acid. On the basis of the anodic sweep, the most active surface appears to be Pt(111), due to the lack of its poisoning.

Although the structure of some of the surfaces used has been checked by LEED, it appears necessary to prove by that technique that the flame-annealing-hydrogen cooling method gives ordered surfaces. The systematic changes of hydrogen adsorption and of HCOOH oxidation with step density indicate on a high probability that the surface structures were well-ordered and well-oriented. Further work involving a fast transfer from the UHV into the cell seems desirable.

Acknowledgments

The authors are indebted to the Research Fund of S.R. Serbia, Yugoslavia and Department of Energy, Washington, USA contract 553 for financial support.

Literature Cited

1. F.G.Will, J.Electrochem.Soc., 112 (1965) 481.
2. A.T.Hubbard, R.M.Ishikawa and J.Katekaru, J.Electroanal.Chem., 861 (1978) 271.
3. E.Yeager, W.E.O'Grady, M.Y.C.Woo and P.Hagans, J.Electrochem.Soc., 125 (1978) 348.
4. P.N.Ross, J.Electroanal.Chem., 150 (1983) 141.
5. D.Aberdam, R.Durand, R.Faure and F.El-Omar, Surf.Sci., 171 (1986) 303.
6. J.Clavilier, J.Electroanal.Chem., 107 (1980) 211.
7. N.Marković, M.Hanson, G.McDougal and E.Yeager, J.Electroanal.Chem., 214 (1986) 555.
8. R.R.Adžić, W.O'Grady and S.Srinivasan, Surf.Sci., 94 (1980) L191.
9. R.R.Adžić, A.V.Tripković and W.O'Grady, Nature (London) 196 (1982) 137.
10. J.Clavilier and G.Sun, J.Electroanal.Chem., 199 (1986) 479.
11. R.R.Adžić, A.V.Tripković and V.B.Vešović, J.Electroanal.Chem., 204 (1986) 329.
12. N.M.Marković, N.S.Marinković and R.R.Adžić, J.Electroanal.Chem., 241 (1988) 309.
13. D.M.Kolb, Zeitschrift fur Phys.Chem. Neue Folge, 154 (1987) 179.
14. H.Angerstein-Kozlowska, B.Conway, A.Hamelin and L.Stoicoviciu, Electrochem.Acta, 31 (1986) 1051; J.Electroanal.Chem., 228 (1987) 429.
15. S.Štrbac, R.R.Adžić and A.Hamelin, J.Electroanal.Chem., in press.
16. R.R.Adžić and S.Štrbac, J.Serb.Chem.Soc., 52 (1987) 587.
17. P.N.Ross, J.Electrochem.Soc., 126 (1979) 67.
18. C.L.Scortichini and C.N.Reilley, J.Electroanal.Chem., 139 (1982) 247.
19. A.N.Tripković and R.R.Adžić, J.Electroanal.Chem., 205 (1986) 335.

20. B.Love, K.Seto and J.Lipkowski, J.Electroanal.Chem., 199 (1986) 259.
21. J.Clavilier, D.Armand, S.Sun and M.Petit, J.Electroanal.Chem., 205 (1980) 267.
22. S.Motoo and N.Furuya, Ber.Buns.Ges, 101 (1987) 624.
23. B.Lang, R.W.Joyner and G.A.Somorjai, Surf.Sci., 30 (1972) 440.
24. P.Blakely and G.A.Somorjai, Surf.Sci., 65 (1977) 419.
25. J.Clavilier and D.Armand, J.Electroanal.Chem., 199 (1986) 187.
26. See e.g. R.R.Adžić, Advances in Electrochemistry and Electrochemical Engineering, vol. 13, H.Gerisher ed., 159-260, J.Willey, New York, (1985).
27. R.Madix, Advances in Catalysis, vol. 29, (1980).
28. A.Razaq and D.Pletcher, J.Electrochem.Soc., 129 (1984) 322.
29. R.R.Adžić and A.V.Tripković, unpublished.
30. A.Bewick and S.Pons In Advances in Infrared and Raman Spectroscopy, Ed. R.E.Nexter and R.Clarke, Heyden and Sun, London 1984.
31. Kunimatsu and H.Kita, J.Electroanal.Chem., 218 (1987) 155.
32. J.Wilsau and J.Heitbaum, Electrochim.Acta, 31 (1986) 8.
33. O.Wolter, J.Wilsau and J.Heitbaum, J.Electrochem.Soc., 132 (1985) 1635.
34. A.Hamelin, Modern Aspects of Electrochemistry, Ed. B.Conway, J.Bockris and K.White, vol. 16, Chapter 1, Plenum Press, New York, (1985).
35. C.Lamy, J.M.Leger, J.Clavilier and R.Parsons, J.Electroanal.Chem., 150 (1983) 71.

RECEIVED May 17, 1988

Chapter 35

Mechanistic Aspects of the Electrochemical Reduction of Carbon Monoxide and Methanol to Methane at Ruthenium and Copper Electrodes

David P. Summers and Karl W. Frese, Jr.

Materials Research Laboratory, SRI International, 333 Ravenswood Avenue, Menlo Park, CA 94025

Carbon monoxide and methanol are reduced to methane. The reduction of carbon monoxide also suffers deactivation by a surface species similar to that for carbon dioxide reduction but which forms at lower temperatures. The reduction of carbon monoxide does appear to proceed via a path similar to that which the reduction of carbon dioxide follows. Rates for methanol reduction are extremely variable. Methanol reduction, like carbon dioxide reduction, both increases in rate with decreasing pH until the surface becomes blocked with surface hydrogen and is also deactivated by increased temperature. For methanol, deactivation does not occur by the formation of the same surface species. Thus the reduction of methanol is not believed to proceed via a mechanism similar to that for carbon dioxide or carbon monoxide. Copper electrodes also reduce carbon monoxide to methane.

The goals of replacing finite world natural gas reserves and producing fuels from inorganic sources and solar energy has been a motivating force for studying the electrochemical reduction of carbon dioxide to methane, especially in light of increasing carbon dioxide concentrations in the atmosphere (1-3). Electrochemical routes are attractive in that they are low temperature processes and can be coupled to solar energy sources. Carbon monoxide and methanol are both possible intermediates in the reduction of carbon dioxide to methane and both are formed as side products in the reduction of carbon dioxide (2-3 and Kim, J. J.; Summers, D. P.; Frese, K. W., Jr. *J. Electroanal. Chem.* in press). Also, whether they are intermediates in carbon dioxide reduction, the reduction of both carbon monoxide and methanol is likely to proceed by similar pathways as the reduction of carbon dioxide and may have a similar intermediates and rate-determining-steps.

The electrochemical reduction of carbon monoxide also offers a route for the production of fuels from inorganic sources. For example, carbon monoxide is formed from coal in gasification

0097–6156/88/0378–0518$06.00/0

schemes. Carbon monoxide is also linked to carbon dioxide by the water gas shift reaction. The reduction of carbon monoxide to methane in the gas phase has been extensively studied but it is only recently that any reports of the electrochemical reduction of carbon monoxide have appeared (5,6 and Kim, J. J.; Summers, D. P.; Frese, K. W., Jr. J. Electroanal. Chem. in press.) There are also only two reports of the electrochemical reduction of methanol to methane (5,7)

The electrochemical reductions of carbon monoxide and methanol to methane (Equations 1 and 2) have potentials, under standard conditions, of +0.019 and +0.390 V vs SCE respectively (or a

$$CO + 6\,H^+ + 6\,e^- \rightarrow CH_4 + H_2O \qquad (1)$$

$$CH_3OH + 2\,H^+ + 2\,e^- \rightarrow CH_4 + H_2O \qquad (2)$$

reversible potential, under electrolysis conditions, of -0.161 and +0.248 V vs SCE respectively at pH 4, 10^{-6} atm methane, 0.1 M CH_3OH). Similarly, the reduction of carbon dioxide to methane has a potential, under standard conditions, of -0.061 V vs SCE. At Ru carbon dioxide is reduced to methane at low overpotentials but with low rates while at copper methane is formed at high rates but with high overpotentials (3 and Kim, J. J.; Summers, D. P.; Frese, K. W., Jr. J. Electroanal. Chem. in press.). The results presented here show that copper electrodes can reduce carbon monoxide, and ruthenium electrodes can reduce both carbon monoxide and methanol, to methane under conditions similar to those for the reduction of carbon dioxide to methane. The data also indicate that there are similarities in the mechanism of reduction to methane between carbon dioxide and carbon monoxide.

EXPERIMENTAL

The electrolysis of carbon monoxide was conducted under 1 atm carbon monoxide and methanol was electrolyzed under a nitrogen atmosphere. Unless otherwise stated, electrolytes were aqueous solutions of either 0.2 M reagent grade sodium sulfate (Ru electrodes) or 0.5 M Na_2HPO_4 at pH 7.6 (Cu electrodes). All solutions were made with distilled deionized water (Millipore). In the experiments with ruthenium the pH was held constant by the addition of reagent grade sulfuric acid using a pH controller and a syringe pump except for the data in Table II in which the pH varied from 4 to 5.5. All electrolysis experiments using Ru electrodes were conducted at ~60°C for 5-6 hrs with reagent grade Na_2SO_4 unless otherwise noted while all Cu experiments were conducted at room temperature and required no pH control.

Ru electrodes were prepared as previously described by plating Ru metal onto spectroscopic carbon rods, except for the electrode used for Auger analysis (before and after carbon dioxide reduction) which was plated on Ti (2). Cu electrodes were prepared from Cu foil as previously described (Kim, J. J.; Summers, D. P.; Frese, K. W., Jr. J. Electroanal. Chem. in press.). Each entry in the tables and figures was obtained on different days with the electrode kept in ordinary laboratory air overnight between runs.

Electrolyses were performed using an Aardvark model PEC-1 potentiostat and Keithley model 616 digital electrometer, and a microcomputer data acquisition system for measuring current as a function of time. A two compartment cell was employed to avoid oxidation of the carbon dioxide reduction products. All electrolyses were carried out using a closed system as previously described (2). The circulated gas was bubbled through the solution causing gentle agitation. The temperature was controlled by placing the entire system with the exception of the circulating pump in a heated enclosure. In all cases electrolyte volumes were 50 ml. Samples were analyzed on a Gowmac model 750 gas chromatograph with a FID detector. Samples were collected for CH_4 analysis from the gas phase over solution. A column of Porapak Q (6ft) followed by Porapak R (3ft) at 50°C was used for CH_4/CO analysis.

Auger spectra were obtained with a Perkin-Elmer PHI Auger spectrometer. Auger samples were removed under potential control, rinsed with water, and allowed to dry before mounting on sample holder.

Results and Discussion

Carbon Monoxide Reduction at Ruthenium. Carbon monoxide can be reduced to both methane and methanol under conditions nearly identical to those for the reduction of carbon dioxide (Table I, All experiments, using one electrode, are presented in the order they were performed).

The rate and faradaic efficiency of methane formation from carbon monoxide appears lower than from carbon dioxide. For comparison, at 60 °C, -0.545 V vs SCE, pH 4, and 60 °C a typical rate for carbon dioxide reductions is 1.5 x 10^{-7} mol cm^{-2} hr^{-1} with a faradaic efficiency of 20-30 %. Carbon monoxide is 40 times less soluble in water than carbon dioxide (at 1 atm). However the surface coverage will depend on the partial pressure which is the same (1 atm) for both carbon monoxide and carbon dioxide. Additionally, carbon monoxide probably adsorbs more strongly than carbon dioxide and so should have a higher coverage than carbon dioxide at the same partial pressure. Indeed a clear anodic stripping peak can be seen at +0.15 V vs SCE (pH 3) for carbon monoxide adsorbed on the Ru surface at 1 atm (60 °C) while exposure to carbon dioxide produces no peak. Since these rates are too slow to be diffusion controlled, transport limitation cannot account for the differences in rate. It is not known if the lower rate from carbon monoxide is simply due to the blockage of surface hydrogen sites necessary for hydrogenation of intermediates to methane or if there is a fundamental difference in rate. The rate of methane formation from carbon dioxide decreases (along with the total current) when carbon monoxide is added (Table II) consistent with the blocking of the surface by more-slowly-reduced carbon monoxide.

When the temperature is raised to 75 °C a decrease in the rate of carbon monoxide reduction is observed with a parallel decrease in the faradaic efficiency. When the electrode is used a second time for carbon monoxide reduction at 60 °C, after it was used for electrolysis at 75 °C, (last entry in Table I) it shows considerable deactivation. The reduction of carbon dioxide also shows a similar

Table I. Rate and Faradaic Efficiency of CH_4 and CH_3OH formation from CO at Electroplated Ru Electrodes [a]

					Methane		Methanol	
Time (hr)	pH	j[b] (μA cm^{-2})	E (V vs SCE)	T (°C)	Eff[c] (%)	Rate[d] x 10^8 (mol cm^{-2} hr^{-1})	Eff[c] (%)	Rate[d] x 10^8 (mol cm^{-2} hr^{-1})
6.3	3	845	-0.54	60	1.3	6.6	N.M.	N.M.
20.3	3	538	-0.54	60	2.1	7.0	~5	~25
18.3	4	101	-0.54	75	5.8	3.7	~15	~14
6.0	4	141	-0.54	60	N.D.	N.D.	N.M.	N.M.

a) All electrolyses were in 0.2 M reagent grade Na_2SO_4.
b) Average current density based on geometrical area.
c) Faradaic efficiency for methane formation.
d) Average rate of methane formation.

Table II. The Effect of CO on the Electrochemical Reduction of CO_2 to Methane at Electroplated Ru Electrodes [a]

CO added (ml)	j^b ($\mu A\ cm^{-2}$)	$Rate^c \times 10^8$ ($mol\ cm^{-2}\ hr^{-1}$)	Eff^d (%)
0	107	9.3	18.6
35	72	4.9	14.7
0	79	7.5	20.2
50	63	3.6	12.4

a) All electrolyses times are 5-6 hrs in 0.2 M reagent grade Na_2SO_4 at 60°C and -0.545 V vs SCE with an initial pH of 4. CO was added displacing an equal volume of CO_2 from the 1.3 l CO_2 reservoir (see experimental).
b) Average current density based on geometrical area.
c) Rate of methane formation.
d) Faradaic efficiency for methane formation.

deactivation though, in this case, a decrease in the rate is not seen until higher temperatures, ~90 °C (2).

There are similarities between deactivation of Ru by carbon dioxide and by carbon monoxide. The Auger electron spectrum of the surface of the electrode (Ru on Ti) deactivated by carbon dioxide reduction at 90 °C is very different from that taken before electrolysis (Figure 1). The primary signals for Ru and C overlap but by comparing the size of the secondary Ru peaks at slightly lower energy it can be seen that the spectrum taken before deactivation shows mostly a Ru signal while the spectrum taken after deactivation shows mostly a C signal and virtually no Ru signal. If the surface is subjected to Ar^+ sputtering all signals attenuate with respect to those for Ru indicating that we are observing a surface carbon species. This has been interpreted as the formation of either graphitic carbon or deactivating C_nH_m species on the electrode surface (2).

The Auger electron spectrum of the surface of the electrode deactivated by carbon monoxide reduction at 75 °C has a spectrum nearly identical to that of the electrode deactivated during carbon dioxide reduction, showing the presence of enough surface carbon to almost totally block the Ru signal (Figure 2). Ar^+ sputtering restores the spectrum of a clean Ru surface, again indicating surface carbon. This implies that the reduction of carbon monoxide proceeds via dissociation to surface carbon atoms just as the reduction of carbon dioxide does. However, the temperature of deactivation is lower for carbon monoxide than for carbon dioxide indicating that it is easier to split the carbon monoxide to carbon.

Methanol Reduction at Ruthenium. The reduction of methanol to methane does occur as shown by the data in Table III. The data for each electrode are presented in the order that they were collected. Rates can be higher for methanol reduction compared to carbon dioxide reduction though faradaic efficiencies are lower. Unlike carbon dioxide reduction, the rate of methane formation is extremely

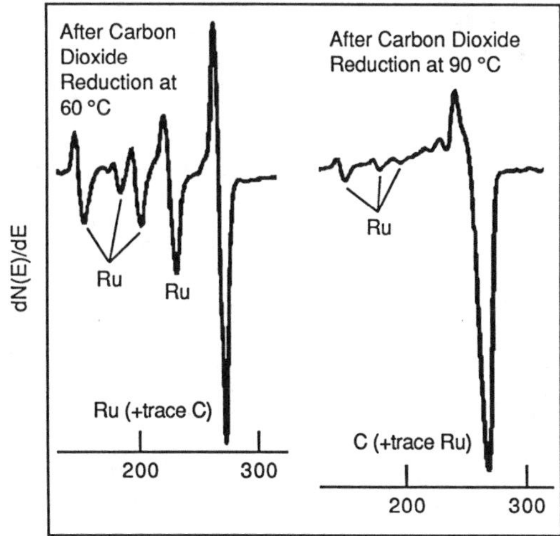

Figure 1. Auger electron spectrum of the surface of a Ru electrode before and after deactivation by reduction of carbon dioxide at higher temperatures (~90 °C in 0.2 M Na_2SO_4 at pH 4 and -0.545 V vs SCE).

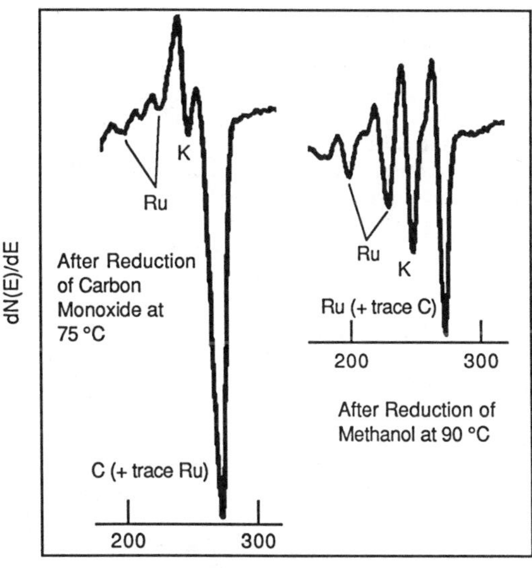

Figure 2. Auger electron spectrum of the surface of two Ru electrodes after deactivation by reduction of carbon monoxide and methanol at higher temperatures (75 and 90 °C respectively in 0.2 M Na_2SO_4 at pH 4 and -0.545 V vs SCE). The presence of K on the surface must result from the adsorption of K^+ ions present as an impurity in the electrolyte.

Table III. Rate and Faradaic Efficiency of CH_4 Formation from CH_3OH at Electroplated Ru Electrodes [a]

experiment	$[CH_3OH]$ (mM)	pH	j^b ($\mu A\ cm^{-2}$)	E (V vs SCE)	T (°C)	Eff^c (%)	$Rate^d \times 10^8$ ($mol\ cm^{-2}\ hr^{-1}$)
electrode #1							
1	75	4	255	-0.54	60	5.0	23.6
2	75	4	220	-0.54	70-75	6.2	25.5
3	75	4	408	-0.54	90	5.0	38.0
4	75	4	147	-0.54	60	0.8	2.2
electrode #2							
1	75	4	255	-0.54	60	1.9	9.0
2	1	4	178	-0.54	60	1.8	5.9
3	500	4	165	-0.54	60	1.6	4.8
4	200	4	121	-0.54	60	1.6	3.6
5	200	5.2	60	-0.61	60	0.5	0.5
6	200	3	312	-0.48	60	1.8	10.3
7	200	1.7	929	-0.41	60	2.5	2.5
electrode #3							
1	1	4	485	-0.54	60	0.3	2.5
2	75	3	907	-0.54	60	0.3	4.6
3	75	4	146	-0.54	22	0.3	0.7
4	75	4	296	-0.70	60	0.8	4.7

a) All electrolyses were in 0.2 M reagent grade Na_2SO_4.
b) Average current density based on geometrical area.
c) Faradaic efficiency for methane formation.
d) Average rate of methane formation.

variable from one electrode to the next. However the activity of each electrode is relatively constant from experiment to experiment. This indicates that the mechanism for methanol reduction is sensitive to some, as yet unknown, surface condition that does not affect the reduction of carbon dioxide which shows much more reproducible rates.

Like the reduction of carbon dioxide to methane, the rate of reduction of methanol to methane increases with temperature (see experiments 1-4, electrode #1 in Table III) (2). Unlike the reduction of carbon dioxide, the faradaic efficiency does not increase indicating that the formation of methane has an activation energy similar to the competing process (H_2 formation). Like carbon monoxide and carbon dioxide reductions, the reaction deactivates when run at excessive temperature. However, in the case of methanol reduction, even at 90 °C the rate is still increasing indicating that methanol reduction is the least prone to deactivation. There is some deactivation since electrolysis at 60 °C, after electrolysis at 90 °C, leads to significantly reduced rates (Table III). An electrode deactivated during methanol reduction does not show the presence of a large amount of surface carbon (Figure 2) indicating a different deactivating species. The lack of surface carbon and the higher temperature of deactivation implies that, contrary to carbon monoxide and carbon dioxide reduction, methanol reduction does not involve dissociation to surface carbon. At 60° C the reduction of methanol also shows a slow deactivation from one experiment to the next (see below) that is not seen for carbon dioxide reductions at the same temperature (2). Species such as COH_{ad} are possible poisons (8).

The first four experiments with electrode #2 (Table III) were run at different methanol concentrations. Dropping the concentration from 75 mM to 1 mM and from 500 mM to 200 mM does lead to a decrease in the rate of reduction, but less than would be expected for such a large change in concentration. Also, there is a steady drop in the rate across all the experiments indicating a slow deactivation of the electrode so that not all of the decrease is due to concentration changes. In increasing the concentration from 1 mM to 500 mM there is even a small decrease in rate. Thus there does not appear to be a strong dependence on methanol concentration implicating the importance of a surface chemical step.

The influence of pH is more prominent than either the effects of deactivation or methanol concentration. The data (experiments 4-7 on electrode #2, Table III) show that there is an optimum pH for methane formation from methanol with the rate increasing with greater acidity until a maximum at pH ~3 is reached whereupon the rate begins to decrease (Figure 3). This is nearly identical to results seen (Figure 4) for carbon dioxide reduction to methane (2). This is interpreted as indicating that the reaction proceeds at a faster rate as the hydrogen coverage increases due to faster hydrogenation of surface intermediates, but that excessive surface hydrogen coverage blocks carbonaceous intermediates (hence the decrease at lower pH). Again the importance of a surface chemical step is implied.

Carbon Monoxide Reduction at Copper. At copper electrodes carbon monoxide is thought to be an intermediate in the reduction of carbon dioxide and is formed as the major product with nitric acid pretreated electrodes (9 and Kim, J. J.; Summers, D. P.; Frese, K. W., Jr. J. Electroanal. Chem. in press.). As the data in Table IV indicates, methane can be formed by carbon monoxide reduction at

Table IV. Rate and Faradaic Efficiency of CH_4 Formation from CO at Cu Electrodes

Electrolyte (0.2 M)	pH	j[a] (mA cm^{-2})	E (V vs SCE)	Eff[b] (%)	Rate[c] x 10^6 (mol cm^{-2} hr^{-1})
Na_2HPO_4	6.1	5	-1.53	3.64	1.13
Na_2HPO_4	6.1	14	-1.72	0.54	0.46
Na_2HPO_4	7.3	5	-1.8	0.61	0.19
Na_2HPO_4	7.4	10	-1.83	0.91	0.56
Na_2HPO_4	7.3	10	-2.0	2.1	1.33
$KHCO_3$	9.5	26	-1.98	1.5	2.43

a. Average current density based on geometrical area.
b. Faradaic efficiency for methane formation.
c. Average rate of methane formation.

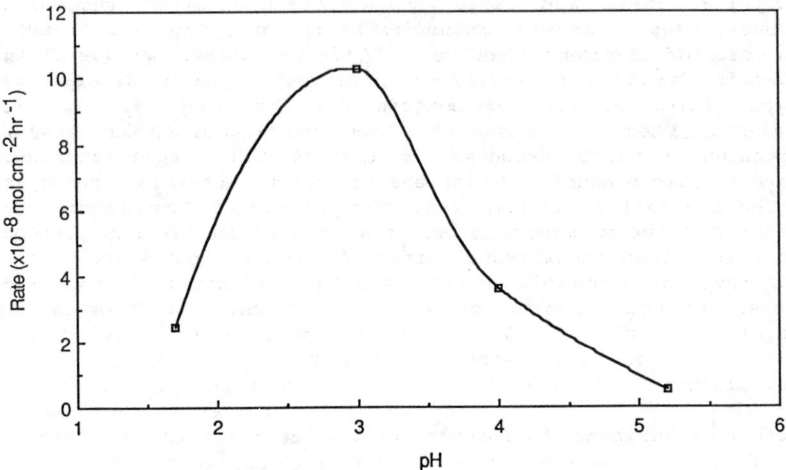

Figure 3. The effect of pH on the average rate of methane formation from methanol. In 0.2 M Na_2SO_4 at 60 °C and at constant over potential (see table 3).

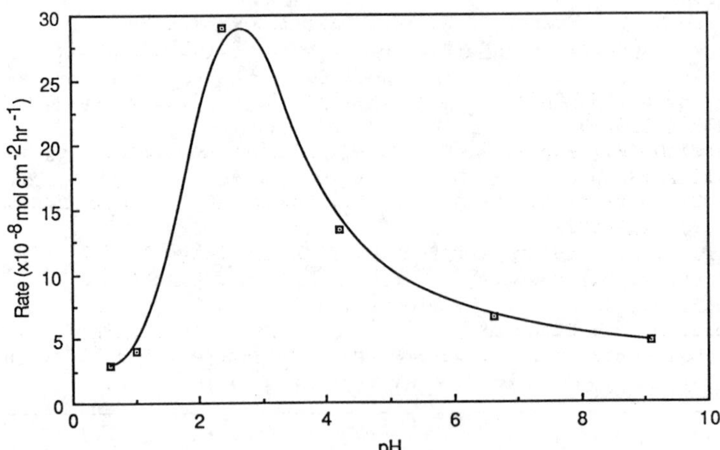

Figure 4. The effect of pH on the average rate of methane formation from carbon dioxide. In 0.2 M Na_2SO_4 at 60 °C and at constant over potential (-0.545 V vs SCE at pH 4).

copper electrodes under similar conditions to those at which methane is formed from carbon dioxide. Just as copper gives higher current densities but poorer overvoltages for carbon dioxide reduction than ruthenium, the reduction of carbon monoxide at copper electrodes shows higher rates and larger overvoltages than at ruthenium electrodes. Again, as with ruthenium electrodes, the reduction of carbon monoxide is slower than the reduction of carbon dioxide along with lower faradaic efficiencies (the major other product is hydrogen) (Kim, J. J.; Summers, D. P.; Frese, K. W., Jr. J. Electroanal. Chem. in press.). Total currents are also lower, consistent with the blocking of the surface with carbon monoxide. However, in this potential region the reduction of carbon dioxide is diffusion controlled and the ratio of the rates of carbon dioxide reduction and carbon monoxide reduction (30-60 in favor of carbon dioxide) is within experimental error of the ratio of solubilities (~40 in favor of carbon dioxide). Thus lower solubility can not be dismissed as the source of lower rates for carbon monoxide reduction.

Acknowledgments

The Authors acknowledge the support of the Gas Research Institute.

Literature Cited

1) Catalytic Activation of Carbon Dioxide; W. M. Ayers, ed.; ACS Symposium Series No. 363; American Chemical Society: Washington, DC, 1988
2) Summers, D. P.; Frese, K. W., Jr. Langmiur 1988,4,51.
3) Hori, Y.; Kikuchi, K.; Murata, A.; Suzuki, S. Chem. Lett. 1986,897.
4) Cooke, R. L.; MacDuff, R. C.; Sammells, A. F. J. Electrochem. Soc. 1987,134,1873.
5) Summers, D. P.; Frese, K. W., Jr. J. Electrochem. Soc. 1988,135,264.
6) Hori, Y.; Murato, A.; Takahashi, R.; Suzuki, S. J. Amer. Chem. Soc. 1987,109,5022.
7) Stenin, V. F.; Podlovchenko, B. I. Elektrokhimiya 1967,3,481.
8) Yasil'ev, Y. B.; Bagotsky, V. S. J. Appl. Electrochem. 1986,16,703.
9) Summers, D. P.; Frese, K. W., Jr. SRI Annual Report, April 30, 1987, GRI Contract No. 5083-260-0922, SRI Project PYU 7142.

RECEIVED May 17, 1988

Chapter 36

Surface Organometallic and Coordination Chemistry of Iridium, Platinum, and Gold Electrocatalysts

Manuel P. Soriaga, G. M. Berry, M. E. Bothwell, B. G. Bravo, G. J. Cali, J. E. Harris, T. Mebrahtu, S. L. Michelhaugh, and J. F. Rodriguez

Department of Chemistry, Texas A&M University, College Station, TX 77843

The interaction of selected organic and inorganic ligands with Ir, Pt and Au electrocatalysts has been studied in parallel to help establish the interfacial organometallic or coordination chemistry of these electrocatalysts in aqueous solutions. As a first study, iodide, hydroquinone, 2,5-dihydroxythiophenol, and 3,6-dihydroxypyridazine were employed as model surface ligands because they are reversibly electroactive and exhibit variable affinities towards the subject electrocatalysts. Under the present conditions, chemisorption has been found to be directly analogous to oxidative addition, and desorption similar to reductive elimination. Consequently, surface coordination of an electroactive center overwhelmingly favors its oxidized state over its reduced form. If the redox center is only indirectly attached to the surface via an electrochemically inert anchor, *substrate-mediated* interactions between the electroactive moieties may arise if the redox center itself interacts strongly with the electrode surface. It seems plausible to view the phenomenon of substrate-mediated adsorbate-adsorbate interactions in terms of mixed-valence metal complexes.

The modeling of intermediates formed from the interaction of organic compounds with transition-metal surfaces based on analogies to homogeneous organometallic complexes has been attempted in the past (1). Although the extension of traditional concepts of molecular coordination chemistry to surface compounds is a sound and logical proposition, early studies were only marginally successful because detailed structural and reactivity information on the

0097–6156/88/0378–0528$06.00/0

surface complexes had been unavailable until recently (2). Experiments based upon modern surface analytical methods have now established that the interaction of inorganic and organic compounds with transition-metal surfaces has several commonalities with the bonding of such ligands to metal centers in molecular and metal-cluster complexes (3). Theoretical treatments currently exist which support the empirical assertions (4), and studies of organometallic chemistry at the gas-metal interface have proliferated (5-7). Unfortunately, similar investigations focused on the electrode-solution interface are meager. Results derived from gas-solid studies cannot be adopted directly to electrocatalytic phenomena in view of experimental parameters which are unique only to the electrochemical environment. Differences in structure and reactivity between surface intermediates formed in vacuum and under electrochemical conditions have been documented (8-9). Surface complexes formed at the electrode-solution interface are more closely related to homogeneous complexes most probably because the electrode surface is initially covered by weakly coordinating electrolyte and solvent molecules (9). Reactions between surface ligands and the electrode surface thus involve displacement reactions very similar to those in molecular coordination chemistry.

Table 1. Surface-Cluster Analogies

Electrode Surfaces	Cluster Complexes
Chemisorption	Synthesis
Adsorbate Orientation	Mode of Coordination
Competitive Chemisorption	Ligand Substitution
Adsorbate Exchange	Ligand Exchange
Adsorbate Reactivity	Ligand Reactivity
Electrocatalysis	Homogeneous Catalysis

Based upon analogies between surface and molecular coordination chemistry outlined in Table 1, we have recently set forth to investigate the interaction of surface-active *and* reversibly electroactive moieties with the noble-metal electrocatalysts Ru, Rh, Pd, Ir, Pt and Au. Our interest in this class of compounds is based on the fact that chemisorption-induced changes in their redox properties yield important information concerning the coordination/organometallic chemistry of the electrode surface. For example, alteration of the reversible redox potential brought about by the chemisorption process is a measure of the surface-complex formation constant of the oxidized state relative to the reduced form; such behavior is expected to be dependent upon the electrode material. In this paper, we describe results obtained when iodide, hydroquinone (HQ), 2,5-dihydroxythiophenol (DHT), and 3,6-dihydroxypyridazine (DHPz), all reversibly electroactive and surface-active, are made to react under identical conditions with smooth polycrystalline Ir, Pt and Au electrodes in aqueous solutions. These results will be viewed in terms of traditional concepts of organometallic and coordination chemistry.

Experimental

Smooth polycrystalline Au, Pt and Ir thin-layer electrodes were utilized (10-11). Electrodes were cleaned between trials by sequential electrochemical oxidation above 1.2 V [Ag/AgCl (1 *M* Cl^-) reference] and reduction below -0.2 V in 1 *M* H_2SO_4; surface cleanliness was verified with the aid of cyclic voltammetry in the same molar sulfuric acid solution. Experiments were carried out in 1 *M* H_2SO_4, 1 *M* $NaClO_4$ buffered at pH 7 and 10, and in 1 *M* NaOH; solutions were prepared with pyrolytically triply distilled water (12). Surface reagents employed were iodide, hydroquinone (HQ), 2,5-dihydroxythiophenol [DHT (13)], and 3,6-dihydroxypyridazine (DHPz).

Coordination of the iodo ligand onto the Ir, Pt and Au surfaces was accomplished by exposure of the clean electrode to 1 m*M* NaI for 180 seconds at the same pH at which subsequent experiments were to be performed. Unadsorbed iodide was rinsed away with supporting electrolyte. We determined the absolute surface coverage of iodine Γ_I (mole cm^{-2}) by means of thin-layer coulometry in 1 *M* H_2SO_4 using two reactions attributable to the surface iodine. (i) Γ_I (mole cm^{-2}) is obtainable from the charge for oxidation of adsorbed iodine to aqueous IO_3^-

$$\Gamma_I = (Q - Q_b)_{ox,I}/5FA \quad (1)$$

where Q is the total charge for oxidation of both iodine and metal surface, Q_b is the charge for oxidation of the metal surface, F is the Faraday, and A is the active surface area (11,14). The use of an *n*-value of 5 in Equation 1 implies that iodine is zerovalent in the chemisorbed state. (ii) Γ_I is also proportional to the charge for reduction of aqueous IO_3^- to aqueous I_2

$$\Gamma_I = (Q - Q_b)_{red,IO_3^-}/5FA \quad (2)$$

where $(Q-Q_b)$ is the corrected cathodic charge. Since the *n*-value of 5 in Equation 2 is independent of the valency of surface iodine, the quantity $|(Q-Q_b)_{red,IO_3^-}|$ must equal $|(Q-Q_b)_{ox,I}|$ if the surface-coordinated iodine is zerovalent.

For the diphenolic compounds, absolute surface-packing-density measurements were based upon differential thin-layer coulometry (9)

$$\Gamma = [(Q - Q_b) - (Q_1 - Q_{1b})]/nFA \quad (3)$$

where (Q_1-Q_{1b}) and $(Q-Q_b)$, respectively, denote the background-corrected faradaic charges for the diphenol-to-quinone redox process for the unadsorbed species after one filling (during which saturation chemisorption of the diphenol occurs) and multiple fillings (during which no additional chemisorption takes place) of the thin-layer cell. The *n*-value for the redox reaction of solution species was determined from the relation

$$n = (Q - Q_b)/FVC^o \qquad (4)$$

where V is the volume of the thin-layer cell and C^o is the bulk concentration of diphenol.

Results and Discussion

Surface Coordination of the Iodo Ligand. The chemisorption of iodine at Au, Pt and Ir surfaces has been demonstrated (15-9). Previous studies with single- and polycrystalline Pt (15-7) showed that aqueous iodide undergoes spontaneous oxidation upon chemisorption to form a monolayer of zerovalent iodine:

$$I^-_{(aq)} + H^+_{(aq)} \rightarrow I_{(ads)} + 1/2H_{2(g)} \qquad (5)$$

This oxidative chemisorption reaction also occurs on Ir and Au since, in addition to evidence described below, it was found that $|(Q-Q_b)_{red,IO_3^-}|$ is equal to $|(Q-Q_b)_{ox,I}|$.

Although rinsing of the pretreated surface with typical solvents at open circuit does not damage the iodine layer, application of sufficiently negative or positive potentials removes the chemisorbed species. Data which illustrate the cathodic stripping of chemisorbed iodine are shown in Figure 1 where cathodic current and Γ_I at Pt are plotted simultaneously as functions of the applied potential. The decrease in Γ_I is clearly correlated with the appearance of the reduction peaks; the multiple reduction peaks signify a coverage-dependent process. Results for Au and Ir are similar to that shown in Figure 1 except that the cathodic stripping process is pH-dependent at Ir and Pt but not at Au. This is demonstrated in Figure 2 which gives plots of $E_{1/2}$, the potential at which $\Gamma_I = 0.5\Gamma_{I,max}$, as a function of pH. In the case of Au, $E_{1/2}$ is -0.50 V, a value which may be identified with the redox potential $E^o_{I(ads)}$ for the surface-coordinated $I_{(ads)}/I^-_{(ads)}$ couple. The pH-independence of $E^o_{I(ads)}$, coupled with results from coulometric measurements, indicates that the reductive elimination of iodine from Au is a simple one-electron process:

$$I_{(ads)} + e^- \leftrightarrow I^-_{(ads)} \leftrightarrow I^-_{(aq)} \qquad (6)$$

In the cases of Pt and Ir, coulometric measurements of the

cathodic peaks yield an *n*-value of 2 for cathodic desorption of iodine. This outcome, coupled with the facts that (i) $E_{1/2}$ at both surfaces is pH-dependent and that (ii) Ir and Pt, in contrast to Au, have strong affinities towards surface hydrogen at sufficiently negative potentials, suggests the following reductive elimination reaction:

$$I_{(ads)} + H^+_{(aq)} + 2e^- \leftrightarrow H_{(ads)} + I^-_{(aq)} \quad (7)$$

It is possible to obtain an estimate of $E^o_{I(ads)}$ at Pt and Ir from the plot of $E_{1/2}$-vs.-pH (20). If Equation 7 is correct, the $E_{1/2}$-vs.-pH plot should yield a straight line with a slope of -0.0296 and an intercept K proportional to $E^o_{I(ads)}$:

$$K = E^o_{I(ads)} + E^o_{H(ads)} + 0.0296\ pI \quad (8)$$

where $pI = -\log C_{I^-(aq)}$, with $C_{I^-(aq)}$ being the solution concentration of the desorbed iodide. At $E_{1/2}$, $C_{I^-(aq)}$ equals $0.5\Gamma_{I,max}(A/V)$. $E^o_{H(ads)}$ is the potential for underpotential deposition of hydrogen, a value not unique for a polycrystalline surface. If $E^o_{H(ads)}$ is simply taken as the potential at which the fractional coverage of chemisorbed hydrogen is one-half, then it is approximately equal to -0.1 V for both Pt and Ir in molar sulfuric acid. Substitution of these and the K values from Figure 2 into Equation 8 yields $E^o_{I(ads)} \approx -0.36$ V for Pt and ≈ -0.32 V for Ir. The slopes obtained for Pt and Ir were -0.028 and -0.029, respectively, in accordance with Equation 7.

The $E^o_{I(ads)}$ values obtained here indicate that, upon surface coordination, the redox potential of the iodine/iodide couple is shifted in the *negative* direction by about 0.90 V on Au, 0.76 V on Pt, and 0.72 V on Ir. These chemisorption-induced redox potential shifts can be employed to estimate the *ratio* of the formation constants for surface coordination of iodine and iodide:

$$RT \ln [K_{f,I}/K_{f,I^-}] = -F[E^o_{(ads)} - E^o_{(aq)}] - \Delta G^o_d \quad (9)$$

where $K_{f,I}$ and K_{f,I^-} are the respective formation constants for surface-coordinated iodine and iodide, $E^o_{(sol)}$ and $E^o_{(ads)}$ are the respective redox potentials for the $I_{2(aq)}/I^-_{(aq)}$ and $I_{(ads)}/I^-_{(ads)}$ couples, and ΔG^o_d is the energy involved in the $I_{2(aq)} \rightarrow 2I_{(aq)}$ dissociation (21). Substitution of experimental values into Equation 9 yields a formation-constant ratio of 2×10^{28} for Au, 1×10^{26} for Pt, and 2×10^{25} for Ir. These exceedingly large values signify overwhelming preference by the subject metals for surface coordination of zerovalent iodine over iodide. Comparison of the $K_{f,I}/K_{f,I^-}$ values indicates that this preferential surface coordination decreases from Au to Pt

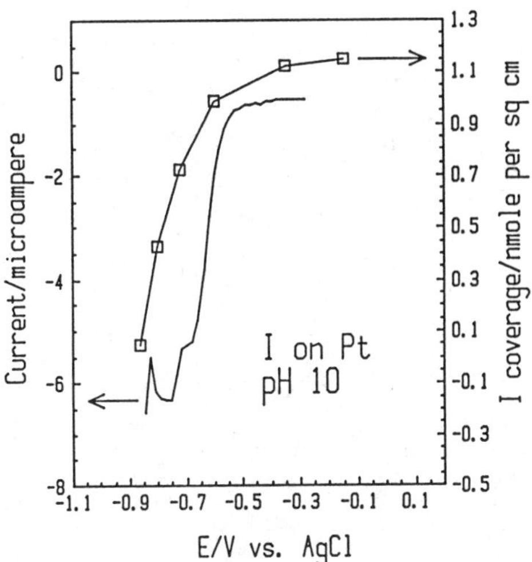

Figure 1. Cathodic current and Γ_I as functions of potential for an I-coated Pt in 1 *M* $NaClO_4$ buffered at pH 10. Area of electrode, A = 1.04 cm^2; sweep rate, r = 2 mV/s; temperature, T = 298 K.

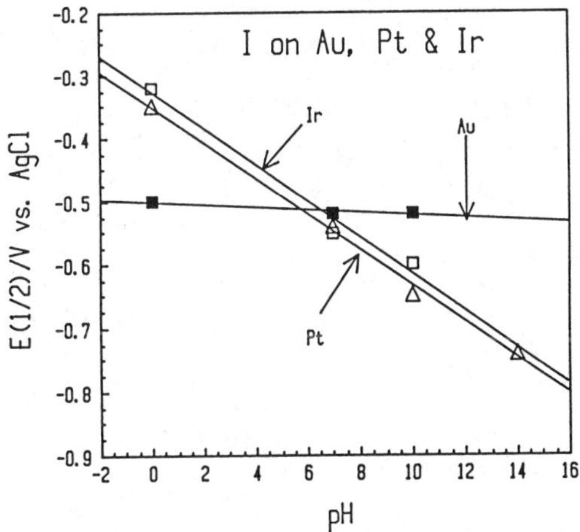

Figure 2. Plot of $E_{1/2}$, the potential at which Γ_I is at half-maximum as a function of pH. A_{Au} = 1.07 cm^2; A_{Ir} = 1.40 cm^2. Other conditions were as in Figure 1. The solid lines represent the linear least squares fit.

to Ir, a trend which may be qualitatively rationalized in terms of the acceptor properties of the zerovalent metal.

Surface Organometallic Chemistry of Hydroquinone. Previous studies (9) have established that the interaction of aqueous HQ with smooth polycrystalline Pt is concentration-dependent. Below 0.1 m*M*, HQ undergoes spontaneous oxidation to form a chemisorbed monolayer of flat-oriented (η^6) benzoquinone (BQ) similar to that in the homogeneous coordination of benzoquinone to tris(triethylphosphine)Pt(0) (9):

$$C_6H_4(OH)_{2(aq)} \rightarrow \eta^6\text{-}C_6H_4O_{2(ads)} + H_{2(g)} \qquad (10)$$

Above 1 m*M*, C-H activation occurs to form a chemisorbed layer of edge-bonded (2,3-η^2) hydroquinone analogous to *o*-benzyne organometallic compounds formed with Pt and Os clusters (9):

$$C_6H_4(OH)_{2(aq)} \rightarrow 2,3\text{-}\eta^2\text{-}C_6H_2(OH)_{2(ads)} + H_{2(g)} \qquad (11)$$

Equations 10 and 11 indicate that the redox potential of the HQ/BQ couple is shifted in the negative direction when η^6-chemisorbed but shifted in the positive direction when 2,3-η^2-bonded. This orientation-dependent shift in redox potential is not unexpected by analogy with molecular organometallic compounds. For example, the redox potential for the reversible, one-electron reduction of duroquinone in acetonitrile is shifted from -0.90 V (vs. SCE) to -0.69 V in bis(duroquinone)Ni(0) and to -1.45 V in (1,5-cyclooctadiene)(duroquinone)Ni(0) (22).

Figures 3 and 4, respectively, show thin-layer current-potential curves for polycrystalline Au and Ir in molar sulfuric acid before and after exposure to a 2 m*M* HQ solution. For smooth Au, no changes in the voltammetric curves are seen. In comparison, a prominent anodic oxidation peak is observed for Ir after pretreatment with HQ. These observations, which indicate that Ir is reactive towards HQ but Au is not, are consistent with what is known from the literature on homogeneous organometallic chemistry (21): Ir and Pt complexes are reactive towards a variety of organic compounds, whereas Au is inert.

Surface Coordination of 3,6-Dihydroxypyridazine. Room-temperature Γ_{DHPz}-vs.-log C curves for DHPz at Au and Pt electrodes in 1 *M* H_2SO_4 and in 1 *M* $NaClO_4$ buffered at pH 7 are shown in Figure 5. It is interesting to note that Γ_{DHPz} values independent of concentration and pH are attained at Pt, whereas pH-dependent isotherms are obtained at Au. At pH 7, the Γ_{DHPz} values at Pt and Au are identical [0.66(6) nmole cm^{-2}]. In 1 *M* H_2SO_4, however, Γ_{DHPz} on Au is profoundly concentration-dependent.

Earlier exploratory studies (9) of the interaction of DHPz with smooth polycrystalline Pt indicated that, at

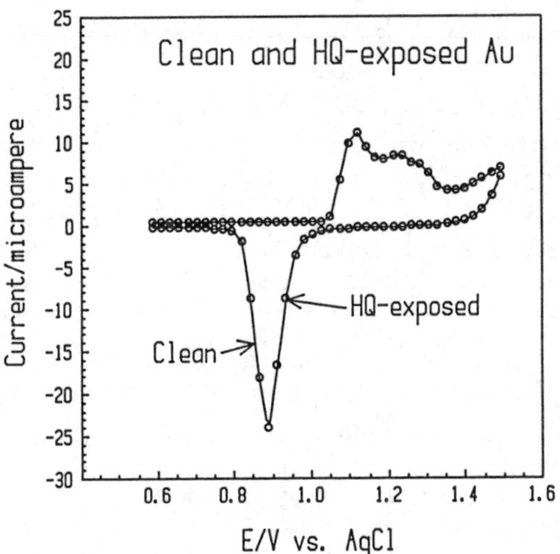

Figure 3. Thin-layer current-potential curves in 1 *M* H_2SO_4 for Au before and after exposure to 2 m*M* HQ. Experimental conditions were as in Figures 1 and 2.

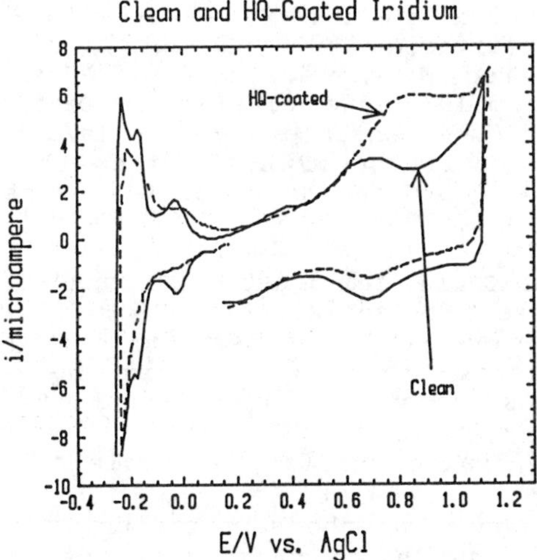

Figure 4. Thin-layer current-potential curves in 1 *M* H_2SO_4 for Ir before and after exposure to 2 m*M* HQ. Experimental conditions were as in Figures 1-3.

room temperature, chemisorbed DHPz is bound to only one nitrogen heteroatom even if the two nitrogens in DHPz are chemically equivalent. This is in agreement with the fact that, in complexes with zerovalent Pt, pyridazine is η^1-N coordinated at ambient temperatures (9).

The pH-independence of Γ_{DHPz} at Pt indicates that the driving force for coordination of the nitrogen heteroatom to the Pt surface is much larger than that for protonation even in molar acid. This behavior is in contrast to that of pyridine, where protonation of the nitrogen heteroatom in molar acid hinders N-coordination to the surface (9). Such a difference in chemisorption characteristics may be related to the fact that the basicity of the nitrogen heteroatom in pyridine (pK_b = 8.8) is much greater than that of the nitrogens in pyridazine (pK_b = 11.8) (23).

The fact that Γ_{DHPz} on Au is identical to that on Pt at pH 7 is evidence that η^1-N surface-coordinated DHPz is also formed on Au at this same pH. Since (i) the DHPz isotherm on Au at pH 0 is not stepwise unlike those exhibited by compounds attached in multiple orientational states, and (ii) it has already been shown above that hydroquinone, the homoaromatic analogue of DHPz, is not chemisorbed on Au, it can be argued that η^1-N surface-coordination of DHPz occurs on Au at pH 0 even at coverages below the saturation value. It can be inferred further that the driving forces for protonation and Au-surface-coordination of the N heteroatom are equally competitive in molar acid.

The unfettered N-coordination of DHPz to the Pt surface even in molar acid clearly indicates that the surface-coordination strength of DHPz is larger on Pt than on Au. This is consistent with the fact that, although analogous DHPz-Pt coordination complexes exist, none has been reported for Au.

Surface Coordination of 2,5-dihydroxythiophenol. Figure 6 shows cyclic current-potential curves for DHT chemisorbed at maximum coverage at Pt and Au in 1 *M* H_2SO_4. These voltammetric curves were obtained in the absence of unadsorbed species; hence, the peaks are due only to redox of the surface-coordinated DHT. The areas under these curves yield Γ_{DHT} values of 0.56(4) nmole cm^{-2} at Pt and 0.54(4) nmole cm^{-2} at Au. This, along with the fact that the redox peaks appear at the same potential region where unadsorbed DHT reacts, signifies a vertical S-η^1-orientation in which the diphenolic group is pendant (Γ_{calc} = 0.57 nmole cm^{-2}). The half-width of the redox peak is 0.21 V at Pt and 0.13 V at Au. Since the modes of attachment and packing densities are identical at both surfaces, the peak broadening at Pt can only be attributed to *substrate-mediated* adsorbate-adsorbate interactions.

Figure 7 shows current-potential curves for Pt and Au precoated with DHT at half coverage. At Pt, no quinone/diphenol redox reaction is observed. Clearly,

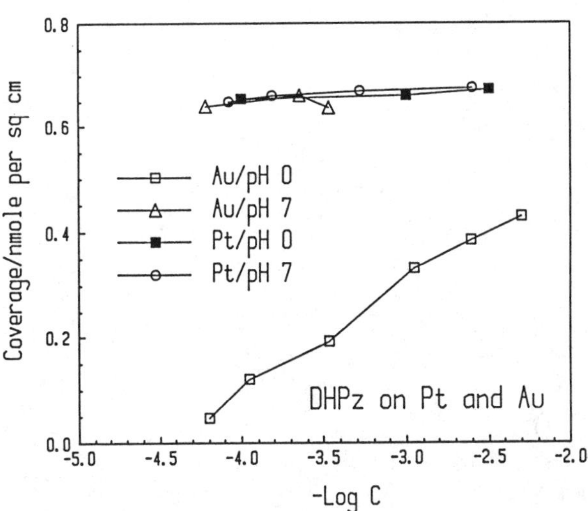

Figure 5. Chemisorption isotherms for DHPz at Au and Pt at pH 0 and pH 7. Experimental conditions were as in Figures 1-3.

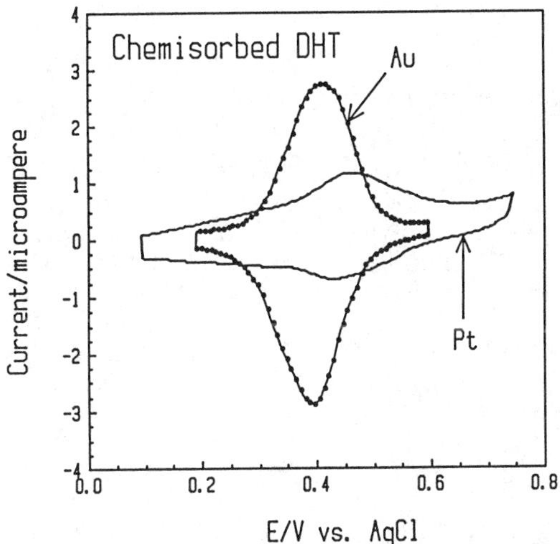

Figure 6. Thin-layer voltammetric curves in *M* H_2SO_4 for the quinone/diphenol redox of DHT chemisorbed at full coverage on Pt and Au. Sweep rates: r_{Au} = 3 mV/s; r_{Pt} = 2 mV/s. Other conditions were as in Figures 1-3.

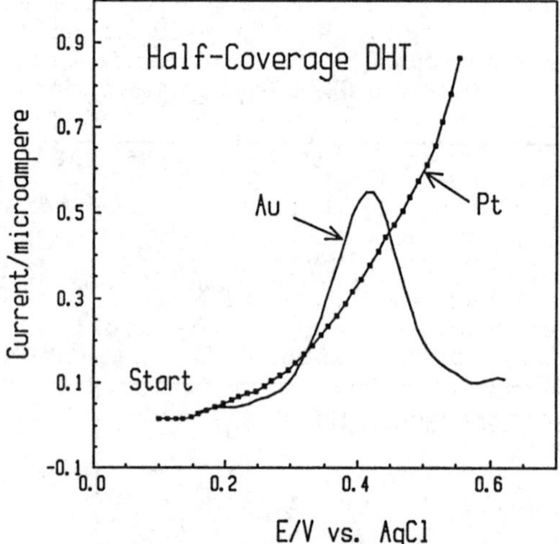

Figure 7. Thin-layer voltammetric curves in 1 *M* H_2SO_4 for the quinone/diphenol redox of DHT chemisorbed at half coverage on Pt and Au. Experimental conditions were as in Figures 1-3.

direct diphenol-Pt interaction has taken place at submonolayer coverages. It may be noted that the peak width at Au is unchanged in going from full to half coverage of DHT. These results indicate that, at submonolayer coverages, DHT remains exclusively S-η^1 on Au but behaves as a surface chelate on Pt in which both sulfur and diphenol groups are bonded directly to the surface. The quinone/diphenol redox activity at Pt starts to disappear at Γ/Γ_{max} ~ 0.8. The tilt angle of the surface-chelated DHT probably varies with coverage.

These results suggest that the critical factor in the substrate-mediated intermolecular interactions which occur within the close-packed DHT layer is the inherent strong reactivity of the diphenolic moiety with the Pt surface. The interaction of adsorbates with each other through the mediation of the substrate is of fundamental importance in surface science. The theoretical treatment, however, involves complicated many-body potentials which are presently not well-understood (2). It is instructive to view the present case of Pt-substrate-mediated DHT-DHT interactions in terms of mixed-valence metal complexes (24). For example, in the binuclear mixed-valence complex, $(NH_3)_5$Ru(II)-bpy-Ru(III)$(NH_3)_5$ (where bpy is 4,4'-bipyridine), the two metal centers are still able to interact with each other via the delocalized electrons within the bpy ligand. The interaction between the Ru(II) and Ru(III) ions in this mixed-valence complex is therefore ligand-mediated. The Ru(II)-Ru(III) coupling can be written schematically as:

$$\text{Ru(II)-bpy-Ru(III)} \leftrightarrow \text{Ru(III)-bpy-Ru(II)} \qquad (12)$$

By analogy with Equation 12, the DHT-DHT interactions mediated by the Pt surface may be represented by the following reaction:

$$\text{Q-S-Pt-Pt-S-H}_2\text{Q} \leftrightarrow \text{H}_2\text{Q-S-Pt-Pt-S-Q} \qquad (13)$$

where Pt-Pt denotes the Pt surface. Only one diphenol (H_2Q) and quinone (Q) substrate-mediated pair is depicted for simplicity. One can now visualize how the coupling between the H_2Q and Q centers can be mediated by the Pt surface.

Acknowledgments

Acknowledgment is made to the Robert A. Welch Foundation and to the Regents of Texas A&M University for support of this research.

MPS wishes to dedicate this article in memory of Professor Rebecca Soriaga-Paredes, late Chairman of the Department of Chemistry, de la Salle University (Philippines).

Literature Cited

1. Ugo, R. Catalysis Revs.1975, 11, 225.
2. Somorjai, G.A. Chemistry in Two Dimensions: Surfaces; Cornell University Press: Ithaca, NY, 1981.
3. Muetterties, E.L. Bull. Soc. Chim. Belg. 1975, 84, 959.
4. Saillard, J.Y.; Hoffman, R. J. Am. Chem. Soc. 1984, 106, 2006.
5. Albert. M.R.; Yates, J.T., Jr. A Surface Scientist's Guide to Organometallic Chemistry; American Chemical Society: Washington, D.C., 1987.
6. Friend, C.M.; Muetterties, E.L. J. Am. Chem. Soc. 1981, 103, 767.
7. Canning, N.D.S.; Madix. R.J. J. Phys. Chem. 1984, 88, 2437.
8. Wieckowski, A.; Rosasco, S.D.; Salaita, G.; Hubbard, A.T.; Bent, B.; Zaera, F.; Somorjai. G.A. J. Am. Chem. Soc. 1985, 107, 21.
9. Soriaga, M.P.; Binamira-Soriaga, E.; Hubbard, A.T.; Benziger, J.B.; Pang, K.W.P. Inorg. Chem. 1985, 24, 65.
10. Hubbard, A.T. Crit. Rev. Anal. Chem. 1973, 3, 201 .
11. White, J.H.; Soriaga, M.P.; Hubbard, A.T. J. Electroanal. Chem. 1984, 177, 89.
12. Conway, B.E.; Angerstein-Kozlowska, H.; Sharp, W.B.A.; Criddle, E.E. Anal. Chem. 1973, 45, 1331 .
13. Alcalay, W. Helv. Chim. Acta. 1947, 30, 578.
14. Rodriguez, J.F.; Mebrahtu, T.; Soriaga, M.P. J. Electroanal. Chem. 1987, 233, 283.
15. Lane, R.F.; Hubbard, A.T. J. Phys. Chem. 1975, 79, 808.
16. Stickney, J.L.; Rosasco, S.D.; Salaita, G.; Hubbard, A.T. Langmuir. 1985, 1, 89.
17. Felter, T.E.; Hubbard, A.T. J. Electroanal. Chem. 1979, 100, 473.
18. Rodriguez, J.F.; Soriaga, M.P. J. Electrochem. Soc. 1988, 135, 616.
19. Rodriguez, J.F.; Bothwell, M.E.; Harris, J.E.; Soriaga, M.P. J. Phys. Chem. 1988, 92, 2702.
20. Soriaga, M.P. J. Electroanal. Chem. 1988, 240, 309.
21. Cotton, F.A.; Wilkinson, G. Advanced Inorganic Chemistry; Wiley: New York, 1980.
22. Nesmeyanov, A.N. Dokl. Akad. Nauk. SSSR. 1976, 230, 1114.
23. Weast, R.C. Handbook of Chemistry; CRC Press: Boca Raton, FL, 1986.
24. Creutz, C. Prog. Inorg. Chem. 1983, 30, 1.

RECEIVED June 27, 1988

Author Index

Affiliation Index

Subject Index

L

M

N

Production by Rebecca A. Hunsicker
Indexing by Deborah H. Steiner
Jacket design by Elizabeth Binamira-Soriaga and Alan Kahan

Elements typeset by Hot Type Ltd., Washington, DC
Printed and bound by Maple Press, York, PA

Recent ACS Books

Biotechnology and Materials Science: Chemistry for the Future
Edited by Mary L. Good
160 pp; clothbound; ISBN 0–8412–1472–7

Chemical Demonstrations: A Sourcebook for Teachers
By Lee R. Summerlin and James L. Ealy, Jr.
Volume 1, Second Edition, 192 pp; spiral bound; ISBN 0–8412–1481–6

Chemical Demonstrations: A Sourcebook for Teachers
By Lee R. Summerlin, Christie L. Borgford, and Julie B. Ealy
Volume 2, Second Edition, 229 pp; spiral bound; ISBN 0–8412–1535–9

Practical Statistics for the Physical Sciences
By Larry L. Havlicek
ACS Professional Reference Book; 198 pp; clothbound; ISBN 0–8412–1453–0

The Basics of Technical Communicating
By B. Edward Cain
ACS Professional Reference Book; 198 pp; clothbound; ISBN 0–8412–1451–4

The ACS Style Guide: A Manual for Authors and Editors
Edited by Janet S. Dodd
264 pp; clothbound; ISBN 0–8412–0917–0

Personal Computers for Scientists: A Byte at a Time
By Glenn I. Ouchi
276 pp; clothbound; ISBN 0–8412–1000–4

Writing the Laboratory Notebook
By Howard M. Kanare
146 pp; clothbound; ISBN 0–8412–0906–5

Principles of Environmental Sampling
Edited by Lawrence H. Keith
458 pp; clothbound; ISBN 0–8412–1173–6

Phosphorus Chemistry in Everyday Living
By Arthur D. F. Toy and Edward N. Walsh
362 pp; clothbound; ISBN 0–8412–1002–0

Chemistry and Crime: From Sherlock Holmes to Today's Courtroom
Edited by Samuel M. Gerber
135 pp; clothbound; ISBN 0–8412–0784–4
